CLINICAL MAGNETIC RESONANCE IMAGING

CLINICAL MAGNETIC RESONANCE IMAGING

Edited by

ALEXANDER R. MARGULIS, M.D.
CHARLES B. HIGGINS, M.D.
LEON KAUFMAN, Ph.D.
LAWRENCE E. CROOKS, Ph.D.

RADIOLOGY RESEARCH AND EDUCATION FOUNDATION
SAN FRANCISCO

Radiology Research and Education Foundation
Third and Parnassus Avenues—Suite C324
San Francisco, California 94143

Library of Congress Catalog Card Number: 83-81823

Printed in the United States of America by the
University of California Printing Department

CONTRIBUTORS

HERBERT L. ABRAMS, M.D., Philip H. Cook Professor of Radiology, Harvard Medical School

ELIAS BOTVINICK, M.D., Associate Professor of Medicine and Radiology, University of California, San Francisco School of Medicine

MICHAEL BRANT-ZAWADZKI, M.D., Associate Professor of Radiology, University of California, San Francisco School of Medicine

ROBERT C. BRASCH, M.D., Associate Professor of Radiology and Pediatrics, University of California, San Francisco School of Medicine

THOMAS F. BUDINGER, M.D., PH.D., Henry Miller Professor of Medical Research, Donner Laboratory and Lawrence Berkeley Laboratory, University of California, Berkeley

NEIL I. CHAFETZ, M.D., Assistant Professor of Radiology, University of California, San Francisco School of Medicine

LAWRENCE E. CROOKS, PH.D., Associate Professor of Electrical Engineering, University of California, San Francisco School of Medicine

CHRISTOPHER CULLANDER, Research Associate, Donner Laboratory and Lawrence Berkeley Laboratory, University of California, Berkeley

PETER L. DAVIS, M.D., Assistant Professor of Radiology, University of California, San Francisco School of Medicine

JACK DEGROOT, M.D., PH.D., Professor of Anatomy and Radiology, University of California, San Francisco School of Medicine

WILLIAM P. DILLON, M.D., Assistant Professor of Radiology, University of California, San Francisco School of Medicine

GORDON GAMSU, M.D., Professor of Radiology, University of California, San Francisco School of Medicine

HARRY K. GENANT, M.D., Professor of Radiology, Medicine, and Orthopaedic Surgery, University of California, San Francisco School of Medicine

HENRY I. GOLDBERG, M.D., Professor of Radiology, University of California, San Francisco School of Medicine

CHARLES A. GOODING, M.D., Professor of Radiology and Pediatrics, University of California, San Francisco School of Medicine

CLYDE A. HELMS, M.D., Assistant Professor of Radiology, University of California, San Francisco School of Medicine

ROBERT J. HERFKENS, M.D., Assistant Professor of Radiology, Duke University School of Medicine

CHARLES B. HIGGINS, M.D., Professor of Radiology, University of California, San Francisco School of Medicine

HEDVIG HRICAK, M.D., Associate Professor of Radiology and Urology, University of California, San Francisco School of Medicine

THOMAS L. JAMES, PH.D., Professor of Chemistry, Pharmaceutical Chemistry, and Radiology, University of California, San Francisco School of Medicine

LEON KAUFMAN, PH.D., Professor of Physics, University of California, San Francisco School of Medicine

PETER LANZER, M.D., Clinical Affiliate Fellow, Department of Cardiology, University of California, San Francisco School of Medicine

PAUL C. LAUTERBUR, PH.D., Professor of Chemistry, State University of New York at Stony Brook

MARTIN J. LIPTON, M.D., Professor of Radiology and Medicine, University of California, San Francisco School of Medicine

ALEXANDER R. MARGULIS, M.D., Professor of Radiology, University of California, San Francisco School of Medicine

BARBARA J. MCNEIL, M.D., PH.D., Professor of Radiology and Clinical Epidemiology, Harvard Medical School

CATHERINE M. MILLS, M.D., Assistant Professor of Radiology, University of California, San Francisco School of Medicine

KIRK L. MOON, JR., M.D., Postdoctoral Fellow, Department of Radiology, University of California, San Francsico School of Medicine

ALBERT A. MOSS, M.D., Professor of Radiology, University of California, San Francisco School of Medicine

DAVID NORMAN, M.D., Associate Professor of Radiology, Neurology, and Neurological Surgery, University of California, San Francisco School of Medicine

JAMES O. PROSEUS, Diasonics, Inc., South San Francisco, California

DAVID D. STARK, M.D., Postdoctoral Fellow, Department of Radiology, University of California, San Francisco School of Medicine

W. RICHARD WEBB, M.D., Associate Professor of Radiology, University of California, San Francisco School of Medicine

CONTENTS

General and Ancillary Considerations

FOREWORD

There can be few better examples of the complex and unanticipated interactions of basic research and technological innovation than the development of nuclear magnetic resonance (NMR) techniques for medical diagnosis and research. This book, written 10 years after the first publication of an NMR image, presents a compelling case for the idea that NMR is giving rise to the most powerful and versatile imaging technique in medicine.

The roots of this seemingly rapid development reach far back, however. In 1924 Pauli proposed the existence of nuclear magnetism to account for the hyperfine structure of atomic spectra. During the next 15 years or so, various methods for measuring the magnetic properties of nuclei were developed, and in 1946 the first successful NMR experiments in solids and liquids were reported. Applications to nuclear physics and to studies of the structures and motions of molecules followed quickly. Within 5 years it was also found that the electronic structures of molecules and solids could be studied by NMR, and its use began to spread rapidly through many areas of physics and chemistry. As improvements in electronics continued, and as digital computers and practical superconducting magnets became available (long after the first observation of superconductivity in 1911), the sensitivity of NMR methods improved dramatically, allowing useful studies to be done on small biochemical samples and on solutions of macromolecules. Increasingly clever methods were also developed for NMR spectroscopy of complex molecules and solids, and even a few observations of biologic tissues were made.

After more than a quarter of a century of applications of NMR to homogeneous samples, the idea of controlling the nonuniformity of magnetic fields so as to identify the spatial coordinates of the nuclei made possible the generation of detailed three-dimensional images. Practical implementation of these methods was aided by the mathematical principles underlying image reconstruction, first worked out as a purely mathematical problem and published by Radon in 1917, and by advances in applied mathematics such as the Cooley-Tukey Fast Fourier Transform algorithm introduced in 1965. When the medical community became aware of the potential usefulness of NMR phenomena, ideas, techniques, and apparatus that had been developing throughout most of the twentieth century were available and ready to support the rapid growth of new technology and its applications to clinical problems.

The successes and promises of NMR in medicine, amply documented and discussed in this book, are still immature. The effects of the human body on the radio-frequency signals and circuitry are not yet well enough known to allow the equipment to be optimized. Interpretations of relaxation time differences in tissues are still almost entirely empirical. Most of the potential for quantitative measurements of flow and motion remains to be explored, theoretically and experimentally. The combining, in human studies, of NMR spectroscopy with imaging—a vast, complex, difficult, and promising prospect—is not yet practical. All of these topics are being investigated by scientists, engineers, mathematicians, and clinicians, drawing upon and extending theory, lore, and expertise developed for other purposes. Not only are these tasks intellectually fascinating in themselves, but this volume demonstrates the usefulness to the medical profession and the benefits to the public of the fruits of long cultivation of many fields.

PAUL C. LAUTERBUR

PREFACE

In the short time since its introduction, magnetic resonance imaging has stirred a great deal of interest in the medical community. As the images have improved in quality and as the amount of information that they can offer has been appreciated, this modality has become the primary topic of discussion among imaging scientists, imaging physicians, hospital administrators, and government officials responsible for funding health care. This book presents the clinical experiences in magnetic resonance imaging gathered predominantly at the University of California, San Francisco, but also reflects the state of the art as of the middle of 1983. Besides giving an introduction to the physics, techniques, and instrumentation of magnetic resonance imaging, the following chapters attempt to present all the anatomic areas in which there has been experience with this technology. The final section deals with the present state of paramagnetic contrast media, biologic and environmental hazards, and siting and economic considerations.

The reason that we even attempted to publish a book at this early stage of the development of this modality is simple: We would like to make life easier for those who are just starting in the field by helping them to avoid the early difficulties that we encountered in obtaining and interpreting images. Our approach at the University of California, San Francisco—having multiple teams develop protocols for establishing clinical experience with magnetic resonance imaging—made it possible to acquire more information and devote more manhours to understanding its applications than if we had only had a small number of physicians working with it.

We hope that we have been successful in this effort and that the book will be helpful for those who are starting their clinical work in this most exciting and challenging field.

ALEXANDER R. MARGULIS
CHARLES B. HIGGINS
LEON KAUFMAN
LAWRENCE E. CROOKS

1

The Role of Magnetic Resonance Imaging in Medicine

ALEXANDER R. MARGULIS, M.D.

Not since the discovery of x-rays before the turn of the century has any diagnostic procedure caused as much excitement and anticipation as has NMR. Since the first images were published in 1973[1] a tremendous amount of progress has been made, so that today's NMR images are surpassing in versatility those of any other modality.[2,3] In addition to providing images in any plane without reformatting, NMR also offers rapid throughput by allowing simultaneous sections through any part of the head or body being imaged.[4,5]

What makes NMR so attractive is that it combines the advantages of the other imaging modalities without sharing their disadvantages.[6,7] Like CT, it offers superb tomography with excellent spatial resolution. Like ultrasound, it offers a great deal of tissue information and is non-ionizing. Like nuclear medicine, it offers the possibility of following metabolic processes and giving information about physiology. Nuclear magnetic resonance does all of this without any known biologic ill effects and with a resolving power considerably exceeding that of x-ray CT. It achieves this by having an acceptable spatial resolution of 0.8 mm at its best, which is comparable to the spatial resolution of much of today's x-ray CT equipment.[5,8] The contrast resolution of NMR, however, exceeds 500% in soft tissue, resulting in outstanding anatomic detail.

Advantages

The advantages of NMR imaging are related to the fact that the signal obtained and imaged is a synthesis of many more parameters than the signal of any other imaging modality. Besides offering data about hydrogen density, it provides information, through its T_1 and T_2 relaxation times, about the chemical and physical environment in which the hydrogen protons reside.[8] In addition, because it takes approximately 50 msec to register the signal from protons residing in a field, information about motion is obtained. Should the hydrogen protons move through the plane of imaging at a faster rate, no signal will be recorded. This in itself can provide extremely valuable information about the status of patency of blood vessels.[9]

The multiplicity of techniques for obtaining the image is staggering.[10] Saturation recovery, spin echo, and inversion recovery techniques and their variations offer multiple ways of acquiring data. The various techniques do not provide the same information about disease processes; for instance, the spin echo data acquisition technique will offer information about processes related to changes in the T_2 as well as the T_1 times, while the inversion recovery technique will usually provide more information about processes in which T_1 times are critical. It is possible, in addition, to obtain different information from images in which the pulse interval (TR) or spin echo delay (TE) is changed (Figs. 1–3). Furthermore, it is possible to vary the number of sequences, the number of averages from which images are derived, the slice thickness, the interval between slices, the plane of imaging, and the size of pixels, and so on. All of this results in an exceedingly large amount of information that must be carefully evaluated and judiciously limited so that it does not overwhelm the examiner and slow the patient throughput and interpretation of studies.

Not all of the possible applications of NMR have been developed yet. The limits of its ability to diagnose and stage disease and determine the best mode of therapy for it have yet to be established. The great promise of totally harmless paramagnetic contrast media will probably be achieved in the very near

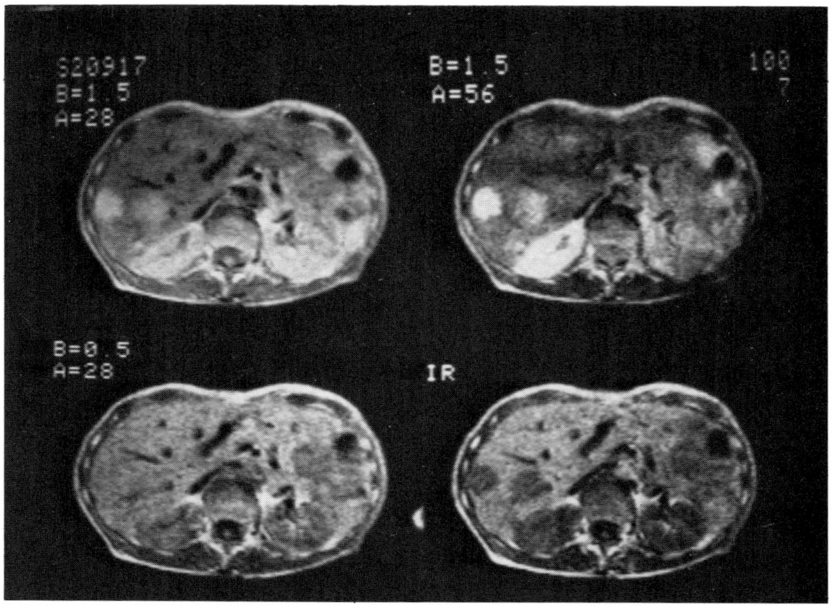

Figure 1. Four spin echo images through the same level of the liver in a patient with multiple metastases from carcinoid. With a short pulse interval (TR) of 0.5 sec and an echo delay (TE) of 28 msec, the image in the left lower corner shows no evidence of metastases. They are much better seen with an echo delay of 56 msec and a pulse interval of 1.5 sec (upper right). The metastases are well seen also with inversion recovery technique (lower right). The image obtained with a pulse interval of 1.5 sec and an echo delay of 28 msec shows the lesions, but not to the best advantage (upper left). Note that the kidney architecture is best seen on the image with a short pulse interval and a short echo delay; this is the image that fails to show the metastases. It is obvious, therefore, that multiple sequences are necessary for the best examination.

future.[11] It is very likely that such contrast media can be attached to metabolites like glucose or amino acids and followed in their passage through various organs.

The evaluation of gallbladder function by NMR has been done by Hricak et al.[12] A normal gallbladder in a fasting patient is able to concentrate bile, which emits a strong signal, whereas a diseased gallbladder is unable to do so and will emit a signal of very low intensity. Similarly, after administration of a cholecystogogue a gallbladder will empty itself of concentrated bile and fill with dilute bile, which images with low intensity. This is a harbinger of many similar physiologic tests that will be soon developed.

Present Status

It appears today that superconducting imagers are superior to those having resistive or permanent magnets. This is particularly noticeable in images of the body, although images of the head are also better when obtained by superconducting imagers.[10]

It appears proven at this time that NMR imaging is superior to that of any other modality in the examination of the brain.[9,10,13,14] This is true whether the diagnosis is that of degenerative disease, ischemic disease, trauma, or neoplasm. Although calcifications are not well seen by NMR, sufficient other information is obtained to compensate for the lack of data about calcification. Nuclear magnetic resonance is also superior to other imaging modalities in the examination of the spinal cord and cancellous bone.[15,16] Another area in which NMR excels is in the examination of the pelvis. Diseases of the prostate, urinary bladder, seminal vesicles, as well as the uterus, ovary, and vagina, are detected and can be studied to excellent advantage.

Nuclear magnetic resonance applications in diseases of the mediastinum,[17] liver,[18,19] spleen,[20] and pancreas[19] are currently being studied in a number of institutions. It appears very likely that in these areas, too, NMR will be superior to CT. It is not well established yet whether NMR can differentiate inflammatory from

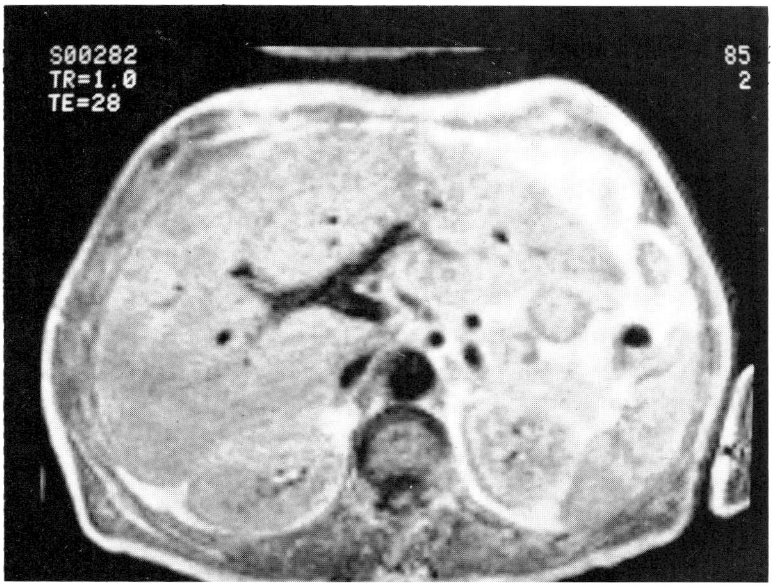

A.

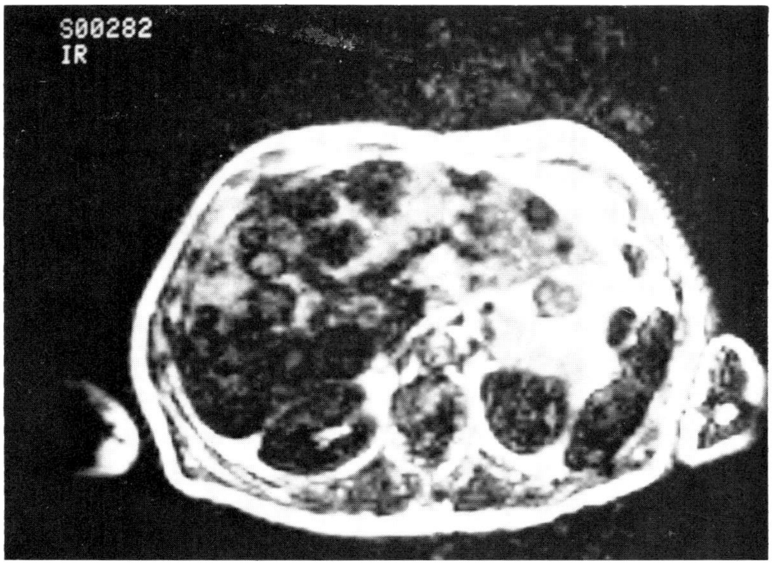

B.

Figure 2.

A. Examination of the liver in a patient with multiple metastases from carcinoma of the colon. With a pulse interval of 1 sec and an echo delay of 28 msec, an unevenness in liver architecture can be seen, raising the suspicion of some type of lesion. The lesions themselves are not seen.

B. With inversion recovery technique, multiple metastatic lesions are seen throughout the liver. The inhomogeneous architecture of the metastases is noted. In many such cases a special technique with different parameters and sequences is necessary to bring out the lesions to the best advantage.

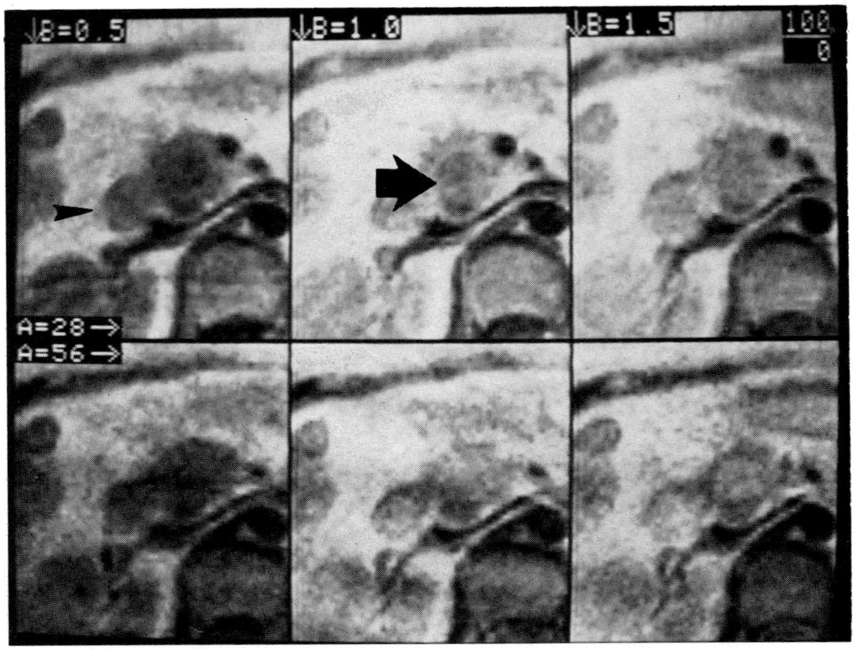

Figure 3. Carcinoma of the pancreas. Six images with three different pulse intervals: 0.5 sec, 1.0 sec, and 1.5 sec from left to right. The upper three images are with an echo delay of 28 msec, and the lower with one of 56 msec. Notice the changes in the appearance of the tumor (large arrow) on the images obtained with different sequences. The tumor is inhomogeneous. The duodenum (small arrowhead) can be differentiated from the tumor even without contrast material when the multiple images are analyzed.

neoplastic nodes in the retroperitoneum and the mediastinum. If this were possible, totally new vistas would be opened in the diagnosis and follow-up of neoplastic disease. Although the adrenals[21] and kidneys[22] also have been successfully studied by NMR, it is too early to declare that NMR is superior to x-ray CT and ultrasound in this respect.

Results of NMR imaging in the evaluation of the stomach, duodenum, and large bowel are promising (see Chap. 13). Contrast media to study these parts of the alimentary tube are still being developed.[23] It is not clear yet whether high-intensity contrast media or air will produce the most satisfactory images there. Neither one, however, can be expected to even approach the detail obtained with barium studies.

The small bowel remains at the moment an area in which NMR imaging is inferior to both CT and particularly to conventional barium sulfate intraluminal radiography. The complicated multidirectional motions of the small bowel preclude good detail in imaging by NMR. Whether or not paramagnetic contrast media combined with the use of drugs such as glucagon, which paralyzes the small bowel, will result in useful NMR images of the peritoneal cavity is still a question to be answered.[21]

All this information has been derived from a relatively small number of centers with clinically active units. At the end of 1982 there were only three imagers in clinical use in Great Britain, one in Germany, one in Holland, and seven in the United States. This number, however, is rapidly increasing, and it is expected that by the end of 1983 there will be over 60 units in use in the United States, Canada, Great Britain, continental Europe, and Japan.

Methods of Interpretation

Nuclear magnetic resonance imaging and interpretation are still in their infancy. As time progresses, more and more is being learned and the applications of NMR are being broadened (Fig. 4).

Unlike x-ray CT, where there are only a limited number of ways of enhancing contrast between tissues (i.e., without intravenous contrast administration, with bolus injection, or with slow intravenous drip

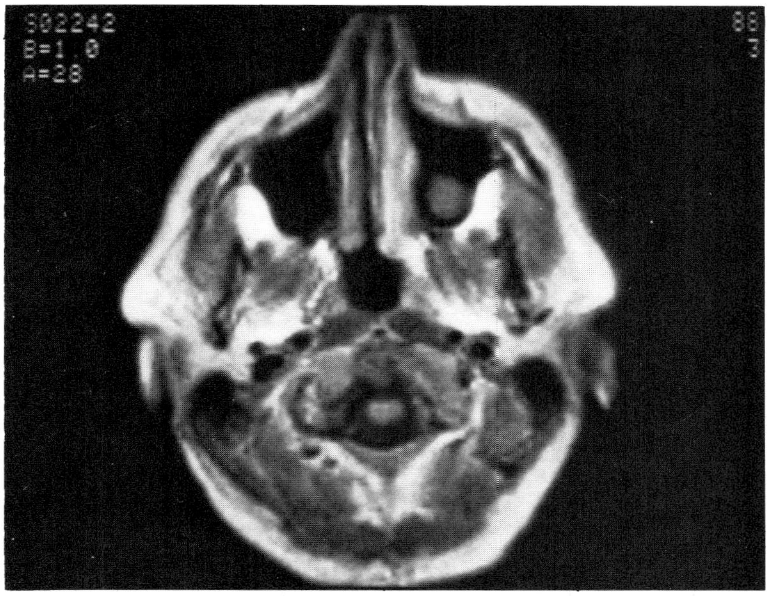

A.

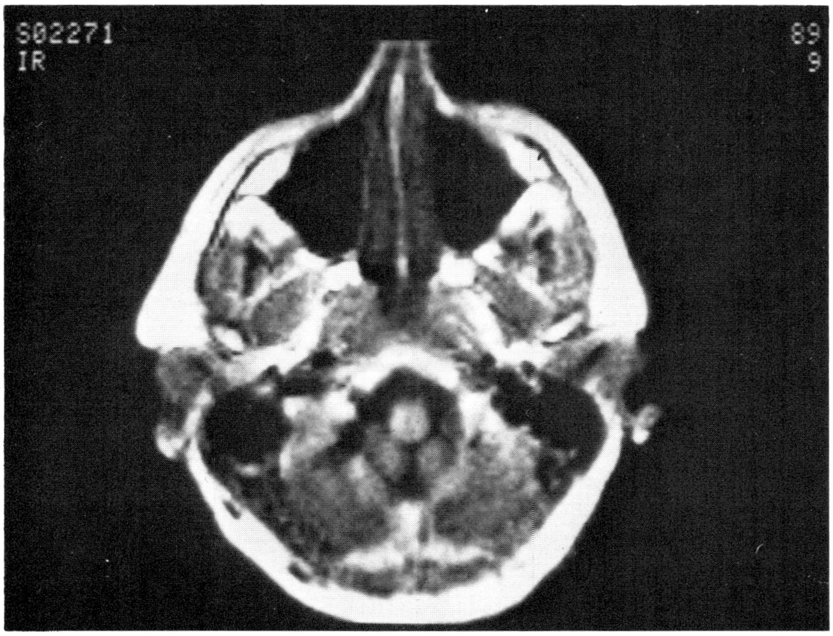

B.

Figure 4. A spin echo image (A) with a pulse interval of 1.0 sec and an echo delay of 28 msec showing a retention cyst in the right maxillary sinus. The inversion recovery image (B) fails to show the lesion.

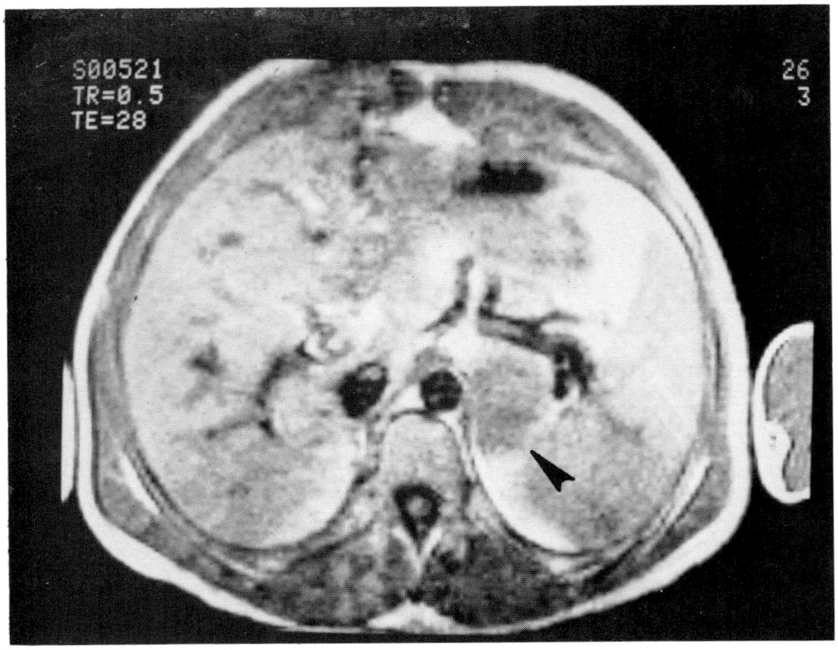

A.

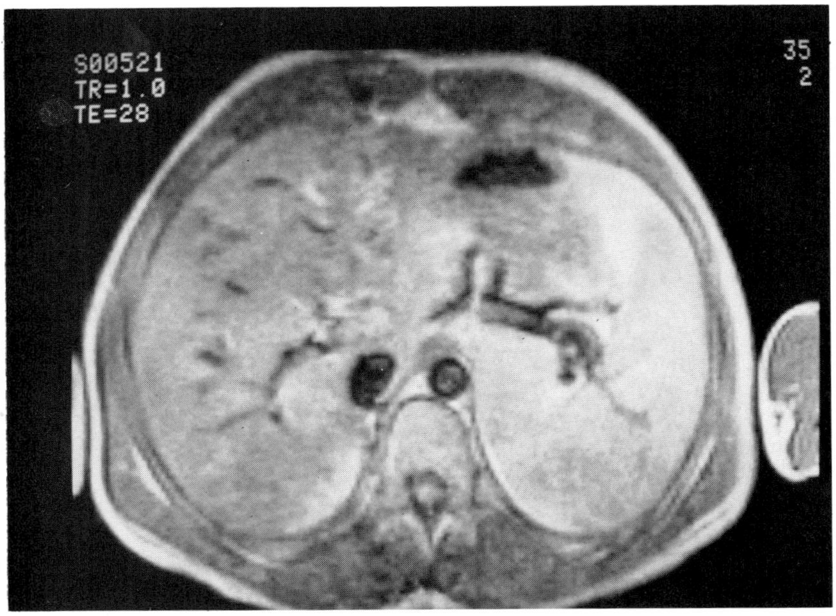

B.

Figure 5. Three images through the liver in a patient with a pheochromocytoma of the left adrenal.

A. The tumor (arrowhead) is best seen in this figure, in which the image was obtained with a pulse interval of 0.5 sec and an echo delay of 28 msec.

B. With a pulse interval of 1.0 sec and an echo delay of 28 msec, the lesion is not well seen.

C. A computer-generated T_1 calculated image shows the lesion to have a long T_1.

(continued on next page)

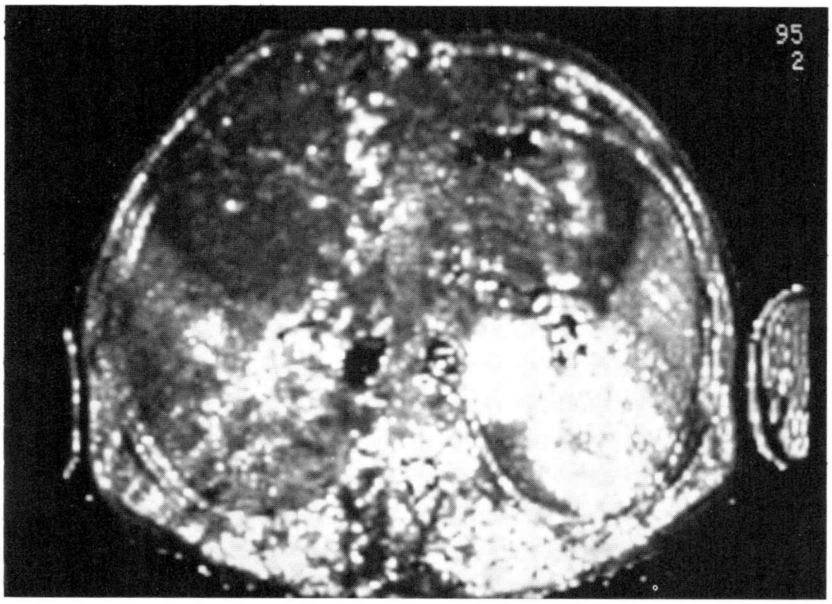

C. A computer-generated T_1 calculated image shows the lesion to have a long T_1.

of contrast material), NMR offers an almost unlimited number of technical approaches to achieve the desired goal. It is possible to devise techniques for studying almost any type of disease producing tissue changes and to evaluate not only gross pathology, but also the physiologic and biochemical changes in the tissues (Fig. 5). In addition to assessing the appearance and signal intensity of a lesion in relation to the organ in which it is located, one can compare the signal intensity with that of other organs. Variations in signal intensity of the lesion are compared with the signal obtained with different techniques. Relaxation-time information about tissues and computer-synthesized images also offer valuable information about the normal and abnormal state of the organs examined. It is on the basis of these multiple approaches that lesions that otherwise would be visualized only after invasive interventions become strikingly obvious by NMR imaging. It becomes clear, however, that the right clinical question has to be asked, and the NMR imaging must be based on the clinical situation. If this is not done, an impracticably complicated imaging protocol, lasting an inordinate amount of time, will have to be employed on every patient.

The study of NMR images requires a different approach from that used in interpreting conventional or even x-ray CT studies. Multiple NMR intensity images obtained with spin echo technique using different pulse sequence intervals and echo delays (Fig. 6) and different, individualized inversion recovery sequence images may be necessary to show lesions. The diagnosis depends on the careful study of all of the images obtained simultaneously on multiple levels. The direction of the scan, whether transverse, sagittal, or coronal, or a combination of them, will also require careful selection based on experience with the modality, the specific NMR imager, and, particularly, the clinical problem. It is essential, therefore, that the radiologist be thoroughly familiar with the patient's clinical status and the techniques best suited to arrive at images providing a maximum of information. A vast amount of information is thus gathered. It takes a long time to interpret an NMR examination because individual sections in various orientations obtained with different sequences need to be studied and assessed individually. Fast access to images on the screen and prompt production of hard copies of good quality are important for acceptable application of NMR imaging to clinical practice.

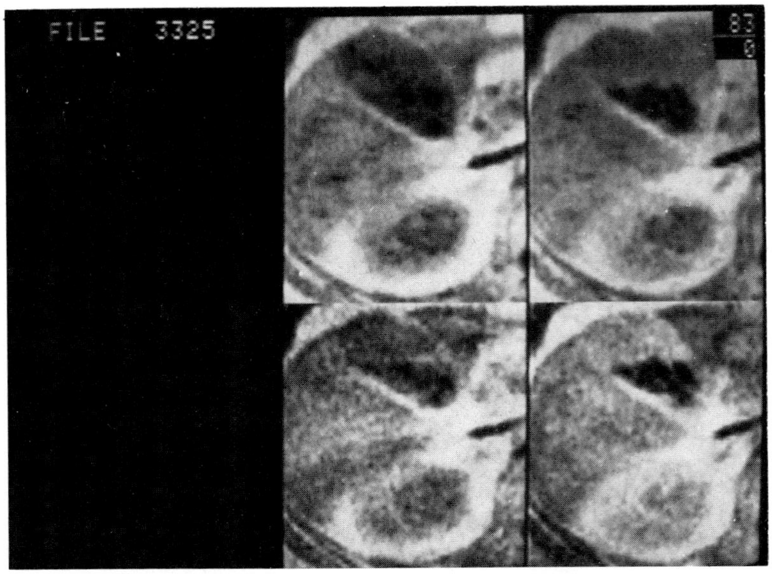

Figure 6. Gallstones in a diseased gallbladder unable to concentrate bile. The four images were obtained with pulse intervals of 0.5 and 1.0 sec going from left to right, and echo delays of 28 msec in the upper row and 56 msec in the lower row. Notice the different intensities of the bile signal, which behaves like water: low intensity with the 0.5-sec pulse interval and the echo delay of 28 msec, much higher intensity because of a relatively long T_2 with an echo delay of 56 msec and the longer pulse interval of 1.0 sec.

Disadvantages

Although NMR imaging is thus far unrivaled in providing valuable morphological and physiologic information without any known harmful biologic side effects, within the range of clinical applications it is not without disadvantages:

1. The slow scanning degrades the images of organs that move in multiple directions (e.g., the small bowel).
2. The long barrel of the magnet induces claustrophobia in about 1%–5% of patients, who then refuse to complete the study.
3. The strong magnetic field of a superconducting imager creates hazards with loose metallic objects that may fly into the bore of the imager, damaging it and injuring the patient. Strict security rules regarding access to the machine must be established; for example, metal detectors should be set up, and combination locks whose code is known only to a limited number of operators should be installed on the examination room door.
4. While it is agreed today that superconducting NMR imagers provide the best resolution and the most information, the optimal magnetic field strength is still an object of debate.
5. Together with the debate over optimal magnetic field strength, claims by manufacturers of NMR imagers with resistive and permanent magnets that their type of machine will supersede the cryogenic imagers creates an atmosphere of uncertainty, which is magnified by the number of manufacturing companies also offering multiple choices of imagers. The comparison with the history of x-ray CT manufacturing companies is too recent to be forgotten.
6. The siting of NMR imagers also presents problems for every individual installation. The strong magnetic fields and the radio-frequency pulses affect the functioning of electronic equipment such as scintillation cameras, x-ray image intensifiers, television monitors, and ordinary cameras. Similarly, elevators, steel beams, and moving trucks in the vicinity can affect the homogeneity of the magnetic field. Most of

these drawbacks in siting require individual consideration and careful planning.[24]

7. Ferromagnetic clips may present dangers, particularly when freshly introduced in the treatment of cerebral aneurysms.[25]
8. Patients with cardiac pacemakers are currently excluded from NMR studies. Although the danger of altering their rhythm is probably exaggerated in the range of clinically applied radio frequency,[26] it is clear that for some pacemakers energies used for NMR imaging can alter function from demand to permanent mode.
9. The cost of the NMR imagers is also a drawback, which will eventually force the health system regulating agencies in the United States and elsewhere to come up with plans for regional distribution not only of imaging devices, but also of hospitals and other medical care providers.

In spite of these minor disadvantages, NMR imaging is the most impressive and productive imaging modality in the early 1980s.

Future Directions

It is apparent that, in the near future, imaging with other elements besides hydrogen protons will take place. It is highly unlikely that such imaging, whether with phosphorus, sodium, or fluorine, will come even close in providing the morphological information supplied by hydrogen protons. Such images promise, however, to supply a wealth of physiologic and biochemical information and will possibly be displayed in different color scales superimposed on a map of hydrogen-based images. Nuclear magnetic resonance imagers of the near future will also be capable of performing NMR spectroscopy of an area selected from the proton image. How soon push-button switching of magnetic field strength, fast conversion from imaging to spectroscopy, and computerized selection of best technique for a given clinical problem will become a reality is a matter of conjecture. However, the dizzying progress from primitive to highly sophisticated images providing an enormous treasure of information permits predicting an accelerating arc of further successes for NMR. The ultimate goal is to unravel the secrets of normal and abnormal metabolism in areas of tissue that are precisely located within fine three-dimensional maps with microanatomic detail.

REFERENCES

1. Lauterbur PC: Image formation by induced local interactions: Examples employing nuclear magnetic resonance. Nature 242:190, 1973.
2. Pykett IL, Newhouse JH, Buonanno FS, et al: Principles of nuclear magnetic resonance imaging. Radiology 143:157, 1982.
3. Pykett IL: NMR imaging in medicine. Sci Am 246:78, 1982.
4. Crooks LE, Ortendahl DA, Kaufman L, et al: Clinical efficiency of nuclear magnetic resonance imaging. Radiology 146:123, 1983.
5. Crooks L, Arakawa M, Hoenninger J, et al: Nuclear magnetic resonance whole-body imager operating at 3.5 kGauss. Radiology 143:169, 1982.
6. Kaufman L, Crooks LE, Margulis AR (Eds): *Nuclear Magnetic Resonance Imaging in Medicine*, pp 11–52. New York: Igaku-Shoin, 1981.
7. Young IR, Bailes DR, Burl M, et al: Initial clinical evaluation of a whole body nuclear magnetic resonance (NMR) tomograph. J Comput Assist Tomogr 6:1, 1982.
8. Herfkens R, Davis PL, Crooks L, et al: Nuclear magnetic resonance imaging of the abnormal live rat and correlations with tissue characteristics. Radiology 141:211, 1981.
9. Crooks LE, Mills CM, Davis PL, et al: Visualization of cerebral and vascular abnormalities by NMR imaging: The effects of imaging parameters on contrast. Radiology 144:843, 1982.
10. Alfidi RJ, Haaga JR, El Yousef SJ, et al: Preliminary experimental results in humans and animals with a superconducting, whole-body, nuclear magnetic resonance scanner. Radiology 143:175, 1982.
11. Brasch RC: Work in progress: Methods of contrast enhancement for NMR imaging and potential applications. A subject review. Radiology 147:781, 1983.
12. Hricak H, Filly RA, Margulis AR, et al: Work in progress: Nuclear magnetic resonance imaging of the gallbladder. Radiology 147:481, 1983.
13. Young IR, Burl M, Clarke GJ, et al: Magnetic resonance properties of hydrogen: Imaging the posterior fossa. AJNR 137:895, 1981.
14. Buonanno FS, Pykett IL, Brady TJ, et al: Clinical relevance of two different nuclear magnetic resonance (NMR) approaches to imaging of a low grade astrocytoma. J Comput Assist Tomogr 6:529, 1982.
15. Moon KL, Genant HK, Helms CA, et al: Musculoskeletal applications of nuclear magnetic resonance. Radiology 147:161, 1983.
16. Brady TJ, Gebhardt MC, Pykett IL, et al: NMR imaging of forearms in healthy volunteers and patients with giant-cell tumor of bone. Radiology 144:549, 1982.
17. Gamsu G, Webb WR, Sheldon P, et al: Nuclear magnetic imaging of the thorax. Radiology 147:473, 1983.

18. Doyle FH, Pennock JM, Banks LM, et al: Nuclear magnetic resonance imaging of the liver: Initial experience. AJR 138:193, 1982.
19. Margulis AR, Moss AA, Crooks LE, et al: Nuclear magnetic resonance in the diagnosis of tumors of the liver. Semin Roentgenol XVIII:2, 1983.
20. Hawkes RC, Holland GN, Moore WS, et al: Nuclear magnetic resonance (NMR) tomography of the normal abdomen. J Comput Assist Tomogr 5:613, 1981.
21. Moon KL, Hricak H, Crooks LE, et al: Nuclear magnetic resonance imaging of the adrenal gland: A preliminary report. Radiology 147:155, 1983.
22. Hricak H, Crooks L, Sheldon P, et al: Nuclear magnetic resonance imaging of the kidney. Radiology 146:425, 1983.
23. Wesbey GE, Brasch RC, Engelstad BL, et al: Gastrointestinal contrast enhancement for NMR imaging using nontoxic oral iron solutions. Radiology (in press).
24. American Hospital Association: Nuclear Magnetic Resonance, Guideline Report. Hospital Technology Series, American Hospital Association, Division of Management and Technology Chicago, Vol 2 (8), 1983.
25. New PFJ, Rosen BR, Brady TJ, et al: Potential hazards and artifacts of ferromagnetic and nonferromagnetic surgical and dental materials and devices in nuclear magnetic resonance imaging. Radiology 147:139, 1983.
26. Pavlicek W, Geisinger M, Castle L, et al: The effects of nuclear magnetic resonance on patients with cardiac pacemakers. Radiology 147:149, 1983.

TECHNICAL CONSIDERATIONS

2

Basic Physical Principles

LAWRENCE E. CROOKS, Ph.D.
LEON KAUFMAN, Ph.D.

Magnetism at the Atomic and Nuclear Level

Magnetism is a result of the motion of electric charges. In wires and the atoms of magnets these charges are electrons. They can produce a magnetic field either by motion along a path (be it in a wire, in the vacuum of a television tube, or in an atomic orbit), or by virtue of their intrinsic spin (which can be visualized as a rotation of the electron around a central axis). The particles that compose the atomic nucleus, collectively called nucleons, also have a spin and magnetic moment. (Because both individual nucleons and nuclei have a charge distribution, rotation or spin of this charge produces a magnetic dipole whose value is called the magnetic moment. The numeric value of the magnetic moment determines the energies of the different orientations of a nuclei in an external magnetic field.) The proton is positively charged and has a relatively large magnetic moment. Although neutral, the neutron also has a net magnetic moment. These magnetic moments exist because the nucleons are composed of subnuclear charged particles called quarks, which can be thought of as having their own orbits. In the neutron, positively charged quarks can be considered as being nearer the axis of rotation than the negative ones, so that there is a net magnetic moment because the difference in orbits prevent cancellation. The neutron's magnetic moment is about two-thirds of the value of the proton's, and points against the axis of spin. When in the nucleus, like nucleons align with their spins pointing against each other (Fig. 1). This phenomenon is called pairing and is favored because it leads to a lower nuclear energy state. Therefore, only the unpaired odd proton, neutron, or both contribute their magnetic moment to the nucleus. As a consequence, only nuclei with odd numbers of protons, neutrons, or both have a magnetic moment.

The magnetic properties of nuclei become important when they are placed in an external magnetic field.

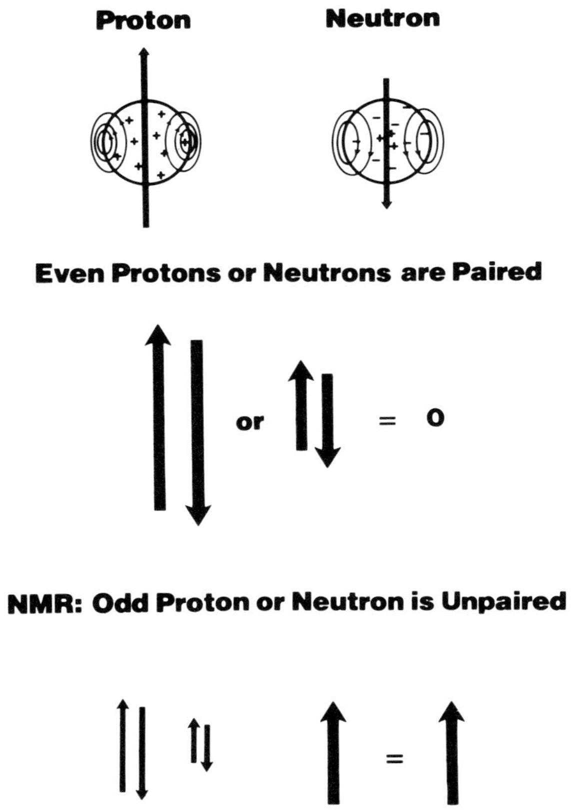

Figure 1. Both the neutron and proton have a magnetic moment. In nuclei, like nucleons pair, canceling their moments. An odd proton or neutron is left unpaired and contributes its magnetic moment to the nucleus.

Just like magnetic compass needles, they will have a tendency to align with the field. Because the randomizing effects of temperature are much stronger than those associated with alignment, magnetization within a substance is very weak, on the order of 10^{-6} to 10^{-5} of the full possible value. For the proton, quantum mechanics allows only two energy states in the magnetic field. The lower energy one is associated with the nuclear magnetic moment pointing along the field (parallel), and the higher one with the antiparallel direction. Given the fixed value of the magnetic moment of each type of nucleus, and a fixed magnetic field strength, the transition between states involves an exact amount of energy; the stronger the field, the higher the energy.

Resonance

Resonance is a familiar phenomenon in the macroscopic world. An example is the excitation of vibration of a tuning fork by sound waves from another similar one. Another, more annoying example is the excitation of vibration modes in household plumbing when the water flow on the pipes induces turbulence with just the right frequency components. Basically, resonance occurs when the amount of energy being fed into a system that is in a certain state, or released by it, matches a transition to another natural energy state of that system.

In quantum mechanics, resonances occur when a packet of energy is absorbed or released, this packet having an energy equal to the difference of energy associated with a transition between states. In the case of the magnetic moment of the nucleus, transitions between parallel (up) and antiparallel (down) states can occur if the right amount of energy is absorbed or released. Because the interaction is with a magnetic element, the necessary energy can be provided by a magnetic field. One way to obtain such a field is by utilizing electromagnetic radiations. These consist of packets (photons) of energy proportional to frequency. To induce resonance, the frequency f has to be proportional to the local magnetic field H_L, the proportionality constant depending on the nucleus involved. The relation between frequency and field is given by

$$f = \gamma H_L / 2\pi \qquad \text{Eq. 1}$$

where γ is magnetogyric ratio of the nucleus.

When the nuclei, originally in equilibrium with the field, are irradiated at the resonant frequency, they can adopt the antiparallel state. When they return to equilibrium, if the field is unchanged, they will radiate emissions of the same frequency. If between excitation and radiation the field strength is changed, they will radiate at a frequency corresponding to the new field value.

At the earth's surface, the magnetic field is approximately 0.5 Gauss (0.00005 T). The field provided by a speaker magnet is in the order of 6–8 kGauss (0.6–0.8 T). Magnets for human NMR operate in the range of 400 Gauss to 20 kGauss. At a field of 3.5 kGauss, resonance for hydrogen nuclei occurs at a frequency of 15 MHz. Because the frequencies involved fall in the radio to television range, the electromagnetic radiations are called radio-frequency radiations, or RF. In free space, a frequency of 15 MHz would correspond to a 20-m wavelength. In the range of field strength given above, wavelengths would range from 3.5–175 m. These wavelengths are clearly too long to resolve fine features within the human body, based on the principles of directionality, focusing, and collimation, which are familiar to diagnostic imaging. Rather, the relationship between frequency and field strength can be used to locate the site of resonating nuclei.

Locating Nuclei in Space

Locating nuclei in space is necessary for the production of images. Consider a conventional car radio. Its antenna is incapable of determining the directionality of the RF reaching it, or even of selecting one direction for reception. Even so, the occupant of the car can use other knowledge to determine where a radio station is located. If he knows that a certain frequency on the dial corresponds to a station located, for instance, in Oakland, on the east side of San Francisco Bay, and another position on the dial corresponds to a station in San Francisco, on the west side, then the occupant can relate position on the dial to geographic location of the transmitter without depending on the directional capabilities of the receiver. A similar process is used to obtain NMR images.

Consider two magnets next to each other, one with a field strength of 2 kGauss and the other 4 kGauss, each with a glass containing different amounts of water (Fig. 2). Place, a few meters away, an antenna that can be made to transmit at frequencies of about 8.6 MHz and 17.1 MHz. Also place some distance away a receiver antenna with the same frequency characteristics. Simultaneous transmissions at both frequencies will excite the hydrogen nuclei in both glasses of water, flipping them from the up to the down alignment with respect to the main magnetic field. If the transmitter is shut off, and the receiver set at 8.6 MHz, only nuclei in the lower field will

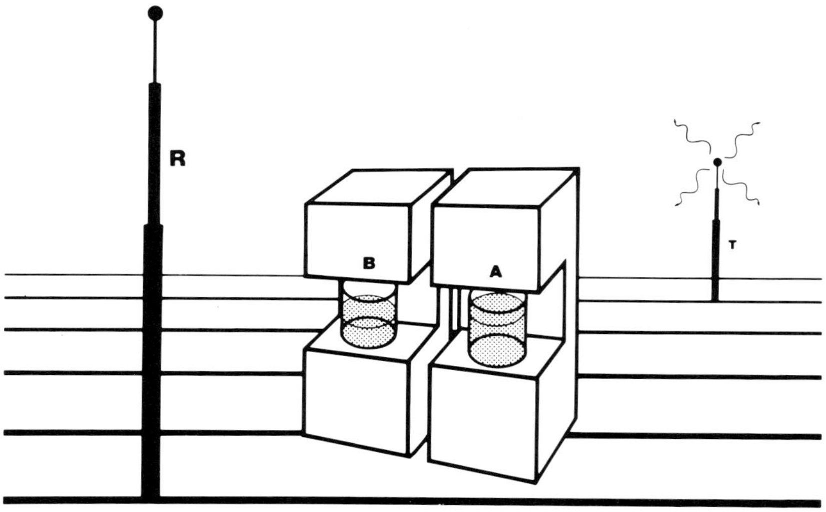

Figure 2. The amount of water in each of two glasses can be measured and located by placing them in magnets of different field strength (A and B). The transmit antenna (T) can be excited by frequencies corresponding to magnet A, so that the receive antenna (R) will detect NMR signals from A only. Conversely, T can transmit at frequencies corresponding to both magnets, and the receiver connected to R can be tuned to receive at a frequency corresponding to either magnet A or B, thus obtaining the amounts of water in each location.

contribute signal. Conversely, we can select the other receiver frequency so that the presence of nuclei in the stronger field is identified. We could also set the transmitter to selectively excite nuclei in one or the other magnet by selecting one of the two frequencies. The second antenna, set to receive both frequencies, will provide information that will enable us to identify which glass of water is radiating if we know the frequency of the transmitter. In each case the strength of the signal will allow us to determine how much water there is in each glass. Note that the location of the antennas does not affect our ability to identify the location of each glass of water.

The discrete, two-position process described above can easily be extended to define a large number of locations. Consider a magnet within which the field can be made to change slightly from one end to the other, and imagine that four glasses of water, each filled to a different level, are placed in the field (Fig. 3). A single antenna surrounds the glasses, and is used to transmit and receive RF. While the main field is uniform, RF magnetic field of a single frequency will excite the nuclei in all four glasses. If immediately after irradiation the field is made to change (this is done by superimposing a field gradient on the main magnetic field), nuclei in the stronger field will radiate at a higher frequency than those in the weaker field. The antenna will sum the signals from the four glasses into a complex wave (called the time domain signal). The position of each glass could be identified by sweeping the receiver frequency through the range of expected frequencies, much as one sweeps the radio dial searching for different stations.

It is more efficient to apply a Fourier transform (FT) to the time domain signal. The FT extracts from the time course of the signal the frequencies that went into forming it. In this case, four frequencies will be found, each with an amplitude or strength dependent on the amount of water in each glass. Since each frequency corresponds to one strength of magnetic field, and each location from end to end has a certain field, frequency and position of the glasses are equivalent parameters, one leading uniquely to the other. This principle, first demonstrated by Lauterbur,[1] is used in all practical applications of NMR imaging to the human body.

Well before imaging, this principle was used by chemists for NMR identification of substances. Rather than varying the external magnetic field, they depended on the small variations of magnetic fields seen by a nucleus in a molecule, since the local field depends on the external field plus the fields produced by neighboring nuclei and electrons. Thus, the ethanol molecule, which contains six hydrogen atoms bound to two different carbon atoms and one oxygen atom,

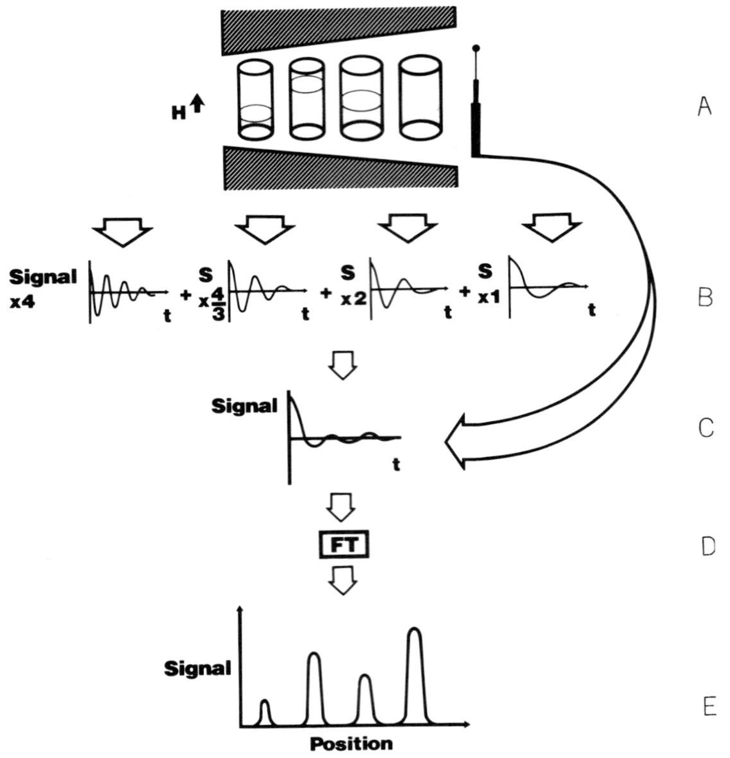

Figure 3. Four glasses of water, filled to different levels (A), will emit NMR signals whose frequency depends on position in the varying magnetic field and whose intensity depends on fluid level. The individual signals (B) are not separated by the antenna, which detects their summed contributions as a time domain signal (C). The Fourier transform (FT) (D) deconvolves the signal into its four frequency components. Since frequency depends on field strength, which depends on position, and amplitude depends on amount of water, the location and content of the water glasses are obtained in the FT process (E).

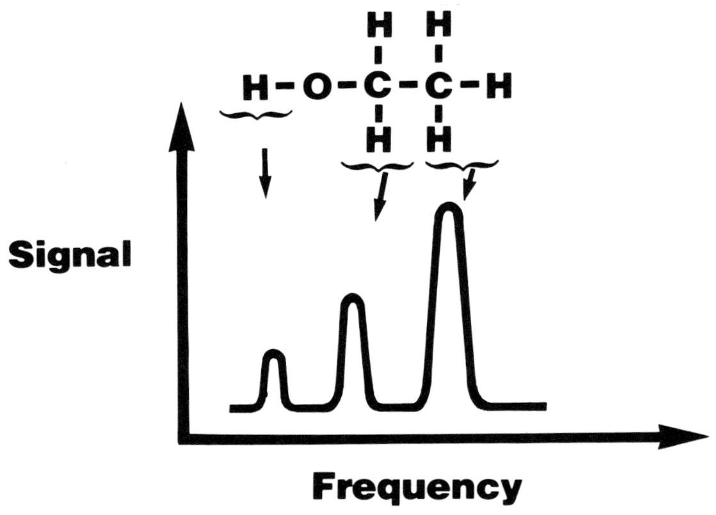

Figure 4. Hydrogen nuclei bound to different atoms in the ethanol molecule produce an NMR spectrum in which the frequency identifies location within the molecule, and signal intensity represents how many hydrogen nuclei are thus bound.

produces a frequency spectrum with three components, each of an amplitude that depends on the amount of hydrogen bound to each carbon or oxygen (Fig. 4). Measuring these different frequencies, or chemical shifts, provides powerful insights into chemical and physical processes in substances.

Behavior of Nuclei Undergoing Nuclear Magnetic Resonance

The NMR process can be described in a quantum mechanical or classical manner. Because the results are identical, the language tends to jump from one model to another. What makes the selection of a descriptive model flexible is that rather than working with a single nucleus, we consider all the nuclei in a sample and describe their behavior by a net magnetization vector M, which characterizes the system by disregarding the state of each nucleus and considering only a net collective effect (Fig. 5).

At room temperature and no magnetic field, the hydrogen in a sample of water has no magnetization, that is, the net magnetization vector is zero. In a magnetic field, the magnetization vector points along the field; its length is proportional to the number of nuclei in the sample and to the field strength, and inversely proportional to temperature. The length and direction of this vector characterize the equilibrium magnetization of the sample, that is, the state that it will revert to after being disturbed if enough time is allowed to pass. This equilibrium magnetization is given by

$$N(up)/N(down) = \exp(\mu H/kT) \quad \text{Eq. 2}$$
$$= 1 + 1.4\times 10^{-6} \text{ at } 37^\circ \text{ C and } 3.5 \text{ kGauss}$$

where μ is the nuclear magnetic moment, k is Boltzmann's constant, and T is the absolute temperature. The vector can be disturbed from equilibrium by application of a second external magnetic field (Fig. 6). If such a field is superimposed on the first one, M will align with the new net field. As M moves to its new direction, energy, stored in the nuclei of the sample, is provided by the second field. When the superimposed field is removed, M returns to equilibrium and the nuclei release the stored energy to the environment, either as heat or RF energy. Note that since the alignment energy is so much smaller than that associated with temperature, significant heating of the sample by NMR signals is not likely. It is apparent from the diagram that the most efficient application of the disturbing field is at 90° to the main one. These two are called, respectively, the transverse and longitudinal fields. The component of M that points along the main field is called the longitudinal magnetization (M_L), and the orthogonal component is called the transverse magnetization (M_T)(Fig. 7). If the transverse field is applied slowly, M tracks it and aligns with the net field; however, if it is switched rapidly, M behaves

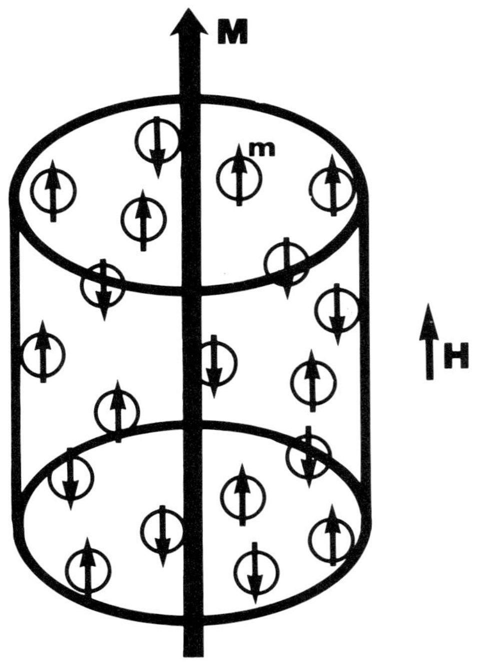

Figure 5. The individual magnetic moments (m) can point the up or down in a magnetic field H, with the equilibrium direction being up. Temperature has a strong randomizing effect, so that the net up-pointing moments are a small fraction of the total. These can be represented by a net magnetization vector M.

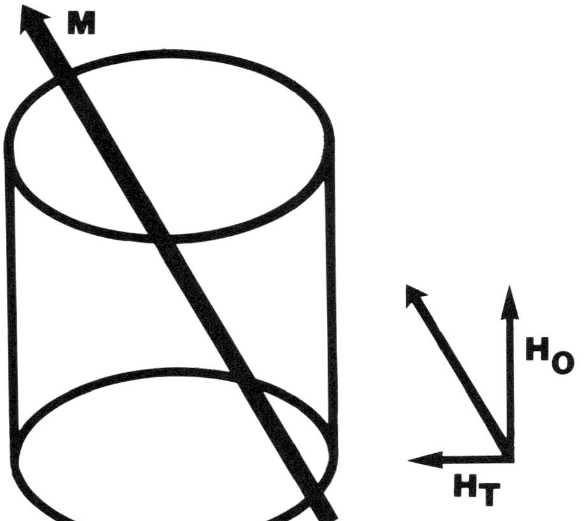

Figure 6. If a small transverse field H_T is applied, M will align with the new net magnetic field.

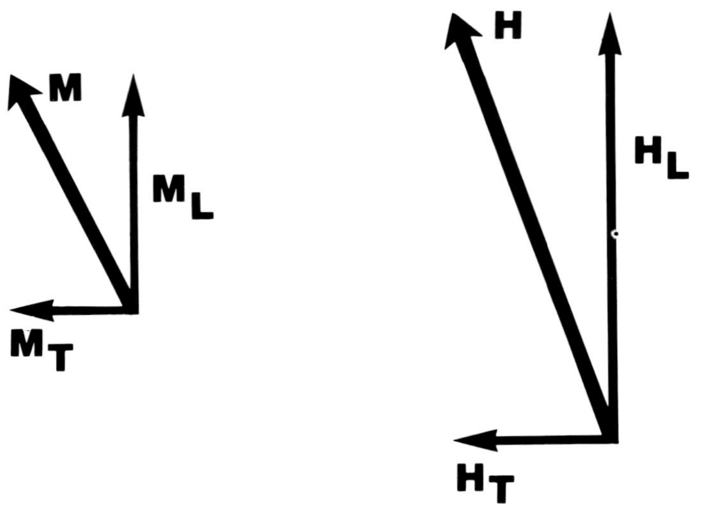

Figure 7. The net magnetic field H results from the sum of the longitudinal and transverse fields, H_L and H_T. M aligns with H, and can be considered as having longitudinal and transverse components, M_L and M_T. In equilibrium, M is given by the vector sum of M_L and M_T.

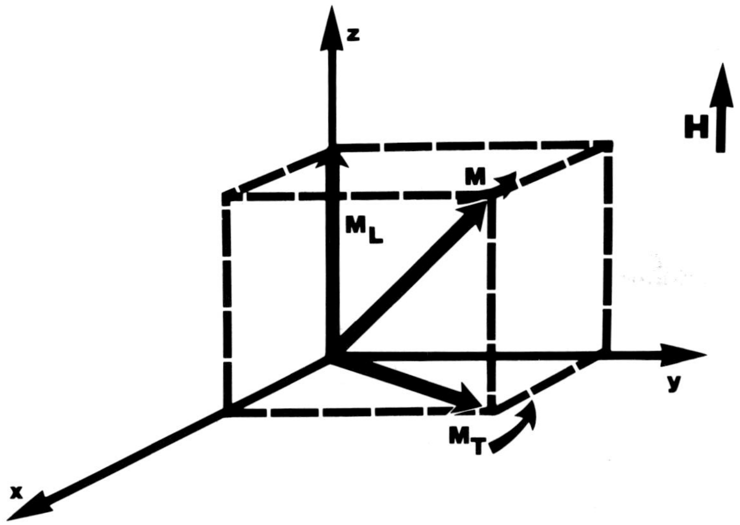

Figure 8. When the field H_T is applied rapidly, the vector M precesses about the z-axis. If H_T is applied repetitively (for instance, by RF oscillation) an equilibrium condition is established where M is the vector sum of M_L and M_T.

as a top, so that as it deviates from the longitudinal axis it precesses about it. If the main magnetic field is defined as being aligned along the z-axis, then M_T rotates in the x,y-plane (Fig. 8). M_L is now reduced from its equilibrium value. In fact, if M is rotated onto the x,y-plane (by what is called a 90° RF pulse), M_L is zero (Fig. 9). Immediately after RF irradiation, M_L begins to grow again towards its equilibrium value M. The growth is exponential with a time constant T_1, so that

$$M_L = M[1 - \exp(-t/T_1)] \qquad \text{Eq. 3}$$

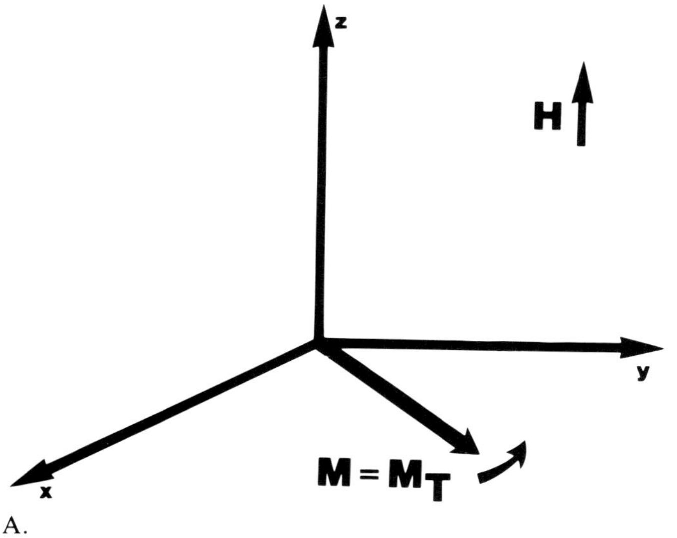

A.

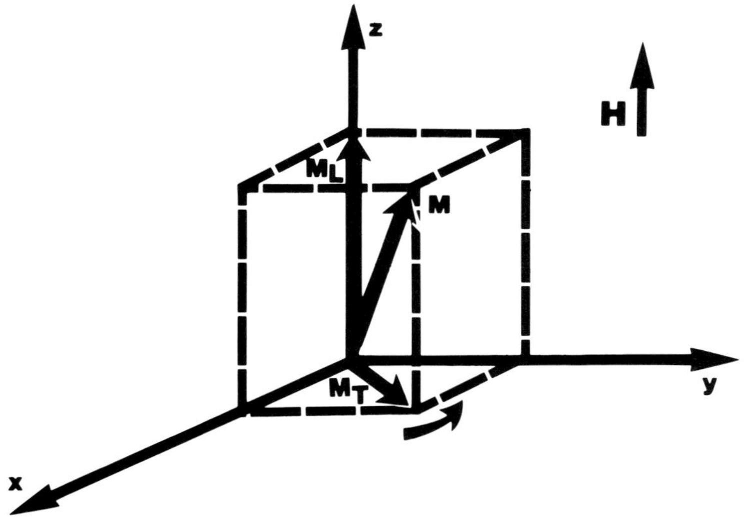

B.

Figure 9. After a 90° RF pulse the magnetization vector is in the x,y-plane (A). Typically, M_T decays more rapidly than M_L grows, so that the length of the vector M changes as it returns to equilibrium (B).

where *t* is the time since irradiation (Fig. 10). During this process M_T decays exponentially with a time constant T_2, so that

$$M_T = M_{To} \exp(-t/T_2) \qquad \text{Eq. 4}$$

where M_{To} is the value of M_T immediately after irradiation, and *t* is the elapsed time (Fig. 11). Note that T_1 and T_2 need not be equal, but T_2 can never be longer than T_1, and that during the process M changes in size. The rotation of M_T produces an oscillating magnetic field, which induces the NMR signal in the antenna. This signal is called a free induction decay, or FID, and is observable immediately after irradiation.

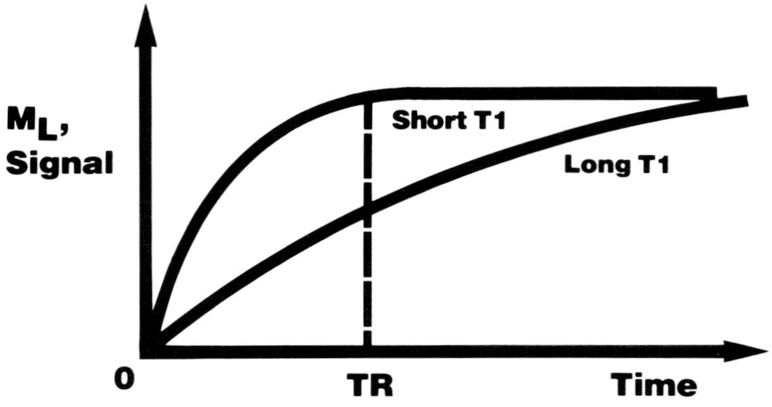

Figure 10. After excitation by an RF magnetic field, M_L and the available NMR signal return to equilibrium value at a rate determined by T_1. If the system is allowed a time TR to recover, contrast between tissues with different T_1 values can be highlighted.

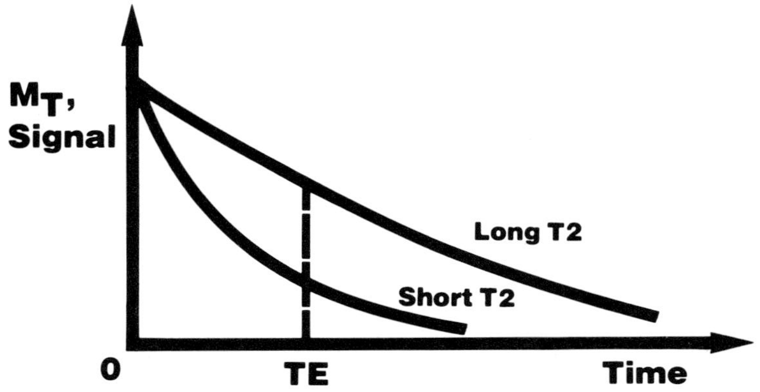

Figure 11. After RF excitation, M_T decays at a rate determined by T_2. The interval TE determines contrast between tissues with different values of T_2.

T_1 Relaxation

The T_1 relaxation time is also called spin lattice or thermal relaxation time. The individual protons exchange fixed amounts of energy when they flip from the down to up alignment in the process of returning to equilibrium. This exchange can occur only at the resonant frequency. A molecule in the lattice surrounding the resonant nucleus appears as an oscillating electric and magnetic field with frequency that depends on its thermal velocity and mean free path. Since both vary over a broad range for any one temperature, of the whole ensemble of molecules only a small fraction provide the right oscillating fields. These then couple to the nucleus and allow the relaxation to occur. As temperature and molecular composition changes, so does the distribution of velocities and mean free paths, thus affecting T_1. For any one substance at a given temperature, field strength also affects T_1, since it changes resonant frequency. In general, T_1 lengthens at the higher fields. In water, it shortens as large molecules are added. Because tissues have different T_1 values, T_1 relaxation can be used to advantage in imaging. Not only does T_1 provide intrinsic contrast between tissues, but this contrast can also be accentuated by proper manipulation of the NMR process.

T_2 Relaxation

The T_2 relaxation time is also called the spin-spin or transverse relaxation time. It occurs as follows: In a perfectly uniform magnetic field all nuclei would resonate at exactly the same frequency, but if the

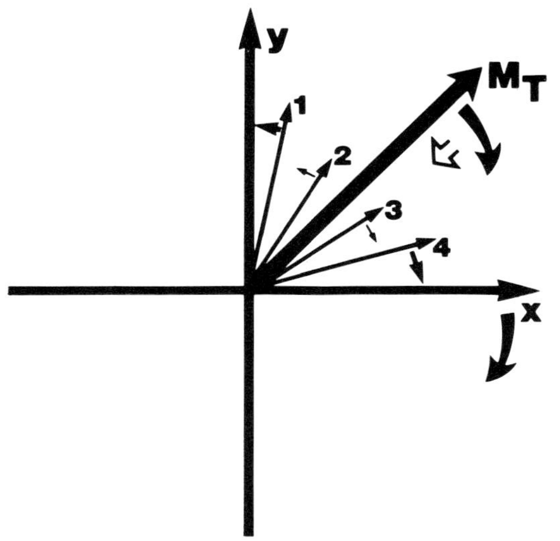

Figure 12. Because individual nuclei find themselves in slightly different fields, their magnetization vectors precess at slightly different frequencies. If we consider the x,y-plane rotating at the average precessional frequency, the individual vectors are fanning apart because some move faster (1 and 4) than others (2 and 3). M_T rotates at the precessional frequency and, because of the spreading of the vectors, shortens with an exponential time constant T_2*.

field is even slightly inhomogeneous, nuclei resonate at slightly different frequencies. Although immediately after RF irradiation they are all in phase (producing the FID), they soon lose coherence, and the signal that is observed decays. In the case of the perfectly uniform field, T_2 would be equal in length to T_1, but the loss of coherence shortens T_2. The effects due to inhomogeneities in the external field produce a rapid decay, characterized by the relaxation time T_2*. This time determines the decay of the FID signal.

In most substances, even if the external field is homogeneous, neighboring nuclei change the local field, so that each nucleus finds itself in a slightly different field. Coherent resonance is lost as a result of the magnetic field changes produced by nearby hydrogen nuclei; hence the name "spin-spin relaxation time." In a very pure fluid whose molecules are characterized by rapid motions, the magnetic field contributions from neighboring nuclei average to zero, and only the main magnetic field determines resonant frequency. T_2 in this case is very long. If the nuclei are in impure liquids, where water molecules attach to proteins and have low rotational speeds, this averaging process is less effective and the resonant frequency is different for each nucleus, making T_2 short.

T_2 relaxation is a collective phenomenon. For any single nucleus the transverse magnetization M_T can

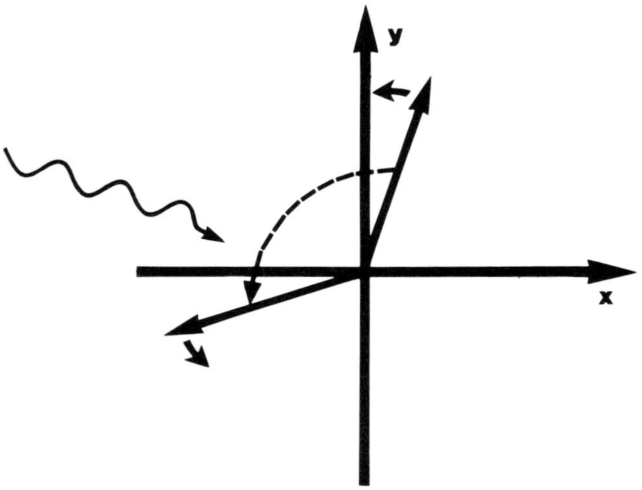

Figure 13. A 180° RF pulse rotates the transverse magnetization vector by that angle, keeping its rotational velocity constant.

be considered as decaying at the same rate as M_L, but since the precessional rotation speeds vary from nucleus to nucleus, the net vector M_T is decaying more rapidly as the individual vectors M_T fan apart (Fig. 12). The portion of the spreading of the vectors that is due to main field inhomogeneity can be refocused by applying an RF pulse whose strength and duration cause the vectors to rotate through a 180° angle about the y-axis (Fig. 13). The differences in precessional velocity remain, but rather than causing a fanning apart of vectors, they now fan together (Fig. 14). When the individual vectors come into phase, they once again provide a strong signal, called

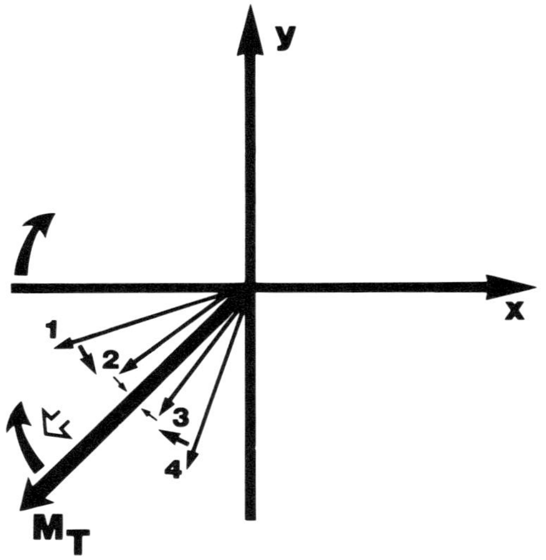

Figure 14. After a refocusing 180° pulse, the individual magnetic moments still have the same rotational direction, but now they are fanning in, as the faster ones (1 and 4) catch up with the slower ones (2 and 3). M_T now passes through a maximum length with an exponential time constant T_2*.

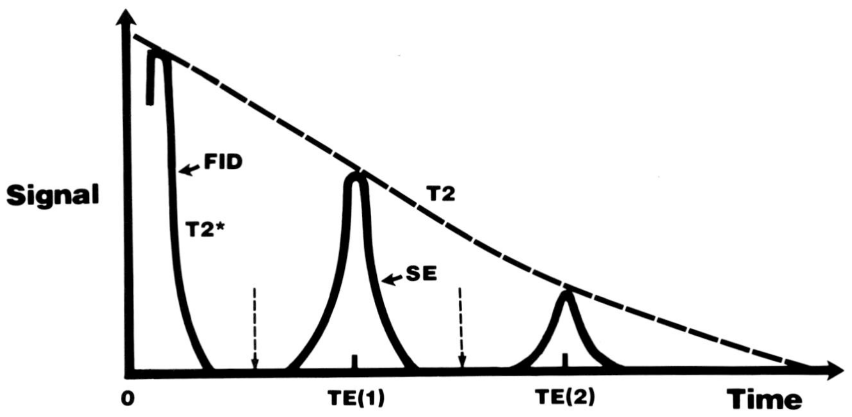

Figure 15. The FID is the signal obtained immediately after RF excitation of the nuclei. Note that in a pure FID the early part of the signal is lost because it overlaps the RF pulse. The FID decays with a rate of T_2*. The application of sequential 180° pulses (dashed arrows) results in spin echoes at times that are integral multiples of TE.

a spin echo. The peak of the spin echo has undergone T_2 decay, but the effects of T_2* have been eliminated by the 180° refocusing pulse. These pulses can be applied at regular intervals, making it possible to obtain multiple echoes from a single 90° RF excitation. Each echo is of progressively lower amplitude as a result of T_2 decay, as given by Equation 4, where the time t is replaced by TE, the time delay between 90° excitation of the sample and reception of the spin echo signal (Fig. 15).

By judicious choice of spin echo parameters, T_2 can also be used to provide intrinsic contrast between tissues. Equation 5 results from combining Equations 3 and 4 after substituting TE for t in Equation 4 and TR for t in Equation 3; thus the intensity I in a spin echo image can be expressed in approximate form as

$$I = N(H)\, f(v) \exp(-TE/T_2)\,[1 - \exp(-TR/T_1)] \qquad \text{Eq. 5}$$

where $N(H)$ is the density of hydrogen nuclei undergoing NMR (a third parameter that produces tissue contrast), TE is as defined above, and TR is the time allowed for recovery of magnetization after the excitation pulse; $f(v)$ denotes that, in general, the motion of nuclei changes their signal. The reason is that the sequence of pulses needed to generate a signal requires 20–100 msec. If, for instance, blood moves so that it traverses the imaged voxels in a time comparable to this, it experiences sequences different from those experienced by stationary tissues. Thus, flow can be considered a fourth endogenous contrast parameter. In the following chapter we shall discuss how different imaging techniques translate those tissue parameters onto an output image.

Impact of Tissue Parameters on Imaging

T_1 Relaxation

As described in the following chapter, most imaging techniques require repeated applications of RF and gradient pulse sequences. After each sequence the magnetization (i.e., signal available) is left at a reduced level or at zero (Fig. 10). Therefore, sequences cannot be repeated too rapidly, since magnetization grows at a rate given by T_1 (Eq. 3), and it is necessary to let a significant interval TR pass between sequences. To the extent that tissue T_1 values range from 0.2–3.0 sec, and that in some clinical applications long TR values are favored (1.0–2.0 sec), T_1 effects slow down the imaging process. Since T_1 increases with magnetic field strength, for equal fraction of the maximum possible response both TR and imaging time need to increase with increasing field strength.

T_2 Relaxation

Clinically, we have found that T_2 relaxation times are as important as or more important than T_1 effects when characterizing normal and pathologic tissues. As previously described, to obtain T_2 information it is necessary to allow an interval TE between excitation of the sample and reception of the spin echo. This interval is short because tissue T_2 values range from 20–300 msec, so that TE can be in the range of tens of milliseconds. The T_2 contrast is gained at the expense of signal strength, which decays as shown in Equation 4 (Fig. 11). This necessitates the selection of TE in such a manner that it balances the need for T_2 contrast and adequate NMR signal levels.

Effects of Blood Flow on Magnetic Resonance Imaging

Imaging sequences apply carefully tailored RF pulses and magnetic field gradients designed to produce optimal amounts of signal from a volume in the subject. It would then appear that, in general, moving nuclei would observe a disrupted sequence and consequently produce little or no signal. In a simple manner, we would expect that signal would disappear if all the nuclei traverse the imaged volume in the time it takes to carry out a single sequence. Consider, for instance, a 70-msec sequence. Nuclei moving across an imaged plane of 7-mm thickness will take 70 msec to traverse this distance if their velocity is 10 cm/sec. For this case, cut-off velocity (the velocity above which the moving nuclei produce no signal) should be near 10 cm/sec.

Consider now a very low-velocity regimen across the section, low enough that few nuclei leave the imaged volume. Loss of signal is not appreciable, and the small fraction of nuclei leaving the volume are being replaced by nuclei that have not undergone previous irradiations. While the stationary nuclei may recover only half of their magnetization, nuclei that enter this volume between successive excitations have not undergone a prior irradiation and are fully magnetized, consequently providing full signal. Thus, if 10% of the flowing nuclei in a pixel are replaced by fresh nuclei providing twice as much signal as the remaining ones, we expect that the net signal should be

$$(0.9 \times 1) + (0.1 \times 2) = 1.1,$$

or 10% higher than for stationary fluid. We call this effect "paradoxical enhancement." The exact behavior of signal intensity is dependent on the relaxation time of the fluid, which determines how much magnetization is lost between excitations and between excitation and signal reception, the interval TR between these excitations, and the delay TE between excitation and signal reception, that is, the length of the sequence. Figure 16 shows the relative NMR intensity as a function of velocity for blood. Curves are shown for different values of TR and TE.

In imaging, not only the quantitative, but also the qualitative behavior during flow can change from that described above if the process has added complexities. Let us consider flow in the plane of the section being imaged. The flowing nuclei are being irradiated at the same rate as the stationary ones, and paradoxical enhancement will not occur. Consider also an imaging method where more than one section is imaged at the same time. This is a desirable imaging mode because it allows for higher utilization of the total imaging time. For instance, if the sequence is 70-msec long and the interval between excitations is 0.5 sec, seven sequences could be fit in each interval. While the same section cannot be irradiated, another six can, so that a total of seven contiguous sections are imaged in one 0.5-sec interval. This has proven to be an effective method of multisection imaging.[2] In this case, only the nuclei entering the first (outside) section have not undergone a previous irradiation and show paradoxical enhancement. As they enter subsequent sections, paradoxical enhancement is decreased or disappears. The effect is illustrated in Figure 17.

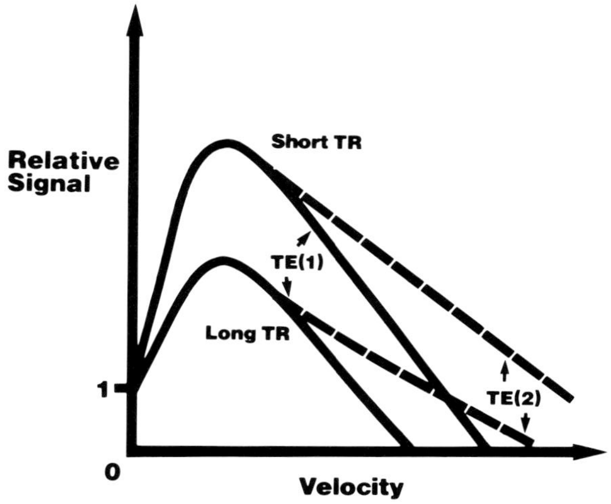

Figure 16. Compared with stationary blood, flowing blood can produce more or less NMR signal as velocity increases. In general, relative intensity is higher for the shorter values of TR. Also, the second echo is of more intensity than the first.

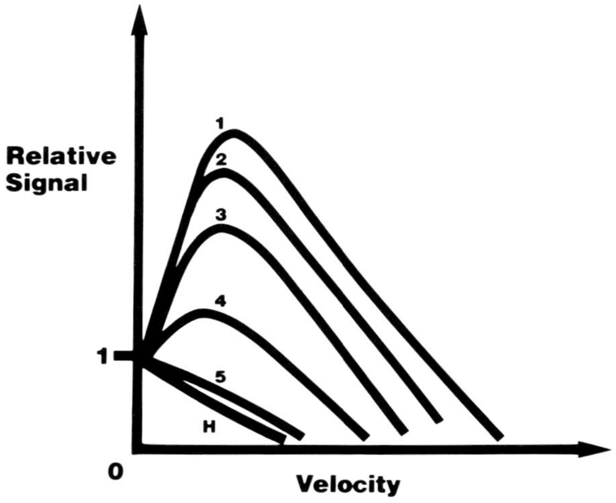

Figure 17. In multisection imaging, paradoxical enhancement is most prominent in the first section encountered by the blood, and decreases as the blood reaches deeper into the imaged volume. The numbers indicate the order in which the blood encounters different sections. No paradoxical enhancement is evidenced for flow in the plane of the image (H).

Although flow effects appear incidentally to the imaging process in NMR images, it is desirable to design sequences specifically for blood flow imaging. Probably the simplest way to highlight blood flow is to decrease the interval TR, so that stationary nuclei lose as much signal as possible when compared with the incoming nuclei. This technique has been demonstrated by Gore et al.[3] More sophisticated techniques can conceivably be used. These derive from Singer's work, and are based on two different principles.[4] In the first, a bolus of excited hydrogen nuclei is produced, and either its disappearance from the site of production or arrival distal to it is used to quantitate flow. A potentially more powerful method to quantitative capillary blood flow involves the generation of "flow graphs."[4] These graphs plot the number of nuclei flowing within a velocity interval and can therefore produce maps of blood flow in terms of ml of blood per gm of tissue per min. Although neither the NMR bolus nor flow graph techniques have been used to map blood flow distributions, this methodology is conceptually capable of doing so, and is likely to be demonstrated in the future.

REFERENCES

1. Lauterbur PC: Image formation by induced local interactions: Examples employing nuclear magnetic resonance. Nature 242:190, 1973.
2. Crooks LE, Arakawa M, Hoenninger JC, et al: NMR whole body imager operating at 3.5 kGauss. Radiology 143:169, 1982.
3. Gore JC: The meaning and significance of relaxation in NMR imaging. In *NMR Imaging,* Witcofski RL, Karstaedt N, Partain CL (Eds), pp 15–23. Winston-Salem: The Bowman Gray School of Medicine in Wake Forest University, 1981.
4. Singer JR: Blood flow measurements by NMR. In *Nuclear Magnetic Resonance Imaging in Medicine,* Kaufman L, Crooks LE, Margulis AR (Eds), Chap 8. New York: Igaku-Shoin, 1981.

RECOMMENDED READING

Abragam A: *The Principles of Nuclear Magnetism.* New York: Oxford University Press, 1961.

Bloch F, Hansen WW, Packard M: Nuclear induction. Phys Rev 69:127, 1946.

Bloembergen M: *Nuclear Magnetic Relaxation.* New York: WA Benjamin, 1970.

Carrington A, McLachlan AD: *Introduction to Magnetic Resonance.* New York: Harper and Row, 1967.

Farrar TC, Becker ED: *Pulse and Fourier Transform NMR.* New York: Academic, 1976.

Kaufman L, Crooks LE, Margulis AR (Eds): *Nuclear Magnetic Resonance Imaging in Medicine.* New York: Igaku-Shoin, 1981.

Purcell EM, Torrey HC, Pound RV: Resonance absorption by nuclear magnetic moments in a solid. Phys Rev 69:37, 1946.

Schumacher TR: *Introduction to Magnetic Resonance.* New York: WA Benjamin, 1970.

Slichter CP: *Principles of Magnetic Resonance.* New York: Harper and Row, 1963.

3

Imaging Techniques

LAWRENCE E. CROOKS, Ph.D.
LEON KAUFMAN, Ph.D.

Nuclear magnetic resonance imaging presents two aspects. One involves the actual techniques by which the spatial distribution of nuclei is mapped. The second, and more exciting and important in the long term, is the type of information that is gathered from each location. Nuclear magnetic resonance imaging depends on at least four tissue parameters: hydrogen density, the T_1 and T_2 relaxation times, and flow. Two other parameters may become clinically useful: nuclear species and chemical shift, that is, location of an atom within a molecule (its chemical binding site). Imaging techniques can be designed so as to highlight any of these, or to eliminate the contributions of some. The "best" techniques can be identified only by clinical trials. Imaging techniques developed in a historical pattern that does not necessarily match what is now in use. Consequently, we will limit our discussion to those techniques that currently appear to offer clinical utility, and some that may be tried in the near future. The general principles leading to image formation are described in the previous chapter. Here we deal directly with the specific aspects of the image formation process. For simplicity, we will separate geometry and signal acquisition modes, something that is not always feasible in practice.

Spatial Localization of Nuclei

Selective Irradiation

A central concept of NMR imaging techniques is that of selective irradiation.[1] In the previous chapter we discussed a situation where an antenna excited nuclei in a glass of water placed in one of two magnets by irradiating the sample with radio waves whose frequency corresponded to the resonant frequency in the field of one of the magnets. Such a principle can be used to select a section of a subject for imaging. In a continuously varying field along the z-direction, every x,y-plane of the subject finds itself at a particular resonant frequency (Fig. 1). If the sample is irradiated

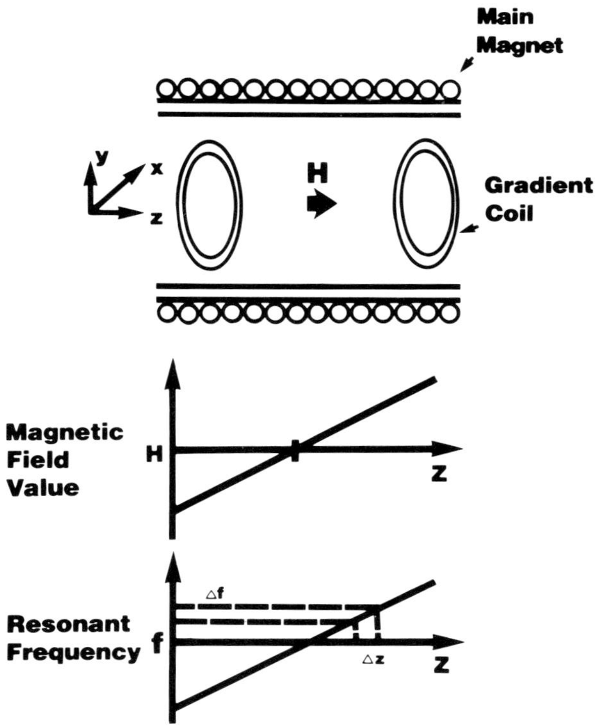

Figure 1. Selective irradiation. In a magnetic field that changes along the z-axis (a z-gradient), the resonant frequency also changes in direct proportion. A narrow band of RF irradiation excites nuclei in a narrow range of z.

with a 90° pulse consisting of a narrow range of frequencies, only the nuclei within a corresponding narrow section will be excited. The thickness of that section will be proportional to the frequency range used, and inversely proportional to the strength of the z-gradient. If the z-axis is defined along the long axis of the body, the sections thus obtained are in the transverse plane. If the gradient is along the x- or y-axis, the sections are obtained in the sagittal and coronal planes, respectively (for a subject who is prone or supine).

Imaging the Plane

Once a plane is excited by a selective irradiation, a gradient can be applied along the x-axis during signal acquisition. As discussed in the previous chapter, the Fourier transform (FT) of the signal thus obtained is the one-dimensional projection of the object onto the x-axis. If the process is repeated with the direction of the projection gradient rotated by 1°, a new projection of the object is obtained. If the process is repeated 180 times, a reconstruction of the object from its projections can be obtained by using one of many conventional back-projection algorithms (Fig. 2). This is the projection reconstruction method, which Lauterbur used to demonstrate NMR imaging.[2]

A second method is used to obtain NMR images of a plane. First, a selective excitation is performed to define the plane, and then a projection is obtained along the x-axis, just as in projection reconstruction. From this point on, all projections are obtained along the x-axis, but a y-gradient pulse of increasing strength is applied to each projection sequence. This gradient makes nuclei lose coherence: The nuclei along the center line of the x,y-plane see no gradient at all, and those farther away see progressively stronger gradients. If the y-gradient is very weak, the loss of coherence is very gradual along the y-axis of the plane, and cancellation of signal occurs between wide bands in the object. The stronger the gradient, the more such bands are produced, until, for a 128-line image, the last projection generates 128 bands among which cancellation of signal occurs. This process encodes y-axis spatial distribution in a form that is mathematically identical to the form in which the time domain signal encodes x-axis information; thus, the y-direction is called the pseudo-time-domain direction. Just as with the x-axis, information along the y-axis is decoded by an FT. The FT has to be performed twice; first, say, along the x-axis, and then along the y-axis. Hence the name 2DFT (two-dimensional FT) for this technique (Fig. 3).[3]

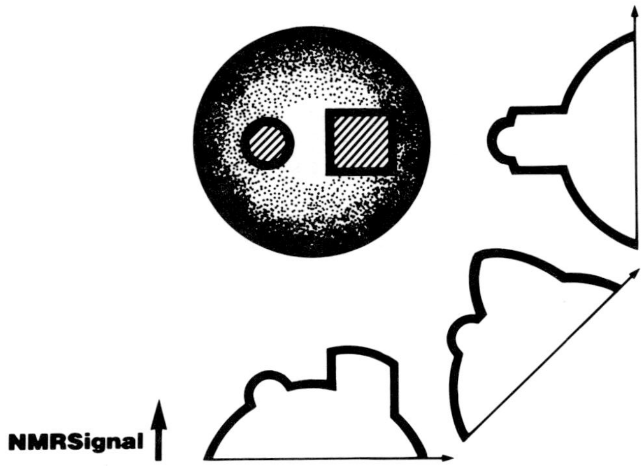

Figure 2. Projection reconstruction. Many angular projections of an object (three are shown) can be used to reconstruct the image by a conventional back-projection algorithm.

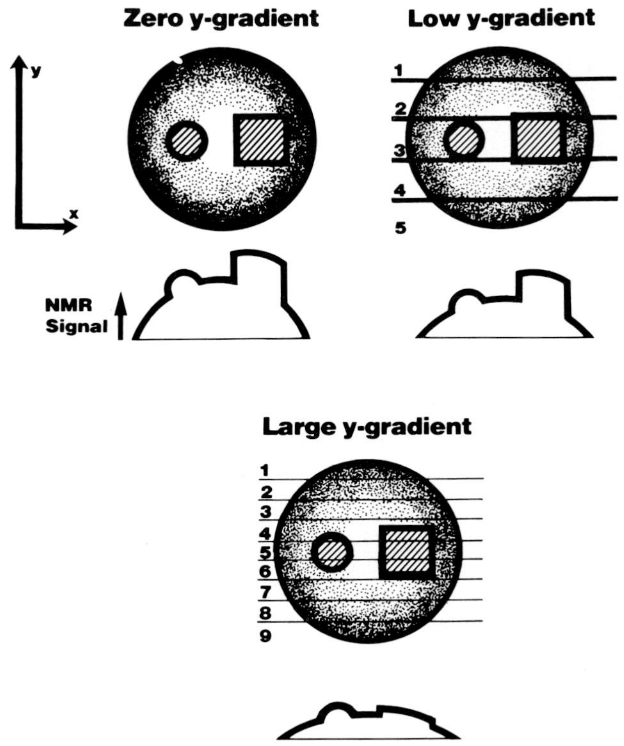

Figure 3. How 2DFT works. A projection of the whole object is obtained with no y-gradient. With a low y-gradient the object provides different polarity signals from a few bands. In this case odd-numbered bands provide positive signals, and even-numbered bands are negative. When the y-gradient increases, the bands become more numerous, so that large-scale features tend to provide signals that cancel out, while smaller features still provide full signal. The object is reconstructed by FTs along x and y.

Imaging with 2DFT has important advantages over projection reconstruction. It is less susceptible to field inhomogeneity, to the presence of magnetic prosthetics, and to motion artifacts; it has the ability to image rectangular fields of view tailored to body shape; and it has some favorable signal-to-noise characteristics.[4] It appears that 2DFT will become the most commonly used imaging technique in commercial NMR imagers.

Time Considerations in Planar Imaging

There are two distinct time scales involved in NMR imaging. The first is related to the time elapsed to acquire one projection. For a free induction decay (FID) method, this time ranges from 90° excitation to the end of the sampling of the FID—typically, some 20–50 msec. In spin echo imaging the time is longer, especially if multiple echoes are obtained. In our case, with a second echo delay of 56 msec, the single-projection time is about 75 msec. The imaging time is governed by a longer time, TR, the interval between the start of contiguous sequences. As we shall see later, clinically significant processes lead us to use TR values between 0.5 and 2.0 sec. The imaging time t is given by the product

$$t = (nN)\ TR \qquad \text{Eq. 1}$$

where N is the number of lines in 2DFT or the number of angular projections in PR (which essentially determines spatial resolution) and n is the number of times the same projection is repeated for signal-averaging purposes. When N is 128 lines, n is 4, and TR is 1.5 sec, then t is 12.8 min. The actual time during which the imaging process is active is given by

$$t' = (nN)\text{sequence time.}$$

in our case 0.64 min, or 5% of the imaging time. We say in this case that the duty cycle of the imager is very low (5%). Note that the speed limitation is not intrinsic to the hardware (as in x-ray CT), but arises from the need to allow TR to elapse.

Volume Imaging

There are two distinct ways of imaging a volume with NMR. These are multisection and three-dimensional (3D) imaging. In 3D imaging the 2DFT process is repeated s times without section selection (s is the number of sections desired), each time with a different value of z-gradient applied prior to readout of the FT data. This process encodes z-axis information in exactly the same way as it is encoded along the y-axis. Reconstruction is by a 3DFT method, where the FT is applied sequentially three times. The imaging time is

$$T = (nNs)\ TR \qquad \text{Eq. 2}$$

The second method is multisection imaging, which cycles through sections of the object during the interval TR. As noted above, the actual time spent carrying out a projection is small (in the previous example, 5% of the imaging time), and 95% of the time is spent waiting for T_1 recovery of magnetization (TR). While a section cannot have a new sequence started for a time TR, an adjacent section can undergo a sequence, and so on, until TR is used up. If, for instance, we cycle through sections in 100 msec, in an interval in which TR is 1.5 sec, 15 sections can be imaged. Fourteen of these are being sampled in a time that would otherwise be lost. Therefore, multisection imaging produces images of different sections in the time it would take to image just one, without sacrificing image quality. The average imaging time per section is fast, on the order of 25–100 sec in our imager,[5] which compares favorably with x-ray CT when tube cooling intervals are included in the imaging time.

Time Considerations in Volume Imaging

Given a certain value of N (spatial resolution), TR (T_1 contrast), and n (averages), the imaging time in 3D imaging mode is s (the number of sections) longer than for a single-section imaging mode. In multisection imaging the number of sections (which in our case is TR/0.1 sec) does not affect imaging time, since s sections are obtained in the same time that one is, without a loss of image quality.[5] The signal-to-noise levels for 3D and multisection imaging are proportional to $\sqrt{(nNs)}$ and $\sqrt{(nN)}$, respectively, giving an advantage of $\sqrt{s}$ to the former, but at the expense of a time longer by s. If both imaging techniques were compared at equal imaging time, s times more averages could be obtained in multisection imaging, leading to a gain of $\sqrt{s}$ in signal-to-noise; that is, for equal time and s, performance is equivalent. Table I compares these techniques with an example in which TR is 1.5 sec, a useful value in imaging the head. The data acquisition and data processing advantages of multisection imaging, even when three views are obtained, are obvious. Three-dimensional imaging can only be practical for short values of TR (e.g., 0.1 sec), which has limited clinical utility.

TABLE I

Comparison of Volume Imaging by Multisection and Three-Dimensional Techniques

Multisection $t = (nN)TR$	*3D* $t = (nNs)TR$
No. of sections = TR/0.1 sec For n = 2 N = 256 TR = 1.5 sec t = 13 min/15 sections	No. of sections = s For n = 1 N = 256 TR = 1.5 sec t = 96 min/15 sections
× 3 for coronal, sagittal, and transverse planes t = 39 min imaging time + 5–6 min returning = 44–45 min	× 1 for all views

Rapid -Frame Imaging

By rapid switching of gradient power supplies, it is possible to obtain very fast spin echoes (1/msec). Since T_2 varies from 20–300 msec, these rapid echoes experience little T_2 decay. Each echo can be used to obtain one of the projections needed for image reconstruction, in this manner accumulating many or all the necessary projections following a single irradiation. The resolution is then limited by T_2 decay, which determines how many adequate spin echoes or projections can be obtained. This method is called echo-planar imaging[6] and has produced cardiac images in animals in 32 msec. Extension to whole-body imaging poses a challenge in terms of gradient power supply capabilities, but, if successful, opens the way to real-time NMR imaging.

Signal Acquisition

At the risk of excessive repetition, the most unique aspect of NMR imaging is that the object representation differs for different signal acquisition techniques. These differences are not the obvious ones that can be obtained by data manipulation. For instance, in x-ray CT all pixels with intensities similar to bone can be set black, white, or in between. This does not mean that the problems associated with image formation in the presence of bone disappear. Similarly, NMR gray scales can be set in an arbitrary manner, but the significant fact is that data can be acquired so as to highlight particular processes. For instance, in a particular technique the pancreas and surrounding fat appear of similar intensity. A different image acquisition technique can make fat the most intense object in the image while giving pancreas a very low intensity. A point that is often missed is that spatial resolution for all signal acquisition techniques can be the same, but that anatomic resolution varies because of the technique-generated contrast among organs. More importantly, pathologic processes evident with one technique can be totally missed with another, so that evaluation of imaging techniques needs to go beyond the search for "good anatomy." The following is a brief description of the imaging techniques now in use. Their clinical applicability is expanded in the following chapters.

Spin Echo Imaging

The relationship between image intensity, tissue, and imager parameters in spin echo imaging was described in Chapter 2. To recount,

$$I = N(H)\ f(v)\ \exp(-TE/T_2)\ [1-\exp(-TR/T_1)] \qquad \text{Eq. 3}$$

where I is the image intensity, $N(H)$ is the hydrogen density, $f(v)$ is the previously discussed velocity effect, TE is delay between RF excitation and the spin echo, and TR is interval between excitations. T_1 and T_2 are the tissue relaxation times.

This technique is probably the single most useful type of NMR imaging mode from the point of view of clinical utility. Spin echo imaging is sensitive to T_1 and T_2 relaxation times, as well as to hydrogen density and flow, and this sensitivity can be easily "tuned" over a wide range of tissue relaxation times. In addition, each tuning range provides image contrast for a significant spread in relaxation time differences between two tissues. This is equivalent to saying that spin echo imaging has a wide dynamic range.

Clinically significant differences in contrast between normal and abnormal tissues are evidenced when TR is changed from 0.5 to 2.0 sec.

Saturation Recovery Imaging

Although saturation recovery imaging is treated as a distinct signal acquisition mode, it is just FID or spin echo imaging in which TR is kept short compared with T_1, for instance, TR is kept in the range of 0.1 to 0.5 sec. In commercial parlance, saturation recovery applies mostly to TR of 0.1 sec. It is used because it permits the acquisition of a large number of lines in a short time. From a technical point of view this advantage has to be weighed against the disadvantages associated with very short TR values: low signal and impairment of multisection capabilities. From a clinical point of view, the more important question involves the diagnostic utility of imaging with TR of 0.1 sec, on which there are limited data.

Free Induction Decay Imaging

Free induction decay imaging can be thought of as spin echo where TE equals 0. The lack of T_2 decay provides strong signals, but no T_2 information. As we discussed in Chapter 2, the FID overlaps in part the RF excitation pulse. As a result, FID imaging is typically carried out with very short values of TE (10–20 msec), making it, in reality, spin echo imaging.

Inversion Recovery Imaging

Compared with the techniques already discussed in this chapter, inversion recovery imaging offers a different set of capabilities. Basically, they are those of a high-contrast film, where within a certain narrow range of T_1 relaxation times the contrast is high. The high contrast is accomplished by first inverting the magnetization with a 180° pulse, and then allowing it to recover, for a time called T_1, before applying a 90° pulse and obtaining a signal. The signal obtained is either an FID image or, more typically, a spin echo image. The NMR signal intensity is made positive whether or not the net magnetization is along or against the main magnetic field. The intensity goes through zero as the magnetization changes from the negative to positive direction (Fig. 4). Tissues with slightly different values of T_1 cross the axis at different times, so that contrast is high for small changes in T_1 time (Fig. 4). The problem is that longer values of T_1 tend to blend with each other, so that differentiation is lost outside of the narrow range for which this technique is designed.

Imaging Elements Other than Hydrogen

A number of elements other than hydrogen that participate in metabolic processes in the body can be imaged by NMR, but with difficulties. These elements are referred to as "X." The difficulties in imaging arise from many basic problems. Because

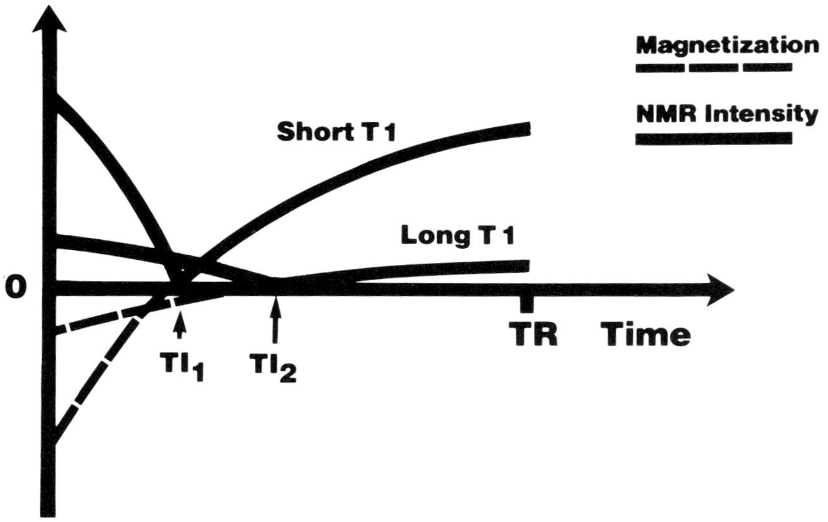

Figure 4. In inversion recovery imaging the magnetization is inverted and allowed to recover for an interval TI. The 90° excitation pulse is applied at TI, and the process is repeated at intervals TR. If the inversion time is TI_1, the longer T_1 values will yield larger NMR intensities (which are always positive). For TI_2, the shorter T_1 value will yield the higher intensities.

the mass of any X is higher than that of hydrogen (H), the sensitivity per unit weight of X is down by at least the atomic mass; for instance, per unit of mass, fluorine is about 20 times less sensitive than H. All have lower concentrations than hydrogen. These elements require higher fields to obtain the same resonant frequency, for instance, 2.5 times higher for phosphorus. Many of them have complex energy states; nitrogen, for instance, can adopt three energy states, making NMR imaging much more difficult. The isotopic abundance of the NMR-sensitive nucleus can be low; for instance, C-13 is just 1% of natural carbon, which is mostly composed of the even-even isotope C-12. Finally, not all of element X may be found with T_1 and T_2 relaxation times that are within an imageable range. In the brain, only one-seventh of phosphorus can be imaged. These factors are considered by Kramer in a more detailed treatment.[7]

These factors lead to some rather straightforward conclusions: The two most attractive nuclei for NMR imaging of endogenous elements are sodium and phosphorus. Sodium is better than phosphorus not only because of its relatively higher concentration in tissues, but also because it has a short T_1, allowing rapid repetition of imaging sequences. Even at high fields their NMR sensitivity (signal per unit volume) is thousands of times less than that of hydrogen. This translates in the image as larger voxels (sensitivity being compensated linearly by volume), longer imaging times (which need to increase as the square of the loss of sensitivity), or both. A compensating factor may be that if variations in X are larger than in H for pathologic tissues, lower signal-to-noise levels may be acceptable. With all factors considered, we may expect that with state-of-the-art technologies, images of X will require long data acquisition times, an assertion supported by recently published sodium images of the cat head, which required hours to obtain.[8]

REFERENCES

1. Garroway AN, Grannell PK, Mansfield P: Image formation in NMR by a selective irradiative process. J Phys[C] 7:457, 1974.
2. Lauterbur PC: Image formation by induced local interactions: Examples employing nuclear magnetic resonance. Nature 242:190, 1973.
3. Kumar A, Welti D, Ernst R: NMR Fourier zeugmatography. J Magnet Reson 18:69, 1975.
4. Crooks LE, Ortendahl DA, Kaufman L, et al: Clinical efficiency of nuclear magnetic resonance imaging. Radiology 146:123, 1983.
5. Ortendahl DA, Kaufman L, Crooks LE: A comparison of projection reconstruction and two-dimensional Fourier transformations in NMR imaging. IEEE Trans Nucl Sci NS-30:692, 1983.
6. Mansfield P, Ordidge RJ, Redzian RR, et al: Real-time dynamic imaging by NMR. J Magnet Reson Med (in press).
7. Kramer DM: Imaging of elements other than hydrogen. In *Nuclear Magnetic Resonance Imaging in Medicine,* Kaufman L, Crooks LE, Margulis AR (Eds), Chap 8. New York: Igaku-Shoin, 1981.
8. Maudsley AA, Hilal SK, Simon HE, et al: Multinuclear NMR imaging. J Magnet Reson Med (in press).

4

Instrumentation

LEON KAUFMAN, Ph.D.
LAWRENCE E. CROOKS, Ph.D.

As discussed in the previous two chapters, the NMR imaging process requires a strong, uniform, and stable magnetic field to magnetize the sample. It also needs rapidly changing, weak magnetic-field gradients along the x-, y- and z-axes, transmission and reception of radio-frequency (RF) fields, and data-processing equipment. The hardware that provides each of its needs can be considered as a set of separate subsystems that are integrated into an NMR imager (Fig. 1). Although for descriptive purposes they can be considered in this manner, in fact there are subtle and not so subtle, but significant impacts of subsystem design on performance. While the magnet is the tree that dominates the view of the forest, it is not necessarily the component that most uniquely determines system performance. In the following section we provide subsystem descriptions and then consider some of their interactions when integrated into an imaging system.

Magnets

The technical requirements of a magnet are that it provide a magnetic field of adequate strength, that over the field of view the uniformity requirements of the imaging techniques in use be satisfied, and that it be stable. Another important consideration is reliability. Other considerations that affect widespread utilization are ease and cost of installation, which will be dealt with in Chapter 20. Although the importance of cost is not to be minimized, we must note that other subsystems can approach or match the cost of the magnet, so that it, by itself, is not necessarily the price-driving component.

Permanent Magnets

The simplest way to generate a magnetic field is to use a permanent magnet. Two problems have hampered the use of this approach. The first is related to aperture and field strength. To increase either of these parameters requires large increases in magnet size, to weights as large as a quarter-million pounds. In addition, the uniformity of the field is determined by exact pole-tip shapes, so that temperature equalization to 0.1°–0.01° F over the volume of the magnet is necessary. The invention or discovery of a material with higher magnetism or lower temperature coefficients, or both, than those now available would greatly aid this kind of approach.

Iron-Core Magnets

A current is surrounded by a magnetic field, which can be used to induce magnetism in iron. Since the iron concentrates and shapes the magnetic field lines, this is a very efficient way to obtain relatively large (up to 20-kGauss) magnetic fields, and has been used very successfully (Fig. 2) for imaging small objects (6–7 cm in diameter). When scaled for human-size imaging, this approach presents serious drawbacks, similar to those of permanent magnets: size and temperature stability become serious issues.

Air-Core Resistive Magnets

Current-carrying conductors can be configured to directly produce magnetic field with the desired uniformity, without depending on iron to guide and shape the magnetic field lines. Resistive conductors dissipate energy at a rate dependent on the square

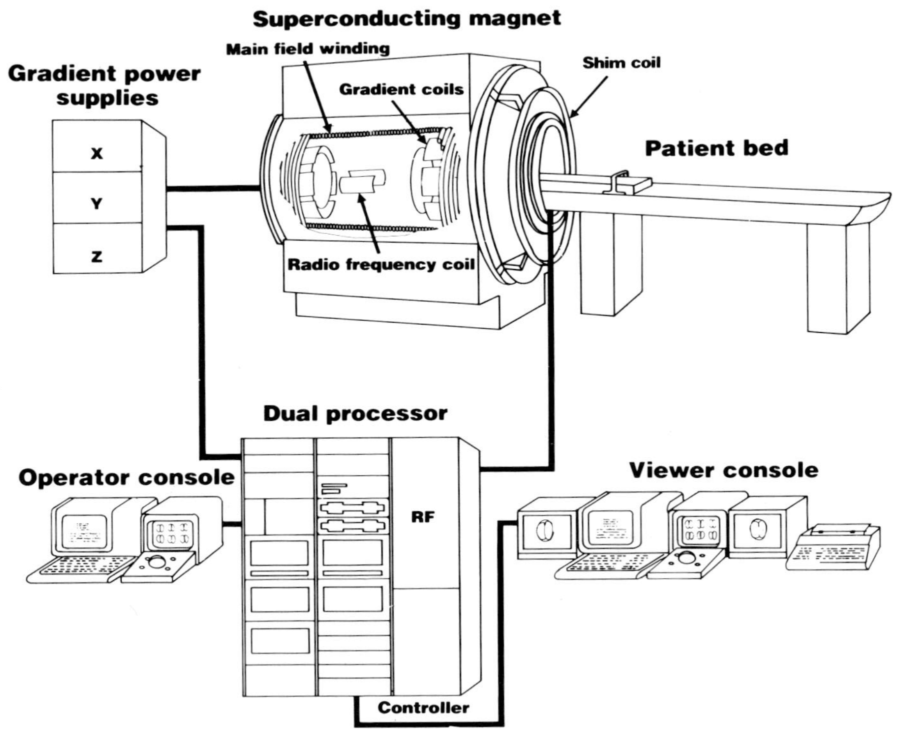

Figure 1. The basic components of an NMR imaging system. The two large coils outside of the magnet are used for field uniformity corrections.

of the current that they carry. Therefore, a doubling of current and the magnetic field it produces necessitates a quadrupling of the power needed to drive the magnet. These power requirements dictate that the desired field be produced as efficiently as possible. The ideal conductor configuration would have wires wound on the surface of a spherical former, the wires running along the "latitudes" (Fig. 3). Inside of this sphere the field is perfectly uniform, pointing along the axis perpendicular to the plane in which current circulates. This geometry is impractical when there is a need to have patients placed in the field, but a reasonable approximation is the "four-coil" magnet, in which a large and smaller pair of coils are wound on the surface of an imaginary sphere (Fig. 4). These magnets can produce fields of up to 2 kGauss, but require power levels on the order of 80 kW or more to do so. They present other problems that make their use in high-quality imagers unlikely: Current-driving requirements necessitate large power supplies and considerable cooling by water, both of which lead to high operating costs and low reliability. Temperature changes and limitations in power supply regulation lead to unstable magnetic fields. To reduce power consumption requirements, the coils are made as small as possible, but this affects field uniformity

Figure 2. H-shaped iron-core NMR magnet uses current to magnetize the iron, and field uniformity is determined by pole tip shape. (H, direction of magnetic vector)

over human-size objects. The attractive aspects of these magnets are that they are easily obtained or produced and have low capital costs. Because of these characteristics, early human NMR imaging programs used four-coil magnets (Fig. 5), but these were quickly superseded by a more sophisticated technology.

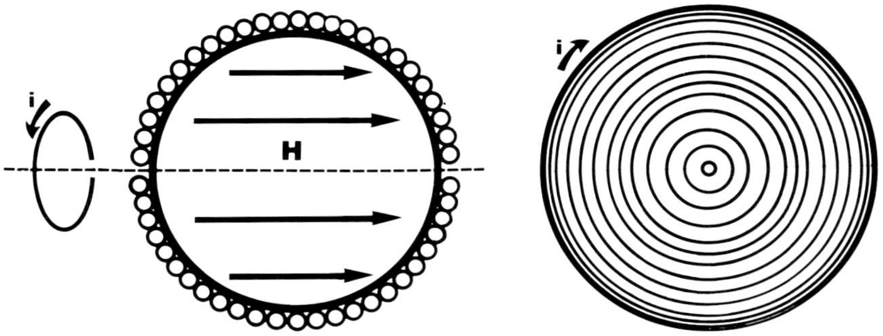

Figure 3. A set of coils along the latitudes of a sphere produce a perfectly uniform field inside of it. View on the left is a cross-section. View on the right shows the current-carrying coils when viewed along the direction of the magnetic field. (H, direction of magnetic vector; I, direction of electric current)

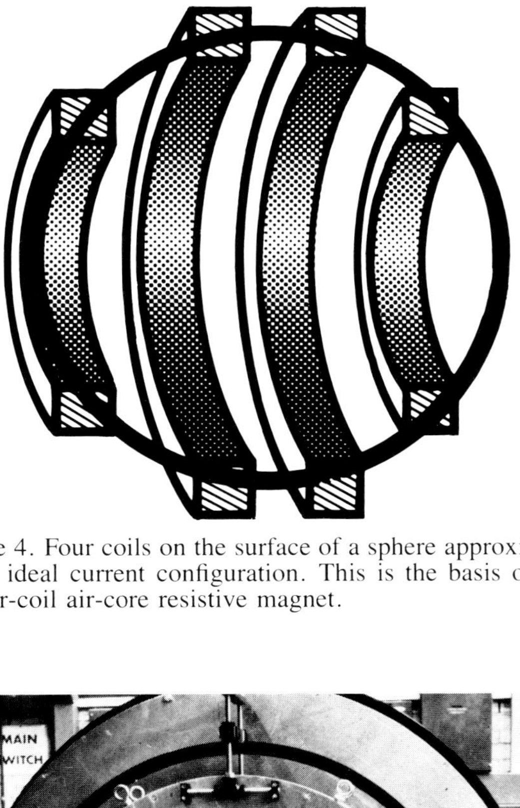

Figure 4. Four coils on the surface of a sphere approximate the ideal current configuration. This is the basis of the four-coil air-core resistive magnet.

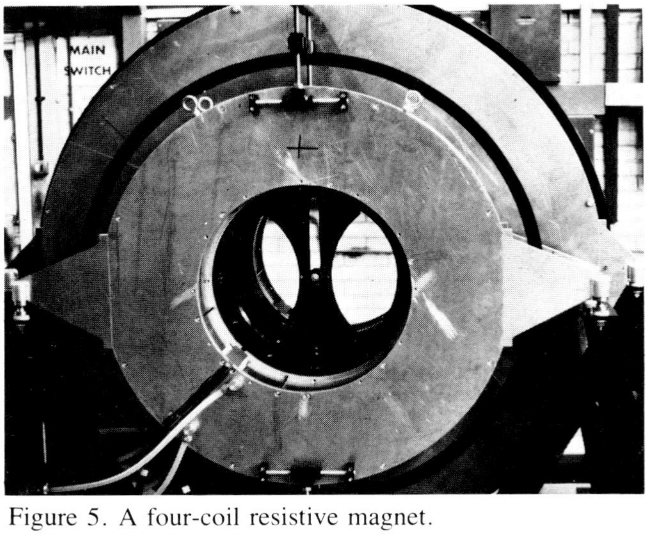

Figure 5. A four-coil resistive magnet.

Superconducting Magnets

A large number of metals and alloys lose all traces of resistance to current flow when cooled to near absolute zero. The most widely used superconducting material is niobium-titanium (Nb-Ti). To form it into wire, billets of Nb-Ti are placed in copper tubes, which are then stacked together. These large billets are drawn many times through progressively smaller dies, until very fine wires are obtained (Fig. 6). The copper serves two purposes: First, it is a ductile material that provides mechanical support for the Nb-Ti. Second, when the Nb-Ti is in the superconducting mode at liquid helium temperatures, the copper (which is the best conductor of Nb-Ti) behaves as an insulator, since it retains some resistance. If the Nb-Ti quenches (ceases to be superconducting), the copper safely dissipates the current carried by the superconductor prior to the quench.

Superconducting magnets offer a number of advantages: Because efficient utilization of current is not a concern in NMR imaging fields, simple solenoidal windings of large diameter provide the needed uniformity over human-size apertures (Fig. 7); magnetic fields can be higher than with other technologies; and stability is excellent. Since energy is dissipated only where there are joints in the wire, the field can be maintained for a long period of time, decaying at a rate of a fraction of a part per million (ppm) per hr. A 0.3 ppm/hr decay rate in a 3.5-kGauss field means a loss of 1 mGauss/hr. For a gradient of 1 Gauss/cm, a 1-mm resolution element spans 100 mGauss, so that over a half-hour procedure the superconducting magnet field drifts over 1/200 of a pixel, an insignificant amount.

The advantages of superconducting magnets over other approaches come coupled to some disadvantages. The magnets are more expensive to manufacture

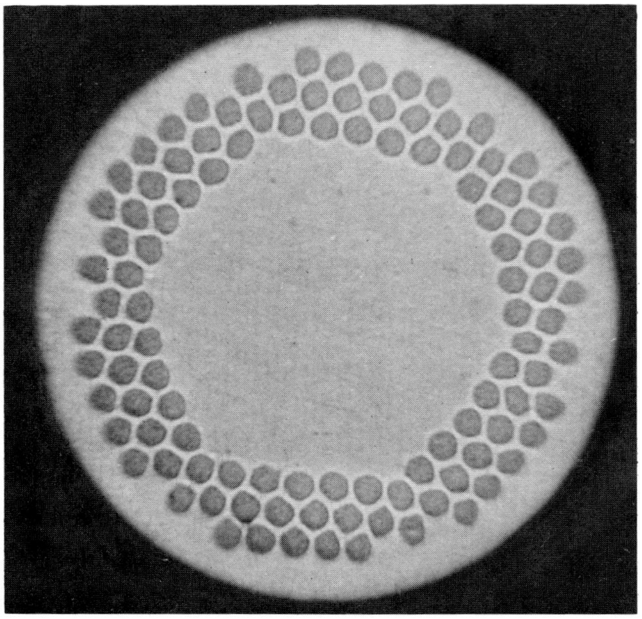

Figure 6. Cross-section of superconducting wire.

than other magnets, raising capital costs for imagers. The savings in electricity and cooling water are offset by the costs of cryogens (the liquid helium to cool the wires and the liquid nitrogen used as a thermal blanket). Their high fields cause concerns regarding

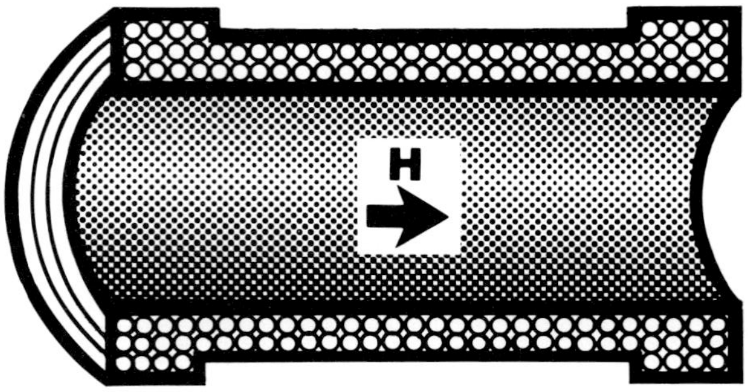

Figure 7. Solenoidal wiring for a superconducting magnet, with end corrections. (H, direction of magnetic vector)

installation needs and hazards. These particular problems are discussed in Chapter 20. Refrigerators that totally eliminate the need for liquid nitrogen, liquid helium, or both are feasible, even though they will raise capital costs further (Fig. 8). Even so, the cost of the magnet constitutes only a fraction of the system cost. All other components being equal, and the installation costs of resistive and superconducting magnets being about the same, the latter costs 25% more than the former. These differentials are in practice larger, since the lower inherent performance of resistive magnets leads to downgrading of the performance of other subsystems, resulting in a cost differential of approximately 50%. As of this writing

Figure 8. Refrigerated NMR imaging superconducting magnet.

Figure 9. Superconducting magnet in a typical installation configuration.

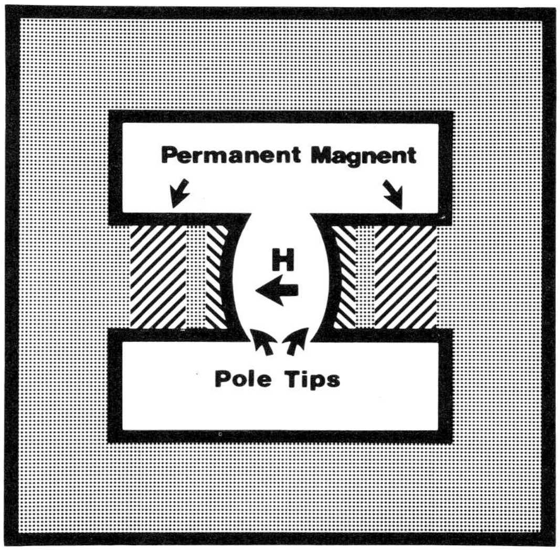

Figure 10. Iron-permanent magnet hybrid configuration. (H, direction of magnetic vector)

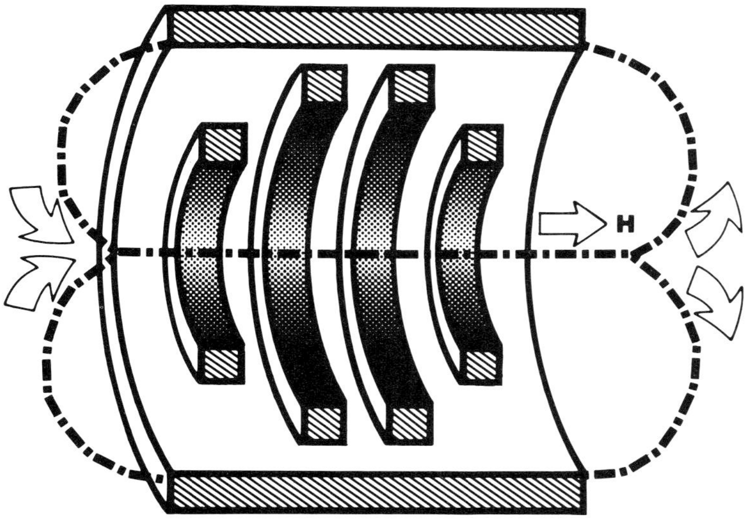

Figure 11. Resistive-iron shielded hybrid configuration. (H, direction of magnetic vector)

it appears that high-performance commercial NMR imagers will be based on superconducting magnets (Fig. 9), with only some companies offering resistive magnets as a lower-cost, lower-performance alternative.

Hybrid Magnets

Given the basic magnet types, hybrid configurations can also be built. For instance, instead of driving an iron-core magnet with a current, it can be magnetized by a block of permanently magnetized material (Fig. 10). Iron cylinders can be used around resistive or superconducting magnets to provide return paths for the magnetic field lines (Fig. 11). Resistive magnets can be cooled by liquid nitrogen rather than chilled water; this drastically reduces power consumption, but requires cryostats and cryogens.

In each case these hybrids can be considered as combining the best of the two technologies they join, or the worst of them, and the desirability of each approach will be demonstrated by practice.

Shimming Coils

Shimming coils are special coils provided to shape the field into desired configurations or to correct for inhomogeneities in the main field. These field-correction coils can be resistive (Fig. 1) or superconducting; the latter are favored when very high uniformity is desired. Iron pieces are also used in shimming. Typically, the fields thus induced are of a very small magnitude.

While the magnet is at the center of attention of the radiologic community, other subsystems can severely limit or degrade the imaging performance of even the most sophisticated magnets. These problems are discussed in the following sections.

Magnetic Field Gradients

As described in the preceding chapter, the imaging process requires that during one sequence there be magnetic field gradient applied along each of the three main axes. Three sets of resistive coils, called gradient coils, are provided for this purpose. A convenient gradient coil set could be two coils, each with a current flowing in a direction opposite to that in the other coil. With the appropriate dimensions, this configuration generates a linear gradient in a volume at the center of the coils. Practical considerations force deviations from this geometry. A set of gradient coils for a superconducting magnet is shown in Figures 1 and 12. Each coil is driven by a gradient power supply. For optimal performance, gradients should be strong (we use greater than 1 Gauss/cm), fast (risetimes of 1 msec or better), and controllable within precise limits. It may be fair to say that the heart of an NMR imager is the gradient power supply, since its ability to provide gradient pulses with fidelity establishes many of the limits of the imager's performance.

Radio-Frequency Antennas

Imagers may use two RF antennas, one as transmitter and one as receiver, or a single antenna switched between the transmit and receive mode. Since the RF magnetic field has to be at 90° to the main field, the antenna shape varies according to the direction of the main magnetic field with respect to the body axis. If the field is perpendicular to the axis, a solenoidal antenna can be used (Fig. 13). If it is along the axis, as in superconductors, saddle-shaped antennas are used (Fig. 14). There are also variations of these shapes.[1] On the basis of an early paper,[2] a folkloric notion developed that in human imaging solenoidal antennas are two or three times more sensitive than saddle-shaped ones. Were this to be the case, magnets that provide fields that cross the body axis would become favored. Our measurements have indicated that for equal antenna aperture and length, both antennas provide equal signal-to-noise levels. Independently of our measurements, another author has explained our results: In NMR the patient is the main source of noise, because a conductive body at

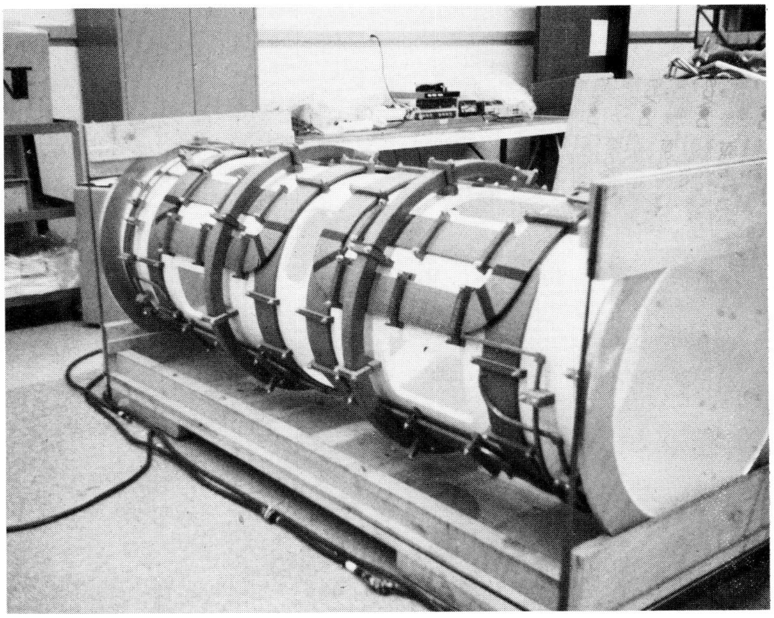

Figure 12. Gradient coils for a superconducting magnet.

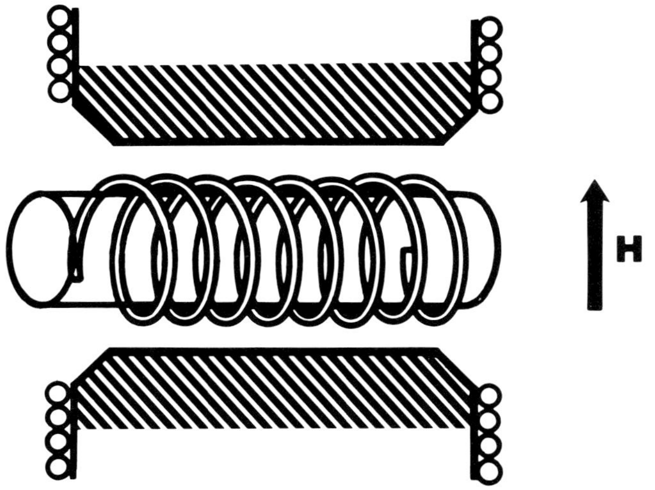

Figure 13. Solenoidal RF antenna, used when the main field is across the body axis. (H, direction of magnetic vector)

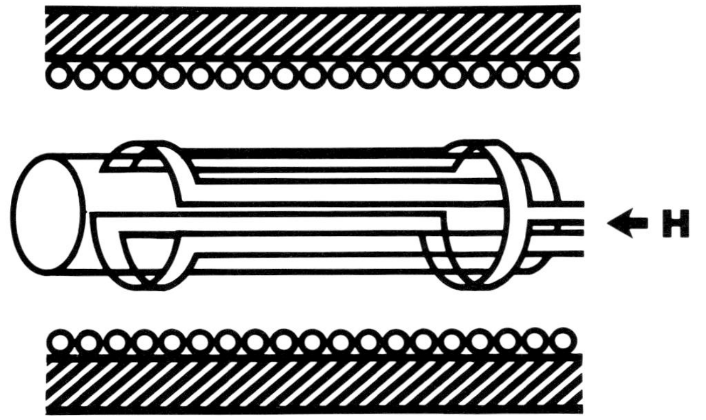

Figure 14. Saddle-shaped RF antenna, used when the main field is along the body axis. (H, direction of magnetic vector)

a finite temperature has random currents that are picked up by the coil as noise.[3] The solenoid is indeed more sensitive to NMR signals from the body, but it is also more sensitive to the (indistinguishable) noise signals from the same body. Therefore the signal-to-noise ratio, the significant parameter, is equal for both antenna shapes except for patients cooled to cryogenic temperatures. Also in use are surface coils, typically circular or elliptical, which are placed on the body, nearest to the organ of interest. The proximity of the antenna and its small, sensitive volume realize a larger signal-to-noise level, at the expense of poor uniformity of response and small imaged volume.

It is important to note that as the strength of the field increases, it becomes more difficult to obtain efficient antennas unless their size is decreased. This has consequences in terms of uniformity of response and simultaneous imaging over significant portions of the human body.

The antennas are driven by suitable electronics capable of RF and amplitude modulation to meet imaging needs.

Computers for Nuclear Magnetic Resonance Imagers

Early NMR imagers used computers devised from CT scanners. Attached to this computer there usually is a sequence-control computer (controller) that generates the magnetic gradient and RF pulses, as well as the signal reception intervals. It quickly became evident that CT computers will be a limiting factor in NMR performance. Consider the following rationale: An x-ray CT scanner generates an image in 1–20 sec. Reconstruction times range from 0–60 sec. The longer processing times would seem unacceptable when compared with the acquisition times, but, in fact, tube cooling intervals between sections are about as long.[4] Therefore, relatively long single-section reconstruction times would not significantly affect throughput. In NMR, multisection, double-echo imaging generates images at a rate of one section (two images at different echo delays) every 26 sec.[5] Because the speed of throughput can be limited by data-flow time (acquisition to archiving), fast reconstructions are a necessity, as is large data-storage capacity. Three-dimensional reconstructions increase computational complexities by orders of magnitude, imposing even more stringent requirements on data-processing hardware. Thus, high-throughput systems have to be carefully configured to maximize performance/cost trade-offs. Another difference between CT and NMR is that color displays are useful in the latter modality.

The consensus is that color has not yet proved useful in either x-ray or nuclear medicine imaging. The use of the color spectrum as an intensity scale is often more confusing than informative because we do not perceive a change in hue as a simple change in intensity. The transition between blue and green, for example, is perceived as more decisive than the transition between two shades of gray. Consequently, in a color image, a distinct edge is perceived whenever a small change in intensity involves a shift in the color spectrum. Therefore, to create meaningful color images, one must have an imaging system in which object contrast is great enough, and spatial resolution is fine enough, for shifts in color to give a fairly reliable indication of morphological features. In CT, contrast between soft tissues is low, and color has been used to try to enhance this contrast. However, the application of the entire color spectrum to an objectively small range of intensities is misleading.

Fractional shifts in the intensity scale are perceived as substantial shifts; edges are perceived where no morphological distinctions exist.

In nuclear medicine imaging, object contrast is large, but spatial resolution is poor. In a black-and-white image, poor spatial resolution will be perceived as blurring, a slow decline in the intensity scale from peak value to background value. In the color image, however, the decline in intensity is indicated by transitions through many colors, each of which the eye perceives as distinct. Thus, what in the black-and-white image would be perceived as inadequate delineation, in the color image is perceived as excessive, and falsely significant, delineation. Nuclear magnetic resonance combines the high contrast of nuclear medicine with spatial resolution comparable to that of x-ray CT; therefore, color is not likely to produce false edges or confusing features and has become a useful technique to highlight the specificity of NMR images.

Another area of potential use for color displays is to synthesize the different parameters provided by NMR images (hydrogen density and T_1 and T_2 relaxation times). If the composite image is in black and white, features that are appreciably enhanced by one or the other of the imaging parameters can be tinted to combine in one image the information provided from several images. Color can be used in a similar fashion to synthesize information from NMR images of other endogenous elements with the hydrogen image. Because of the variety of information that NMR images can provide, it is important to develop color display methods that allow for the concise and meaningful transmission of this information.

Whatever limitation computers introduce will soon be negated, since computer technology will soon produce thoughtfully designed NMR computers that will be upgradable on a component or addition basis, rather than by total replacement.

Examples of Subsystem Interaction

In light of the foregoing discussion, there are obvious cases of subsystem interactions: Computer capacity can negate multisection imaging, weak gradient fields may not be able to overcome main-field inhomogeneities produced by magnet design, fabrication tolerances, or iron in the environment. This then forces more stringent magnet designs, or more complex shimming (field-correcting) assemblies, both driving costs upward. Another example of interaction is that gradient pulses can induce currents on shimming coils, negating their utility. Also, RF pulses can couple into gradient supplies and trigger them at the wrong time.

There are more subtle examples as well. Gradient pulses induce eddy currents in the magnet cryostat, increasing cryogen consumption. This consumption becomes dependent on imaging technique, since multisection imaging has many more gradient pulses per unit time than comparable single-section imaging. Also, high-resolution imaging, obtained by increasing gradient strength, raises consumption.

Moreover, higher fields produce increased NMR signals, but also lengthen T_1 relaxation times, which results in less signal in a given imaging time. Higher fields also increase stresses in gradient coils, decrease antenna efficiency, and result in higher absolute field inhomogeneities.

This brief list of interactions is not meant to be all-inclusive, or even to indicate the most significant ones. Rather, it is hoped that it conveys a sense of the complexity of trade-offs involved in the design of NMR imagers and in imaging technique.

REFERENCES

1. Bottomley PA: Instrumentation for whole-body NMR imaging. In *NMR Imaging,* Witcofski RL, Karstaedt N, Partain CL (Eds), pp 25–32. Winston-Salem: The Bowman Gray School of Medicine, Wake Forest University, 1981.
2. Hoult DI, Richards RE: Signal-to-noise ratio of nuclear magnetic resonance experiment. J Magnet Reson 24:71, 1976.
3. King K: "Signal-to-Noise Rates in NMR Imaging" (dissertation). Madison: Physics Department, University of Wisconsin, 1982.
4. Cann CE: Advanced diagnostic imaging techniques in musculoskeletal computed tomography. In *NMR, Interventional Radiology, and Diagnostic Imaging Modalities,* Moss AA (Ed), pp 85–94. San Francisco: Department of Radiology, University of California, San Francisco, 1983.
5. Crooks LE, Ortendahl DA, Kaufman L, et al: Clinical efficiency of nuclear magnetic resonance imaging. Radiology 146:123, 1983.

RECOMMENDED READING

Becker JJ: Permanent magnets. Sci Am 223:92, 1970.

Garrett MW: Axially symmetric systems for generating and measuring magnetic fields. J Appl Phys 22:1091, 1951.

Geballe TH, Hulm JK: Superconductors in electric power technology. Sci Am 243:138, 1980.

Hulm JK, Kunzler JE, Matthias BT: The road to superconducting materials. Phys Today: 34, 1981.

Hulm JK, Matthias BT: High field, high-current superconductors. Science 290:881, 1980.

Kolm HH, Freeman AJ: Intense magnetic fields. Sci Am 212:66, 1965.

Kinzler JE, Tanenbaum M: Superconducting magnets. Sci Am 206:60, 1962.

Montgomery DB: *Solenoid Magnet Design: The Magnetic and Mechanical Aspects of Resistive and Superconducting Systems*. New York: Wiley-Interscience, 1969.

Terman FE: *Radio Engineer's Handbook*, pp 73–90. New York: McGraw-Hill, 1943.

5

Basic Principles and Biologic Research Applications

THOMAS L. JAMES, Ph.D.

In little over 3 decades since the discovery of the phenomenon of NMR, its use has spread from physics to chemistry, biochemistry, physiology, and, most recently, medicine. Part of the reason for the proliferation of NMR in so many areas is the diversity of information that can be obtained from performing different types of NMR experiments. As with many other physical methods, some knowledge may be gained empirically with NMR, for example, by comparison of results obtained with small variations in experimental conditions. However, better insight in planning and interpreting NMR experiments (especially if something unexpected occurs) can be derived from a deeper understanding of the theory underlying the experiments. This chapter is designed as an introduction to the principles of NMR as a nondestructive technique capable of revealing some of the biochemical details of intact cells, tissue, and organisms. More detailed explanations exist. The monographs of Abragam[1] and Slichter[2] are recommended for a rigorous, mathematically oriented discussion of NMR. The book by James[3] describes the principles of NMR physically, including the useful mathematical expressions without detailed derivations of those equations, and covers the use of NMR in several biochemical areas. Fukushima's and Roeder's text[4] describes many of the practical aspects of NMR. Gadian[5] has recently published a monograph that emphasizes the biologic applications of NMR, principally P-31 NMR.

Magnetic Resonance Properties

Magnetic Properties

Chapter 2 covers the basic physics of NMR. For consistency of description, we briefly summarize here some basic aspects of NMR. About half of the known atomic nuclei act as though they were spinning. A nucleus is positively charged. Anything that has a charge and spins creates a magnetic field. Thus the charged, spinning nucleus is equivalent to a tiny bar magnet with its orientation coincident with the spin rotation axis. One often hears the term *nuclear spin* applied to this tiny bar magnet. Imagine that this tiny magnet is placed in a strong magnetic field. The orientation of that small magnet will no longer be random; rather, the strong magnetic field will orient the nuclear magnet. There will be one most probable orientation. However, if the tiny magnet could be oriented *exactly* 180° in the opposite direction from the most probable orientation, it would be able to maintain that orientation as well. This less favorable orientation can be regarded as a higher energy state, while the more favorable orientation is a lower energy state. (This simplified analogy is reasonable for nuclei possessing a spin quantum number $I = \frac{1}{2}$, which includes most nuclei of interest in biologic studies, for example, H-1, P-31, C-13, F-19, and N-15; the texts mentioned above should be consulted for a more rigorous description.) The extent to which one orientation (energy state) is favored over the other depends on the strength of the tiny nuclear magnet and the strength of the strong magnetic field in which it is placed (*vide infra*, Eqs. 2 and 3).

Magnetic Resonance

It is possible to effect a change in the tiny nuclear magnet from one orientation to the other by applying energy of the appropriate frequency. The energy required (ΔE) is just the energy difference between the two nuclear orientations and is shown in Figure 1 to depend on the strength of the magnetic field in which the nucleus is placed:

$$\Delta E = \gamma h B_o / 2\pi \quad \text{Eq. 1}$$

where h is Planck's constant. The Bohr condition ($\Delta E = h\nu$) enables the frequency of the nuclear transition to be written as:

$$\nu_o = \gamma B_o / 2\pi \quad \text{Eq. 2}$$

Equation 2 is often referred to as the Larmor resonance equation. The gyromagnetic ratio γ is a constant for any particular type of nucleus and is directly proportional to the strength of the tiny nuclear magnet. Table I lists the gyromagnetic ratios for several nuclei of biologic interest. At magnetic field strengths typically used in NMR experiments, the frequencies necessary to fulfill the Larmor condition are in the radio-frequency (RF) range; for example, in a magnetic field of 2.35 T, the transition frequency for H-1 is 100 MHz, for P-31 is 40.5 MHz, and for C-13 is 25.1 MHz. The implications of Equation 2 are very important for imaging and for chemical shifts as described below and in other chapters.

Boltzmann Distribution and Sensitivity

In reality, we do not place one nucleus in a magnetic field. Rather a tremendous number (approaching Avogadro's number) of nuclei are in the sample that is placed in a magnetic field. The distribution of nuclei in the different energy states (i.e., orientations of nuclear magnets) under conditions in which the nuclear spin system is unperturbed by application of any RF energy is given by the Boltzmann equation:

$$N_{upper}/N_{lower} = \exp(-\Delta E/kT) = \exp(-h\nu/kT) \quad \text{Eq. 3}$$

where N_{upper} and N_{lower} represent the population (i.e., number) of nuclei in upper and lower energy states, respectively, k is the Boltzmann constant, and T is the absolute temperature. For hydrogen nuclei in a 2.35-T magnetic field ($\nu = 100$ MHz) at thermal equilibrium at room temperature, the population ratio will be 0.999984. That means for every 1,000,000 nuclei in the upper energy state there are 1,000,016 nuclei in the lower energy state. Without this small excess number of nuclei in the lower energy state, the NMR phenomenon could not exist.

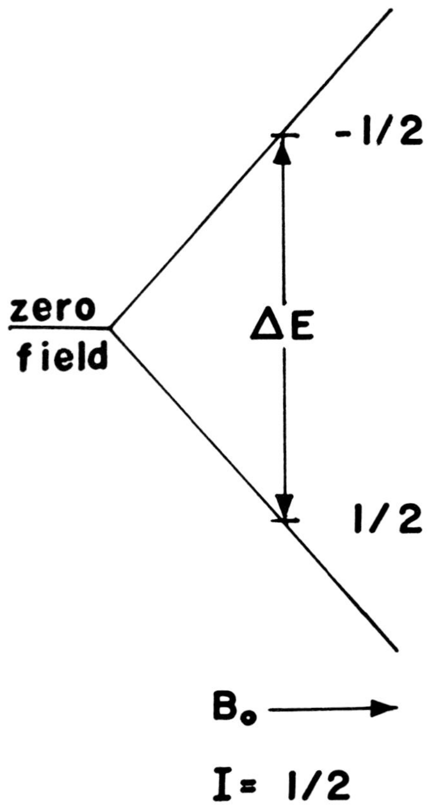

Figure 1. Splitting of the nuclear energy levels is dependent on magnetic field strength B_o.

Such a small population difference presents a significant sensitivity problem for NMR since only the difference in populations (i.e., 16 out of 2,000,016 nuclei) is detected, the others effectively canceling one another. The low sensitivity of NMR, which has its origin here, is probably its greatest limitation for applications to biologic systems. As seen from Equations 2 and 3, stronger B_o magnetic fields will increase the population ratio and, consequently, the sensitivity. Different nuclei have different inherent sensitivities; the relative sensitivities are listed in Table I.

Other factors influence sensitivity. As implied above, the signal-to-noise ratio in an NMR experiment will be enhanced as the number of nuclei in the lower energy state relative to the upper energy state increases. This increase can be achieved in either of two ways: by raising the concentration or by increasing the size of the sample detected.

For most types of experiments, the magnetic field strength B_o should be uniform across the sample; to the extent that it is not, the different nuclei in a sample will achieve the Larmor condition (Eq. 2) at different frequencies, leading to a broader signal in the NMR spectrum with a lower signal-to-noise ratio. The exact geometry of the receiver coil used

TABLE I

Properties of Selected Nuclei*

Nucleus	*Spin Quantum Number (I)*	*Natural Abundance (%)*	*Gyromagnetic Ratio (γ) (/10^7 rad/T-sec)*	*Relative Sensitivity for Equal Number of Nuclei at Constant Field (% vs. H–1)*	*Electric Quadrupole Moment (Q) (/e· 10^−24 cm^2)*
H-1	1/2	99.9844	26.752	100	N.A.**
H-2	1	0.0156	4.1067	0.965	2.77×10^{-3}
C-13	1/2	1.108	6.7265	1.59	N.A.**
N-14	1	99.635	1.9325	0.101	7.1×10^{-2}
N-15	1/2	0.365	−2.7108	0.104	N.A.**
F-19	1/2	100	25.167	83.3	N.A.**
Na-23	3/2	100	7.0764	9.25	0.1
P-31	1/2	100	10.829	6.63	N.A.**

* Modified, with permission, from Appendix 2 in James TL: *Nuclear Magnetic Resonance in Biochemistry: Principles and Applications*. New York: Academic, 1975.
**Not applicable.

in detecting the NMR signal also has an effect. These factors have been discussed by Fukushima and Roeder.[4] For biologic samples, the high dialectaic constant leads to additional signal loss. This has been discussed previously,[6] but it might be noted that a recent circuit design tends to minimize these losses.[7]

Other factors influencing sensitivity, which include the spin lattice relaxation time T_1, the spin-spin relaxation time T_2, and the length of time allowed to collect the NMR signal will be discussed later in this paper.

Nuclear Magnetic Resonance Relaxation

As already described by the Boltzmann distribution, at equilibrium there will be a slightly greater number of nuclei in the lower energy state. If RF energy is applied to the nuclear spin system at the resonance frequency, (see Eq. 2), the probability of an upward transition is equal to that of a downward transition. Since there are a greater number of nuclei in the lower energy state, there will be more transitions from the lower energy state to the upper energy state than vice versa, resulting in a non-equilibrium distribution of nuclear spins. To return to the example of protons in a 2.35-T field that possess a difference of 16 out of 2 million nuclei at equilibrium, when energy at the resonance frequency is applied, the greater number of transitions from the lower energy state to the upper energy state reduce that difference to something less than 16. In fact, it is possible for that difference to be zero when the populations of the two states are equal. Under these circumstances, the NMR signal disappears and the spin system is said to be "saturated."

There are relaxation processes, which neither emit nor absorb radiation, that permit the nuclear spin system to redistribute the population of nuclear spins. Some of these processes lead to the point where the non-equilibrium spin distribution ($N_{lower} - N_{upper}$) exponentially approaches the equilibrium distribution $(N_{lower} - N_{upper})_{equil}$:

$$N_{lower} - N_{upper} = (N_{lower} - N_{upper})_{equil} (1 - \exp[-t/T_1]) \qquad \text{Eq. 4}$$

where the time constant for the exponential relaxation is T_1, the spin lattice relaxation time. Such a relaxation curve is illustrated in Figure 2 for the example used above. The lattice is the environment around the nucleus, including other molecules in the sample as well as the remainder of the molecule containing the nucleus of interest.

Additional relaxation processes adiabatically redistribute any absorbed energy among the many nuclei in a particular spin system. This also occurs exponentially and has a time constant T_2, the spin-spin relaxation time. The linewidth of an NMR signal at half-height, $W_{1/2}$, can be related to T_2 by

$$W_{1/2} = 1/\pi T_2 \qquad \text{Eq. 5}$$

under some circumstances. However, in most biologic samples other factors contribute to make the NMR signal broader than T_2 might imply.

The relaxation processes occur by interaction of the nuclear spin with fluctuating magnetic fields produced by magnetic dipoles (e.g., other nuclei, paramagnetic ions, etc.). The magnetic fields fluctuate as a result of random molecular motions, both rotational and translational. The nature and the rate of

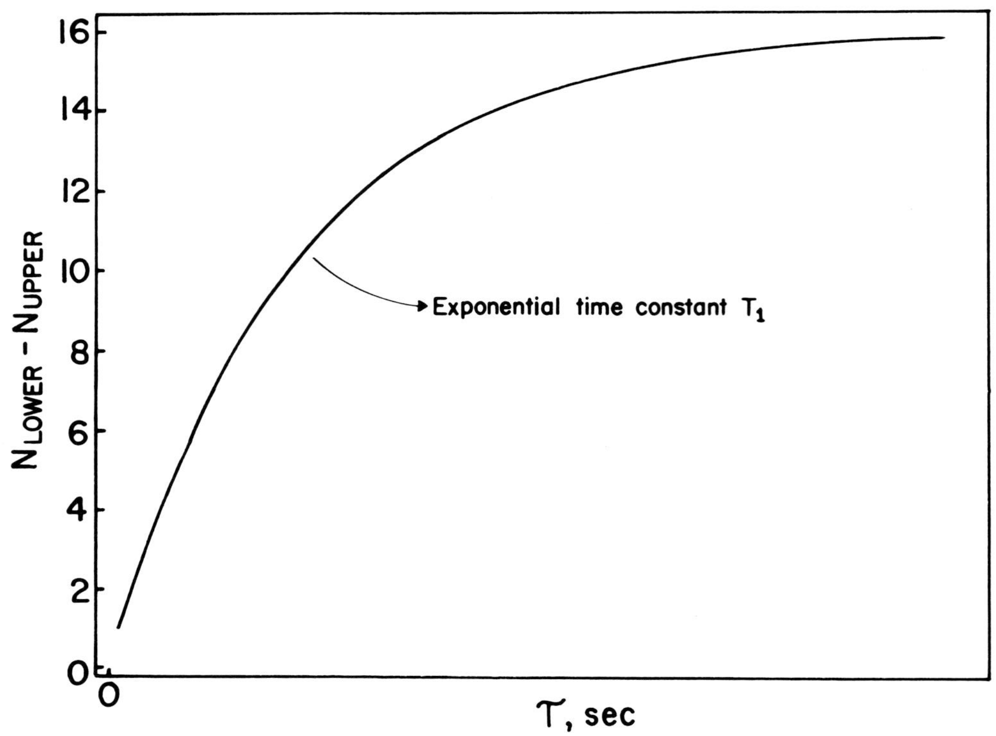

Figure 2. Spin-lattice relaxation of a group of 2,000,016 hydrogen nuclei in a magnetic field of strength 2.35 T following saturation of the spin system. The time constant is T_1 for the exponential relaxation of the spin system toward the value of 16, the population difference for that group of nuclei at thermal equilibrium.

the molecular motions affect the T_1 and T_2 relaxation times. Molecular motions that occur at a rate comparable with that of the resonance frequency ν for the nucleus are most effective in promoting spin lattice relaxation, that is, yield the lowest values for T_1. T_2 values can be decreased even further as the molecular motion becomes slower than ν, but T_1 values will begin to increase. These relationships can be expressed generally in terms of the following equations:

$$1/T_1 = \gamma^2 H^2\{\tau_c/[1 + (2\pi\nu\tau_c)^2]\} \qquad \text{Eq. 6}$$

and

$$1/T_2 = \gamma^2 H^2\big(\tau_c + \{\tau_c/[1 + (2\pi\nu\tau_c)^2]\}\big) \qquad \text{Eq. 7}$$

where H^2 is a measure of the strength of the interaction between the nuclear spin and the fluctuating magnetic fields and the τ_c is the correlation time. The correlation time is a quantitative measure of the rate of a molecular motion. For rotational motion, it is the length of time required to rotate through an angle of 33°.

Log-log plots of Equations 6 and 7 are shown in Figure 3. It is evident that as τ_c becomes larger, that is, the molecular motion becomes slower, the value of T_2 becomes smaller than T_1. Typically, the correlation time is 10^{-12} to 10^{-11} sec for small molecules, 10^{-9} to 10^{-6} sec for macromolecules (or small molecules bound to macromolecules), and 10^{-6} to 10^{-3} sec for some motions in membranes. Clearly, from Figure 3, we expect T_1 and T_2 to be equal for small molecules but to diverge for macromolecules (or small molecules bound to them). Faster internal mo-

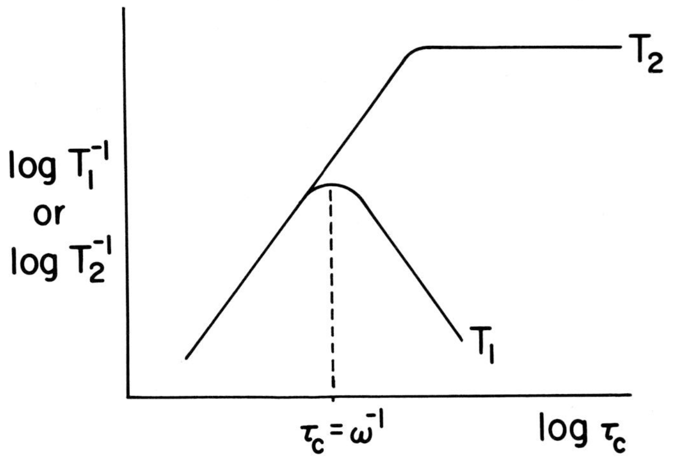

Figure 3. Dependence of $1/T_1$ and $1/T_2$ on the correlation time τ_c. ω is equal to $2\pi\nu$ where ν is the spectrometer frequency (see Eq. 2).

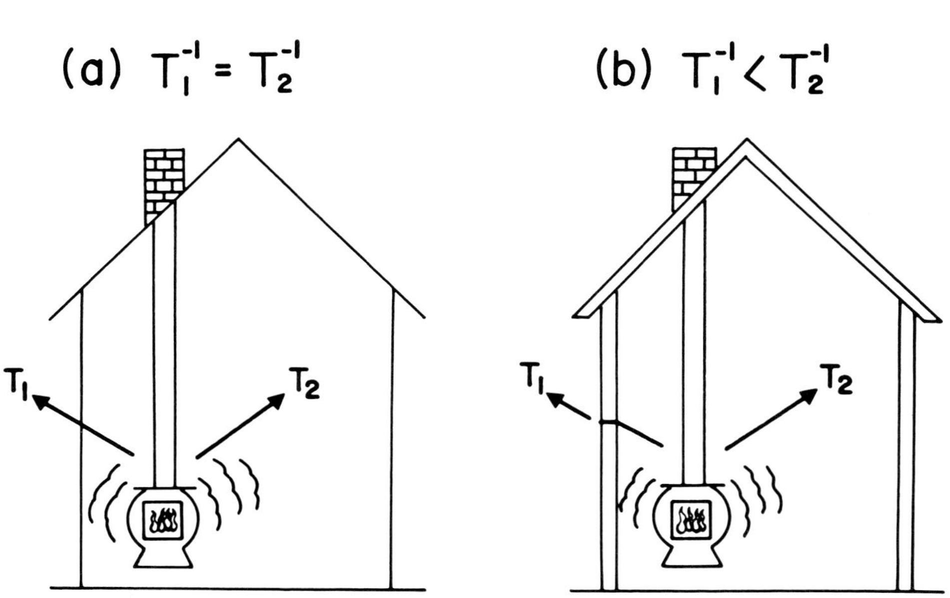

Figure 4. Stove-in-house analogy to T_1 and T_2 relaxation. In (a), the house is not insulated, so the energy emitted by the stove is lost to the surroundings environment (lattice) as fast as it is dispersed throughout the house. In (b), the house is insulated so that the energy from the stove is not dissipated to the surroundings nearly as fast as it equilibrates in the house.

tions can exist in a macromolecule or membrane, which can make the situation somewhat more complicated. One discussion of internal motions and NMR can be found in an article by James.[8] One important result of this discussion is that we might anticipate that NMR signals arising from macromolecules (or bound ligands) will be much broader than signals for small molecules in solution (see Eqs. 2 and 7). In fact, some signals may be too broad to detect.

We may summarize this discussion of T_1 and T_2 with the rough analogy depicted in Figure 4. Energy in the form of heat from the stove is put into the system (house). Relaxation processes will then tend to return the house to its equilibrium condition (cold). Adiabatic T_2 processes will distribute the energy (heat) throughout the system (house), but T_1 processes will dissipate the energy to the out-of-doors (lattice). The uninsulated house on the left of the figure represents the situation with small molecules, that is, where $1/T_1$ equals $1/T_2$. The house on the right is insulated, so that $1/T_2$ is greater than $1/T_1$ and consequently represents macromolecular systems. (Warning: Do not try to push the analogy any further!)

Other Nuclear Magnetic Resonance Parameters

Chemical Shift

It is evident from Equation 2 that the nuclei of different elements, having different gyromagnetic ratios, will yield signals at different frequencies in a particular magnetic field. However, it also turns out that nuclei of the same type can achieve the resonance condition at different frequencies. This can occur if the local magnetic field experienced by a nucleus is slightly different from that of another similar nucleus; for example, the two P-31 signals of ADP occur at different frequencies since the local field that each phosphorus experiences is different.

The basis for the variation in local magnetic fields can be understood by reference to Figure 5. If a molecule containing the nucleus of interest is put in a magnetic field B_o, simple electromagnetic theory indicates that the B_o field will induce electron currents in the molecule in the plane perpendicular to the applied magnetic field. These induced currents will then produce a small magnetic field opposed to the applied field that acts to partially cancel the applied field, thus shielding the nucleus. In general, the induced opposing field is nearly a million times smaller than the applied field. Consequently, the magnetic field perceived by the nucleus will be *very* slightly modified from the applied field; therefore, the resonance condition of Equation 2 will need to be modified:

$$\nu = (\gamma/2\pi)B_{local} = (\gamma B_o/2\pi)(1 - \sigma) \quad \text{Eq. 8}$$

where B_{local} is the local field at the nucleus and σ is a nondimensional screening or shielding constant.

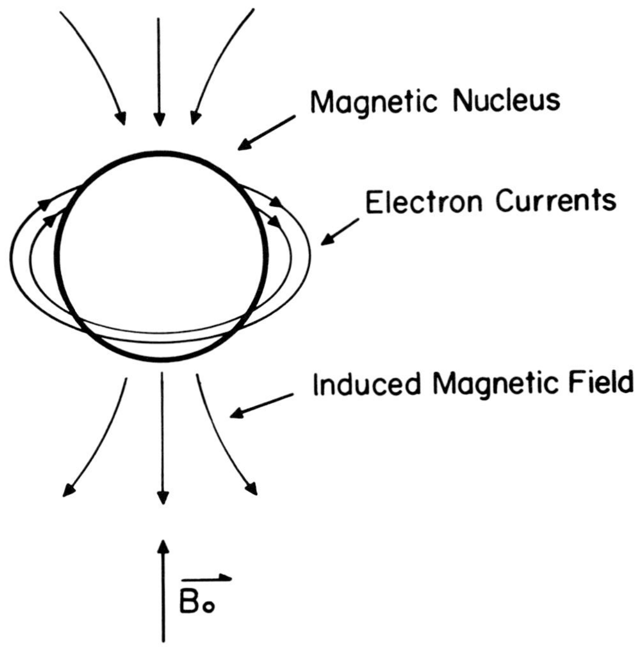

Figure 5. Electron currents around a nucleus are induced by placing the molecule in a magnetic field B_o. These electron currents, in turn, induce a small magnetic field opposed to the applied magnetic field B_o.

In actuality, there will be more electronic currents induced in the molecule than just those directly around the nucleus. In fact, some of those currents will tend to increase B_{local} (see Figure 6). Consequently, the shielding and the resulting resonance frequency will depend on the exact characteristics of the electronic environment around the nucleus.

As mentioned, the induced magnetic fields are typically more than a million times smaller than the applied magnetic field. Therefore, if the Larmor frequency ν_o (see Eq. 2) is on the order of several MHz, differences in resonance frequencies for two different phosphorus nuclei, for example, will be on the order of Hz. Although we cannot easily determine absolute RFs to an accuracy of ± 1 Hz, we can determine the relative positions of two signals in an NMR spectrum to an accuracy of better than ± 1 Hz. Therefore, a reference is chosen, and the difference between the position of the nuclear signal of interest and that of the reference is called the chemical shift.

The chemical shift is usually expressed in terms of parts per million (ppm), actually a dimensionless number, by the following expression:

$$\Sigma = [(\nu_{ref} - \nu_{sample})/\nu_{\text{instr}}] \times 10^6 \qquad \text{Eq. 9}$$

where the difference between the resonance frequency of the reference and the sample ($\nu_{ref} - \nu_{sample}$) measured in Hz (e.g., 50 Hz) divided by the instrument operating frequency (e.g., 200 MHz) gives the chemical shift (e.g., 0.25 ppm).

Typical ranges in chemical shifts for biologically important samples are: H-1, 15 ppm; C-13, 200 ppm; N-15, 400 ppm; and P-31, 35 ppm.

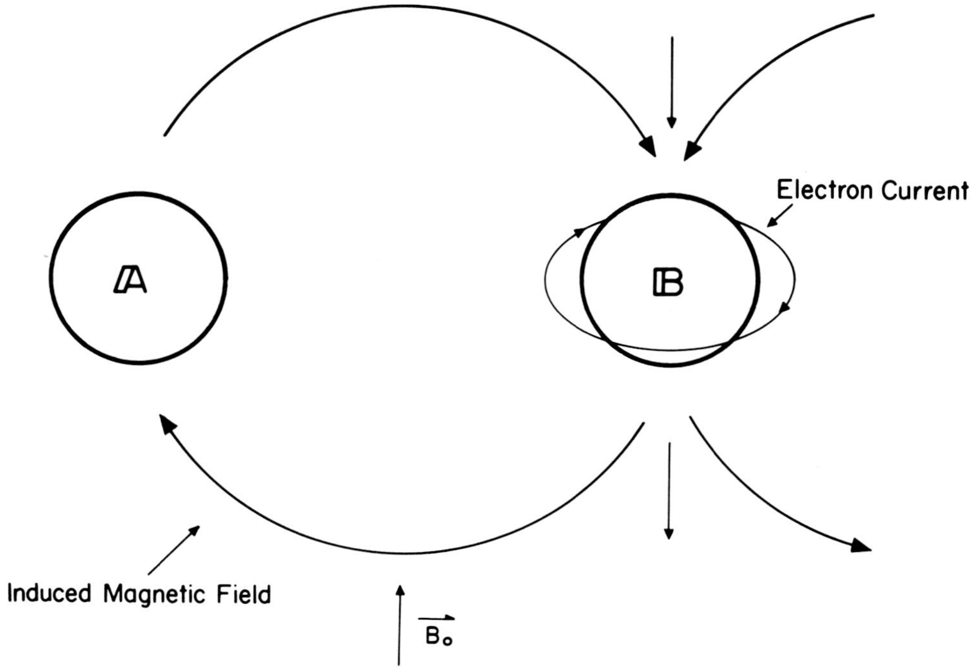

Figure 6. Effects at nucleus A caused by the secondary magnetic field arising from induced electronic currents at nucleus B.

Spin-Spin Splitting

Nuclei that have a spin can interact with other nuclear spins, causing a mutual splitting of the signals from each set of nuclei into multiplets. Indeed the number of the components in the resonance multiplet of a nucleus is $2nI + 1$, depending on the spin quantum number I (see Table I) and the number n of the other nuclei interacting with the nucleus. For example, a nucleus (e.g., C-13 or H-1) interacting with three methyl protons will give rise to a quartet. The relative intensities of the multiplets to a first approximation are given by binomial coefficients: 1:1 for a doublet, 1:2:1 for a triplet, and 1:3:3:1 for a quartet. The difference between any two adjacent components of a multiplet is the same and yields the value of the spin-spin coupling constant J in Hz.

One important feature of spin-spin splitting is that it is independent of magnetic field strength. Therefore, increasing the magnetic field strength will increase the chemical shift difference between two peaks in Hz (not ppm!—see Eq. 9), but the coupling constant J will remain unaltered.

In many biologic samples, the linewidth for one reason or another may be broader than the splitting because interactions between P-31 and H-1 or H-1 and H-1 vitiate a determination of the coupling constants. It should also be noted that rapidly exchanging -OH and -NH protons (*vide infra*) do not cause splitting. If coupling constants can be determined, they can reveal details about molecular geometry (see Chap. 4 in James[3]).

In order to simplify a spectrum and to improve the signal-to-noise ratio, decoupling (usually of protons) is often employed— especially with C-13 and N-15 NMR. Strong irradiation of the protons at their resonance frequency will cause a collapse of the multiplet in the C-13 or N-15 resonance into a singlet.

Peak Intensity

The area under an NMR peak (the peak intensity), but *not* the height (the peak amplitude), is directly proportional to the number of nuclei contributing to the signal under suitable experimental conditions. Those conditions are that a delay between pulses $\geq 5T_1$ must be employed in signal-averaging experiments (*vide infra*). Consequently, if the concentration of nuclei is known for a particular peak, it can be used as a standard.

General Features of Nuclear Magnetic Resonance Spectrum

Various NMR spectral parameters already mentioned are summarized in Figure 7.

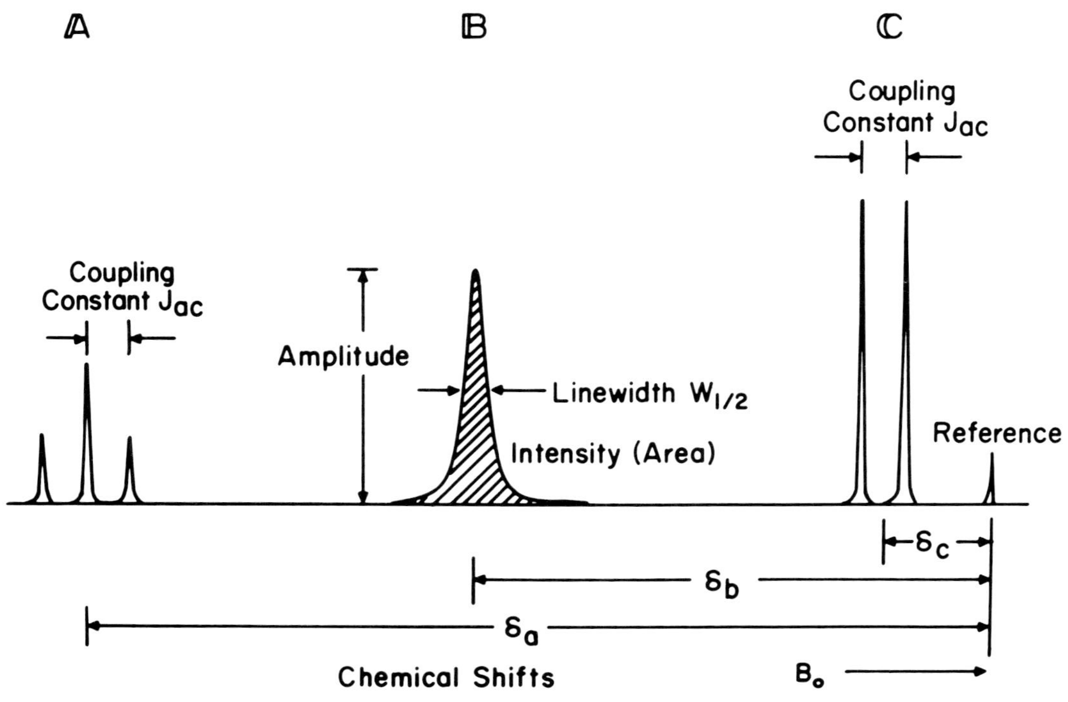

Figure 7. Nuclear magnetic resonance spectral parameters.

Chemical Exchange

Nuclear magnetic resonance parameters can be affected by the monitored nucleus spending part of its time in one environment and part of its time in another environment. If the exchange between the two environments is sufficiently rapid, the observed NMR parameter will be a weighted average of the parameter values in each of the different environments or sites, the weighting taking account of the number of nuclei in each of the sites. One often encountered example of exchange is provided by the observation of H-1 NMR of molecules containing hydroxyl or amino groups in aqueous solution. Generally, the hydrogens on these functional groups exchange so quickly with water hydrogens that only a single averaged resonance signal is seen for these "exchangeable" protons; since water is usually present in great excess, the properties of this averaged resonance are weighted such that water protons dominate. It should be noted, however, that in proteins and nucleic acids, the complicated structure precludes much of that exchange, thus permitting signals from these exchangeable protons to be observed. Another example of exchange is that of a small molecule that may spend part of its time free in solution and part bound to a macromolecular structure.

In the case of rapid exchange between two sites, a single resonance peak results that is centered at frequency

$$\nu = f_A \nu_A + f_B \nu_B \qquad \text{Eq. 10}$$

with linewidth $W_{1/2}$ ($= 1/\pi T_2$) such that

$$1/T_2 = (f_A/T_{2A}) + (f_B T_{2B}) \qquad \text{Eq. 11}$$

and

$$1/T_1 = (f_A/T_{1A}) + (f_B T_{1B}) \qquad \text{Eq. 12}$$

where f_A and f_B represent the fractions of nuclei at sites A and B, and the subscripts *A* and *B* refer to the values of those parameters when the rapidly exchanging nuclei are at sites A or B. Of course, $f_A + f_B = 1$. For rapid exchange to occur, the exchange rate must be much larger than (a) the chemical shift difference ($\nu_A - \nu_B$), (b) $1/T_{2A}$ and $1/T_{2B}$, and (c) $1/T_{1A}$ and $1/T_{1B}$. The other limit is that of slow exchange, in which case two separate resonance peaks occur for nuclei at sites A and B; the conditions for this are just the inverse of those for rapid exchange. When exchange rates are on the same order as ($\nu_A - \nu_B$), $1/T_2$, or $1/T_1$, the situation is more complicated, with line broadening occurring. This intermediate exchange situation permits exchange rates to be measured.[9]

With rapid exchange, it can be easily seen from Equations 11 and 12 that the measured T_2 (or linewidth) and T_1 values of a small molecule can be significantly affected if that small molecule spends part of its time bound to macromolecular structures. For example, if only 1% of the rapidly exchanging small molecule is bound to the macromolecule, the linewidth will double if the molecular motion is 100 times slower in the bound state; in practical situations, the molecular motion may be up to 10^2 to 10^6 times slower.

Nuclear Magnetic Resonance Experiment

It is possible to generate a proton NMR spectrum by slowly sweeping either the magnetic field or RF (the other remaining constant) such that the resonance condition expressed by Equation 2 is successively achieved by all the protons of various chemical shifts in the sample. This method is still used in some low-cost, routine proton NMR spectrometers. However, most spectrometers utilize RF pulses at the resonance frequency.

Pulsed Nuclear Magnetic Resonance

The signal-to-noise ratio of an NMR signal is dramatically improved by signal averaging, the signal-to-noise ratio improving as $\sqrt{n}$, where *n* is the number of signals averaged. The fastest way to obtain an NMR signal is to detect the signal following a strong RF pulse applied at the resonance frequency; this signal is the free induction decay (FID). This signal can be quickly followed by another pulse and FID, which can be acquired by a computer and averaged with the first FID to enhance the signal-to-noise ratio. Limitations on the rate of pulsing are set by the acceptable linewidth ($W_{1/2} \geqslant$ pulse rate) and T_1.[3,4] As mentioned before, the pulse rate should be less than $1/(5T_1)$ to obtain quantitative results from peak areas.

Pulsed NMR, by using various pulse sequences, provides the best means of measuring relaxation times. Several different pulse sequences have been utilized to measure T_1 and T_2.[3,4] The most common (and probably best) technique of determining T_1 is to employ the inversion recovery sequence ($180° - \tau - 90°$). After an initial 180° pulse inverts the spin populations, the spin system begins relaxing toward thermal equilibrium as discussed earlier. After time τ, a 90° pulse is applied, and the FID following the pulse is acquired. The value of T_1 may be determined by measuring the magnitude *M* of the FID as a function of τ using the following expression:

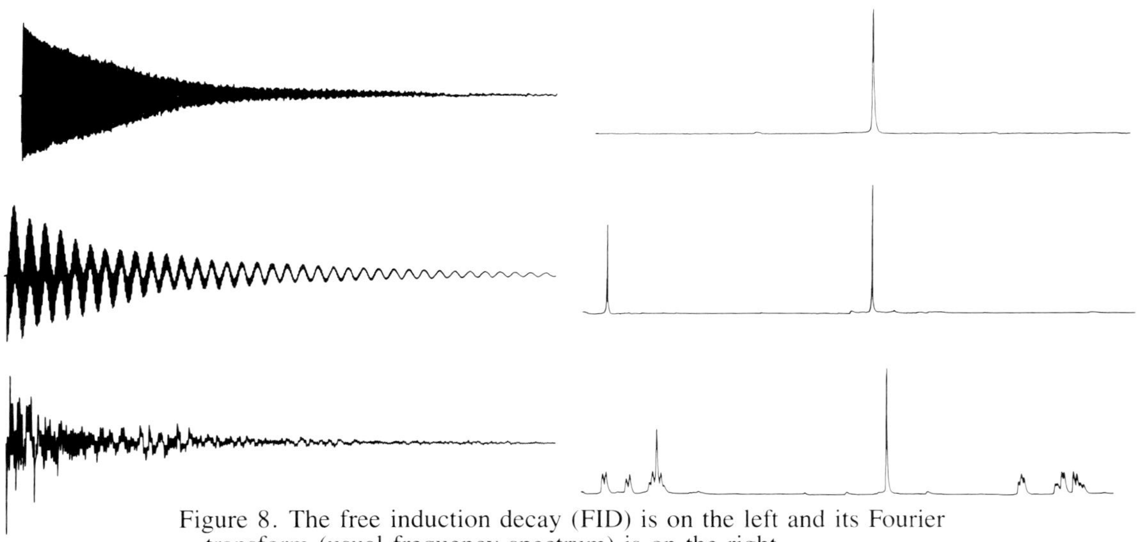

Figure 8. The free induction decay (FID) is on the left and its Fourier transform (usual frequency spectrum) is on the right.
(Reproduced, with permission, from James TL: *Nuclear Magnetic Resonance in Biochemistry: Principles and Applications*. New York: Academic, 1975.)

$$M = M_\infty[1 - 2\exp(-\tau/T_1)] \qquad \text{Eq. 13}$$

where M_∞ is the magnitude of the FID at

τ = infinity

(i.e., at thermal equilibrium). A fast inversion recovery method of T_1 determination has been described.[10]

Determination of T_2 is best carried out using a Carr-Purcell pulse sequence with the Meiboom-Gill modification.[3,4] This sequence is initiated with a 90° pulse following by a series of 180° pulses. The result is a train of spin echoes, signals that follow each 180° pulse exactly halfway between the 180° pulses. The magnitude of the echo decays exponentially as the train of echoes progresses, so that the echo magnitude can be expressed as

$$M = M_\infty[\exp(-t/T_2)] \qquad \text{Eq. 14}$$

where t is the length of time from the 90° pulse to the top of the echo. Obviously, T_2 can be determined from the decay of the echoes.

Fourier Transform Nuclear Magnetic Resonance

The foregoing discussion on pulsed NMR is sufficient if there is only one frequency for the nuclei of interest, for example, monitoring Na^+-23 in a sample or the proton signal of H_2O in a tissue (since it is present in great excess over other observable proton signals). However, the FID is a time-domain signal that contains the same information as the frequency-domain spectrum. The usual frequency spectrum can be obtained by computing the Fourier transform (FT) of the signal-averaged FID. Modern computers associated with NMR spectrometers perform the transform quickly. The signal-averaged FIDs on the left side of Figure 8 yield the frequency spectra on the right side after FT.

The aforementioned pulse sequences for T_1 and T_2 determination are also applicable to FT NMR. In this case, the peak amplitudes in the Fourier-transformed spectrum are monitored as a function of interpulse spacing τ rather than FID or spin echo amplitudes.

Biologic Applications

There is not sufficient space for a complete review of the biologic applications of NMR. Rather, a few recent reviews will be cited, with some indications of future directions from a technological viewpoint. Gadian's monograph gives a good overview of biologic applications.[5] Recent reviews have appeared on P-31 NMR of intact tissue,[11] tissue metabolism,[12] studies of several nuclei in biologic samples,[13] in vivo P-31 and C-13 NMR,[14] a popular account of P-31 and C-13 NMR in living cells,[15] and an edited conference transcript including P-31 NMR in human arm muscle and H-1 imaging.[16] Nuclear magnetic resonance imaging will not be considered here since it is already extensively covered in this volume.

Studies to date have emphasized observation of P-31, C-13, and H-1 NMR signals from small molecules in whole cell or intact tissue because of the relatively narrow linewidths of these signals. A few studies of macromolecules in whole cells have appeared: C-13 NMR of hemoglobin in normal erythrocytes and sickle cells,[17] C-13 NMR of glycogen in perfused rat livers,[18] N-15 NMR of bacterial cell wall components in whole cells,[19] and F-18 NMR of RNA with cells incorporating 5-fluorouracil.[20] It is anticipated that we will see more work on macromolecular systems in cells and tissue in the future. The mobility of some of these macromolecular systems is likely to be too limited for the usual solution NMR techniques to work, so the solid-state NMR techniques of dipolar decoupling, magic angle sample spinning, and cross-polarization[21] are likely to be employed. Indeed, initial experiments have been carried out on seeds, pods, and leaves.[22]

Nitrogen-15 NMR suffers from the poor sensitivity of detecting N-15, low natural abundance, and long T_1 values. These deficits undoubtedly account for its lack of use in following N-15 metabolites in cellular systems. However, these problems can be overcome, and it is anticipated that N-15 experiments in biologic systems will prove useful, as C-13 NMR has. Nitrogen-15-enrichment is less expensive than C-13 enrichment. Furthermore, population transfer techniques can be employed with small molecules, permitting signal-to-noise increases by a factor approaching the ratio of proton-to-nitrogen gyromagnetic ratios (≤ 10) as well as faster pulse rates based on H-1 relaxation rather than N-15 relaxation.[23] Nitrogen-15 NMR signals have already been detected from soluble amino acids in cell suspensions of microorganisms even without the significant advantages of the population transfer techniques.[19,24]

Pulsed NMR techniques are also available for studying diffusion and flow.[3] Some studies have been made, but, considering the importance of transport processes in physiology, for the most part these techniques have not been exploited in biologic tissue. Singer[25] has described some work on blood flow. The technique developed by Hemminga and deJager,[26] which permits the detection of fluid flow in the presence of a stationary fluid, may very well prove to be of considerable utility for in vivo flow studies.

REFERENCES

1. Abragam A: *Principles of Nuclear Magnetism*. London: Oxford University Press, 1961.
2. Slichter CP: *Principles of Magnetic Resonance*, (ed 2). Berlin: Springer-Verlag, 1978.
3. James TL: *Nuclear Magnetic Resonance in Biochemistry: Principles and Applications*. New York: Academic, 1975.
4. Fukushima E, Roeder SB: *Experimental Pulse NMR: A Nuts and Bolts Approach*. Reading: Addison-Wesley, 1981.
5. Gadian DG: *Nuclear Magnetic Resonance and Its Applications to Living Systems*. Oxford: Clarendon, 1982.
6. Hoult DI, Lauterbur PC: The sensitivity of the zeugmatographic experiment involving human samples. J Magnet Reson 34:425, 1979.
7. Murphy-Boesch J, Koretsky AP: An in vivo NMR probe circuit for improved sensitivity. (Submitted for publication.)
8. James TL: Conformational flexibility of nucleic acids investigated by NMR relaxation. Bull Magnet Reson 4:119, 1983.
9. Jackman LM, Cotton FA (Eds): *Dynamic Nuclear Magnetic Resonance Spectroscopy*. New York: Academic, 1975.
10. Kowalewski J, Levy GC, Johnson LF, et al: A three-parameter non-linear procedure for fitting inversion recovery measurements of spin-lattice relaxation times. J Magnet Reson 26:533, 1977.
11. Burt CT, Cohen SM, Bar̆any M: Analysis of intact tissue with ^{31}P NMR. Ann Rev Biophys Bioeng 8:1, 1979.
12. Gadian DG, Radda GK: NMR studies of tissue metabolism. Ann Rev Biochem 50:69, 1981.
13. Kuchel PW: Nuclear magnetic resonance of biological samples. In *Critical Reviews in Analytical Chemistry*, Campbell B, Meites L (Eds). Boca Raton: CRC Press, 1981.
14. Shaw D: In vivo chemistry with NMR. In *Nuclear Magnetic Resonance Imaging in Medicine*, Kaufman L, Crooks L, Margulis AR (Eds), p 147. New York: Igaku-Shoin, 1981.
15. Shulman RG: NMR spectroscopy of living cells. Sci Am 248:86, 1983.
16. Bradbury EM, Radda GK, Allen PS: Nuclear magnetic resonance techniques in medicine. Ann Intern Med 98:514, 1983.
17. James TL, Matthews R, Matson GB: Hemoglobin aggregation in oxygenated sickle cells studied by carbon-13 rotating frame spin-lattice relaxation in the presence of an off-resonance radiofrequency field. Biopolymers 18:1763, 1983.
18. Sillerund LO, Shulman RG: Structure and metabolism of mammalian liver glycogen monitored by carbon-13 nuclear magnetic resonance. Biochemistry 22:1087, 1983.
19. Lapidot A, Irving CA: Dynamic structure of whole cells probed by nuclear overhausser enhanced nitrogen-15 nuclear magnetic resonance spectroscopy. Proc Natl Acad Sci USA 74:1988, 1977.
20. Gochin M, James TL, Shafer RH: In vivo ^{19}F NMR of 5-fluorouracil incorporation into RNA and metabolites in E. coli cells. (Unpublished results.)
21. Schaefer J, Stejskal ED: High resolution ^{13}C NMR of solid polymers. In *Topics in Carbon-13 NMR Spectroscopy* (Vol 3), Levy GC (Ed), p 283. New York: Wiley-Interscience, 1979.

22. Schaefer J, McKay RA, Stejskal ED: Double cross-polarization NMR of solids. J Magnet Reson 34:443, 1979.

23. Bolton PH: Enhancement of signals from insensitive nuclei by magnetization transfer via J couplings. J Magnet Reson 41:287, 1980.

24. Kanamori K, Legerton TL, Weiss RL, et al: Nitrogen-15 spin-lattice relaxation times of amino acids in Neurospora crassa as a probe of intracellular environment. Biochemistry 21:4916, 1982.

25. Singer JR: Blood flow measurements by NMR. In *Nuclear Magnetic Resonance Imaging in Medicine*, Kaufman L, Crooks L, Margulis AR (Eds), p 128. New York: Igaku-Shoin, 1981.

26. Hemminga MA, deJager PA: The study of flow by pulsed nuclear magnetic resonance: II. Measurement of flow velocities using a repetitive pulse method. J Magnet Reson 37:1, 1980.

6

Tissue Characterization

PETER L. DAVIS, M.D.
LEON KAUFMAN, Ph.D.
LAWRENCE E. CROOKS, Ph.D.

The application of NMR to humans makes it important to study how tissue characteristics influence the NMR image. Although the etiology of the T_1 and T_2 relaxation times is well understood in theory, in practice T_1 and T_2 can only be predicted for simple materials. In particular, tissues are much too complex to predict their T_1 and T_2 relaxation times from theory.

It is important to keep in mind that, as discussed in previous chapters, in NMR imaging the T_1 and T_2 relaxation times are not measured or imaged directly, but are calculated from the image data. It is also important to recognize that the T_1 and T_2 relaxation times are inherent properties of how the hydrogen protons in the tissue in a magnetic field behave when perturbed by the appropriate radiofrequency (RF) field. In turn, these properties combine with the imaging technique to determine the NMR signal intensity received from the tissue and depicted in the image. Only if there is a difference in NMR intensity (contrast) will two tissues, normal or abnormal, be differentiated. This chapter will review the physical basis of the T_1 and T_2 relaxation times, describe how they interact with the imaging technique, and use a series of experiments performed at the University of California, San Francisco to elucidate the NMR properties of tissues in vivo.

T_1 and T_2 Relaxation Times

The etiology of the T_1 and T_2 relaxation times will only be briefly reviewed here to demonstrate how the microenvironment influences the relaxation time. The basis of relaxation of the proton is that NMR is not a spontaneous process but a stimulated process—something has to give the excited proton a "push" to stimulate it to relax and emit its radio signal.[1] The "something" is a fluctuating magnetic field that is generated in the tissue. The source of the fluctuating magnetic field is hydrogen (usually in the form of water) and other nuclei and molecules with magnetic moments. Because of thermal motion, these nuclei move, tumble, and collide, which from the viewpoint of a proton causes their magnetic field to appear to fluctuate. When the frequency of the fluctuation is close to the Larmor frequency of the proton, the proton relaxes.[2] Hence, there is an optimal motion by the molecules to induce relaxation in protons and a shortest T_1 time. Small molecules tumbling faster or large molecules tumbling slower will have changing magnetic fields at frequencies correspondingly greater or less than the Larmor frequency and will not as frequently cause relaxation as molecules tumbling mainly at the Larmor frequency, although there will be overlap (Fig. 1). Therefore, anything that affects thermal molecular motion, such as temperature, viscosity, composition, or state, will affect the T_1 relaxation time.

The T_2 relaxation time is also dependent upon the internal magnetic fields in the tissue because they determine in part the Larmor frequency at which an unrelaxed molecule relaxes.[2] The more nonuniform and fixed the fields seen by the protons, the larger the range of Larmor frequencies of the protons. This in turn causes them to emit NMR radiosignals of different frequencies, lose coherence faster, and, hence, have a short T_2 relaxation time. If the molecules are moving rapidly, however, the different magnetic fields will tend to average to a uniform field, the spectrum of Larmor frequencies becomes narrower, and T_2 becomes longer (Fig. 1).

Water is an appropriate example of a molecule whose T_1 and T_2 relaxation times vary as its rate of tumbling varies. In liquid water the water molecules tumble very rapidly. As a result, the frequencies of their changing magnetic fields are usually much higher than the typical Larmor frequency of protons.[3] Since these high frequencies stimulate relaxation infrequently, T_1 will be long. However, as water is cooled, the molecules tumble slower, and the frequencies of their changing magnetic fields are closer to the Larmor frequency.[1] Therefore the colder water will stimulate relaxation more frequently, and the T_1 times of the molecules will be shorter.

In solid ice, the slowly tumbling hydrogen protons in the water molecule experience varied and fixed magnetic fields from their relatively fixed neighbors. As a result, they have a wide range of Larmor frequencies, and their T_2 relaxation times are short. In liquid water, the molecules tumble extensively and rapidly enough that in the time span of the NMR phenomena the various fluctuating magnetic fields average to a uniform field. Therefore, the unrelaxed protons experience the same average magnetic field and have a narrow range of Larmor frequencies and a long T_2 relaxation time. In fact, for liquid water the T_2 time can become as long as the T_1 time. The T_2 time can never be longer than T_1, and is usually less than T_1.

This discussion leads to the theory of mobile and immobile protons. The theory is that NMR signal can only be detected from mobile protons. Mobile protons are protons that are tumbling fast enough that they experience similar magnetic fields, have similar Larmor frequencies, and, as a result, have collectively a T_2 relaxation time adequate to allow detection of the NMR signal. Immobile protons are protons that are somehow fixed, so that they experience widely varying magnetic fields, have different Larmor frequencies, and, as a result, collectively have a T_2 relaxation time that is too short to permit detection of the NMR signal. What exactly constitutes mobile and immobile protons in biologic tissue is still much debated; in particular, questions focus on hydrogen in water molecules that are part of the hydration shell of other molecules. This issue will be discussed further in the next section.

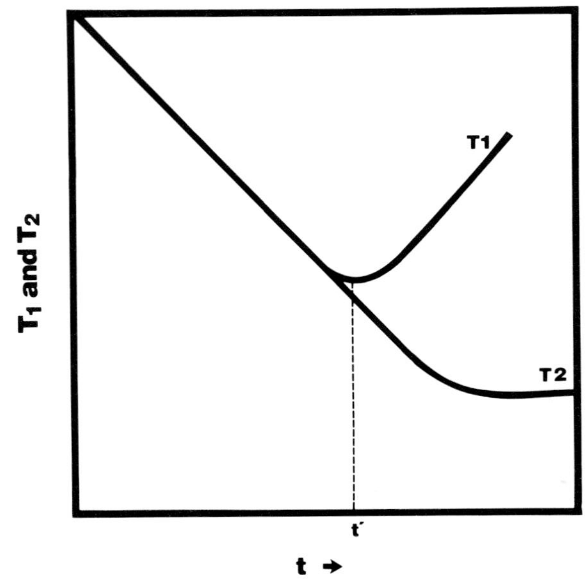

Figure 1. A log-log plot of T_1 and T_2 versus the time t required for a molecule to complete a single rotation. The shortest T_1 occurs at t′, which is the time span of a single wavelength of the Larmor frequency of the imaging system.

The mechanism of the effect of paramagnetic materials on the T_1 and T_2 relaxation times is the same as that described above, except that the fluctuating magnetic fields caused by the paramagnetic material are usually much stronger than those usually present in the tissues.[2] These strong fluctuating fields usually cause shortening of the T_1 and T_2 relaxation times. The final intensity change is dependent on the relative changes in T_1 and T_2 and the imaging technique.

Water in Biologic Systems

Since the body is 70%–80% water, it is important to look closer at how water behaves in different situations. Pure water tumbles once in a time of 2–3 psec at 25° C.[4] This is quite shorter than the tumbling times of about 0.1 to 1 msec that will cause the shortest T_1 times at imaging magnetic field strengths. Hence, pure water has a relatively long T_1 time. In addition, at this rate of tumbling the internal fields average out, and the T_2 time is long.

When some protein is dissolved in this water, some of the adjacent water molecules (how much is a matter of much debate) are influenced by the surface of the protein molecule. At the protein surface are positive and negative polar groups, hydrogen bonding sites, and hydrophobic sites. These tend to hold the water molecule slightly and slow its tumbling to a time of 10–100 psec. These slower tumbling times will tend to decrease the T_1 time. This wider range of tumbling times will also tend to decrease the T_2 time.

Proteins such as serum albumin or hemoglobin tumble typically at about 0.1 msec, which is near the time that would cause short T_1 times at imaging magnetic field strengths. It appears that there are some fluctuating fields in solutions caused by these faster tumbling times. However, it is unclear whether these fields are caused by the constituents of the proteins themselves, by water molecules bound very tightly to the proteins and tumbling at the same speed, or by different water molecules that periodically sample and are influenced by the protein tumbling.

Hence, there are at least three sources of fluctuating magnetic fields in a protein solution that can stimulate relaxation.[3] First, the bulk of the water, which is not influenced by the proteins and therefore will tumble very rapidly and generate fluctuating fields far from the Larmor frequency, will not stimulate many protons to relax; hence, the adjacent protons will have a long T_1 time. Second, a much smaller fraction of the water, which is influenced by the protein surface to tumble somewhat slower and therefore will generate fluctuating fields closer to the Larmor frequency, will stimulate more protons to relax; hence, the adjacent protons will have a shorter T_1 time. Third, a very small fraction of the water, which acts as though it is tumbling at the very slow rate of the protein molecules and generates fluctuating fields near the Larmor frequency, will stimulate many more protons to relax; hence, the adjacent protons will have a short T_1 time.

All the different T_1 components contribute to a final observed T_1 according to the following equation:

$$1/T_{1_{obs}} = X_a(1/T_{1_a}) + X_b(1/T_{1_b}) + \ldots \quad \text{Eq. 1}$$

where $T_{1_{obs}}$ is the observed T_1, X_a is the fraction of water with a T_1 of T_{1a}, X_b is the fraction of total water having T_{1b}, and so on.[4] Therefore, even small fractions of water molecules with short T_1 times may cause considerable shortening of the observed T_1 time. T_2 behaves similarly.

Cellular systems are more complex than simple protein solutions. Unlike protein solutions, cells have both structure and components such as membranes, microtubules, chromatin, and filaments. These can all interact with water to various degrees. In addition, these components can change as a result of a pathologic process and normal cellular metabolism. A few examples can be illustrative. Kuntz and Zipp measured the effects of cellular dehydration by increasing the plasma osmolarity of packed red cells.[5] As the osmolarity increased, the T_1 and T_2 relaxation times decreased. This would be expected since the increased osmolarity should cause the cell volume to decrease and decrease the water tumbling rates to nearer the Larmor frequency.

However, the T_1 time can be changed without changing the cell water content. Beall et al. increased the T_1 time of HeLa cells by condensing the chromatin. This did not change the water content of the cells significantly.[6] Hence, the relaxation times can be influenced by configuration changes of the macromolecules.

Relaxation Times and Imaging Techniques

The observed NMR intensity results from a combination of the T_1 and T_2 relaxation times and the hydrogen density. How these are combined is determined by the imaging tecnique—the manner in which the protons are perturbed in the magnetic field.[7] Each imaging technique does it differently. The equation describing the spin echo technique is

$$I = Hf(v)[\exp(-TE/T_2)][1 - \exp(-TR/T_1)] \quad \text{Eq. 2}$$

Because of the complexity of Equation 2, it is possible for several tissues with vastly different T_1 and T_2 relaxation times to have the same intensity on a spin echo image. Therefore it may not be possible to differentiate two normal tissues or a normal and abnormal tissue on a particular spin echo image.

However, by changing the TE or TR parameters of the images, it should be possible to change the relative intensities of the two tissues. Other tissues may then become isointense, however.

This effect can be studied by rearranging Equation 2 to

$$F = I/Hf(v) = [\exp(-TE/T_2)][1 - \exp(-TR/T_1)] \quad \text{Eq. 3}$$

Since $Hf(v)$ is relatively constant for most soft tissues, F is essentially proportional to I. Equation 3 can then be rearranged so that all combinations of T_1 and T_2 that will result in a certain F can be calculated and plotted for specific values of TE and TR. The resulting graphs can be considered to be isointensity lines. They would be isointensity lines if $Hf(v)$ was identical for all tissues. For the following discussion, therefore, $Hf(v)$ will be assumed to be identical for all tissues.

A set of these graphs calculated for some of the values of TE and TR we currently use are shown in Figure 2. Each line represents a series of tissue T_1

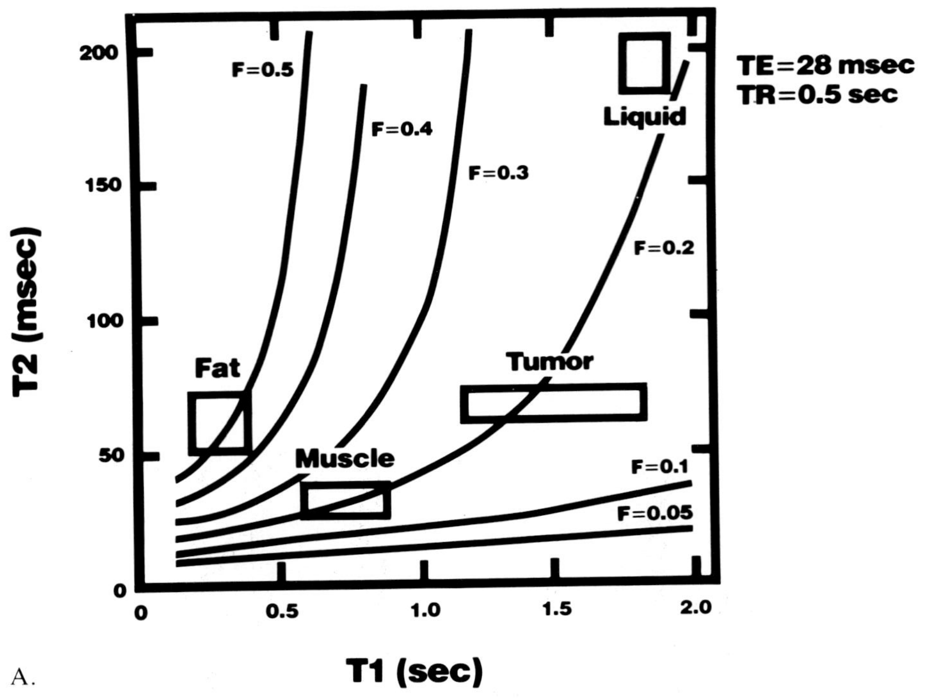

A.

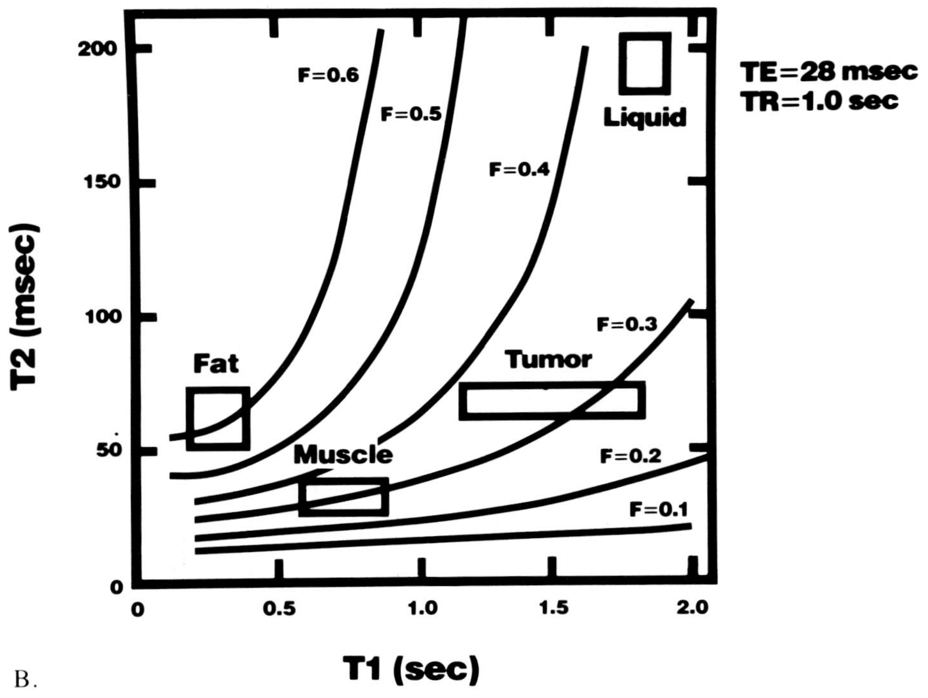

B.

Figure 2.
A. T_1 versus T_2 graph plotting isointensity lines calculated from Equation 3 for TE of 28 msec and TR of 0.5 sec.
B. T_1 versus T_2 graph plotting isointensity lines calculated from Equation 3 for TE of 28 msec and TR of 1.0 sec.

and T_2 combinations that will all have the same value of F. Several trends are evident. First, NMR signal intensity will increase as tissue T_2 increases, and decrease as tissue T_1 increases. Second, as TR increases, signal intensity for any specific T_1 and T_2 combination tends to increase; and conversely, as TE increases, signal intensity for any combination tends to decrease.

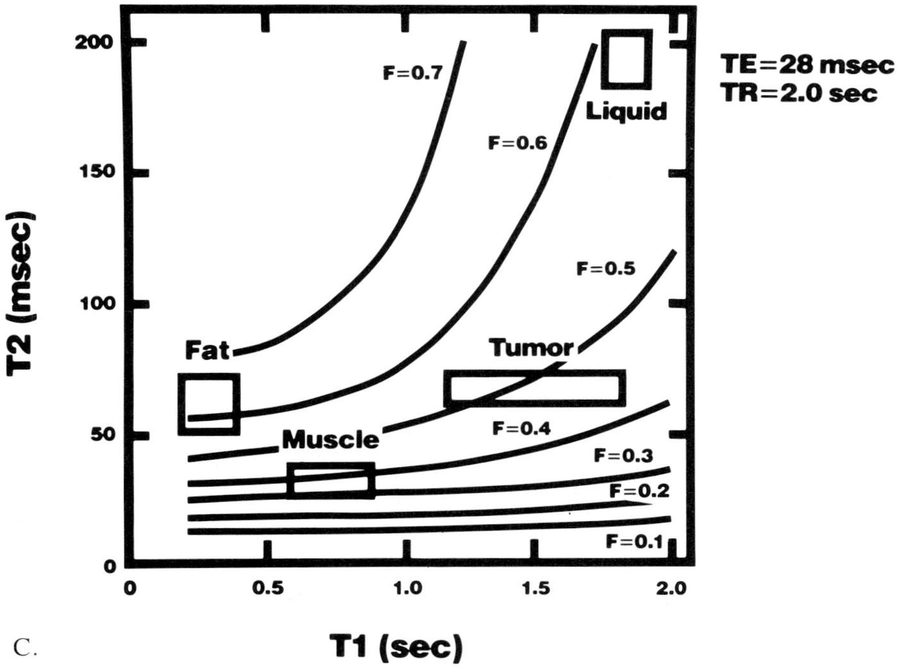

C.

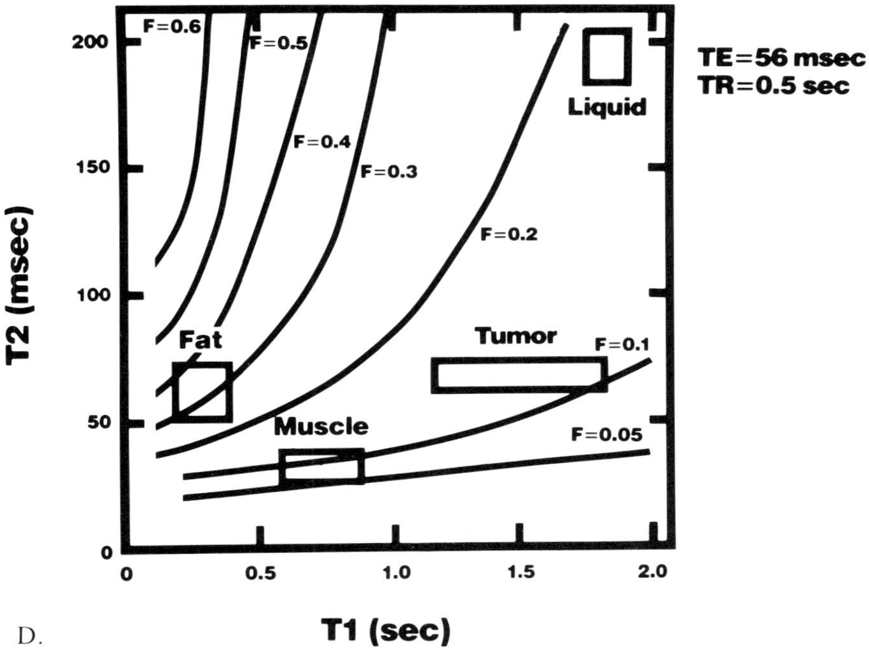

D.

C. T_1 versus T_2 graph plotting isointensity lines calculated from Equation 3 for TE of 28 msec and TR of 2.0 sec.
D. T_1 versus T_2 graph plotting isointensity lines calculated from Equation 3 for TE of 56 msec and TR of 0.5 sec.

However, when we look at an image, we usually do not consider absolute intensity, but rather the relative intensity differences (or contrast) between two or more tissues. To see how the intensity differences will change with different values of TE and TR, the typical T_1 and T_2 ranges for fat, muscle, a tumor model, and an impure liquid have been superimposed on each graph. For TE of 28 msec and

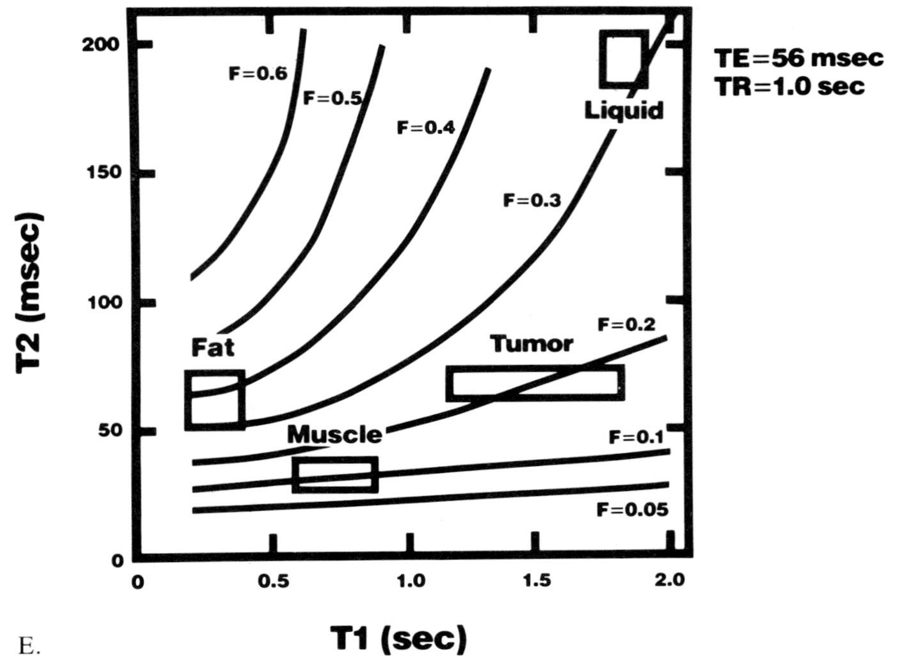

E.

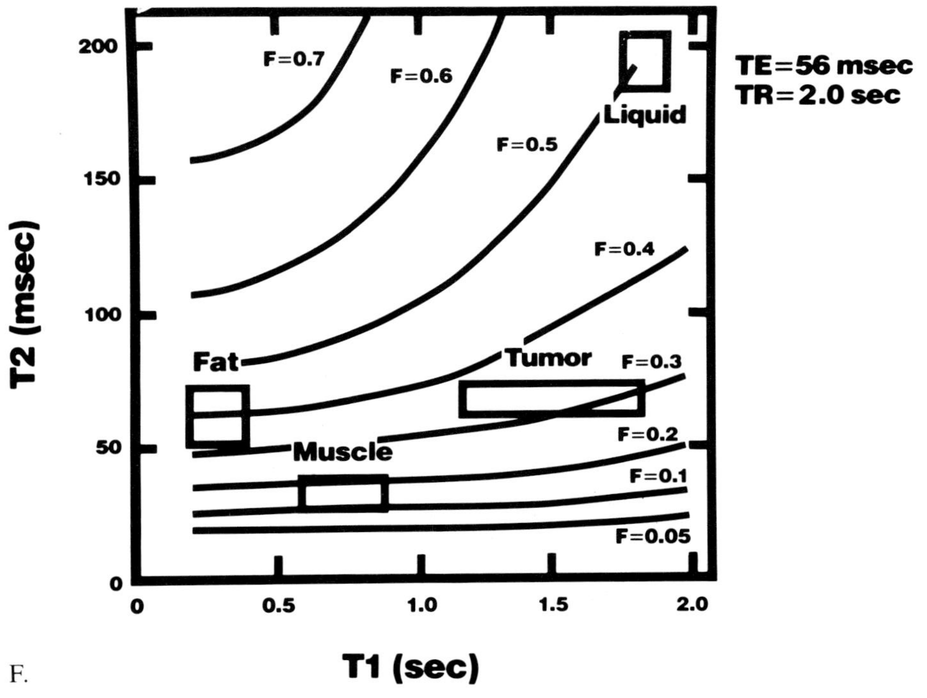

F.

E. T_1 versus T_2 graph plotting isointensity lines calculated from Equation 3 for TE of 56 msec and TR of 1.0 sec.

F. T_1 versus T_2 graph plotting isointensity lines calculated from Equation 3 for TE of 56 msec and TR of 2.0 sec.

TR of 0.5 sec, the muscle, tumor, and liquid will all have similar intensities, but the fat will have a higher intensity. For TE of 28 msec and TR of 2.0 sec, all four tissues will have different intensities: fat, the brightest; the liquid, next; then the tumor; and muscle, the darkest. For TE of 56 msec and TR of 2.0 sec, again all four tissues will have different intensities, but their relative order has changed: the liquid, now the brightest; fat, next; then tumor; and, finally, muscle.

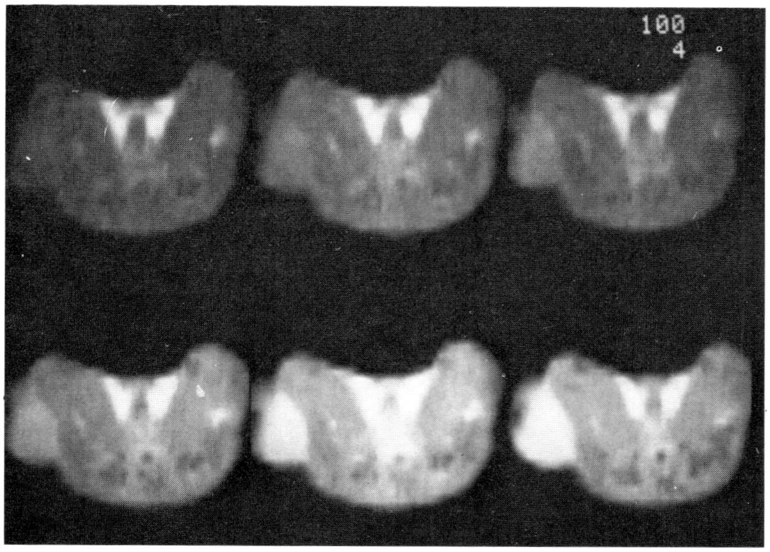

Figure 3. Transverse spin echo images of a dead rat with an implanted sarcoma in the right flank. The TR parameter was increased in steps of 0.25 sec from 0.25 sec to 1.5 sec increasing from left to right, top and bottom. See how the intensity of the tumor changes with respect to the muscle and fat as TR increases.
(Reproduced, with permission, from Hansen G, Crooks LE, Davis PL, et al: In vivo imaging of the rat with nuclear magnetic resonance. Radiology 136: 695, 1980.)

In summary, the difference in intensity between tissues is not only dependent on the difference between their T_1 and T_2 times, but also on the imaging technique and the timing parameters used when imaging with that technique (Fig. 3). While the spin echo technique is represented by Equation 2, other imaging techniques such as inversion recovery are not, but have their own different equations. As a result, the relative intensities for tissues studied with each technique can be different.

In theory, if two tissues differ in T_1, T_2, or both, then their difference should appear on the T_1 or T_2 images, or both. Unfortunately, these T_1 and T_2 images, which are calculated from at least two images of the same plane with different values of TE or TR, are usually noisier than the original images. Therefore, fine detail and small changes can be frequently missed on them. As a result, we frequently end by visually comparing several spin echo images. This requires the viewer to become familiar with how the intensities of normal tissues vary with changes in TE and TR. This is probably the most difficult aspect of becoming "conversant" with NMR imaging.

Experiments

Ideally, a study of the relaxation times should be performed in human subjects. However, humans are too variable and experimental flexibility is too limited to be able to conduct well-controlled studies. There are several advantages in using animals, particularly small ones such as rats. First, at a reasonable cost very uniform strains of animals can be obtained; their near-uniformity allows measurement of T_1 and T_2 times in nearly identical normal tissues. Second, a wide variety of well-understood reproducible pathologic conditions can be induced in animals. Third, a large number of animals with identical pathologic conditions can be rapidly studied to enable measurement of the relaxation times and their variabilities. Fourth, the animals can be sacrified when desired so that physiologic and pathologic correlations can be made.

Our initial experiment involved normal animals. From these we learned to recognize normal tissues, determined their relative NMR intensity, and measured their relaxation times. Next, several benign diseases were similarly studied. Finally, since early in vitro studies suggested that there were significant changes in the relaxation times of malignant tissues, we performed three experiments that evaluated three different malignancies. The results of these experiments allowed us to evaluate the sensitivity and specificity of the relaxation times and compare normal tissue relaxation times to benign disease times and malignant disease times. In particular, we wanted to know if these three malignancies could be differentiated by their relaxation times and if different diseases within a single class could be similarly differentiated.

As part of these experiments, the water and lipid contents of tissue were measured and correlated to the relaxation times. Since significant correlation was found between the water content of tissue and relaxation times, it was decided to study several disease entities in which significant water changes in tissue were known to occur. These included brain edema and infarction and muscle infarction.

Imaging Equipment and Technique

All experiments using small animals were performed in a resistive magnet with a 6.5-cm aperture operating at 3.5 kGauss.[8] All experiments using larger animals were performed in a superconducting magnet with a 55-cm aperture, also operating at 3.5 kGauss. At this magnetic field strength, hydrogen resonates at 15 MHz. The selection of the imaging technique was dictated by the requirement that both T_1 and T_2 relaxation times be obtainable. This could be easily done with the spin echo imaging technique described in Chapters 2 and 3.

Initially, data were collected by the line technique, and later, the two-dimensional Fourier transform (2DFT) technique. No significant changes in relaxation times were associated with this change, as expected, but we obtained the advantage of improved imaging speed and quality. In the small-animal magnet five contiguous planes are imaged simultaneously in 6 minutes. The spatial resolution is usually 1 mm × 1 mm × 5 mm thick, although resolution up to 0.5 mm × 0.5 mm × 5 mm can be obtained. For each plane four images with different combinations of TE of 28 or 56 msec and TR of 0.5 or 1.0 sec are obtained (Fig. 4). From these four images T_1 and T_2 relaxation times can be calculated.[9] If this calculation is done pixel by pixel, T_1 and T_2 relaxation time images can be generated in which the longer times are represented as brighter intensities. If a region of interest is drawn, the T_1 and T_2 times for that region can be calculated and used for statistical purposes.

Normal Animals

Several normal rats were imaged to determine the NMR appearance of normal anatomy and its gray scale.[10] The rats were imaged by NMR and x-ray CT (GE 7800) at 1-cm intervals through the entire length of their bodies. They were then frozen and sliced at 1-cm intervals. The slices were photographed and radiographed. All images were compared (Fig. 5). The NMR images demonstrated all major organs except for the heart, which did not image well because of its rapid motion. A relative gray scale was determined for the spin echo technique. Fat produced the most intense signal, followed by brain and spinal cord, which are also high in lipid content. Next in intensity were most of the viscera, including liver,

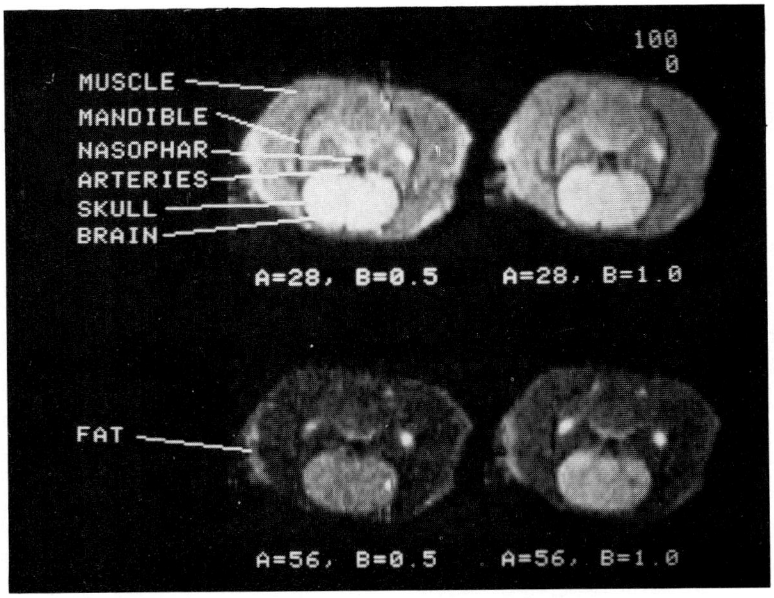

Figure 4. Four transverse spin echo images through the same plane of the head of a supine live rat. Only the TE (A) and TR (B) times differ between the images.

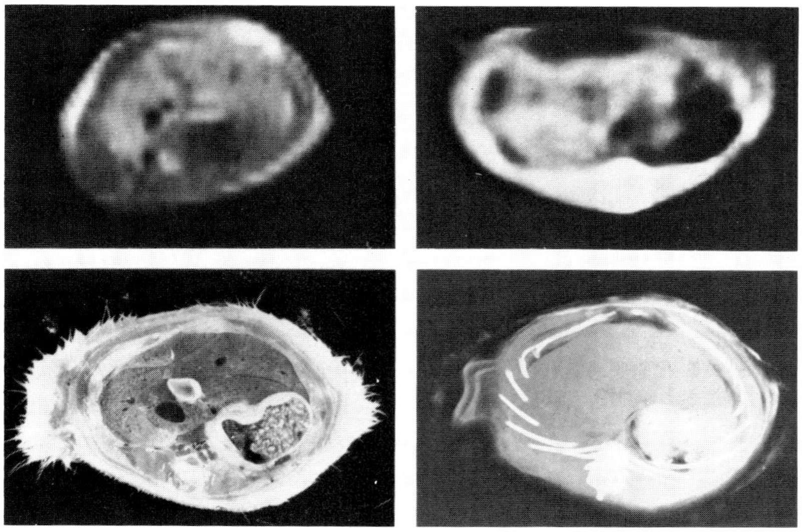

Figure 5. Cross-sections through the upper abdomen of a rat. Top left: NMR image. Top right: CT image. Bottom left: anatomic section. Bottom right: radiograph of anatomic section.
(Reproduced, with permission, from Hansen G, Crooks LE, Davis PL, et al: In vivo imaging of the rat with nuclear magnetic resonance. Radiology 136: 695, 1980.)

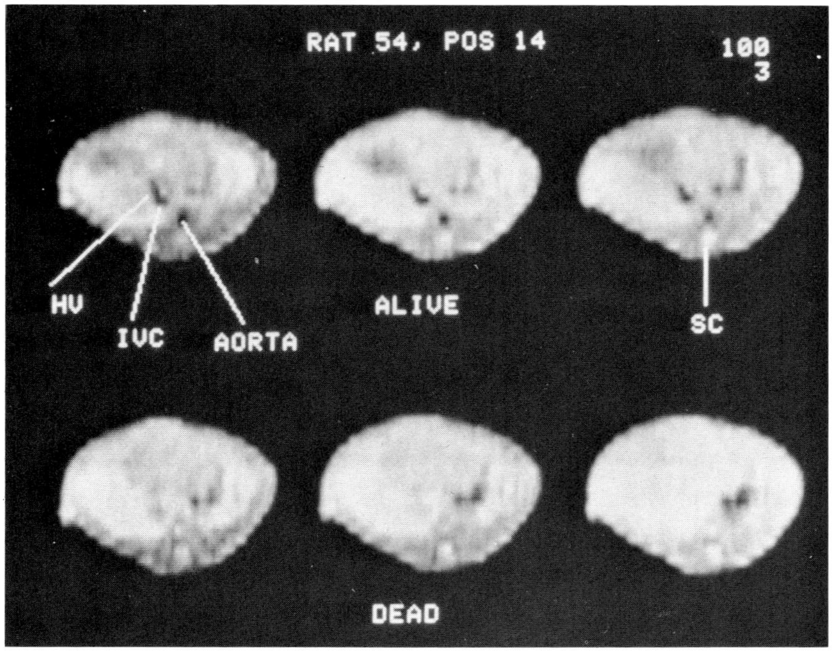

Figure 6. Transverse NMR images through the liver of a rat. Images in the top row were taken while the rat was alive. Blood flow is seen causing low intensity in the aorta, inferior vena cava (IVC) and hepatic vein. Images in the bottom row were taken after the rat was killed. The vessels filled in and the liver becomes uniform.

kidney, and muscle. Water, urine, and cerebral spinal fluid were darker, and lung, bone, and air were the darkest. These latter three materials are of low intensity because of their low hydrogen content. Water and fat have similar hydrogen densities, but fat is brighter because it has a shorter T_1 time than water. Like the heart, rapidly flowing blood also produces low intensity signals, causing blood vessels to image as dark lumina. Effects of blood flow could also be seen in the liver, which became more uniform after the animal was killed and blood flow ceased (Fig. 6). Still blood imaged relatively intensely.

T_1 and T_2 relaxation times were obtained for brain, liver, fat, and muscle. Each tissue's values segregated well (Table I). Fat had a short T_1 and long T_2, muscle had a long T_1 and short T_2, liver had intermediate values, and brain had both long T_1 and T_2.

Benign Pathologic Conditions

A variety of benign pathologic conditions were induced in rats.[9] These included sterile abscesses induced by injecting turpentine, pyogenic abscesses by injecting fecal material, and hematomas by injecting blood (Figs. 7, 8). In addition, several renal lesions were produced by ligating the renal artery, vein, or ureter (Fig. 9).[11] All lesions were confirmed by pathologic examination. All these pathologic conditions were discernible on at least one of the four spin echo images obtained, and most were on several. However, the contrast of the pathologic tissues in relation to their normal surrounding tissue often varied dramatically with different values of TE and TR.

Except for some hematomas and the renal lesions, the T_1 and T_2 times for the pathologic conditions were different from those for normal tissues (Table I). Most pathologic conditions had prolonged T_2 relaxation times. The T_1 and T_2 times for some of the hematomas overlapped those for normal brain. The renal lesion results were inconclusive because of the small size of the lesion, tissue inhomogeneity, and partial volume averaging artifacts. However, the appearance of collecting structures (urine being dark because of long T_1) and their sizes could be used to differentiate renal vein, renal artery, and ureter ligation. When T_1 and T_2 times for each condition were contrasted against those of its normal surroundings, most differed in both T_1 and T_2, but some only differed in one or the other. By having both relaxation times available, the examination was more sensitive and specific.

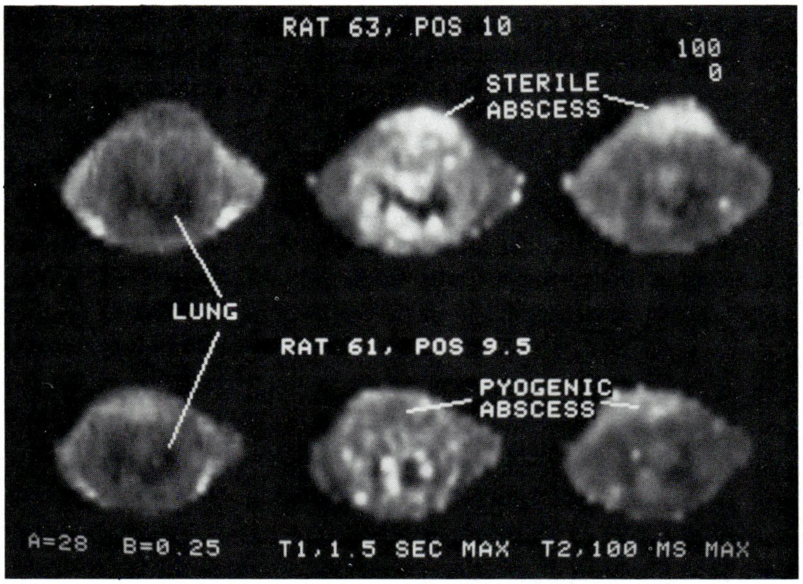

Figure 7. Nuclear magnetic resonance appearance of sterile abscess (top row) and pyogenic abscess (bottom row). The abscesses (in the nape of the neck of prone rats) are difficult to see on the spin echo images (left). They are seen better in the T_1 and T_2 images, in the center and on the right, respectively. A characteristic feature is that the T_1 time of sterile abscesses is consistently longer than that of pyogenic abscesses. Lesions were confirmed on pathologic examination. (TE = A, TR = B)

(Reproduced, with permission, from Herfkens R, Davis PL, Crooks LE, et al: NMR imaging of the abnormal live rat and correlation with tissue characteristics. Radiology 141:211, 1981.)

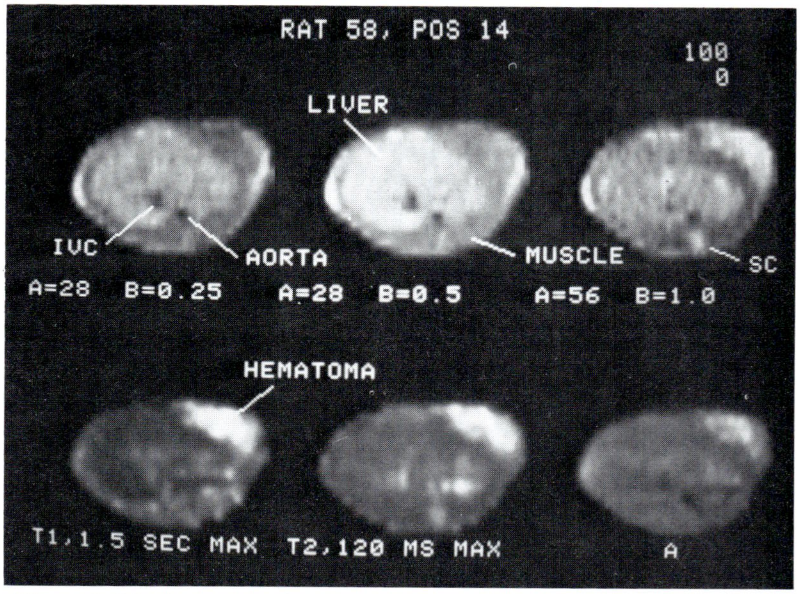

Figure 8. Nuclear magnetic resonance images through the upper abdomen of a supine rat with a subcutaneous hematoma. Calculated images (T_1, T_2, asymptote [A in bottom right]) show the hematoma with long T_1 and T_2 times. The lesion is also seen in the spin echo image (top right). (IVC = inferior vena cava; SC = spinal cord, TE = A, TR = B)

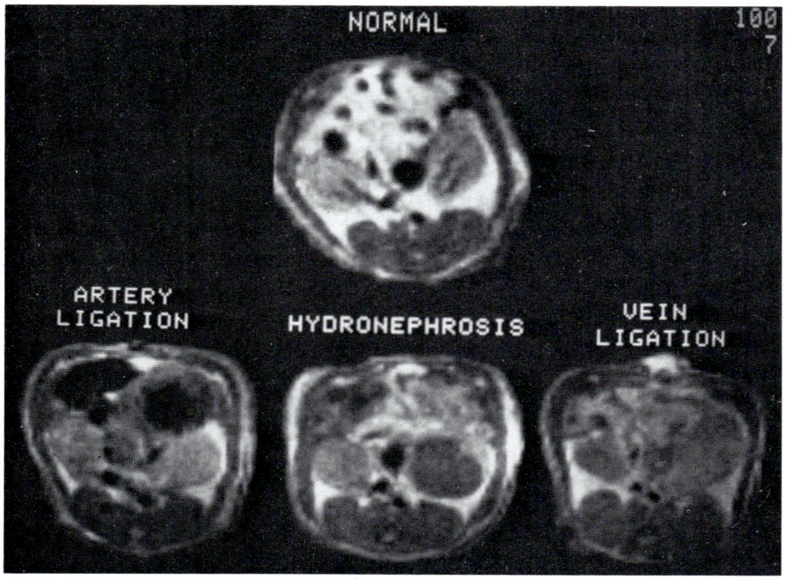

Figure 9. Transverse spin echo images through the kidneys of rats that were normal, had the left renal artery ligated, the left renal ureter ligated, or the left renal vein ligated. The normal kidneys always demonstrated urine in the collecting structures, but usually at levels different from the pathologic kidneys. The left kidney with the renal artery ligation is normal in size, but no urine is detected in the pelvocaliceal system at any level. The left kidney with the ureter ligation shows enlargement of the pelvocaliceal system and thinning of the parenchyma. The left kidney with the renal vein ligation demonstrates gross enlargement of the kidney, but no urine is identified in the collecting structures.

TABLE I

T_1 and T_2 Relaxation Times of Various Rat Tissues

Tissue	No. of Specimens	T_1 Mean (msec) ± SD	T_1 Range	T_2 Mean (msec) ± SD	T_2 Range
Fat	501	305 ± 49	136–468	54.1 ± 4.8	33–97
Muscle	556	707 ± 101	350–1041	29.5 ± 2.3	23–44
Liver	48	426 ± 52	291–504	38.9 ± 6.3	32–64
Brain	6	675 ± 80	625–849	54.8 ± 2.1	51–57
Abscess (Pyogenic)	4	575 ± 66	479–665	58.3 ± 8.0	45–66
Abscess (Sterile)	3	830 ± 142	714–1029	63.0 ± 1.6	61–65
Hematomas	2	827 ± 46	781–872	66.0 ± 16.0	50–82
Mammary Adenocarcinoma	267	977 ± 166	561–1770	58.1 ± 5.5	45–78
Fibrosarcoma	177	1544 ± 388	810–3899	68.0 ± 4.7	53–82
Hepatomas	28	832 ± 187	399–1450	57.3 ± 10.9	37–79

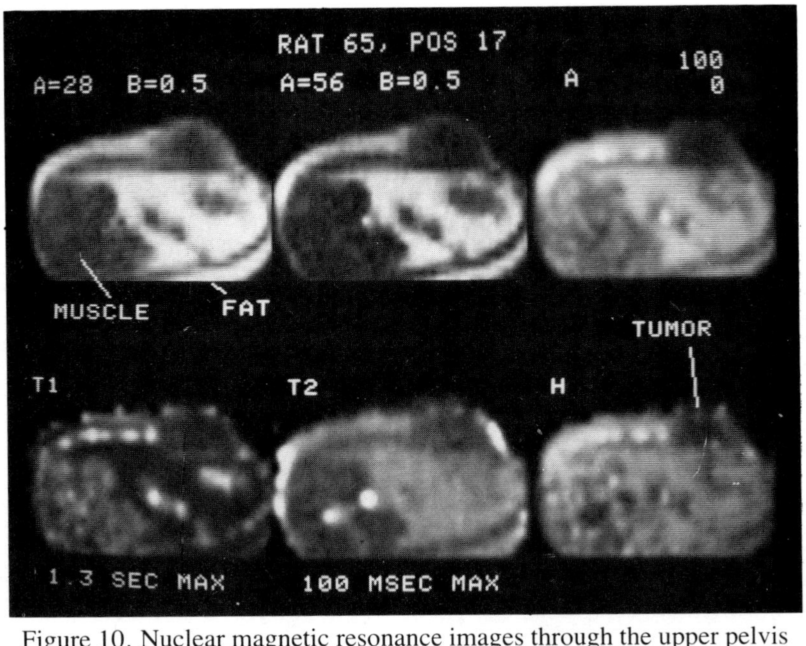

Figure 10. Nuclear magnetic resonance images through the upper pelvis of a supine rat with a spontaneous tumor. The lesion was identified as a fibroma on pathologic examination. Shown are two NMR intensity images and the asymptote (A in top right), T_1, T_2, and hydrogen (H) images. Legends under T_1 and T_2 images show the T_1 or T_2 time corresponding to maximum intensity. The tumor appears dark (low intensity and short relaxation time) on all images.

(Reproduced, with permission, from Herfkens R, Davis PL, Crooks LE, et al: NMR imaging of the abnormal live rat and correlation with tissue characteristics. Radiology 141:211, 1981.)

Benign Tumors

The only benign tumors that we could obtain were two spontaneous fibromas in rats (Fig. 10). They had low NMR intensities and short T_1 and T_2 relaxation times.

Malignant Tumors

Three large series of malignant tumors were studied in rats. These were a hepatoma implanted in the liver, a mammary adenocarcinoma implanted in calf muscle, and a fibrosarcoma implanted in calf muscle.[12–14] Each series consisted of about 25 implanted rats and five control rats. The tumors were imaged weekly for 6 to 8 weeks. Each week several rats were sacrificed for pathologic correlation and, for the latter two series, for water and lipid content.

All tumors greater than 8 mm^2 (3 mm in diameter) were detected in the spin echo images (Figs. 11–13). No false positives occurred. The tumors were usually seen best in the images with the longer TE and TR times (Figs. 14, 15). In some of the spin echo and T_1 images internal structures could be recognized in the tumors; at pathologic examination these structures usually correlated to areas of increased tissue density (which imaged with higher intensity) or to areas of necrosis (which imaged with lower intensity) (Figs. 16, 17). For all the tumors there was good correlation between the NMR image area of the tumor and its corresponding pathologic section area (Fig. 18).

The mean T_1 and T_2 relaxation times of the tumors were long (Table I). The mean T_2 times of all three tumor series were similar. The mean T_1 times of the hepatomas and the mammary adenocarcinomas were similar, but shorter than the mean T_1 time of the fibrosarcomas. However, there was considerable overlap of individual T_1-T_2 pairs. In addition, there was a substantial overlap of tumor relaxation times with the relaxation times of benign pathologic conditions. Necrotic regions of the tumors and bladder contents tended to have very long T_1 and T_2 relaxation times.

The tumors could be differentiated from their normal surrounding tissues by their T_1 or T_2 relaxation times, or both. Most of the hepatomas differed in both T_1 and T_2 from normal liver. The mammary adenocarcinomas and the fibrosarcomas had T_1 times similar to those of muscle, but different T_2 times, and T_2 times similar to those of fat, but different T_1 times. Therefore, for these latter two tumors, T_1 and T_2 times and an imaging technique by which both could be obtained (such as the spin echo) were necessary for best differentiating these two tumors from normal surrounding tissues.

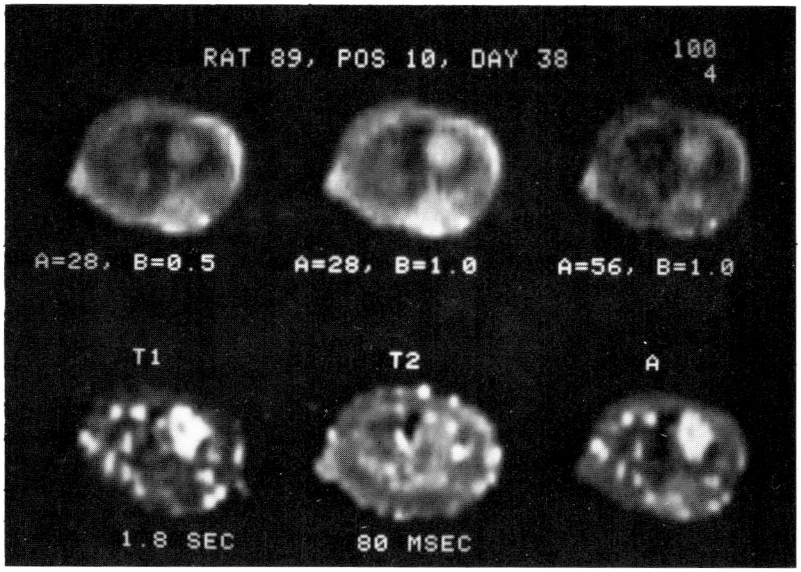

Figure 11. Nuclear magnetic resonance images through lower chest of a rat showing an extrahepatic tumor in the base of the left lung. Top row shows spin echo images. Bottom row shows T_1, T_2, and asymptote (A) images with relaxation time gray scale ranging to the maximum time indicated. (A = TE, B = TR)

(Reproduced, with permission, from Davis PL, Kaufman L, Crooks LE, et al: Detectability of hepatomas in rat livers by nuclear magnetic resonance imaging. Invest Radiol 5:354, 1981.)

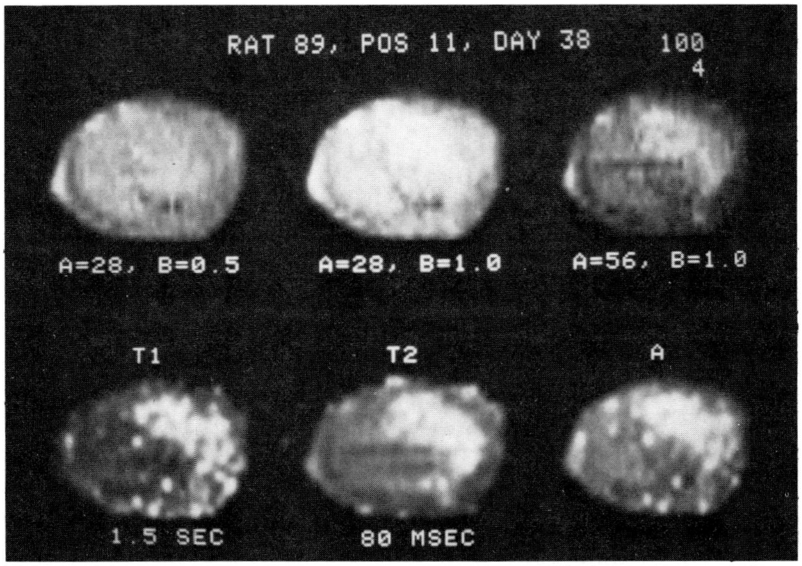

Figure 12. Nuclear magnetic resonance images through liver of a rat. Intense area on right side of image is a large hepatoma.
(Reproduced, with permission, from Davis PL, Kaufman L, Crooks LE, et al: Detectability of hepatomas in rat livers by nuclear magnetic resonance imaging. Invest Radiol 5:354, 1981.)

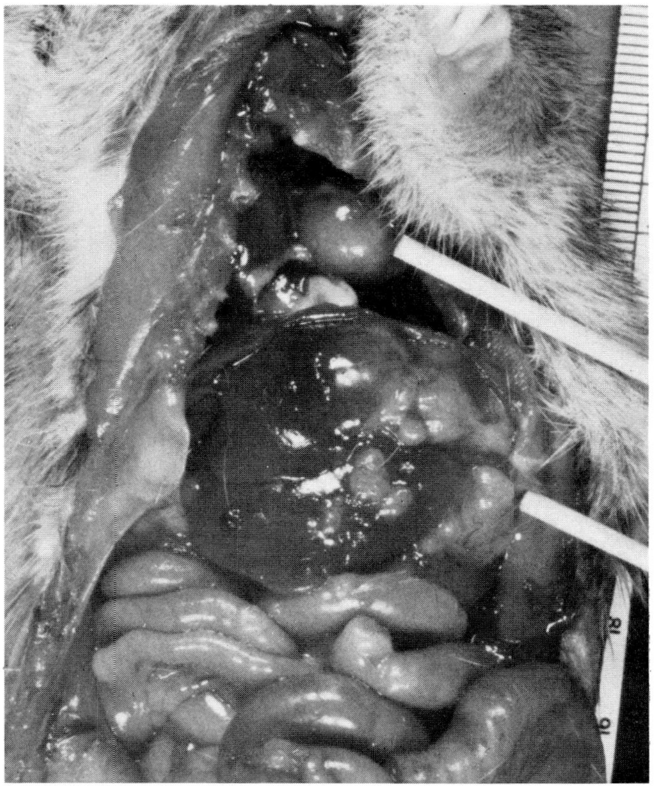

Figure 13. Photograph of dissected rat imaged in Figures 11 and 12 shows both tumors as indicated by pointers.
(Reproduced, with permission from Davis PL, Kaufman L, Crooks LE, et al: Detectability of hepatomas in rat livers by nuclear magnetic resonance imaging. Invest Radiol 5:354, 1981.)

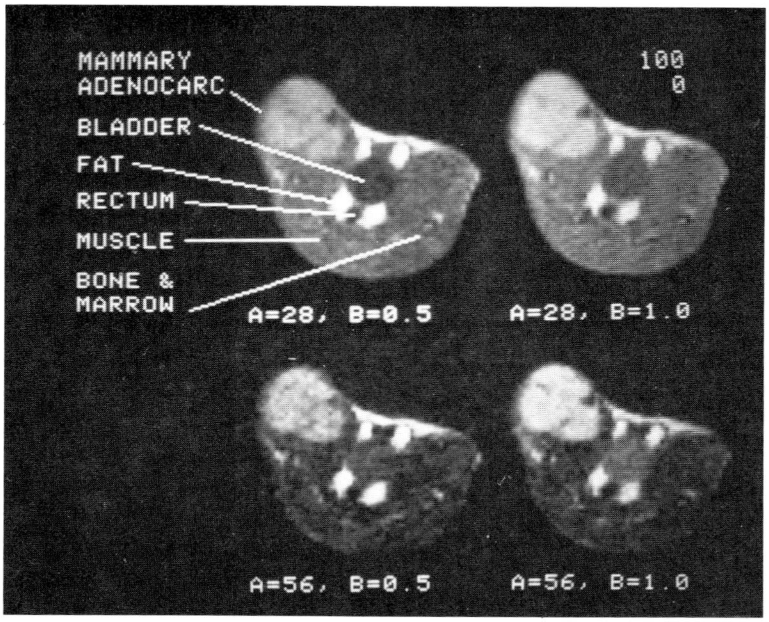

Figure 14. Transverse spin echo images through the pelvis of a rat with a mammary adenocarcinoma implanted in the right calf. The tumor is seen best in the images with the longer TE and TR times. (TE = A, TR = B)

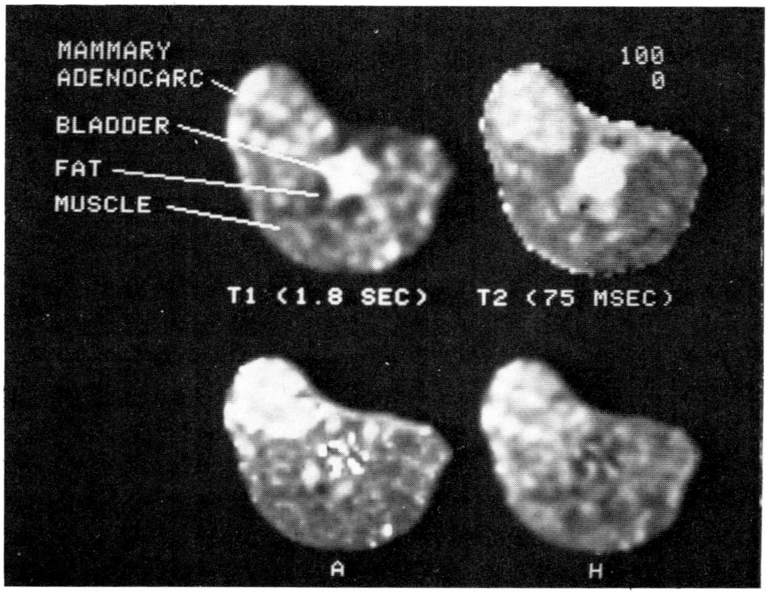

Figure 15. T_1, T_2, asymptote (A), and hydrogen (H) images calculated from the data in Figure 14. The T_1 and T_2 times of the tumor are prolonged.

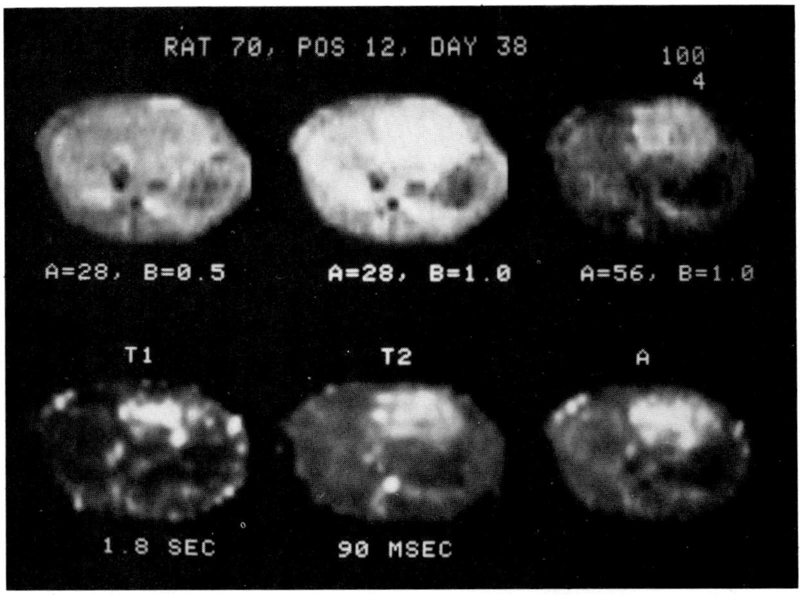

Figure 16. Nuclear magnetic resonance image through liver of a rat demonstrates tumor in upper left of liver. Both T_1 and asymptote (A in bottom right) images demonstrate a more intense core in the tumor.
(Reproduced, with permission, from Davis PL, Kaufman L, Crooks LE, et al: Detectability of hepatomas in rat livers by nuclear magnetic resonance imaging. Invest Radiol 5:354, 1981.)

Figure 17. Pathology section shows tumor as a white core surrounded by a slightly darker region.
(Reproduced, with permission, from Davis PL, Kaufman L, Crooks LE, et al: Detectability of hepatomas in rat livers by nuclear magnetic resonance imaging. Invest Radiol 5:354, 1981.)

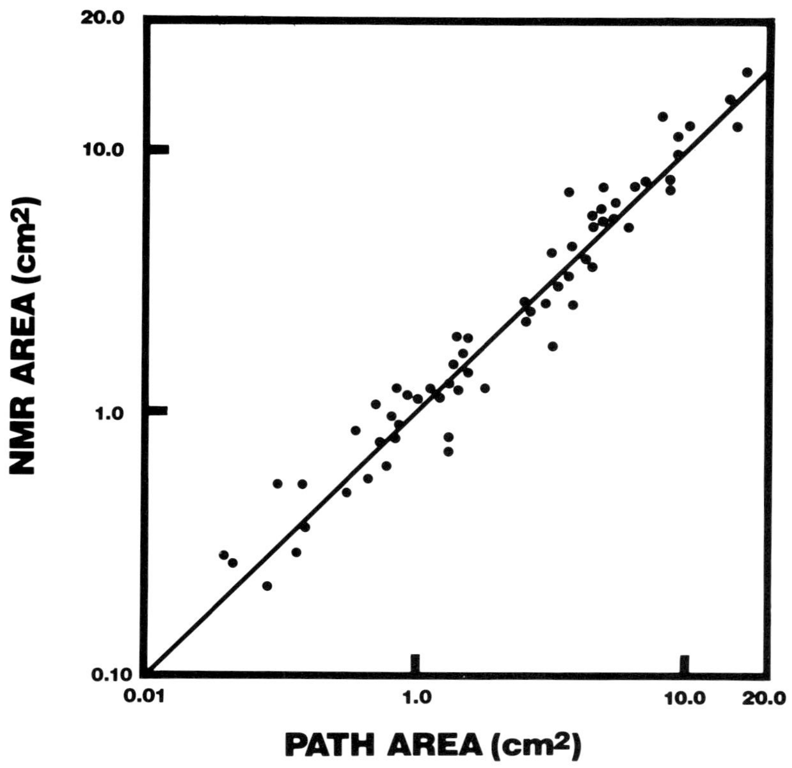

Figure 18. Comparison of tumor areas as obtained from NMR images and pathology section. Line is line of identity.

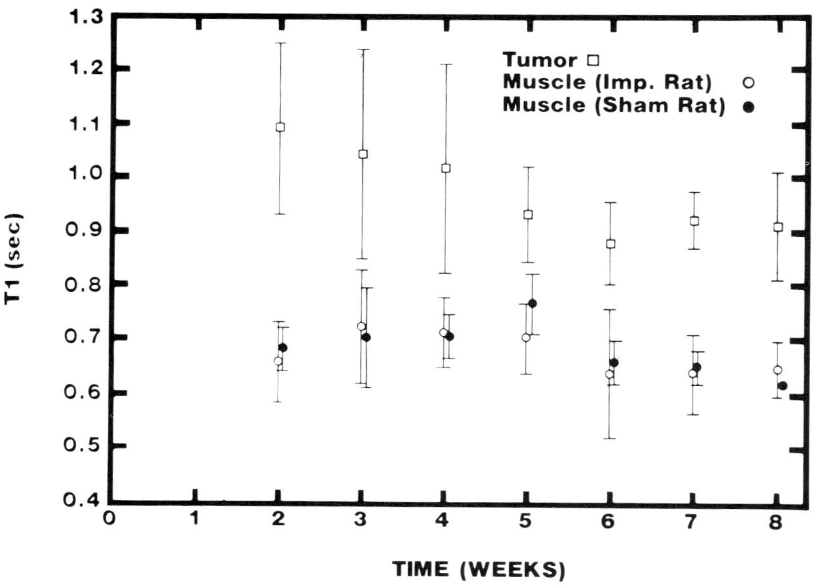

Figure 19. The mean T_1 relaxation times of the mammary adenocarcinomas were quite uniform over the 6-week period. No significant differences were noted between the relaxation time of muscle in rats with implanted tumors and that in sham-implanted rats.
(Reproduced, with permission, from Davis PL, Sheldon P, Kaufman L, et al: Nuclear magnetic resonance imaging of mammary adenocarcinomas in the rat. Cancer 51:433, 1983.)

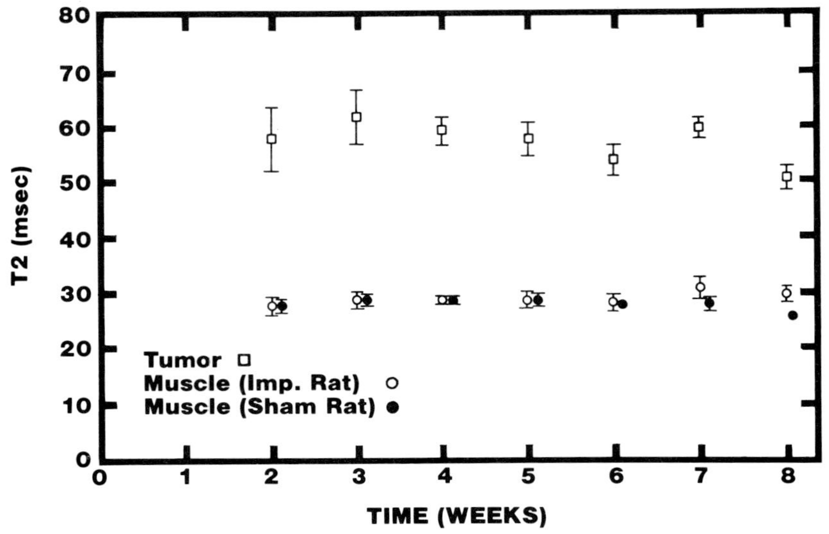

Figure 20. The mean T_2 relaxation times of the mammary adenocarcinomas was quite uniform over the 6-week period. No significant differences were noted between the relaxation time of muscle in rats with implanted tumors and that in sham-implanted rats.
(Reproduced, with permission, from Davis PL, Sheldon P, Kaufman L, et al: Nuclear magnetic resonance imaging of mammary adenocarcinomas in the rat. Cancer 51:433, 1983.)

The relaxation times of the tumors and, in the latter two experiments, of muscle both in the tumor-bearing and the control rats were measured repeatedly as the tumors grew (Figs. 19, 20). These measurements demonstrated that the relaxation times of the tumors did not change significantly as the tumors grew. In addition, the relaxation times of the muscle in the tumor-bearing rats did not change and did not differ significantly from the times of the muscle in the control rats. Thus, these models failed to demonstrate the systemic effects reported elsewhere in the literature.[15]

Brain Edema and Dehydration

Schmidley et al. induced cellular brain edema by administering vasopressin and water, which caused plasma hypo-osmolality.[16] Brain water content increased to 80.1% ± 0.2% (normal, 78.3% ± 0.1%) (Table II). T_2 increased significantly to 54.5 ± 2.1 msec (normal, 49.8 ± 1.8 msec, $P<0.01$). T_1 increased to 1021 ± 81 msec (normal, 953 ± 104 msec), but this increase was not significant. Brain dehydration was induced by administering 2-M sucrose. Brain water decreased to 75.2% ± 1.0%. T_1 and T_2 decreased significantly to 779 ± 85 msec ($P<0.01$) and 45.2 ± 3.5 msec ($P<0.01$), respectively.

Brain Infarction

Mano et al. induced acute ischemic infarctions in one hemisphere of 31 gerbils by occluding their right common carotid artery (gerbils possess an incomplete circle of Willis).[17] Animals were imaged at 3–7, 12–16, or 24–28 hr following carotid occlusion. The ischemic lesion could be detected by NMR at 3 hr after the infarction, the earliest time that these infarcts were imaged. While both T_1 and T_2 increased linearly over the 24-hr period, the change in T_2 became significant during the 3-hr to 7-hr period. The change in T_1, however, did not become significant until the 24-hr to 28-hr period (Table II).

Muscle Infarction

Herfkens et al. infarcted the lower extremity of 11 rats by injecting absolute alcohol into the external iliac arteries.[18] Immediately after occlusion there was no significant change in the T_1 or T_2 relaxation times (Table II). However at 6 and 24 hr significant increases ($P<0.001$) were measured in both T_1 and T_2. The water contents of the infarcted and control extremities were measured. The mean water content of the infarcted extremities (83.0% ± 2.0%) at 24 hr had increased significantly compared to the control extremities (72.0% ± 2.0%) ($P<0.001$).

TABLE II

Relaxation Times of Several Experimental Pathology Models

	T_1	T_2	Water
Model	*Mean(msec) ± SD*	*Mean(msec) ± SD*	*Mean (%) ± SD*
Brain water			
Edema	1021 ± 81	54.5 ± 2.1*	80.1 ± 0.2
Dehydration	779 ± 85	45.2 ± 3.5*	75.2 ± 1.0
Control	953 ± 104	49.8 ± 1.8	78.3 ± 0.1
Brain infarction			
3–7 hr infarction	930 ± 42	62.9 ± 1.7*	
3–7 hr control	850 ± 30	57.9 ± 1.0	
12–16 hr infarction	1030 ± 54	66.4 ± 1.7*	
12–16 hr control	880 ± 54	56.6 ± 0.7	
24–28 hr infarction	1413 ± 90*	70.8 ± 4.2*	
24–28 hr control	1040 ± 90	56.5 ± 2.1	
Muscle infarction			
Immediate infarction	596 ± 177	34.2 ± 3.8	
Immediate control	596 ± 200	32.8 ± 4.2	
24–hr infarction	1079 ± 134*	45.8 ± 4.2*	83.0 ± 2.0
24–hr control	620 ± 127	29.2 ± 1.7	72.0 ± 2.0
Myocardial infarction			
Infarction	728 ± 95	48.4 ± 2.4*	78.6 ± 0.7
Control	650 ± 87	42.1 ± 1.2	76.2 ± 0.6

* Significant at least at $P<0.05$ vs. control.

Myocardial Infarction

Higgins et al. created myocardial infarctions in six dogs by ligating the left anterior descending coronary artery.[19] Twenty-four hours later the hearts were excised and NMR imaging performed. The NMR intensity of the infarcted area was increased in all six dogs. The area of infarction on the NMR image correlated well to the gross pathology findings. The T_1 times of the infarcted heart (728 ± 95 msec) was not significantly longer than the times of the normal heart (650 ± 87 msec) (Table II). Although there was overlap between the two sets of times, the T_1 time of the infarcted myocardium was always longer than the normal myocardium for each individual animal. However, the mean T_2 time of the infarcted heart (48.4 ± 2.4 msec) was significantly longer than the T_2 time of the normal heart (42.1 ± 1.2 msec) ($P<0.01$). There was no overlap of the T_2 times between normal and infarcted myocardium.

The water content of the infarcted myocardium (78.6% ± 0.7%) was significantly greater than that of normal myocardium (76.2% ± 0.6%) ($P<0.01$). Although there was no significant linear correlation between T_1 and water content, there was a good correlation ($r = 0.90$, $P<0.01$) between the T_2 times and water content.

Discussion

Effects of Water and Lipid on T_1 and T_2

Since it is too difficult to derive the relaxation times of complex tissues from theory, we can measure the relaxation times of many different tissues and try to correlate these times to differences in the composition and structure of the tissues. Although the composition and structure of the tissues are complex, we can begin simply. The most common reservoirs of hydrogen in soft tissues are, first, water and, second, lipid.

In the animal experiments, as the water content of tissues increased, the T_1 and T_2 relaxation times of these tissues increased, except for fat. Fat has a much longer T_2 time than would be expected by its water content. With this exception there is some correlation between the rates $1/T_1$ and $1/T_2$ and the water content of the tissues (Figs. 21, 22). Most of the induced pathologic conditions had increased water content and correspondingly increased T_1 and T_2 times. The only condition with decreased water content compared with normal was the dehydrated brain, and there the T_1 and T_2 times were decreased. Some tumors also showed this effect.[9]

There does not appear to be any definitive correlation between T_1 and T_2 time and the lipid content

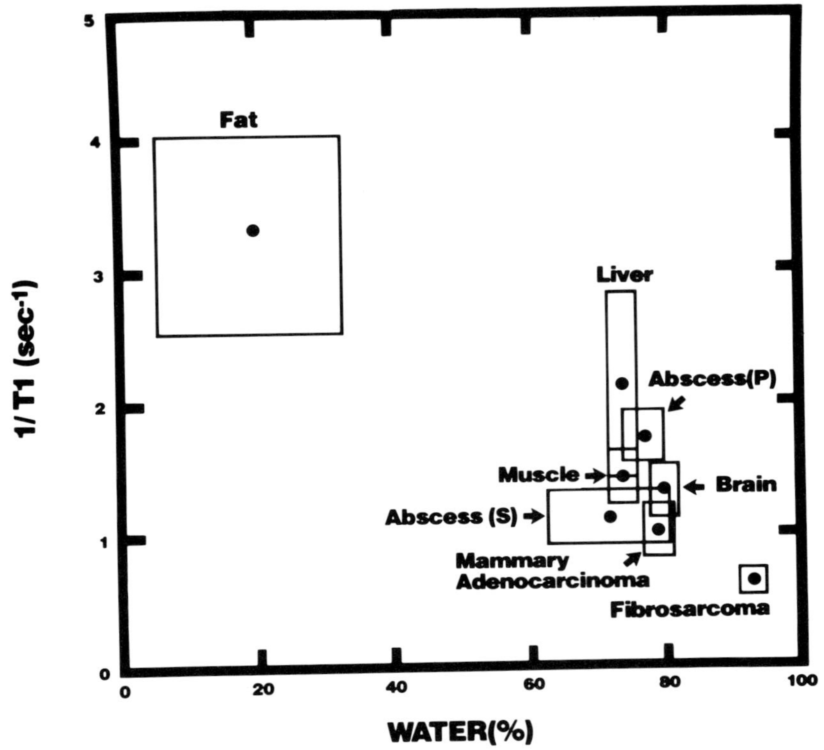

Figure 21. Rate $1/T_1$ as a function of water content of tissue. The width and height of each square indicate 1 standard deviation.

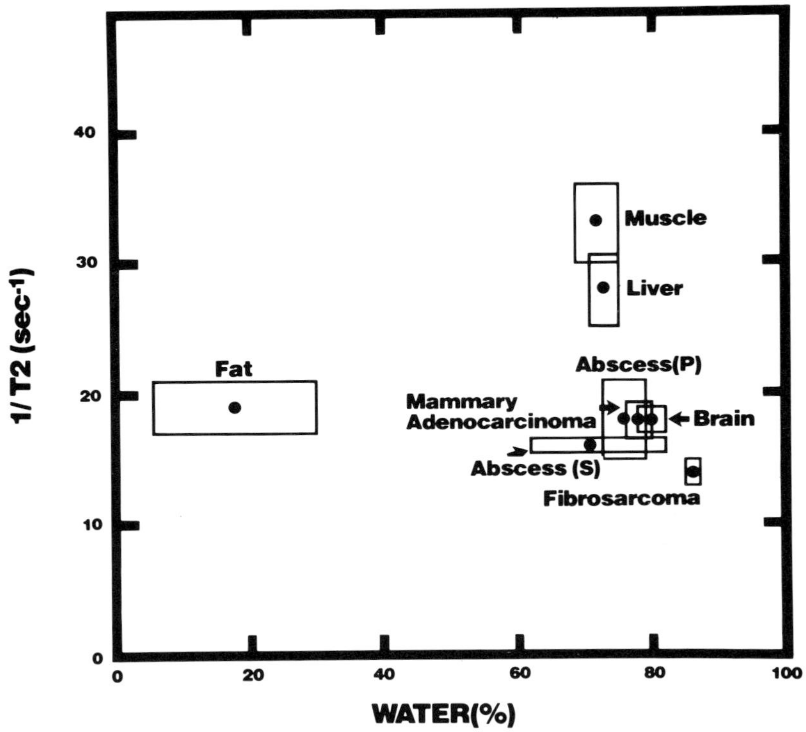

Figure 22. Rate $1/T_2$ as a function of water content of tissue. The width and height of each square indicate 1 standard deviation.

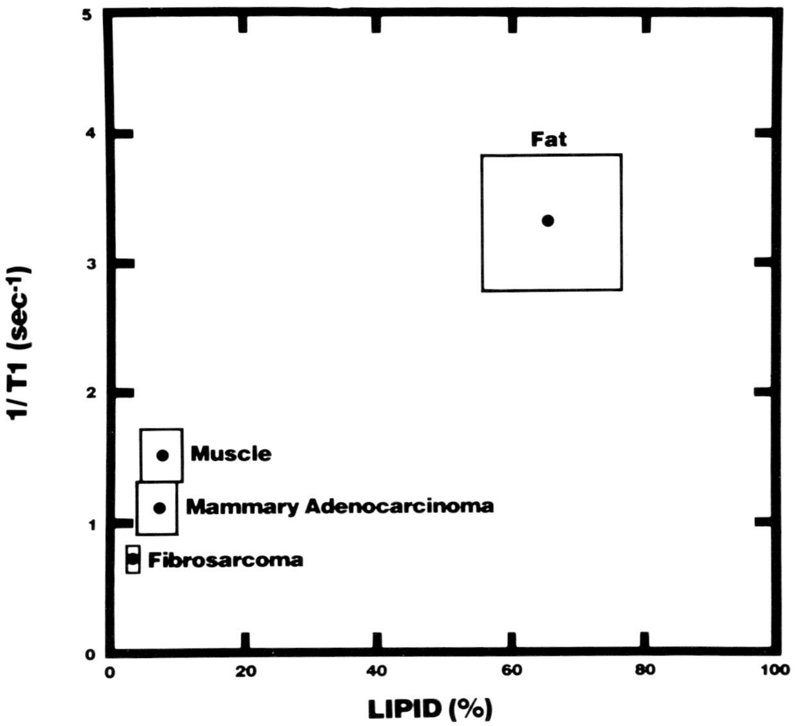

Figure 23. Rate $1/T_1$ as a function of total lipid content of tissue.

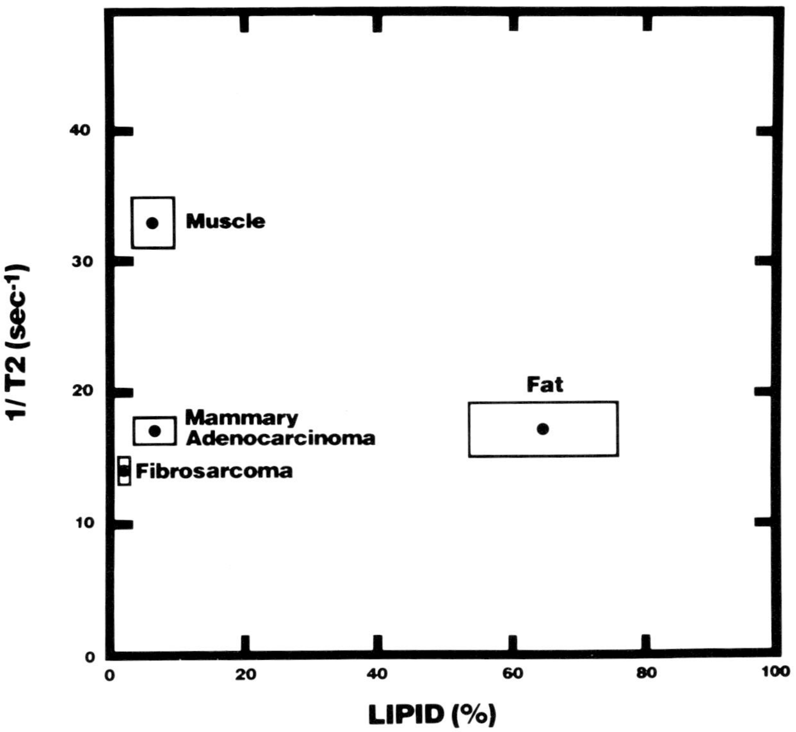

Figure 24. Rate $1/T_2$ as a function of total lipid content of tissue. The long T_2 time of fat is probably due to its high lipid content. Muscle and tumor, with essentially identical lipid content, have very different relaxation times.

of a tissue. In the animal tumor experiment, for example, the muscle and tumor usually had similar lipid content but different T_1 and T_2 times (Figs. 23, 24). The differences in the NMR behavior of these tissues are probably due to the major reservoir of their hydrogen protons. Fat consists of approximately 50%–80% lipid and 10%–30% water, while the other soft tissues consist of approximately 70%–85% water and 5%–10% lipid. Apparently, lipid content accounts for the long T_2 of fat, while water does for nonfatty tissues, tumors, and other pathologic conditions.

Undoubtedly, much more work will be necessary before the etiology of the T_1 and T_2 relaxation times of tissues is fully understood. Our work shows that they are related to the water and fat content of tissue. But, as previously discussed, water in tissue can be in many different compartments—intracellular, interstitial and vascular—that do not always freely communicate. In addition, the water may be free or associated to various degrees with other molecules or structures. All these factors will affect the relaxation times and how they change with different pathologic conditions.

When one considers that pure water has T_1 and T_2 relaxation times of about 3–6 sec, a lot of interesting things must be happening in nonfatty tissues to shorten T_1 and T_2 times to about 0.5–2 sec and 0.03 and 0.15 msec, respectively.

Sequential Changes in T_1 and T_2 of Lesions

It is often important to ascertain how a pathologic condition changes over time. This frequently has prognostic implications. It would be very helpful if changes in the T_1 and T_2 relaxation times of a pathologic condition could provide this information.

The T_1 and T_2 relaxation times of the infarcted gerbil brains increased progressively over the 24-hr period after the infarctions. The water content of the infarcted tissue is also known to increase during this time.

The relaxation times of the three implanted tumors did not change significantly over the several-week period of each experiment. The water content of the mammary adenocarcinoma remained quite uniform until the tumors became necrotic.

Sensitivity and Specificity of T_1 and T_2

How sensitive and specific the T_1 and T_2 relaxation times are for the definitive diagnosis of disease is one of the unanswered questions about NMR. This issue dates back to 10 years ago, when it was first discovered that at least some pathologic conditions had T_1 and T_2 times different from those of normal tissues.[20] Since then an extensive literature has been published on both sides of the debate.[21–32]

Our results derive from nearly 2,000 measurements, mainly in rats, of T_1 and T_2 relaxation times of normal and well-defined abnormal tissues.[9,12–14] To avoid partial volume averaging problems, only tissues greater than a certain volume were measured. These were usually fat, brain, muscle, liver, urinary bladder contents, and the pathologic tissues. The results are shown in Figure 25. The values of the rectangle borders are the limits of the first standard deviation. In theory, if T_1 and T_2 times were completely independent, the rectangles should only contain about 36% of the T_1-T_2 pairs. However, there is so much correlation between the two times that the rectangles usually contain greater than 70% of the T_1-T_2 pairs.

The pathologic conditions that we examined could usually be differentiated from the normal surrounding tissues if both T_1 and T_2 times were used. For example, although the T_1 times of muscle and the mammary adenocarcinoma overlapped slightly, and the T_2 times of fat and the adenocarcinoma overlapped significantly, each of these three tissues could be uniquely defined with no overlap when both T_1 *and* T_2 information was taken into account. This was also true for the fibrosarcoma. The other pathologic conditions occasionally showed some overlap, but this could usually be further minimized where comparison to control tissues within the same animal was obtained. Hence, the sensitivity and specificity of the T_1 and T_2 times for differentiating normal from abnormal tissues are quite good.

The specificity of the T_1 and T_2 times for differentiating different pathologic conditions is quite poor, however. As shown in Figure 25, there is considerable overlap between the T_1 and T_2 times of the different conditions. Although some T_1 and T_2 pairs uniquely exist for the hematomas, hepatomas, and fibrosarcomas, many of their pairs overlap those of other conditions. Some conditions, such as abscesses and mammary adenocarcinomas, have no unique T_1-T_2 pairs.

The in vivo relaxation times for human tissues measured by NMR imagers show a similar trend. For example, Doyle et al. have measured the T_1 times of many liver diseases in 32 subjects.[33] Their two cases of hepatomas could be differentiated from most other pathologic conditions. The T_1 times in their three cases of metastases overlapped those of the four cases of dilated bile ducts. A variety of inflammatory and metabolic diseases such as cirrhosis and hepatitis had overlapping times different from

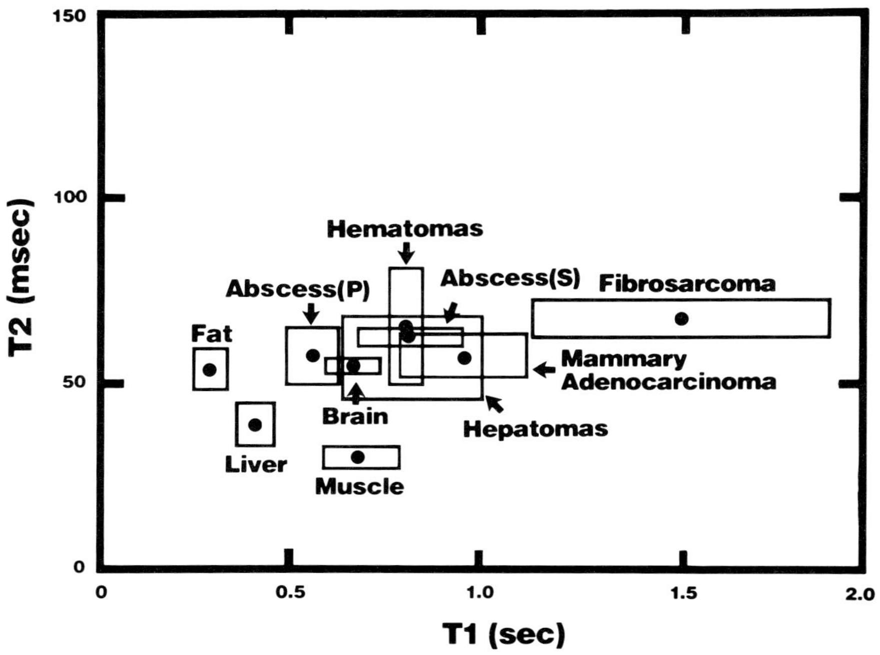

Figure 25. Plot of T_1 or T_2 relaxation times for various tissues. Rectangles are the mean, plus and minus the first standard deviations. The rectangles usually contain greater than 70% of the T_1-T_2 pairs.

the groups just described. Some of these times overlapped the range of normal liver T_1 times.

Smith et al. have measured the T_1 times of a variety of normal organs and a few pathologic conditions for each.[34,35] Most of the carcinomas had little or no overlap with the normal tissue T_1 times, but overlapped considerably between tissue types. For example hepatomas, cholangiocarcinomas, liver metastases, and pancreatic carcinomas had overlapping T_1 times. However, within a single organ, certain disease could be differentiated: for example, cirrhosis from liver tumors, pancreatitis from pancreatic carcinoma. Simple cysts had T_1 times much longer than those of any other tissue.

Ross et al. measured the T_1 times of a variety of breast diseases.[36] While normal and cancerous breast tissue can be differentiated, the T_1 of both types was overlapped by that of benign breast disease.

Currently there is a lack of data obtained from NMR imagers published on in vivo T_2 relaxation times in humans. The reason is that it is difficult to implement the spin echo technique rapidly in humans because of the current and stability requirements of the gradient coils. If the TE time is increased to compensate for this, the NMR intensity drops severely. Approximately 65% of the signal is lost in a single time period equal to the T_2 relaxation time of the tissue being imaged. However, these design problems have been overcome at the University of California, San Francisco and T_2 relaxation times in humans are now being measured and will be reported shortly.

In conclusion, the T_1 and T_2 relaxation times appear to be quite sensitive indications of pathologic conditions. They may even be specific for some. However, a great many will overlap. Within a specific organ, certain groups of diseases may be differentiable by the relaxation times. Having both the T_1 and T_2 times available improves the accuracy of the technique.

Accuracy and Reproducibility of T_1 and T_2

The ranges of normal T_1 relaxation time of normal liver from several groups are shown in Table III. They vary significantly. Although some of the variations may result from nutritional difference (especially "normal" alcohol intake), most of it is

TABLE III

T_1 Relaxation Times of Normal Liver from Three Groups

Location of Group	*T_1 range (msec)*
Aberdeen (1981)[34]	140–170
Hammersmith (1982)[33]	210–270
San Francisco (1983)[38]	228–956

probably from other factors. The most important factor is that NMR imagers are designed to produce a good image. Many elements in this design degrade the ability of the imager to accurately measure T_1 and T_2 relaxation times compared with NMR machines specifically designed to measure these times. Specifically, the long RF pulses and the magnetic gradients used in human imaging can lead to inaccuracies. In addition, in the case of T_1 measurements, two complete imaging cycles, often separated by minutes, are usually necessary to obtain the minimal data required to calculate T_1. During this time the subject can move. The T_1 and T_2 time calculations are noise-sensitive. Relaxation time machines make many measurements to decrease the noise effects; most imagers only make two measurements.

Moreover, all calculations are based on an equation describing the imaging technique, similar to how Equation 2 describes the spin echo technique. Unfortunately, these equations are approximations in two specific ways.[37] First, the imagers do not implement exactly the necessary manipulations to fulfill the requirements of equations. Second, the equations themselves are based on assumptions that are not necessarily true; for example, that the T_1 and T_2 decays are simple exponentials. Other factors, such as inhomogeneities of the tissue being measured and temperature instability, complicate the situation.

However, simply because the times measured by imagers are not accurate does not mean that they are not useful. To the contrary, the previous section showed they can be quite useful. What is more important is that for a specific system, they remain reproducible. The T_1 and T_2 relaxation times obtained at different times over the past 2 years from our 6.5-cm aperture NMR images have been compared. These times have not changed significantly despite changes made in the imager's hardware and software during this period. The relaxation times of muscle measured in the tumor experiments remained very stable over the 6 to 7 weeks of the experiments (Figs. 19, 20). These results demonstrate that the relaxation times of NMR imagers can be very reproducible.

It should be expected that the relaxation times will vary between installations since the times are sensitive to changes in magnetic field strength and imaging technique. Each institution may need to compile its own table of relaxation times of normal and abnormal tissues. However, the trends of increase and decrease in the times should remain fairly consistent. Whether two identical imagers will produce identical results must be determined experimentally. We have compared the relaxation times of rats in the small imager with those of humans in the large imager, and they are similar. However, more imagers will need to be installed and tested before significant conclusions can be made.

Conclusions

The wide variation of the T_1 and T_2 relaxation times is what gives NMR imaging its superb soft-tissue contrast. As will be shown in the following chapters, this is what makes NMR imaging useful now. Even if the relaxation times are not specific to pathologic conditions, they will provide much diagnostic information when combined with the patient's clinical presentation and the image information such as location, size, contour, and homogeneity of the lesion. However, there is the prospect that the relaxation times may provide much more information when their etiology and interactions with the imaging techniques are better understood.

Acknowledgments

The authors would like to thank the following researchers who allowed us to use their data: Kirk Moon, David London, Robert Herfkens, James Schmidley, Robert Levy, Isamu Mano, and Charles Higgins. Special thanks to Phillip Sheldon and Antonio Brito for their extensive technical assistance.

Dr. Davis is supported by a Research Career Development Award from the USPHS National Cancer Institute (CA00836-01). Additional support for many of the experiments came from the Radiology Research and Education Foundation, Diasonics Inc., and USPHS grant from the National Cancer Institute (CA32850).

REFERENCES

1. Glasel JA: Nuclear magnetic resonance studies on water and ice. In *Water: A Comprehensive Treatise,* Franks F (Ed), Vol 1, Chap 6. New York: Plenum, 1972.
2. Becker ED: *High Resolution NMR* (ed 2), pp. 184–275. New York: Academic, 1980.
3. Mathur-DeVre R: The NMR studies of water in biological systems. Prog Biophys Molec Biol 35:103, 1979.
4. Mansfield P, Morris PG: *NMR Imaging in Biomedicine,* Chap 2. New York: Academic, 1982.
5. Kuntz ID, Zipp A: Water in biological systems. N Engl J Med 297:262, 1977.
6. Beall PT, Hazelwood CF, Rao PN: Nuclear magnetic resonance patterns of intracellular water as a function of HeLa cell cycle. Science 192:904, 1976.
7. Crooks LE, Mills CM, Davis PL, et al: Visualization of cerebral and vascular abnormalities by NMR imaging: The effects of imaging parameters on contrast. Radiology 144:843, 1982.
8. Crooks LE, Hoenninger JC, Arakawa M, et al: Tomography of hydrogen with nuclear magnetic resonance. Radiology 136:701, 1980.

9. Herfkens R, Davis PL, Crooks LE, et al: NMR imaging of the abnormal live rat and correlation with tissue characteristics. Radiology 141:211, 1981.
10. Hansen G, Crooks LE, Davis PL, et al: In vivo imaging of the rat with nuclear magnetic resonance. Radiology 136:695, 1980.
11. London DA, Davis PL, Williams RD, et al: Nuclear magnetic resonance imaging of induced renal lesions. Radiology 148:167, 1983.
12. Davis PL, Kaufman L, Crooks LE, et al: Detectability of hepatomas in rat livers by nuclear magnetic resonance imaging. Invest Radiol 5:354, 1981.
13. Davis PL, Sheldon P, Kaufman L, et al: Nuclear magnetic resonance imaging of mammary adenocarcinomas in the rat. Cancer 51:433, 1983.
14. Moon KL Jr, Davis PL, Kaufman L, et al: Nuclear magnetic resonance imaging of a fibrosarcoma tumor implanted in the rat. Radiology 148:177, 1983.
15. Hollis DP, Saryan LA, Economou JS, et al: Nuclear magnetic resonance studies of cancer: V. Appearance and development of a tumor systemic effect in serum and tissues. J Natl Cancer Inst 53:807, 1974.
16. Schmidley JW, Brito A, Chan PH, et al: Using nuclear magnetic resonance (NMR) imaging to study changes in brain water (abstract). Neurology 33: (Suppl 2):151, 1983.
17. Mano I, Levy RM, Crooks LE, et al: Nuclear magnetic resonance imaging of acute experimental cerebral ischemia. Invest Radiol 147:761, 1983.
18. Herfkens RJ, Sievers R, Kaufman L, et al: Nuclear magnetic resonance imaging of infarcted muscle: A rat model. Radiology 147:761, 1983.
19. Higgins CB, Herfkens R, Lipton MJ, et al: Nuclear magnetic resonance imaging of acute myocardial infarctions: Alterations in magnetic relaxation times. Am J Cardiol (in press).
20. Damadian R: Tumor detection by NMR. Science 171:1151, 1971.
21. Damadian R, Zaner K, Hor D, et al: Human tumors detected by NMR. Proc Natl Acad Sci USA 71:1471, 1974.
22. Medina D, Hazlewood CF, Cleveland G, et al: NMR studies on human breast dysplasias and neoplasms. J Natl Cancer Inst 54:813, 1975.
23. Eggleston JC, Saryan LA, Czeisler JL, et al: NMR studies of several experimental and human malignant tumors. Cancer Res 33:2156, 1973.
24. Hazlewood CF, Cleveland G, Medina D: Relationship between hydration and proton NMR relaxation times in tissues of tumor bearing and non-tumor bearing mice. J Natl Cancer Inst 52:1844, 1974.
25. Inch WR, McCredie JA, Knispel RR, et al: Water content and proton spin relaxation times for malignant and nonmalignant tissues from mice and humans. J Natl Cancer Inst 52:353, 1974.
26. Weisman ID, Bennett LH, Maxwell LR, et al: Recognition of cancer in vitro by NMR. Science 178:1288, 1972.
27. Hollis DP, Saryan LA, Eggleston JC, et al: NMR studies of cancer. J Natl Cancer Inst 54:1469, 1975.
28. Barriolhet LE, Moran PR: NMR relaxation spectroscopy in tissues. Med Phys 2:191, 1975.
29. Hollis DP, Economou JS, Parks LC, et al: Nuclear magnetic resonance studies of several experimental human malignant tumors. Cancer Res 33:2156, 1973.
30. Ranade SS, Sha S, Korgaonkar KS, et al: Absence of correlation between spin-lattice relaxation times and water content in human tumor tissues. Physiol Chem Phys 8:131, 1976.
31. Ling GN, Tucker M: Nuclear magnetic resonance relaxation and water contents in normal mouse and rat tissues and in cancer cells. J Natl Cancer Inst 64:1199, 1980.
32. Hazlewood CF: Water content and proton spin relaxation time for malignant and nonmalignant tissues from mice and humans. J Natl Cancer Inst 52:625, 1974.
33. Doyle FA, Pennock JM, Banks LM, et al: Nuclear magnetic resonance imaging of the liver: Initial experience. AJR 138:193, 1982.
34. Smith FW, Mallard JR, Reid A, et al: Nuclear magnetic resonance tomographic imaging in liver disease. Lancet 1:963, 1981.
35. Smith FW, Reid A, Hutchison JMS, et al: Nuclear magnetic resonance imaging of the pancreas. Radiology 142:677, 1982.
36. Ross RJ, Thompson JS, Kim K, et al: Nuclear magnetic resonance imaging and evaluation of human breast tissue: Preliminary clinical trials. Radiology 143:195, 1982.
37. Kinslaw WS, Lent AA: An introduction to NMR imaging: From the Block equation to the imaging equation. Proc IEEE 71:338, 1983.
38. Moss AA, Goldberg HI, Stark D, et al: Nuclear magnetic imaging of hepatic tumors. (Submitted for publication.)

CLINICAL APPLICATIONS

7

Anatomic Approach to Nuclear Magnetic Resonance Images of the Brain

CATHERINE M. MILLS, M.D.
JACK DEGROOT, M.D., Ph.D.

Nuclear magnetic resonance offers the unique combination of an imaging technique that is capable of superb morphological definition and freedom from any detectable biologic hazards.[1-4] The anatomy of the brain is depicted in exquisite detail without ionizing radiation.[5-13] Direct multiplanar images can be obtained without degradation by beam-hardening artifacts. Blood vessels are especially well defined, even in the absence of any contrast agents.[14-19] The signal intensity from blood within the vessel lumen varies with the sequence of gradient and radio-frequency (RF) pulses used to generate the image, as well as with the velocity of blood flow. Appropriate imaging techniques can optimize anatomic detail, distinguish slowly from rapidly flowing blood, and serve to identify marked impairment or complete obstruction of flow in an artery or a vein.

Anatomy

Soft Tissues of the Skull

Skin and Subcutaneous Tissues. The high proportion of fat in the skin and subcutaneous tissues results in an intense NMR signal with all imaging techniques (Figs. 1-3). Tissues that are primarily fat are seen as bright areas with NMR as a result of the short T_1 of fat. Variations in contrast due to differences in T_1 are highlighted by changing the pulse sequence interval (TR) in the spin echo technique. The magnetization of the sample is zero immediately following an RF pulse. With repetitive RF perturbations at short intervals, the signal intensity is low because insufficient time has elapsed for the sample to regain magnetization. For small values of TR, relatively higher signal is noted from tissues with a short T_1 as compared with tissues with a long T_1. As the pulse sequence interval is increased, the recovery of magnetization increases, and, consequently, the NMR signal becomes larger. Images obtained at the second echo delay, at 56 msec, also show fat as high signal intensity; however, its relative intensity, compared with the adjacent soft tissues, has decreased.

Muscles. The musculature of the head and neck has a low NMR signal intensity (Fig. 1). The long T_1 relaxation time of muscle does not permit full recovery of magnetization between repetitive excitation sequences. Images obtained with a longer TR show that the relative intensity of muscle increases; however, it remains less intense than surrounding tissues.

Superficial Vasculature. The superficial vasculature in the soft tissues of the skull is easily identified as a result of the low signal intensity produced by

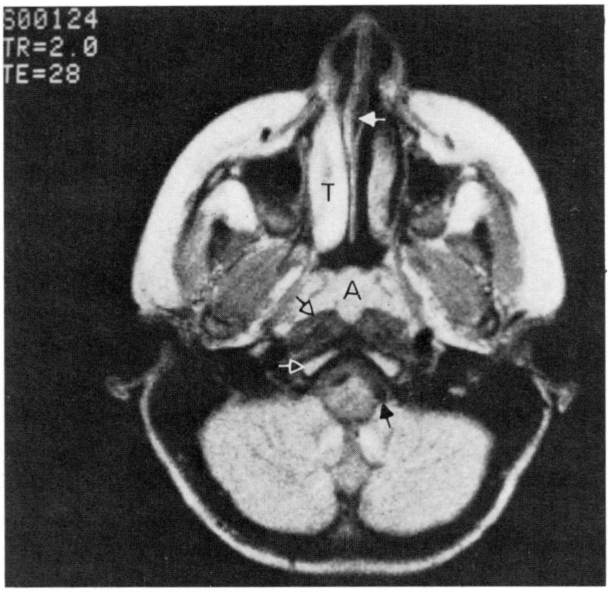

Figure 1. Spin echo (TR of 2.0 sec, TE of 28 msec). The high soft-tissue contrast results in a clear distinction between the adenoid tissue (A) and the longus colli and capitis muscles (open black arrow). (nasal septum, solid white arrow; vertebral artery, solid black arrow; clivus, open white arrow; T, swollen turbinate)

flowing blood in NMR images obtained at a short echo delay of 28 msec (Fig. 3). The variation in NMR signal intensity produced by blood flowing at different velocities allows the differentiation of arterial and venous structures.

Pinna. The cartilaginous structure of the pinna is visualized as a tissue of low signal intensity (Fig. 2). It is surrounded by the skin, which has a higher signal intensity.

Nose. The skin and nasal cartilage produce a lower intensity signal than the subcutaneous tissue does and are generally similar in intensity to muscle (Fig. 1). The nasal turbinates are brighter in NMR intensity than the adjacent musculature, but not as intense as fat. Air within the nasal cavity is visualized as the lowest intensity within the image.

Skull Vault and Base

Inner and Outer Tables. The cortical bone of the inner and outer tables of the cranial vault has a very low signal intensity (Figs. 1-9). The primary cause of the lack of signal is the low mobile proton density. Variations of the pulse sequence interval (TR) and the echo delay (TE) do not result in any variation of the signal intensity of the cortical bone.

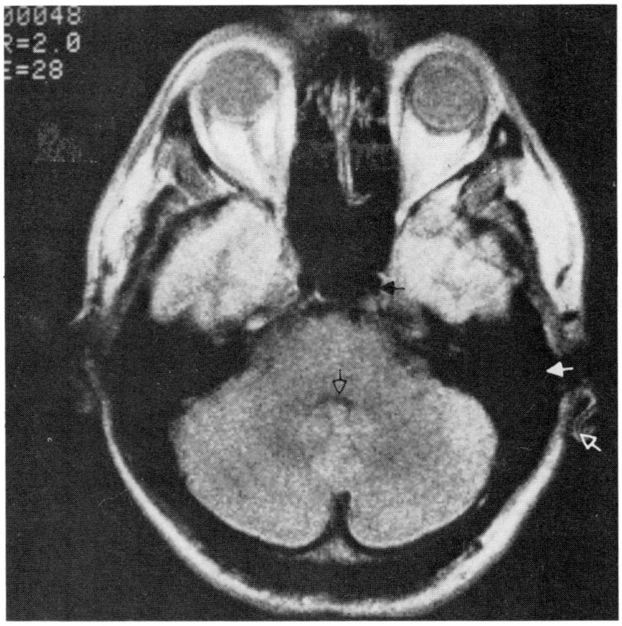

Figure 2. Spin echo (TR of 2.0 sec, TE of 28 msec). The structures of the posterior fossa are sharply defined, without degradation from bone artifacts. (internal carotid artery, solid black arrow; fourth ventricle, open black arrow; pinna, open white arrow; mastoid air cells, solid white arrow)

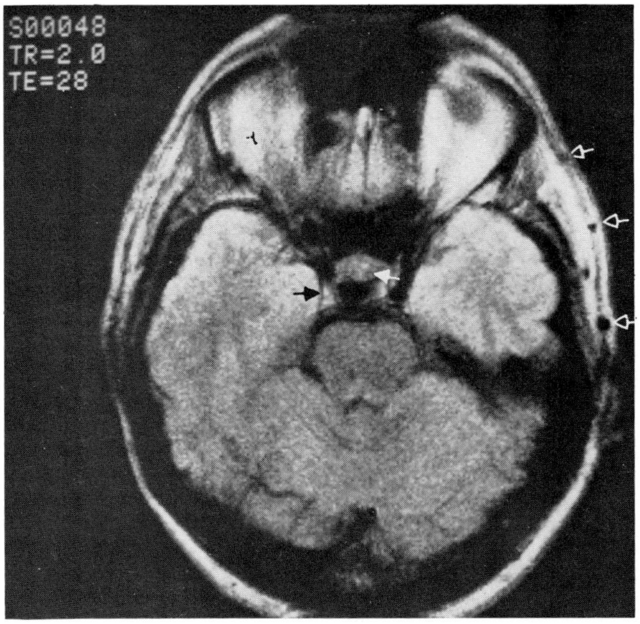

Figure 3. Spin echo (TR of 2.0 sec, TE of 28 msec). The superficial vasculature (open white arrows) has low signal intensity as a result of relatively rapid blood flow. (Pituitary gland, solid white arrow; tentorium, solid black arrow; levator-superior rectus complex, open black arrow)

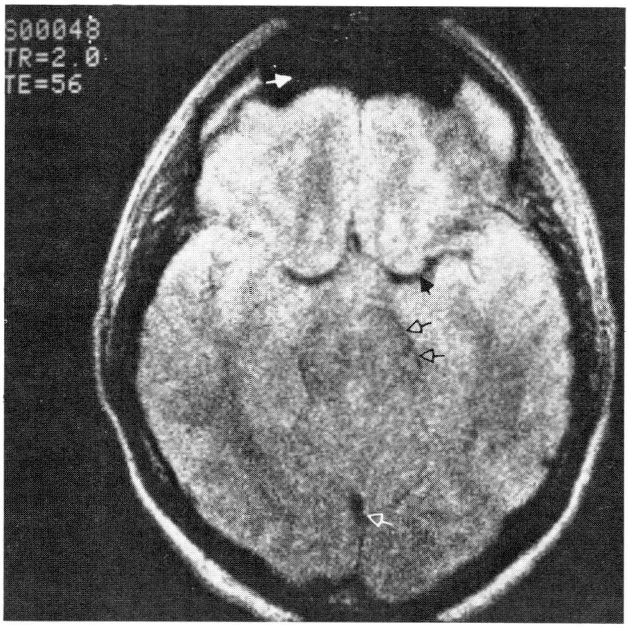

Figure 4. Spin echo (TR of 2.0 sec, TE of 56 msec). Cerebrospinal fluid increases in signal intensity in the second echo image when the repetition time is long, allowing the definition of the vascular structures comprising the circle of Willis. (Frontal sinus, solid white arrow; posterior cerebral artery, open black arrows; straight sinus, open white arrow; middle cerebral artery, solid black arrow)

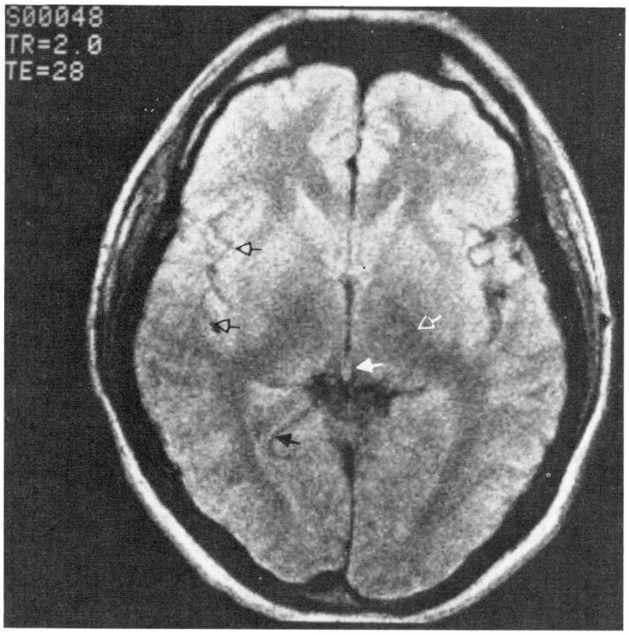

Figure 5. Spin echo (TR of 2.0 sec, TE of 28 msec). The signal intensity of the gray matter is higher than that of the white matter when the repetition time is long, reflecting its higher water content. The branches of the major cerebral arteries are shown clearly. (Internal capsule, open white arrow; internal cerebral vein, solid white arrow; calcarine branch, solid black arrow; insula with middle cerebral branches, open black arrows)

Diploic Space. The medullary cavity is easily identified by its high NMR signal intensity (Fig. 9). The bright signal is the result of two distinct factors: the short T_1 relaxation time of the fat component of the marrow, and the slowly flowing blood within the diploë.

Petrous Bones. The petrous bones are primarily represented as black, signal-free regions of the image because of the low intensity of cortical bone and the aeration of the mastoid air cells (Fig. 2). Definition of the contours of cortical bone with an adjacent air interface is difficult because of the similarity in NMR signal intensity. The neurovascular bundle can usually be identified entering the internal auditory canal. The jugular bulb is frequently identified as a bright spot in the petrous bone that increases in signal intensity with longer echo delays; the most likely explanation for this finding is the presence of turbulence within the bulb. Turbulence produces a subpopulation of protons that experience repeated excitation within the image plane and thus increase their signal.

Clivus. The clivus is primarily identified as a result of the high intensity signal produced by the marrow (Fig. 1). The adjacent cortical bone is recognized by its absence of signal.

Brain

Gray and White Matter. The NMR intensity of gray and white matter varies with the technique of signal acquisition and with alterations of the instrument parameters. Inversion recovery images show superb contrast between gray and white matter, with gray matter having a relatively lower signal intensity than white matter (Fig. 6). Contrast in the inversion recovery image is primarily dependent on the contribution of the T_1 relaxation time, although it is also influenced to a small degree by T_2 relaxation. The particular TR and TI chosen will accentuate a range of T_1 values as the primary determinants of contrast. The contrast seen in the spin echo image varies from a relatively featureless low-contrast image when the echo delay and the pulse interval are small to a higher contrast image when these parameters are increased. The intensity difference between gray and white matter reflects their differing compositions; gray matter is composed of more water and less lipid than white matter is. The increased water results in a prolongation of the relaxation parameters. The contrast due to T_1 relaxation is accentuated by altering the pulse sequence interval. The intensity of gray matter increases as the pulse sequence interval is

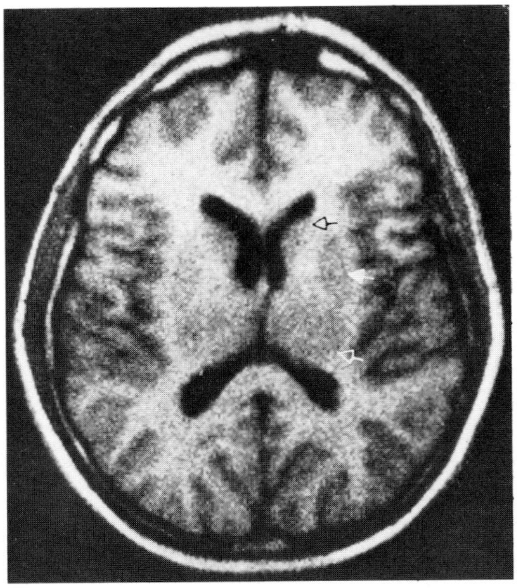

Figure 6. Inversion recovery (TR of 1.0 sec, TI of 0.42 sec). Gray matter has a low signal intensity as compared with adjacent white matter with the inversion recovery technique of signal acquisition. (caudate nucleus head, open black arrow; lentiform nucleus, solid white arrow; thalamus, open white arrow)

prolonged. The accentuation of the tissue with the increased water content by long pulse sequence intervals can be used to advantage to define abnormalities such as infarction, demyelinated plaques, and edema (Figs. 8, 9). The characteristic response of brain to a variety of insults is nonspecific in the production of edema. Indeed, most pathologic processes in the central nervous system are best defined in the image obtained with the long pulse sequence interval.

Pituitary Gland. The pituitary gland is visualized with an NMR intensity slightly greater than that of white matter (Figs 3, 10). The infundibulum often demonstrates a high signal intensity that increases on the second echo image in a manner similar to slowly flowing blood. This rather characteristic feature may represent visualization of capillary or venous blood flow in the pituitary portal system of vessels.

Cerebrospinal Fluid. The T_1 and T_2 relaxation times of cerebrospinal fluid (CSF) are both long. These relaxation characteristics reflect the composition of CSF, a relatively pure liquid whose primary component is water. The NMR intensity of CSF changes dramatically with changes in the echo delay time and pulse sequence interval (Figs. 7, 8, 10). As the repetition time, or TR, is increased, CSF can be made to vary in signal intensity from black to light gray. When the echo delay, or TE, is increased to 56 msec, the NMR intensity of CSF increases further

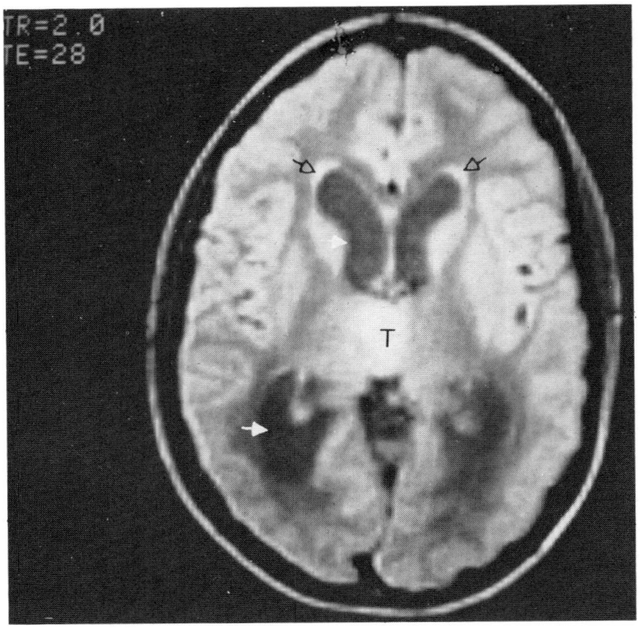

A.

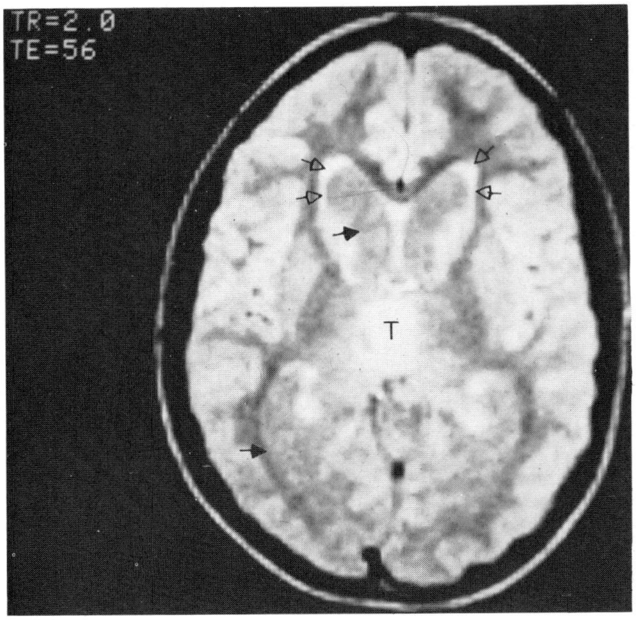

B.

Figure 7. Spin echo images. In A, TR of 2.0 sec, TE of 28 msec. In B, TR of 2.0 sec, TE of 56 msec. Obstruction of the cerebral aqueduct by a tumor (T) in the region of the third ventricle results in ventricular dilatation and transependymal flow of cerebrospinal fluid (CSF) as evidenced by high signal intensity foci adjacent to the anterior horns of the lateral ventricles. The intensity of the CSF (solid arrows), the periventricular edema (open arrows), and the tumor (T) increases with the longer echo delay.

because of the influence of the long T_2 relaxation time. This variation permits the differential visualization of structures outlined by CSF. For example, the brain stem will be seen best when the TR is

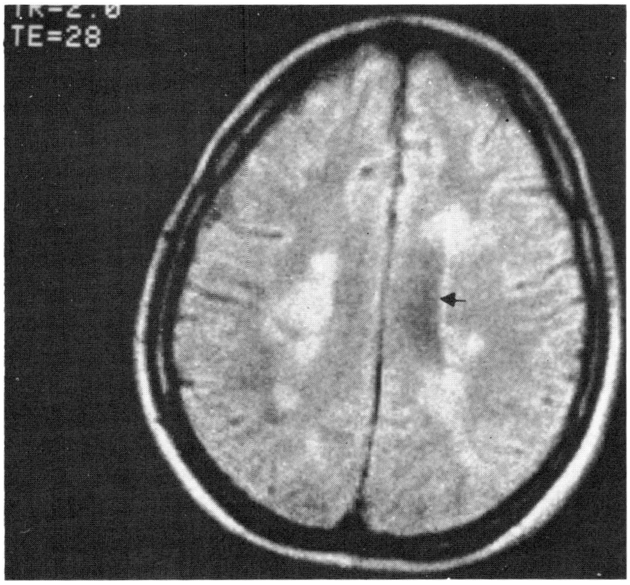

A.

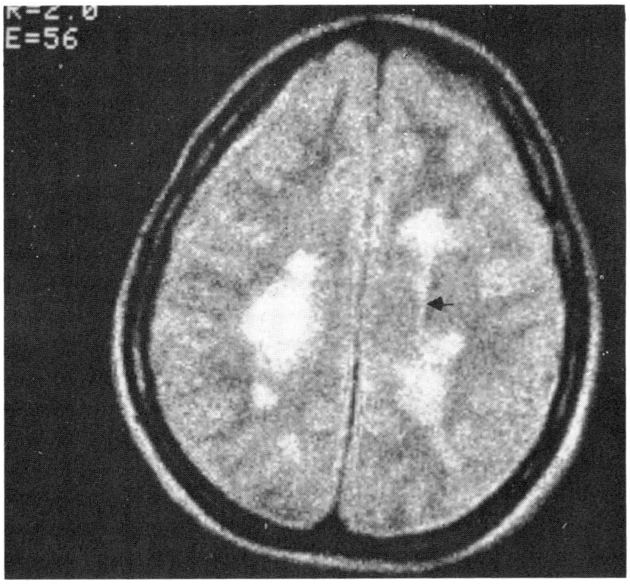

B.

Figure 8. Spin echo images from a patient with multiple sclerosis. In A, TR of 2.0 sec, TE of 28 msec. In B, TR of 2.0 sec, TE of 56 msec. Multifocal periventricular high-intensity foci are present. The intensity of the lesions and the CSF (arrow) is augmented with the longer echo delay.

short, resulting in low intensity CSF, while the basilar artery will be visualized to better advantage when the TR is long and the CSF has increased density (Fig. 10).

Dura. The dura adjacent to the cranial vault is not visualized because of its inherently low signal intensity and its proximity to cortical bone, which also has low signal intensity. The dura of the falx and the tentorium is identified as a structure of low signal intensity silhouetted by the adjacent CSF and

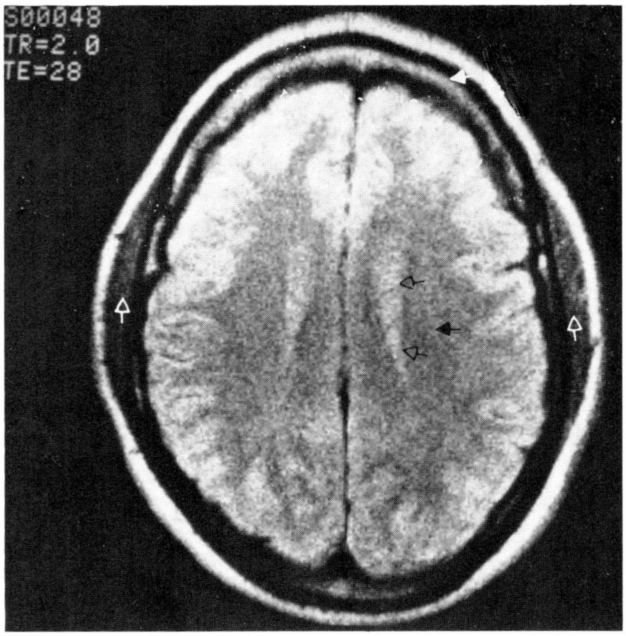

Figure 9. Spin echo (TR of 2.0 sec, TE of 28 msec). The body of the caudate nucleus (open black arrows) is differentiated from each adjacent centrum semiovale (solid black arrow) by its high signal intensity. (bone marrow, solid white arrow; temporalis muscles, open white arrows)

brain (Fig. 3). The low signal from this dense connective tissue may be due to its relative lack of water and, hence, hydrogen.

Vasculature: Effects of Motion on the Nuclear Magnetic Resonance Signal

The intensity of the NMR signal is influenced both by T_1 and T_2 relaxation times, as well as by the motion of the hydrogen nuclei being imaged. For example, if the hydrogen nuclei pass through the imaged volume faster than the time it takes to perform an imaging sequence, the signal will be zero. The signal intensity from flowing blood is a function of the percentage of moving hydrogen nuclei and their velocity. The signal intensity is therefore dependent on temporal aspects of the imaging technique.[8,17,18,20] Relatively low-velocity blood flow has a higher NMR signal intensity in images obtained at the second echo delay as compared with the first echo delay for a given pulse sequence interval (Fig. 10B). High-velocity blood flow in arterial structures produces no detectable signal (Figs. 11-14).

The intensity of flowing blood is affected not only by the instrument parameters, but also by the position of the section in the multisection sequence. The enhancement of the NMR signal is greatest in the sections first entered by the blood. These sections receive protons that have not experienced a prior RF or

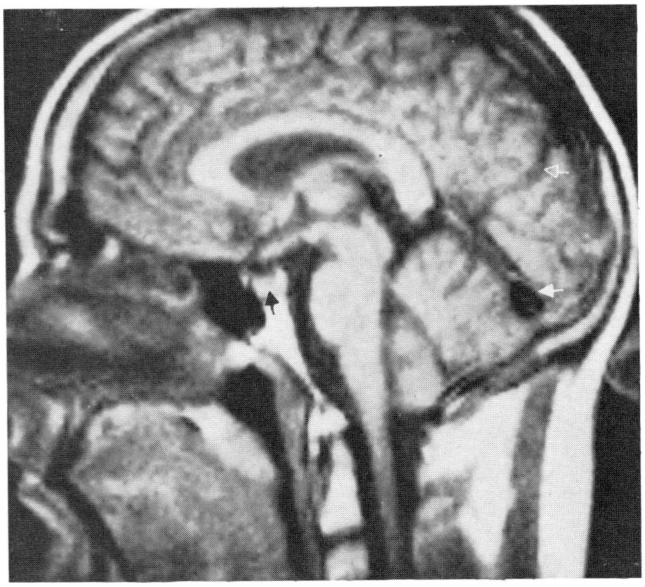

A.

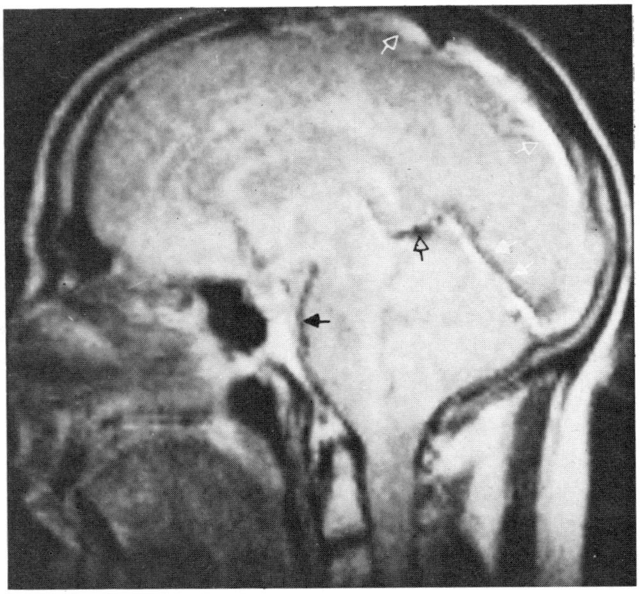

B.

Figure 10.

A. Spin echo (TR of 0.5 sec, TE of 28 msec). The CSF has low signal intensity in images with a short TR, resulting in clear definition of the brain stem. (pituitary gland, solid black arrow; parieto-occipital fissure, open white arrow; torcular, solid white arrow)

B. Spin echo (TR of 1.5 sec, TE of 56 msec). The intensity of the CSF increases with longer TR and TE, permitting the basilar artery (solid black arrow), which remains low in signal intensity, to be defined. The superior sagittal sinus (open white arrows) and straight sinus (solid white arrows) have increased signal intensity. (great vein of Galen, open black arrow)

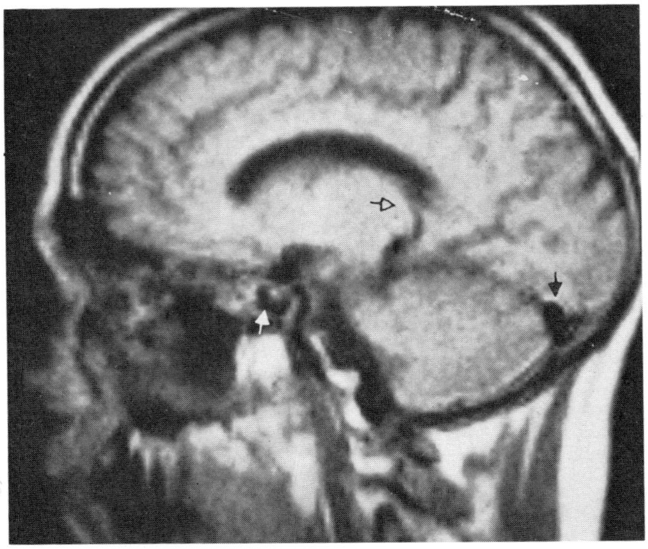

Figure 11. Spin echo (TR of 0.5 sec, TE of 28 msec). The rapid blood flow in the internal carotid artery (no signal, white arrow) contrasts with the slow flow in the surrounding cavernous sinus (high signal). The phenomenon of boundary layer flow is illustrated in the region of the torcular (solid black arrow). (pulvinar, open black arrow)

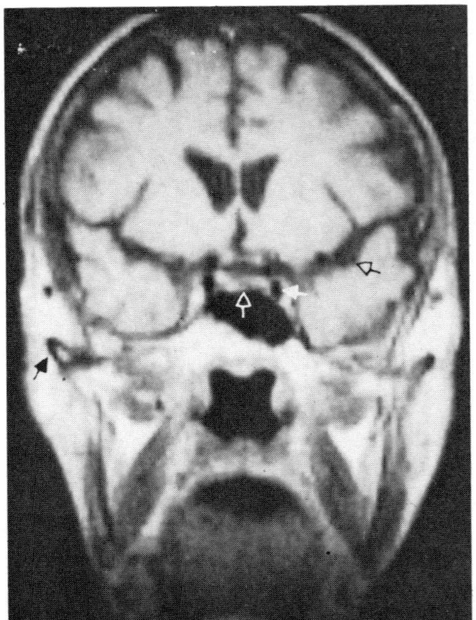

Figure 12. Spin echo (TR of 0.5 sec, TE of 28 msec). The low intensity of the cavernous internal carotid artery (solid white arrow) is silhouetted by the higher intensity of the surrounding cavernous sinus. The pituitary gland (open white arrow) has slightly higher intensity than the white matter. (sylvian cistern with middle cerebral branches, open black arrow; zygomatic arch, solid black arrow)

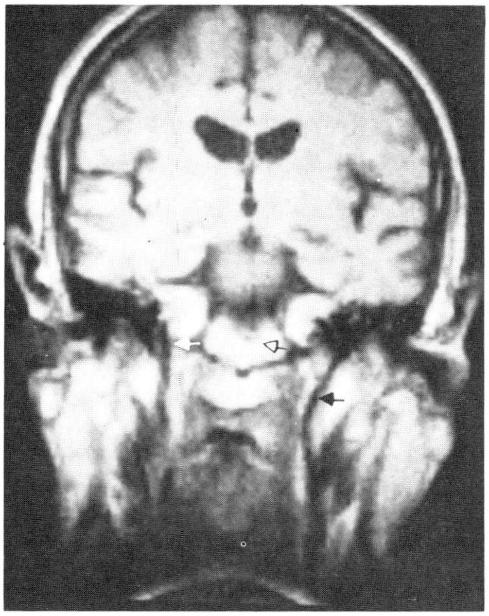

Figure 13. Spin echo (TR of 0.5 sec, TE of 28 msec). Portions of the internal jugular vein (solid black arrow) and the internal carotid artery (solid white arrow) are of low signal intensity because of relatively rapid flow (clivus, open black arrow)

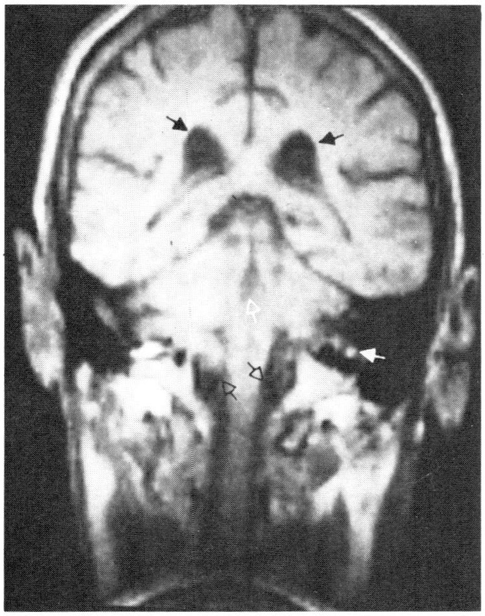

Figure 14. Spin echo (TR of 0.5 sec, TE of 28 msec). The absence of signal from the petrous bones allows the identification of the neurovascular bundle (solid white arrow). Posterior fossa structures are seen without bone artifact. (fourth ventricle, open white arrow; atria of lateral ventricles, closed black arrows; vertebral arteries, open black arrows)

gradient pulse excitation and will, therefore, provide maximal signal. The signal in the subsequent sections is decreased or absent because the protons were perturbed in the previous section. A multisection sequence oriented from cephalad to caudad will demonstrate enhancement of arterial blood in the last section for neck images and in the first section for body images. Venous blood will show the opposite pattern, with enhancement in the first section for neck images, and in the last section for body images. These enhancement patterns reflect the opposing directions of flow of arterial and venous blood.

The size of the volume irradiated affects flow phenomena. If the excited volume is large compared with the sections being imaged, stationary and flowing blood will be represented with a similar intensity since nuclei outside and inside the imaged volume undergo the same sequence of perturbation.

Another NMR flow phenomenon is the ring, or boundary layer pattern (Fig. 11). The higher-velocity blood is centrally located and exhibits no discernable NMR signal. Slowly flowing blood, near the vascular wall, interacts with the RF and gradient pulse sequence and has definable signal. Turbulence will also alter the signal intensity of flowing blood. The effect of turbulence is to maintain a subpopulation of protons within the imaged plane and thus increase the signal.[8]

Orbit

Bone. The dense bony margins of the orbit, like the cortical bone elsewhere in the body, produces no NMR signal because of the low proton density. However, the marrow does produce a high-intensity signal as a result of the fatty component, which shortens T_1.

Orbital Fat. The abundant orbital fat produces a high signal intensity regardless of the imaging technique chosen. There is no apparent difference between fat within the orbit and adjacent subcutaneous adipose tissue, in accordance with its unremarkable histologic features. The strong signal from orbital fat is the result of a very short T_1. T_2 is also short, as demonstrated by the slight decrease in relative signal intensity of orbital fat in images with a longer echo delay. However, the potential reduction in signal due to the short T_2 is made insignificant when compared with the signal's intensification by the very short T_1.

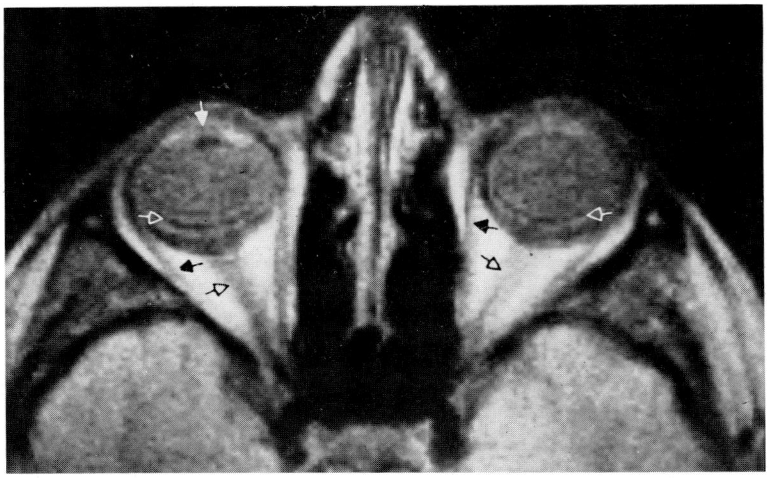

A.

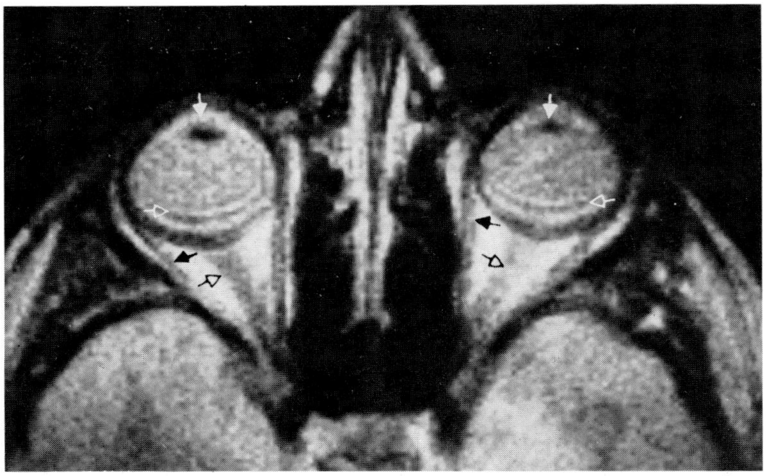

B.

Figure 15. Spin echo images. In A, TR of 2.0 sec, TE of 28 msec. In B, TR of 2.0 sec, TE of 56 msec. There is excellent visualization of the intraorbital structures. The intensity of the lens is lower with a 56-msec echo delay. Motion of the globe during the imaging sequence produces semicircular artifacts (open white arrows). (optic nerve, open black arrows; lens, solid white arrows; extraocular muscles, solid black arrows)

Extraocular Muscles. The extraocular muscles, like muscles elsewhere, have a rather low intensity signal on both spin echo and inversion recovery images, because of their moderately long T_1 and short T_2 (Fig. 15).

The individual tendons of the muscles may be identified as a reduction in signal intensity at the transition from muscle to tendon. The tendon of the superior oblique muscle is particularly well identified as a result of the difference between the high signal from fat and the low signal from fibrous tissue (Fig. 16).

Optic Nerve. The optic nerve has medium signal intensity but appears as a relatively low-intensity structure as compared with orbital fat; it has an NMR intensity similar to those of myelinated white matter (Fig. 15). The delineation of the optic nerve is complicated by the small size of nerve and its dural sheath, which together measure 3 to 4 mm, and by the inherent difficulty in visualizing a small low-intensity structure surrounded by high intensity. The margins of the optic nerve may also be obscured as a result of the movement that occurs during the signal acquisition sequence.

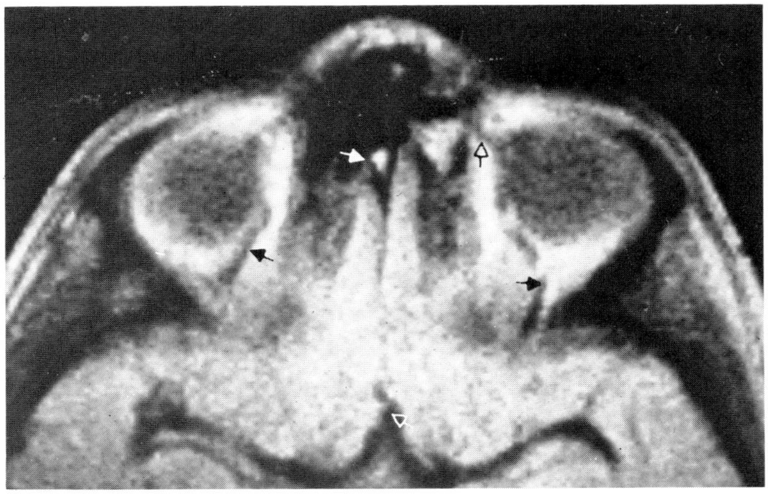

A.

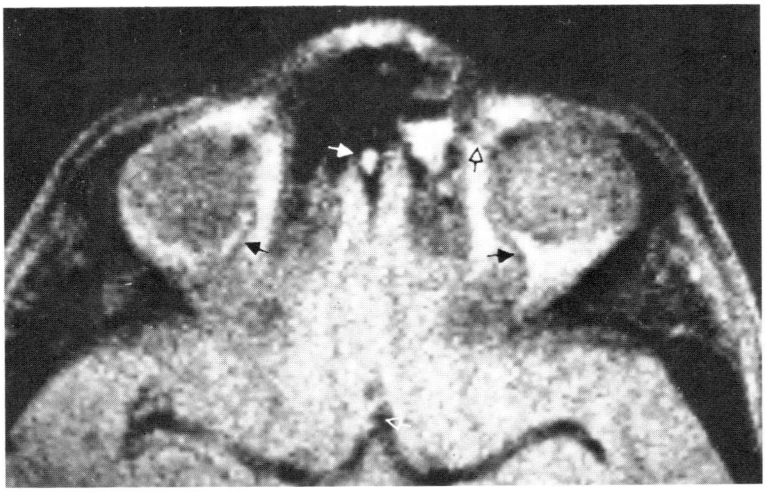

B.

Figure 16. Spin echo images. In A, TR of 2.0 sec, TE of 28 msec. In B, TR of 2.0 sec, TE of 56 msec. The superior ophthalmic vein (solid black arrows) is identified as a low signal intensity structure with a short echo delay. Its intensity increases with a 56-msec echo delay. (tendon of superior oblique muscle [posterior to angular vein], open black arrows; crista galli, solid white arrows; anterior communicating artery, open white arrows)

Sclera and Cornea. The definition of the sclera and cornea is best on images with a long TR, where they are seen as a rim of low intensity signal surrounding the vitreous and separating it from the orbital fat. The sclera consists of dense bundles of collagen fibers with many elastic fibers and may be partially calcified in older patients. Studies performed both in vitro and in vivo have shown the T_1 relaxation time to be long in such intensely fibrous tissue. In addition, the presence of calcification would decrease signal intensity.

Vitreous and Aqueous. The vitreous and aqueous show very low signal intensity when the TR and TE are short. However, the signal is increased somewhat by increasing TR or TE. The T_1 of the vitreous is long, as is the T_2, explaining increasing signal with longer TR and TE.

Lens. The signal intensity of the lens is relatively strong in inversion recovery images and in spin echo images with a short TR. In spin echo images with a long TR, the intensity of the lens decreases with increasing echo delays (Fig. 15). The T_1 of the lens is substantially longer than that of fat, but also much less than that of water, while the T_2 of the lens is very short.

Vasculature. The superior ophthalmic vein is clearly identified in spin echo images of the orbit (Fig. 14). Because of the thickness of the sections used in this study, an area of reduced signal is seen secondary to the superimposition of vein and optic nerve within the orbital fat. With longer TE values, the intensity of the vein increases.

General Considerations

Nuclear magnetic resonance images of cranial morphology provide detailed anatomic information in any desired plane, without degrading artifacts, in a safe and rapid manner. Normal anatomy of the brain is seen in exquisite detail, while the major cranial vessels are demonstrated without contrast administration. The spin echo technique provides tissue characterization that is especially helpful when pathologic processes are present.

REFERENCES

1. Wolff S, Crooks LE, Brown P, et al: Tests for DNA and chromosomal damage induced by nuclear magnetic resonance imaging. Radiology 136:707, 1980.
2. Davis PL, Crooks LE, Arakawa M, et al: Potential hazards in NMR imaging: Heating effects of changing magnetic fields and RF fields on small metallic implants. AJR 137:857, 1981.
3. Davis PL, Kaufman L, Crooks LE, et al: Detectability of hepatomas in rat livers by nuclear magnetic resonance imaging. Invest Radiol 16:354, 1981.
4. Saunders RD: Biologic effects of NMR clinical imaging. Applied Radiol 11:43, 1982.
5. Holland GN, Hawkes RC, Moore WS: Nuclear magnetic resonance (NMR) tomography of the brain: Coronal and sagittal sections. J Comput Assist Tomogr 4:429, 1980.
6. Bydder GM, Steiner RE, Young IR, et al: Clinical NMR imaging of the brain: 140 cases. AJR 139:215, 1982.
7. Crooks L, Arakawa M, Hoenninger J, et al: Nuclear magnetic resonance whole-body imager operating at 3.5 kGauss. Radiology 143:169, 1982.
8. Crooks LE, Mills CM, Davis PL, et al: Visualization of cerebral and vascular abnormalities by NMR imaging: The effects of imaging parameters on contrast. Radiology 144:843, 1982.
9. Davis PL, Crooks LE, Margulis AR, et al: Nuclear magnetic resonance imaging: Current capabilities. West J Med 137:290, 1982.
10. Kaufman L: Reading the NMR image: As the signal changes, so do image features. Diagnostic Imaging, 1982.
11. Young IR, Bailes DR, Burl M, et al: Initial clinical evaluation of a whole body nuclear magnetic-resonance (NMR) tomograph. J Comput Assist Tomogr 6:1, 1982.
12. Brant-Zawadzki M, Davis PL, Crooks LE, et al: NMR demonstration of cerebral abnormalities: Comparison with CT. AJNR 4:117, 1983.
13. Hawkes RC, Holland GN, Moore WS, et al: Nuclear magnetic resonance (NMR) tomography of the brain: A preliminary clinical assessment with demonstration of pathology. J Comput Assist Tomogr 4:577, 1980.
14. Morse OC, Singer JR: Blood velocity measurements in intact subjects. Science 170:440, 1970.
15. Partain CL, James AE Jr, Watson JT, et al: Nuclear magnetic resonance and computed tomography. Radiology 136:767, 1980.
16. Alfidi J, Haaga JR, El Yousef SJ, et al: Preliminary experimental results in humans and animals with a superconducting, whole-body, nuclear magnetic resonance scanner. Radiology 143:175, 1982.
17. Kaufman L, Crooks LE, Sheldon PE, et al: Evaluation of NMR imaging for detection and quantification of obstructions in vessels. Invest Radiol 17:554, 1982.
18. Kaufman L, Crooks L, Sheldon P, et al: The potential impact of nuclear magnetic resonance imaging on cardiovascular diagnosis. Circulation 67:251, 1983.
19. Mills CM, Brant-Zawadzki M, Crooks LE, et al: Nuclear magnetic resonance imaging: Effect of blood flow. AJNR (in press).
20. Crooks LE, Sheldon P, Kaufman L, et al: Quantification of obstructions in vessels by nuclear magnetic resonance (NMR). IEEE Trans Nucl Sci NS-29:1181, 1982.

8

Central Nervous System

MICHAEL BRANT-ZAWADZKI, M.D.
CATHERINE M. MILLS, M.D.
DAVID NORMAN, M.D.

General Considerations

As was the case with CT, the early clinical evaluation of NMR imaging has stressed the utility of this technique in the diagnosis of central nervous system (CNS) disorders.[1-8] Nuclear magnetic imaging has already proved itself comparable to CT in the depiction of brain anatomy. But this quality alone might not have fueled its further development. The fact that NMR obviates the use of ionizing radiation and harbors no known hazards for biologic tissue, and the fact that direct multiplanar cross-sectional images can be obtained with NMR without bone streak artifact are distinct advantages of this modality with respect to CT.

Yet even the above combination of features would have been unlikely to permit NMR to supplement or even complement CT's diagnostic repertoire were it not for the fact that NMR offers clearly superior sensitivity to soft-tissue contrast differentiation compared with CT. Although the final analysis of CT's tissue characterization capabilities must await full development of its dual-energy potential and other hardware refinements, it seems safe to say at the present time that the potential pre-eminence of NMR in the field of medical imaging, especially in the brain and spinal cord, rests on NMR's superior characterization of soft-tissue composition and its alteration by pathologic processes. It has already been established that NMR offers superior sensitivity in the detection of certain CNS disorders when compared with CT.[7,8] The ability to separate one pathologic entity from another requires not only optimal characterization of tissue constituents, but also a knowledge of anatomic and clinical aspects of disease processes.

With the advent of NMR and its broader perspective on CNS constituents and their physiologic and pathologic alteration, the radiologist is pressed to expand his or her basic knowledge of anatomy, its distortion by disease, and especially those fundamentals of pathophysiology first imaged with CT. The electron density and atomic number of tissue and their effects on x-ray attenuation are not relevant to the imaging technology of NMR. The radiologist must now learn how normal and abnormal processes affect the hydrogen density, T_1 and T_2 relaxation times, and blood flow in CNS compartments. Also, an understanding of how the variable NMR imaging parameters affect the appearance of tissues and the quantification of their signal intensity must be garnered.[6] The latter topic is covered in the first section of this volume. The subject of this discussion, namely, the illustration of the current clinical applicability of NMR in the CNS, might best be prefaced by a brief overview of the normal constituents of and pathology peculiar to the brain and spinal cord.

Normal Brain Constituents

Like most organs in the human body, the brain's major single component is water. However, the proportion of water in the two major types of brain

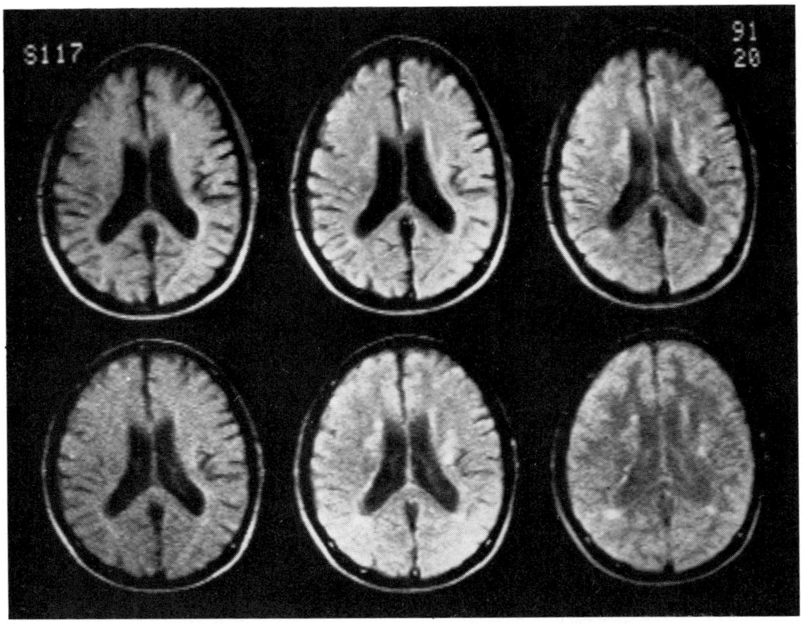

Figure 1. Variation in gray-white matter contrast with different spin echo parameters. Patient with multiple sclerosis. These axial sections at the level of the lateral ventricles were all obtained at the same time. The images on the top were all obtained with the first echo sampling of the signal intensity. The lower row images were the second echo sampling. The images on the left were obtained with a pulse interval of 0.5 sec, those in the middle with pulse interval of 1.0 sec, the right-sided images were obtained with a pulse interval of 1.5 sec. Note that on the short-interval images, the gray matter shows a lower intensity when compared with the white matter. However, the images obtained with the long pulse interval (1.5 sec) show a reversal of gray-white matter contrast, the low intensity of the white matter relative to the higher intensity gray matter being best appreciated on the second echo (lower right). Note that the multiple sclerosis plaques appear as high-intensity foci within the white matter. Demyelination renders the white matter composition more similar to that of the gray matter since the water content is elevated and the fat content is decreased following demyelination. Finally, note that on the lower right image the intensity of the cerebrospinal fluid (CSF) has risen relative to the surrounding brain. This is due to the long T_2 relaxation of CSF, that is, a long decay in signal intensity relative to the surrounding brain.

tissue, the gray and white matter, differs. By virtue of its greater content of fat, a substance that is hydrophobic, the white matter contains approximately 12% less water than the gray matter, whose water content approaches 83% of its net weight.

The greater water content of the gray matter prolongs that tissue's T_1 and T_2 relaxation times. Such prolongation tends to increase the relative intensity of the gray matter's NMR signal obtained with the spin echo technique using radio-frequency (RF) pulse intervals of 1 sec or more. This is especially true when sampling the second echo delay on such images (Fig. 1). Other variables may play a role in gray-white matter differentiation with NMR, such as the differential fat content, or possibly the greater blood volume of the gray matter, but the contribution of such factors is yet unexplored.

The greater fat content of white matter in the adult brain with respect to gray matter (approximately 16% as opposed to 6% wet weight) is due to the presence of myelin. Myelin makes up close to one-half the white matter's dry weight and is highly hydrophobic, having less than half of its own wet weight made up by water. Fat exhibits very short T_1 relaxation times compared with other biologic substances. The intensity of the NMR signal from tissue with short T_1 relaxation is generally high, on both spin echo and inversion recovery images. The straightforward

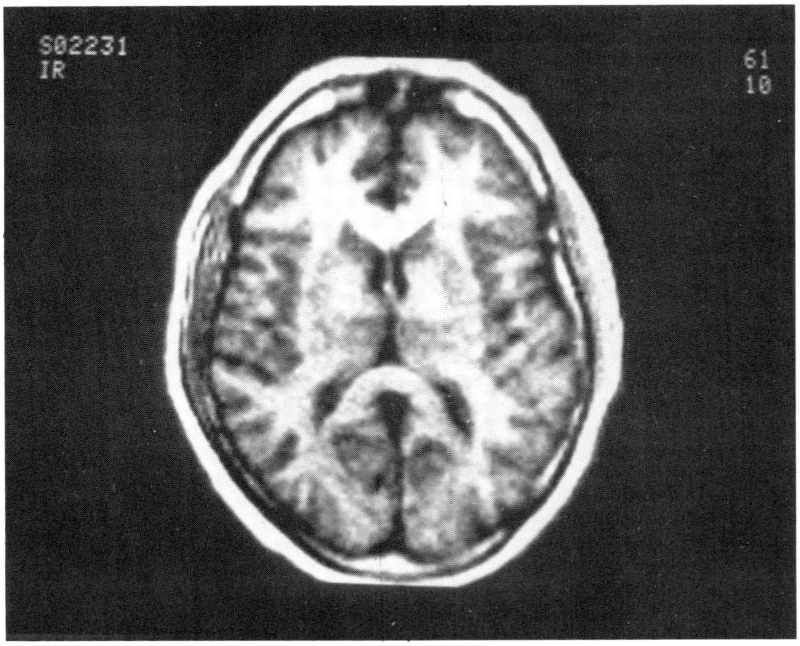

Figure 2. Normal volunteer, inversion recovery. Inversion recovery technique highlights tissue contrast differences based on T_1 relaxation. Note that the short T_1 relaxation of white matter produces very high signal intensity compared with the gray matter on this image. A similar high-intensity signal originates from the subcutaneous fat in the scalp.

explanation that the white matter will yield high intensity on NMR images obtained with those techniques that emphasize the shortened T_1 of fat in the signal (Fig. 2) oversimplifies the multivariate composition of that signal and neglects the influence of the relatively low water content.

Other chemical and physiologic variables of the CNS that enter the equation that yields NMR signal intensity include blood flow and hydrogen density. Since hydrogen is the focus of current NMR imaging, signal intensity differences between gray and white matter would be obviously affected were the density of hydrogen in the two tissues different. It is not different in the normal brain,[9] but the influence of pathology on hydrogen content has received little attention. Blood flow, on the other hand, is known to decrease from the gray to the white matter in the normal brain, a gradient that is accentuated with advancing age.[10] The influence of capillary flow differences on NMR signal intensity deserves intensive study in the near future, since a noninvasive method of accurate quantification of regional cerebral profusion would be a breakthrough in the diagnosis and management of cerebral ischemia. Nuclear magnetic resonance has already shown its ability to detect differences between rapid and slow flow in larger vessels, based on signal intensity (see below).[6,11,12]

The cerebrospinal fluid (CSF), being a relatively pure liquid, has much longer T_1 and T_2 relaxation times than brain tissue. It will show low intensity except with imaging techniques that prolong the interval between RF pulsation to the point that magnetization of the CSF's protons "catches up" with that of more solid tissue. Viewing the CSF on second echo images brings in more influence from its relatively prolonged T_2 relaxation properties; thus the intensity of CSF is greater on such images (Fig. 1).

As the foregoing discussion shows, the multiple physiologic and chemical factors that influence the intensity and decay characteristics of the NMR signal make the relationship between image appearance and tissue characteristics of the gray and white matter and CSF a complex one. Separation of the signal intensity into its various components and their quantification are the source of NMR's superior tissue characterization capability. Let us now briefly illustrate this capability in the context of some basic general pathologic alterations of the brain and spinal cord.

Common Pathologic Changes

Edema

A regional increase in water content of CNS tissue occurs with a wide variety of pathologic processes that affect the brain and spinal cord. The sensitivity of NMR in detecting CNS disease is therefore strongly related to its ability to detect edema. The compartmentalization, location, and progression of brain edema, along with the appearance and characterization of any associated lesion, should help the specificity of its differential diagnosis with NMR. The compartment in which the excess water is found is related to its cause and is used for its classification. Three types of brain edema are generally recognized: cytotoxic edema, vasogenic edema, and interstitial edema.[13]

Cytotoxic edema is that found in the intracellular space and is caused by abnormal cellular metabolism, such as can be induced with hypoglycemia or anoxia. In such cases, the production of high-energy phosphates such as ATP ceases, and cellular membrane homeostasis suffers. Potassium leaks out of the cell, while sodium and water enter from the extracellular space. This in turn induces an osmotic egress of water from the capillary bed to the extracellular space. The amount of water accumulation and size of the resulting mass effect depend on the number of cells affected. Cerebral infarction is a classic cause of cytotoxic edema. It usually affects the gray matter in the distribution of a major vessel. Prolongation of T_1 as a result of the edema induces a relative loss of intensity on inversion recovery images. The lengthening of T_2 relaxation caused by the gray matter edema is best seen with the spin echo technique and lends it even greater intensity than normal gray matter on the second echo images, since by the first echo the discrepancy in T_2 relaxation between the normal and abnormal tissue is not always apparent (Fig. 3).

Vasogenic edema is the second type of edema classically described, and occurs after breakdown of the blood-brain barrier (BBB). This barrier includes the capillary endothelial cells and their transport mechanism, the tight junction between these cells, and the foot processes of the nearby glial cells on the parenchymal side of the vessel wall. Tumors characteristically produce vasogenic edema, since neoplastic vessels do not possess the BBB. As the number of such vessels increases, or when normal cerebral vessels become compromised by disease processes, normal hydrostatic pressure drives water through the "leaky" capillary wall and into the extracellular space. This form of edema spreads by bulk flow, generally along white matter tracts, driven by the force of the hydrostatic pressure. Again, the prolongation of T_2 is seen as high intensity on the second echo sampling with the spin echo technique (Fig. 4). Such edema in the white matter has been found to cause demyelination as well, and the loss of fat with its short T_1 relaxation may contribute to the observed prolongation of T_1 relaxation times within edematous white matter.

Demyelinating processes such as multiple sclerosis produce the same effect, but presumably it is the initial loss of fat and its hydrophobic characteristics that then allow an accumulation of water at the site of the lesion in such instances (Fig. 5). Of course, this focal increase of water at the site of a plaque generally does not spread by bulk flow. The differential impact of increase in water and loss of fat on the NMR signal intensity of edematous white matter will be the subject of analysis in the near future, since studies aimed at staging the early and late lesions in multiple sclerosis have begun.

Finally, the third form of edema is the so-called interstitial edema associated with hydrocephalus. Because the brain is devoid of lymphatics, normal extracellular fluid accumulation is balanced by both the osmotic-hydrostatic relationship at the venous side of the capillary bed and, deep in the brain, by flow into the ventricle. When the ventricular system becomes obstructed, the extracellular fluid in the periventricular space accumulates. Indeed, with sufficient intraventricular pressure, the ependymal lining of the ventricles may tear, allowing CSF to escape into the parenchyma.[14] The same T_1 and T_2 effects are seen as described for vasogenic edema affecting the white matter. The prolongation of T_1 relaxation is best seen on images emphasizing T_1 differences (Fig. 6A). The images obtained with short RF pulse repetition intervals (shorter than the T_1 relaxation time) will emphasize low intensity in the demyelinated white matter adjacent to the ventricles. The longer repetition rate permits the lengthier T_1 of demyelinated white matter to "catch up" with that of the normal white matter, thus losing the ability to differentiate the two on the first echo. As with other forms of edema, periventricular edema of hydrocephalus is best appreciated on images obtained with the second echo sampling of the signal generated with long pulse intervals (Fig. 6B). This is because the long T_2 relaxation of edema is emphasized and high intensity of the edematous region results. This high intensity is easy to separate from the low intensity

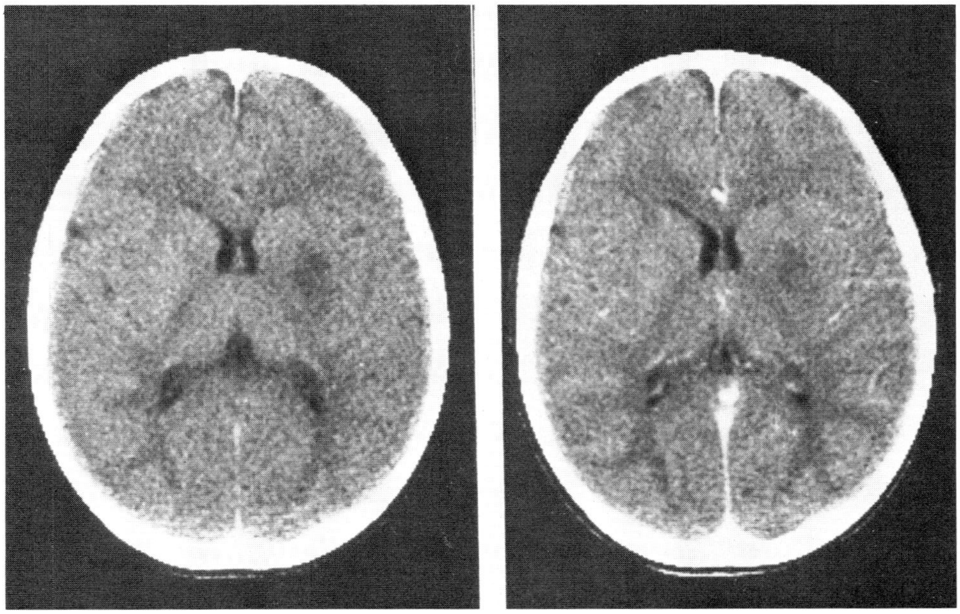

A.

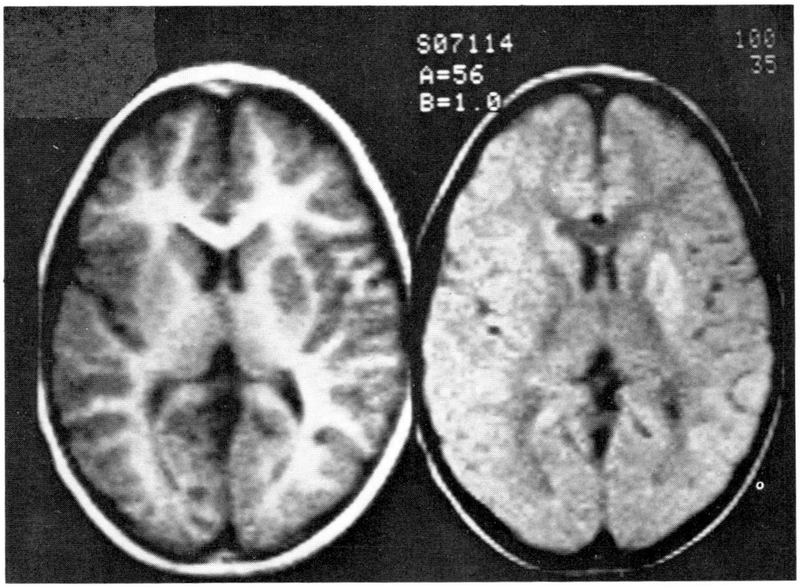

B.

Figure 3. Acute cerebral infarction.

A. Computed tomographic scans before (left) and after (right) administration of intravenous contrast show a focal area of low attenuation in the left globus pallidus.

B. Nuclear magnetic resonance sections obtained with the inversion recovery (left) and spin echo (right) technique at the level of the infarct. The inversion recovery technique, which highlights contrast differences based on T_1 relaxation, shows a low-intensity focus in the infarct, indicating a long T_1. The spin echo technique, with the second echo sampling, highlights T_2 relaxation differences. The infarct is of high intensity as a result of prolongation of T_2 relaxation (slower signal decay) when compared with the surrounding brain.

(Reproduced, with permission, from Brant-Zawadzki M, Davis PL, Crooks LE, et al: NMR demonstration of cerebral abnormalities: Comparison with CT. AJNR 4:117, 1983. © 1983, American Roentgen Ray Society.)

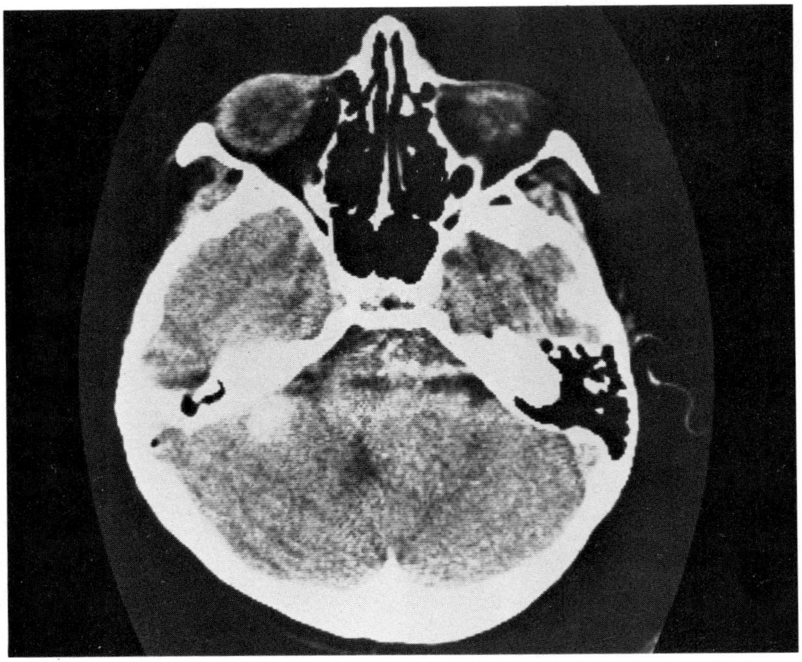

A.

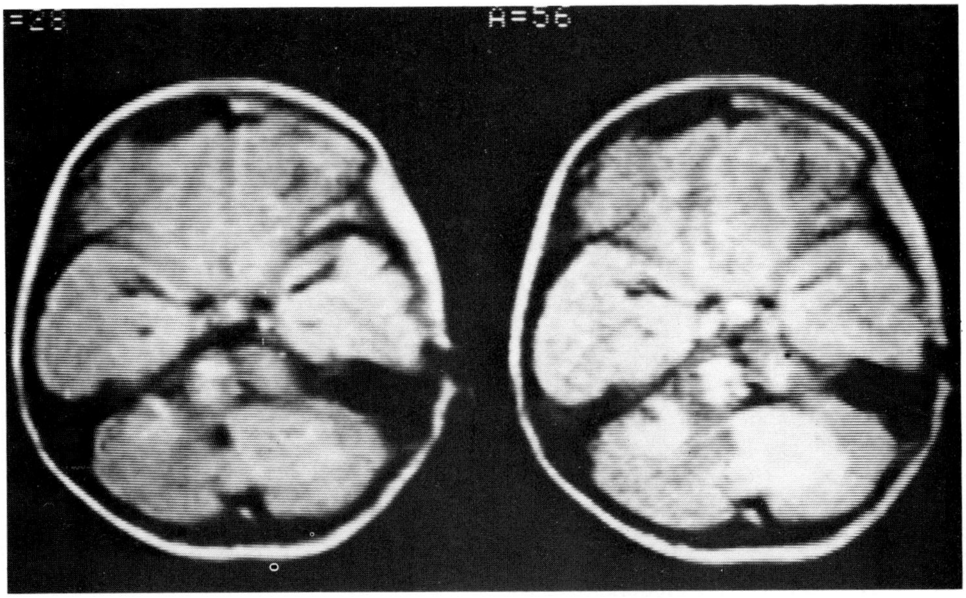

B.

Figure 4. Brain edema secondary to tumor. Patient with a brainstem glioma.

A. Axial CT section at the level of the fourth ventricle shows midline shift of the fourth ventricle from left to right, with abnormality in the brainstem obscured by bone streak artifact.

B. Spin echo NMR sections obtained at the level of the fourth ventricle. The image obtained with sampling of the first echo shows midline shift of the fourth ventricle to the right. Slight elevation of intensity is noted within the brainstem at the site of the tumor. The image on the right is the same section sampled at 56 msec (second echo) and highlights the edema in the left cerebellar hemisphere.

(Reproduced, with permission, from Brant-Zawadzki M, Davis PL, Crooks LE, et al: NMR demonstration of cerebral abnormalities: Comparison with CT. AJNR 4:117, 1983. © 1983, American Roentgen Ray Society.)

of the ventricular CSF space. With techniques that emphasize T_1 characteristics, the edema and CSF both have low intensity, and the abnormality of white matter adjacent to the ventricles may easily be misinterpreted as an artifact of partial volume (Fig. 6).

Nuclear magnetic resonance is extremely sensitive in detecting CNS edema of any sort. This is a major virtue in view of the great spectrum of pathologic entities that can produce such edema. In our experience, edematous CNS is best seen on the second echo images obtained with repetition intervals longer than the T_1 relaxation of the abnormal tissue.

Hemorrhage

Another nonspecific tissue change in the CNS is bleeding. Cerebrovascular accidents including infarct, hypertensive bleed, aneurysmal rupture, venous thrombosis, trauma, and even tumor and infection can result in bleeding. The markedly shortened T_1 of hemorrhage, which lends high intensity to both spin echo and inversion recovery images, makes it easily recognized. The shortened T_1 separates the

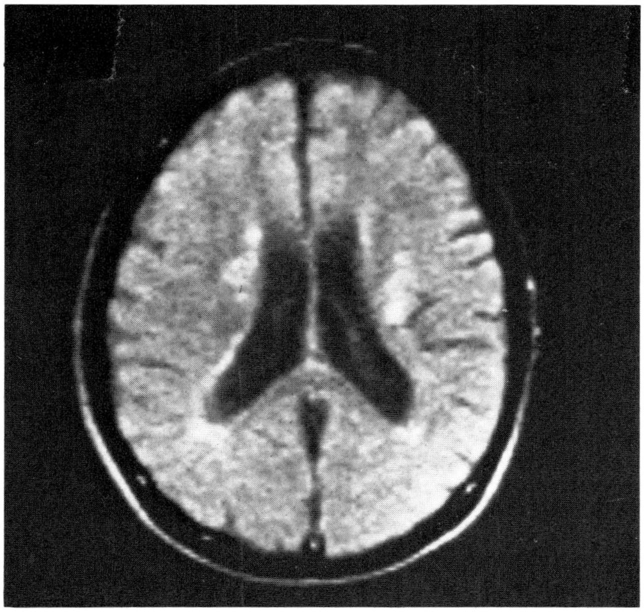

Figure 5. Multiple sclerosis. Typical focal high-intensity plaques are seen around the lateral ventricles in this patient with documented multiple sclerosis (see Fig. 1).

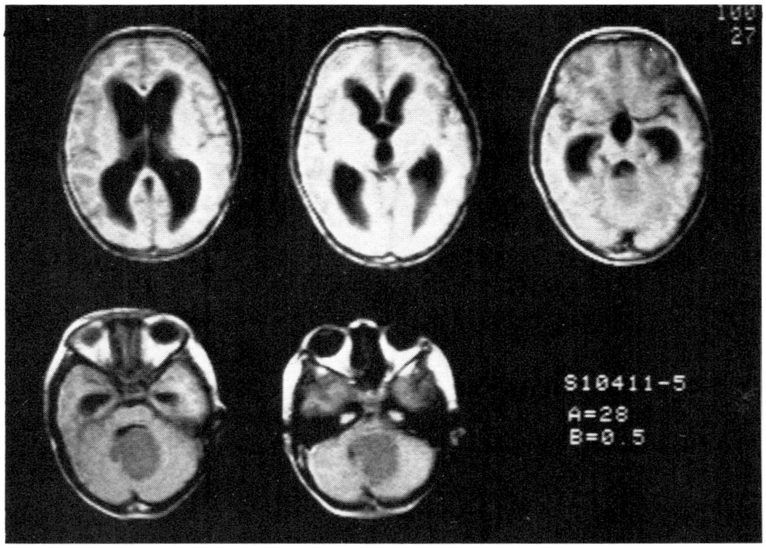

A.

Figure 6.

A. Interstitial edema due to hydrocephalus. Sections obtained with spin echo technique, short pulse interval (0.5 sec), and first echo sampling (28 msec). This technique highlights differences in contrast based on T_1 relaxation. Note that the upper sections, which show ventricular dilatation, reveal a gray "cap" around the ventricles. The normal white matter is of high intensity. Thus, the white matter immediately adjacent to the ventricles is demyelinated, raising its water content, prolonging its T_1 relaxation. Note that the tumor in the posterior fossa on the lower images also shows low intensity and, indeed, had very prolonged T_1 relaxation properties.

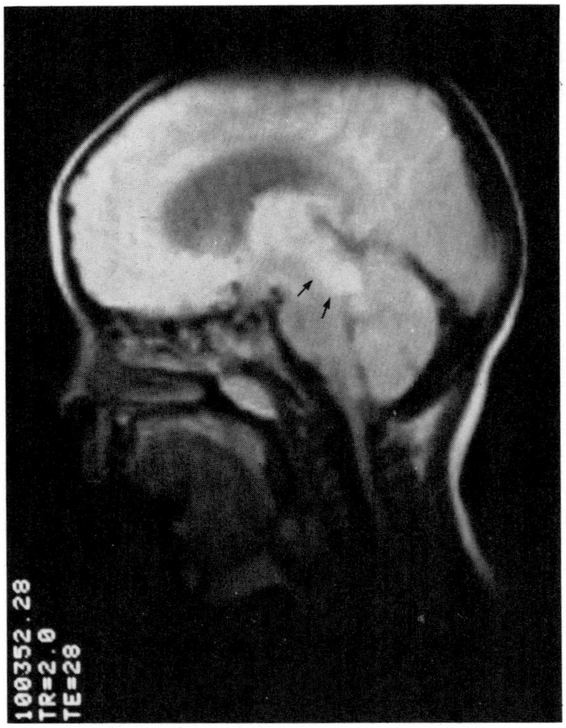

B.

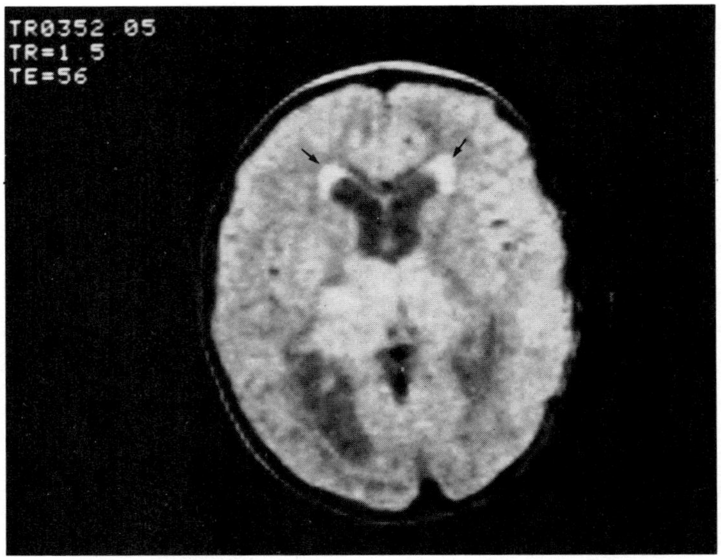

C.

B and C. Brain stem glioma producing hydrocephalus and interstitial edema. Sagittal (B) NMR image depicts a focal tumor (arrows) in the pons and midbrain with compression of the cerebral aqueduct. The axial section (C) shows tumor spreading into both thalami. Note the "caps" of interstitial edema (arrows) atop the dilated frontal horns. Note the easier detection of this edema with the longer TR and TE (1.5 sec, 56 msec) of this study as compared with Figure 6A, where the first echo sampling of the signal obtained with short pulse intervals was used.

(Reproduced, with permission, from Crooks LE, Mills CM, Davis PL, et al: Visualization of cerebral and vascular abnormalities by NMR imaging: The effects of imaging parameters on contrast. Radiology 144:843, 1982.)

high intensity of hemorrhage from that of edema, in which the T_1 is relatively long.

The explanation of the shortened T_1 within areas of CNS hemorrhage is still speculative, but may have to do with the known magnetic properties of hemoglobin or other proteins in clot.[15] Moderate prolongation of T_2 relaxation has also been noted within hemorrhage in our case material (Fig. 7). Computed tomography exhibits hemorrhage in the brain as a focus of high density, since the proteinaceous component of clot introduces a relatively high attenuation of the x-ray beam passing through the involved region. With age, the protein denatures, and the attenuation properties of the clot resemble those of normal brain, or even decrease to the same point that edematous brain exhibits. For this reason, CT may miss the presence of an older hemorrhage, or even misinterpret it as a focus of brain edema (Fig. 7C). However, NMR detects the presence of old clot much more sensitively, presumably because the altered magnetic environment that it produces shortens the T_1 relaxation values of the involved region (Fig. 7D), maintaining the high intensity of the abnormal brain.

Accumulation of Fat

Abnormal accumulations of lipid are rare in the CNS, and generally are seen within teratomatous growths. The characteristically shortened T_1 of lipomatous tissue helps its differentiation from edema, while its T_2 relaxation time is similar to that of normal brain (Fig. 8), unlike the prolongation of T_2 associated with bleeding. Because of the shortened T_1 relaxation of fat, its high intensity diminishes on images obtained with progressively longer RF pulse intervals. If one is interested in visualization of the orbital contents,

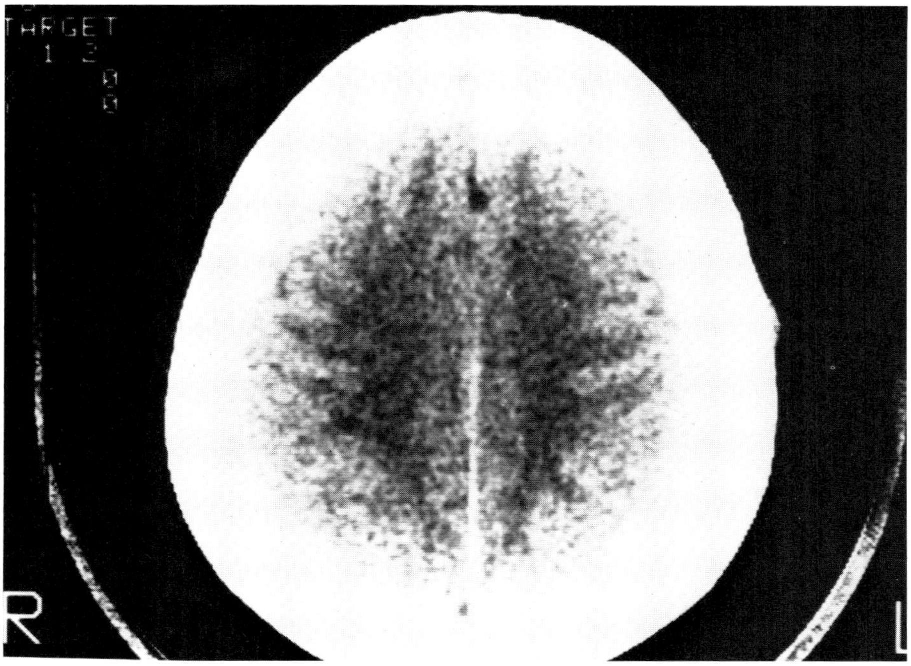

A.

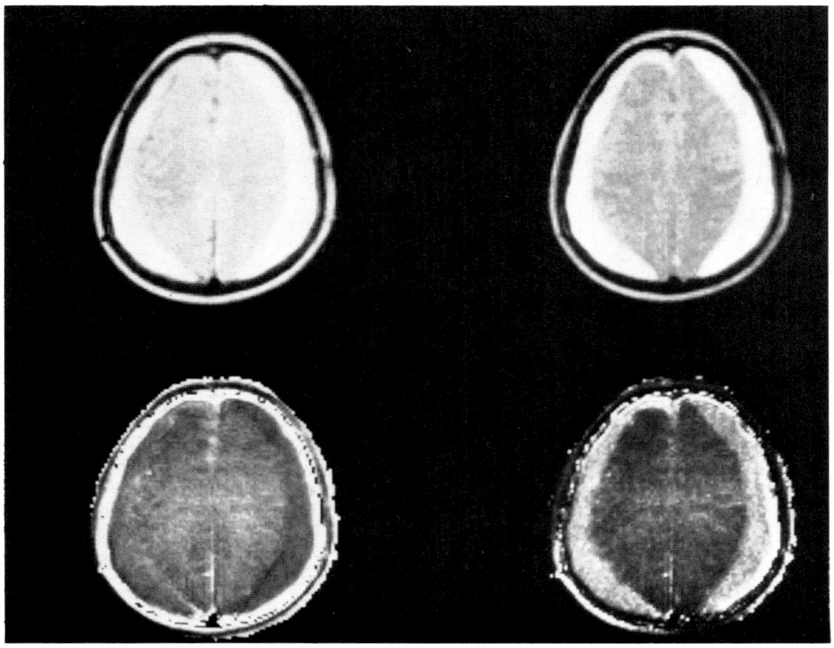

B.

Figure 7.

A. Isodense subdural hematoma. This elderly woman with right hemiparesis showed no significant abnormality on her CT study other than effacement of sulci on lower sections.

B. Spin echo NMR images revealed bilateral subdural hematomas (top row). The calculated images (bottom row) showed a shortened T_1 (left, lower row) and a lengthened T_2 (right, lower row) relaxation time relative to the brain. Such alteration of relaxation values is indicative of hemorrhage.

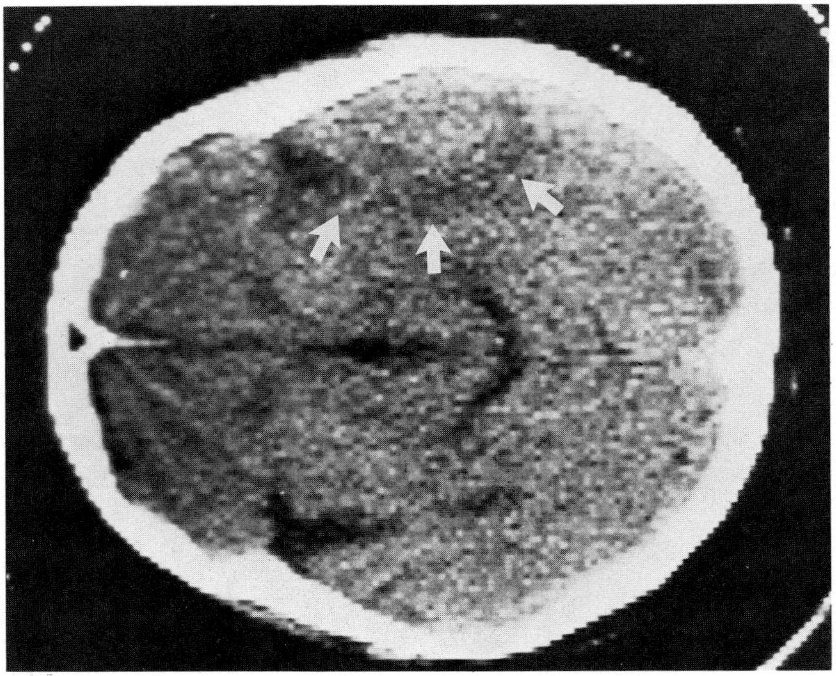

C.

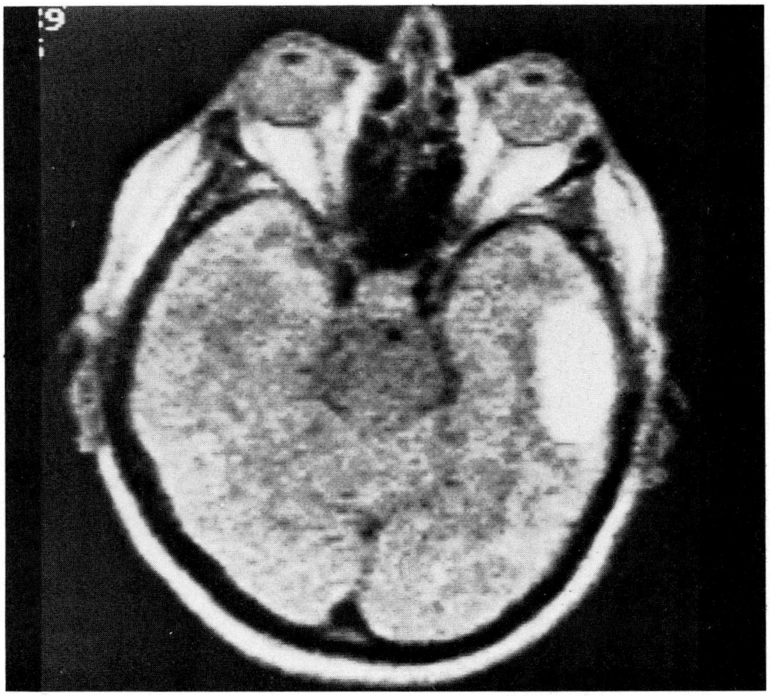

D.

C. A CT scan obtained 8 weeks following a temporal lobe hemorrhage in a middle-aged man shows a vague area of low attenuation (arrows). The initial CT study depicted obvious blood.

D. An NMR image 1 week later shows an oval lesion with intensity characteristic of blood.

one must compensate for orbital fat's high intensity, which tends to obscure the optic nerve and muscles embedded within. Manipulation of imaging technique, or "windowing," can accomplish this.

Pathologic Calcification

Since calcium has little hydrogen, its NMR signal is low, and punctate calcification can be missed. However, large areas of calcium are seen as "holes" of low intensity on the images. Bone also shows no signal, but normal bone marrow gives a strong NMR signal and can provide evidence of bony erosion or infiltration by tumor (Fig. 8).

Calcification in the brain can occur in a variety of disease processes, including infection, tumor (Fig. 9), vascular abnormalities (such as arteriovenous malformations and aneurysms), and other pathologic entities that damage tissue and predispose it to dystrophic calcium deposition. Metastatic calcification in the brain (due to abnormally elevated serum calcium levels) occurs less often than in other organs. Normally calcified structures such as the pineal gland and choroid plexi also must be considered.

Nuclear magnetic resonance is less sensitive to punctate calcific foci in the brain than CT is, and thus some specificity in diagnosis may be lost. However, calcification can be seen in a variety of disorders, as mentioned, and even CT may not offer a pathognomonic appearance.

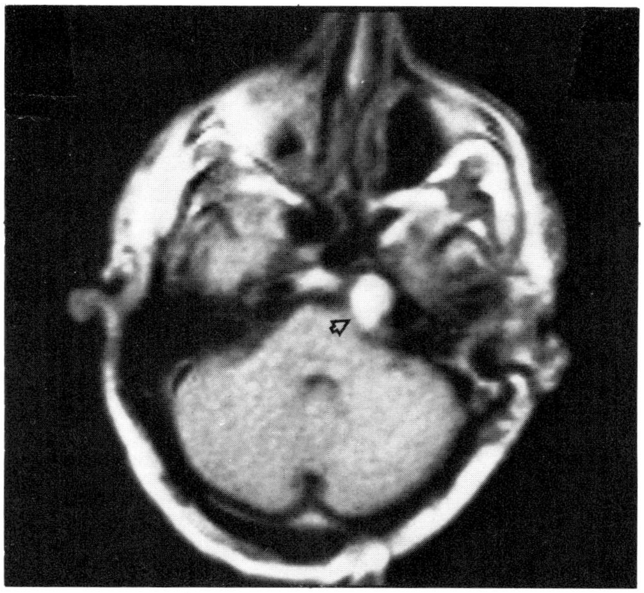

Figure 8. Left petrous teratoma, lipomatous component. A high intensity focus is seen in the left petrous tip (arrowhead), where biopsy proved the existence of a teratoma with a lipomatous component. Note that the high intensity of the tumor matches that of the subcutaneous fat and exceeds that of the high intensity signal from the bone marrow in the adjacent clivus.
(Reproduced, with permission, from Brant-Zawadzki M, Davis PL, Crooks LE, et al: NMR demonstration of cerebral abnormalities: Comparison with CT. AJNR 4:117, 1983. © 1983, American Roentgen Ray Society.)

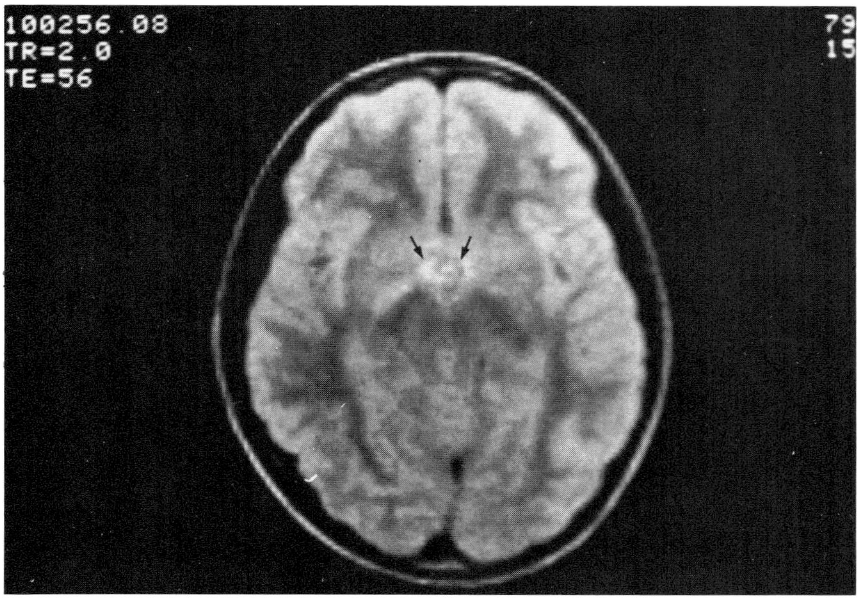

Figure 9. Suprasellar teratoma with calcification. Computed tomography identified a small tumor with calcification in the suprasellar cistern of this patient. A diagnosis of craniopharyngioma was rendered. Nuclear magnetic resonance shows the lesion as a high-intensity focus with internal signal drop-out (arrows), the latter consistent with foci of calcium. At surgery, a benign teratoma with calcium was identified.

Specific Disease Processes

Degenerative Disease

Our initial experience with NMR included a disproportionate number of patients with chronic CNS processes, since such patients were not acutely ill and could be studied in an outpatient setting at the imaging laboratory. Patients with multiple sclerosis figured prominently both in our initial study population and in that of other centers involved in early NMR development.[7] Computed tomography has been suboptimal in depicting the pathologic process of this demyelinative disorder, in that often flagrant clinical symptomatology would be evidenced by relatively subtle abnormalities on the CT study. Another source of frustration in this patient population is the difficulty encountered by the neurologist in establishing the diagnosis in early cases. The initial symptoms of multiple sclerosis may be nonspecific and often subtle; the diagnosis may not be definitely established for some time, adding to the anxiety of both the patient and the clinician. The need for early definitive diagnosis has not been met by CT. This technique's relative insensitivity even to later stages of disease makes monitoring of therapy on the basis of imaging difficult. Such an ability would be of potentially significant importance, since new approaches to therapy of this frustrating disease entity are being developed.

Because of the difficulty in establishing the diagnosis in early cases, the capabilities of NMR in the diagnosis of this disease process had to be assessed in a selected group of patients. Only those with a well-established diagnosis were evaluated, and their NMR and CT studies were compared. The lesions were classically distributed in the periventricular white matter on NMR images. Their focal high intensity was easily seen with spin echo technique.

Again, it was the longer pulse intervals that best depicted the lesions, with the second echo sampling being the most useful. It is likely that this set of parameters proved most sensitive because of the increased water content in the foci of demyelination. The contribution to signal intensity alteration produced by the loss of fat is unknown at present. Computed tomography was not nearly as helpful in these patients, showing no focal lesions in six patients and fewer lesions in another six, whereas NMR depicted multiple

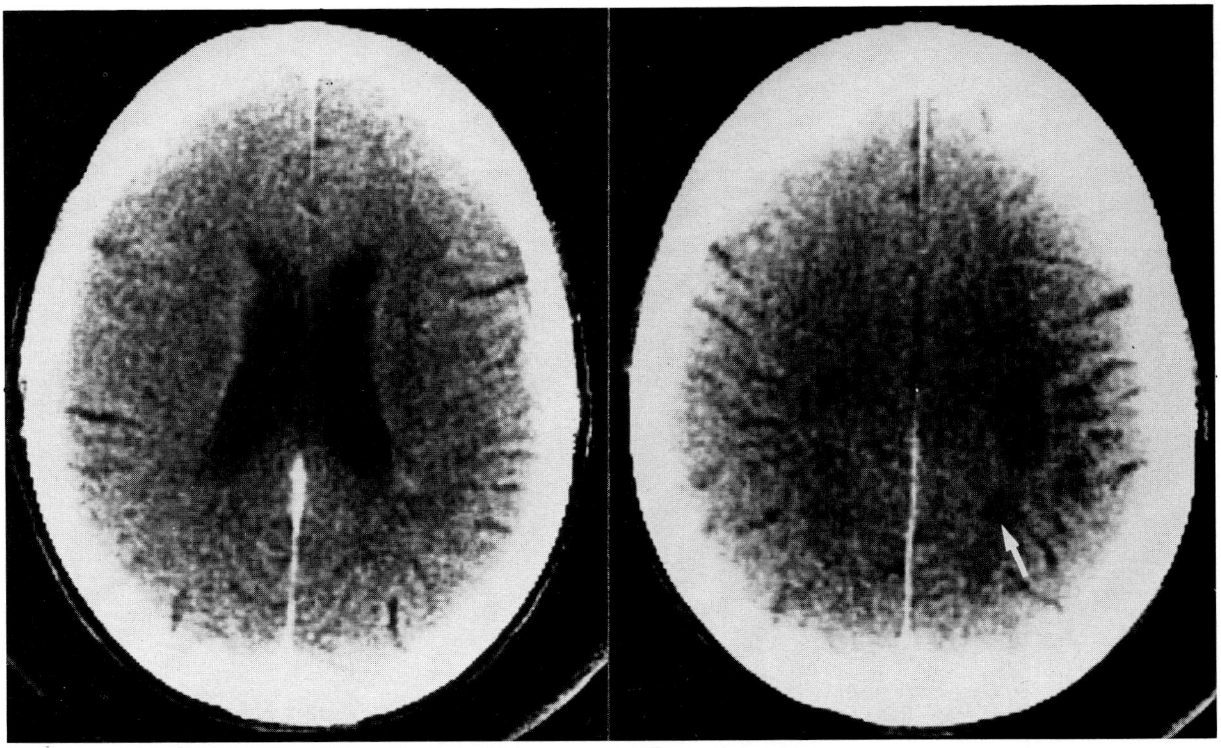

A. B.

Figure 10. Multiple sclerosis.
A. A CT study at the level of the lateral ventricles showed no lesions in this patient with multiple sclerosis.
B. Section at the level of the centrum semiovale showed a faint focal lucency in the white matter (arrow).

C.

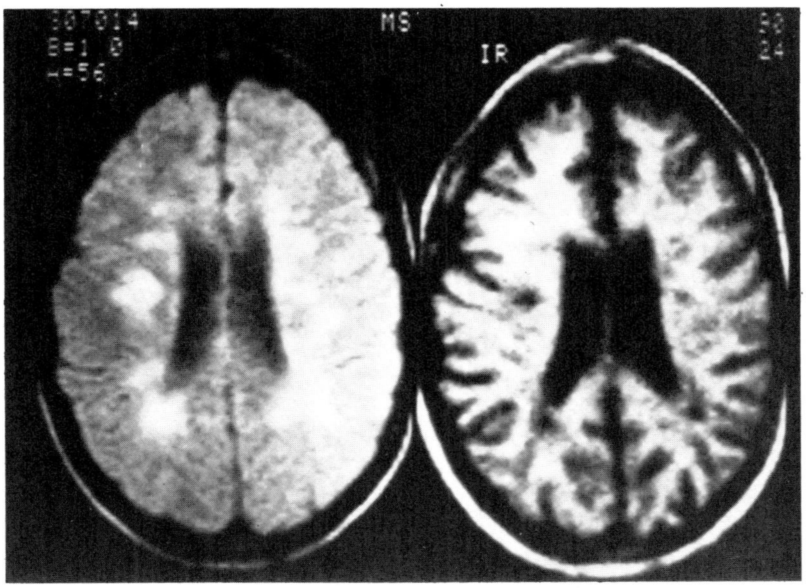

C. Spin echo (left) and inversion recovery (right) NMR sections in this patient show multiple plaques in the periventricular white matter. Note the greater sensitivity to disease of the spin echo technique.

(Reproduced, with permission, from Brant-Zawadzki M, Davis PL, Crooks LE, et al: NMR demonstration of cerebral abnormalities: Comparison with CT. AJNR 4:117, 1983. © 1983, American Roentgen Ray Society.)

lesions in all 12 patients involved in our early comparative study (Fig. 10). It should be stated that the CT studies were not all done with double doses of contrast and delayed scanning, a technique that provides greatest sensitivity with CT.[16] Nevertheless, CT also had problems with beam hardening artifact in the posterior fossa, where NMR showed several lesions.

Degenerative diseases of the gray matter were also studied, but were manifested predominantly by regional or global atrophy (Fig. 11). The question of whether T_1 or T_2 changes occur in the early phases of these diseases is a subject for further research.

Tumors

The appearance of brain neoplasms in NMR studies in our patients was influenced by their size, location, production of surrounding edema, and tissue composition. A germinoma as small as 5 mm was depicted as a result of its difference in intensity from adjacent structures (Fig. 12). A small metastatic focus was seen by virtue of high intensity, which presumably resulted from edema (Fig. 13). Larger intra-axial tumors could be visualized because of intensity differences or, when these were absent, distortion of surrounding anatomy and production of edema (Figs.

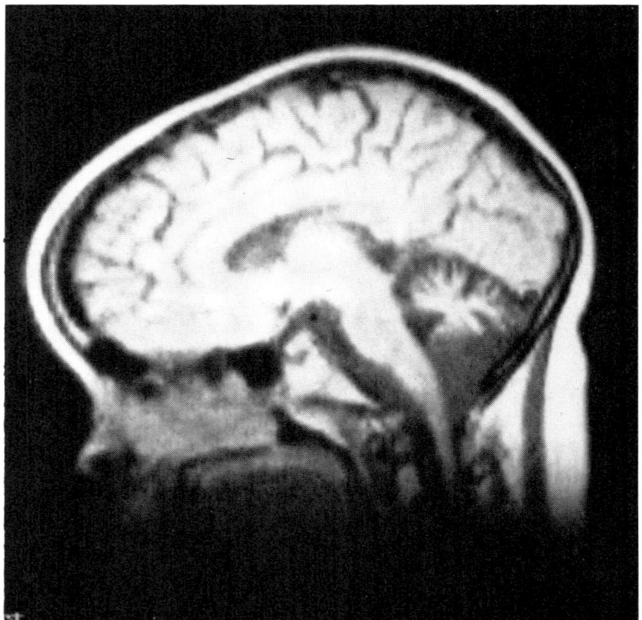

Figure 11. Posterior fossa atrophy (spinocerebellar degeneration). Midsagittal section in this child with spasticity shows a small pons, with enlargement of the CSF spaces including that of the fourth ventricle. Note the small size of the cervicomedullary junction at the level of the foramen magnum.

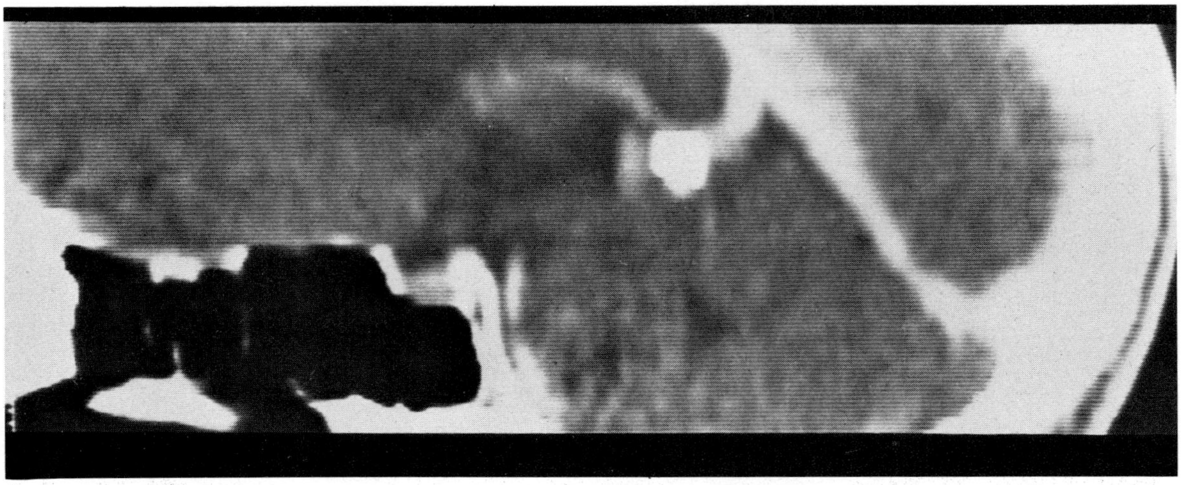

A.

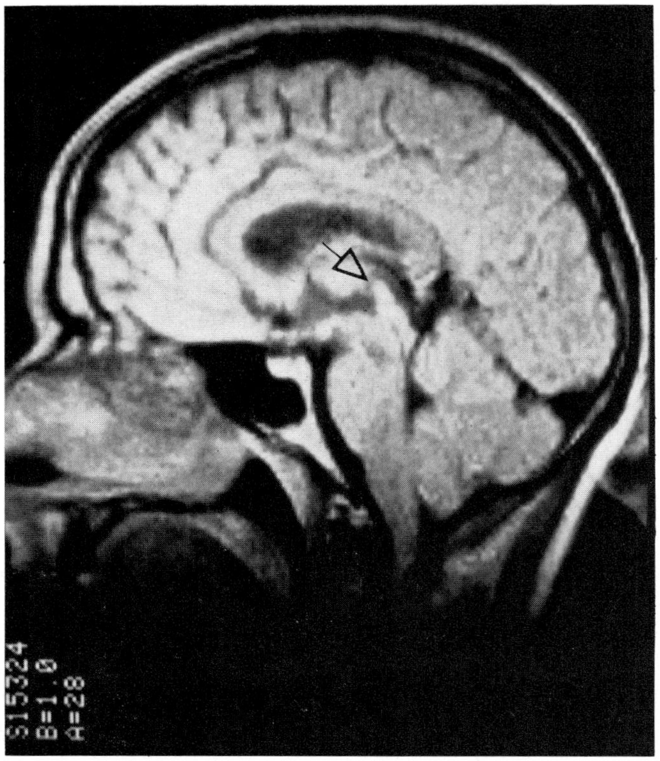

B.

Figure 12. Prepineal germinoma.

A. Reformation of 1.5-mm axial CT images shows an abnormal focus of enhancement immediately anterior to the calcified pineal gland.

B. Direct sagittal NMR image shows a high-intensity focus on top of the superior colliculus (arrowhead). Note the compression of the underlying cerebral aqueduct. Incidentally, note that the calcified pineal gland gives no signal on the sagittal NMR image; thus the small germinoma is immediately anterior to an area of very low signal.

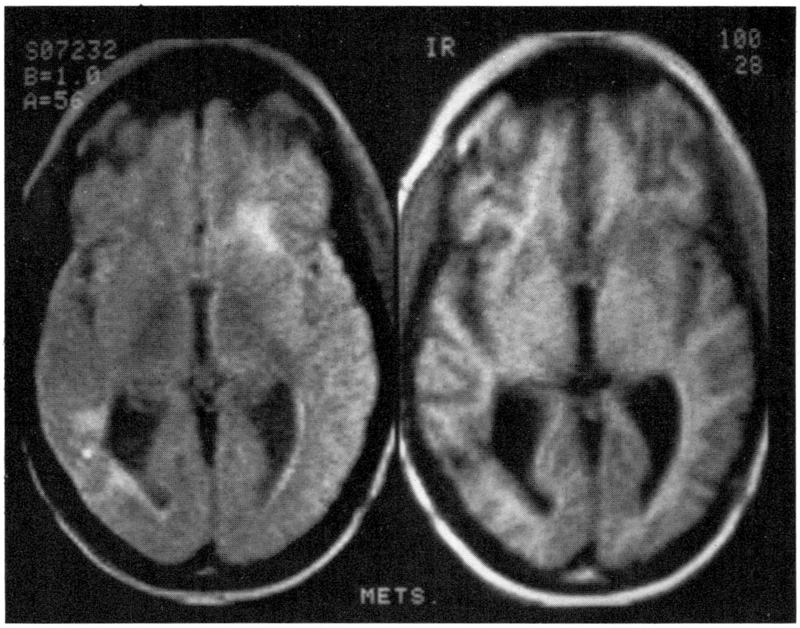

Figure 13. Metastatic disease. Two foci of metastatic disease are seen on the spin echo NMR image (left). The high-intensity foci are noted in the low left frontal lobe and in the posterior right temporal lobe. Note that the inversion recovery image (right) shows the lesions as low-intensity foci, which are difficult to distinguish from the adjacent gray matter and CSF spaces.

4, 6). Certain characteristic components, such as fat, made the diagnostic specificity of NMR increase (Fig. 8).

Tumors in the spinal cord were well seen by virtue of both intensity changes and enlargement of cord size (Fig. 14). Syrinx could be detected in some cases of nonspecific cord enlargement as the responsible cause (Fig. 15).

A major issue in the evaluation of tumors with NMR is the modality's potential for sufficiently specific characterization of histology to permit differentiation of malignant from benign neoplasia, and tumor from the surrounding edema, both of which may show similar intensity alterations of the NMR image (Fig. 16). Initial enthusiasm helped fuel the proposition that the T_1 and T_2 relaxation values in the affected region may permit specific identifications of malignancy.

It has been stated that T_1 prolongation occurs with all malignancies, but several exceptions to this "rule" have already been seen in our relatively early experience and in that from other institutions.[7] Although short T_1 relaxation can be seen in benign lesions (Figs. 8, 17) and long T_1 relaxation can be found in malignant ones (Figs. 6A, 18) when compared with normal brain parenchyma, accumulating evidence indicates that abnormalities in relaxation values provide a great deal of sensitivity but are of limited use in differentiating one lesion from another. In our experience, neoplasia is more likely associated with T_2 prolongation (Figs. 19, 20) since T_1 prolongation occurs any time abnormal accumulation of relatively pure fluid is seen (as with benign cysts) or unusual enlargement of a CSF space develops (Fig. 20). It must be remembered that T_1 and T_2 relaxation values are relative values, and as such must be related to the particular subject's normal tissues. Comparing the values from one subject's tissue to another's is not always reliable. As experience accumulates, variations in the typical findings of relaxation time alterations will be found (Fig. 21).

Further experience is needed to determine the potential of NMR in characterization of tumor tissue, its blood flow characteristics, and possibly response to various chemotherapeutic agents. It is hoped that such experience will improve the specificity of NMR's evaluation of brain and spinal cord neoplasms.

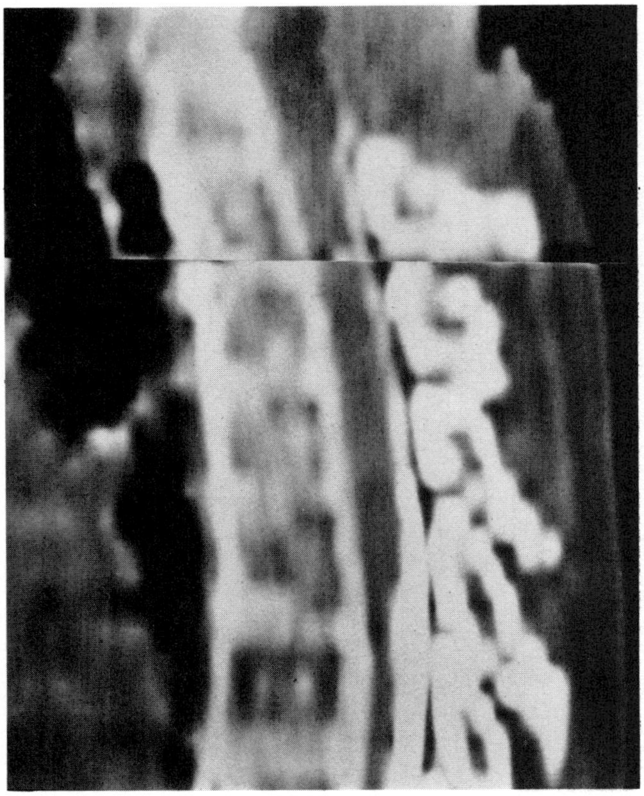

A.

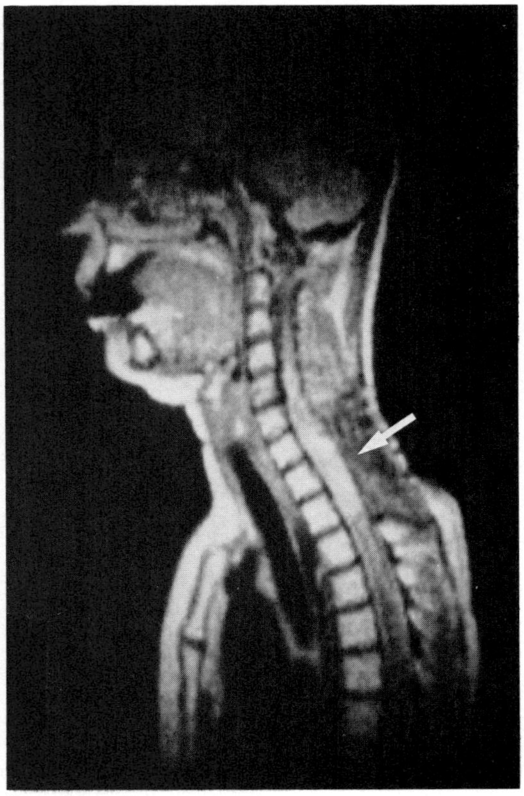

B.

Figure 14. Spinal ependymoma.

A. Midsagittal reformation of 5-mm axial CT images shows an enlarged cord at the cervicothoracic junction surrounded by metrizamide. Note the missing spinous processes following previous laminectomy. The horizontal line represents manual reregistration of the reformation necessitated by patient motion.

B. Spin echo NMR image. Direct sagittal section shows the tumor at the cervicothoracic junction (arrow).

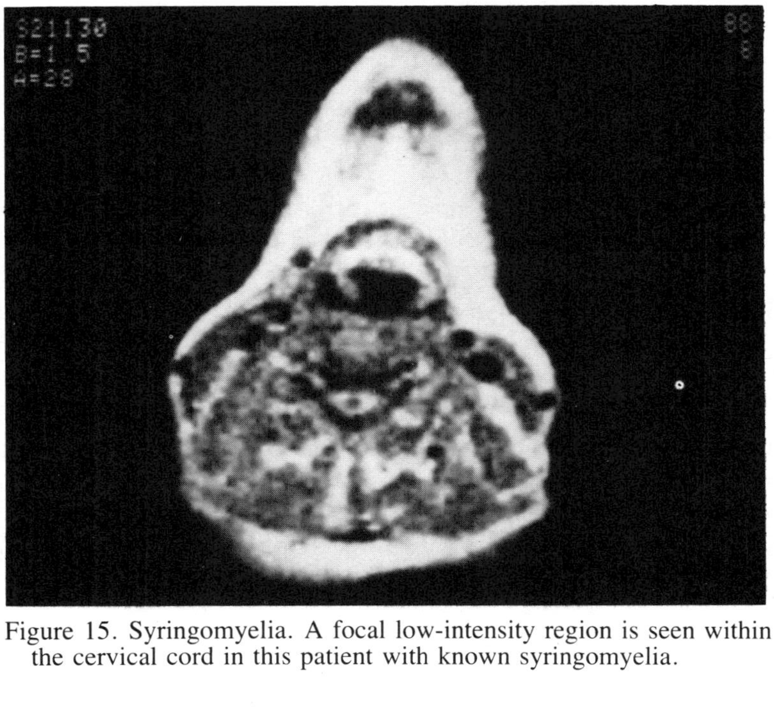

Figure 15. Syringomyelia. A focal low-intensity region is seen within the cervical cord in this patient with known syringomyelia.

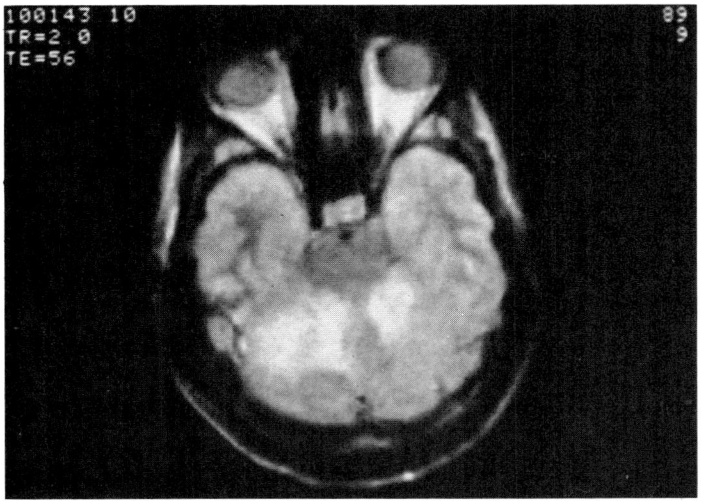

Figure 16. Metastatic testicular carcinoma. Young man with posterior fossa symptoms shows bilateral cerebellar lesions of high intensity compressing the perimesencephalic cisterns. The presence of abnormality is definitely established, but separation of the tumor nidus itself from surrounding edema was impossible.
(Reproduced, with permission, from Brant-Zawadzki M, Davis PL, Crooks LE, et al: NMR demonstration of cerebral abnormalities: Comparison with CT. AJNR 4:117, 1983.)

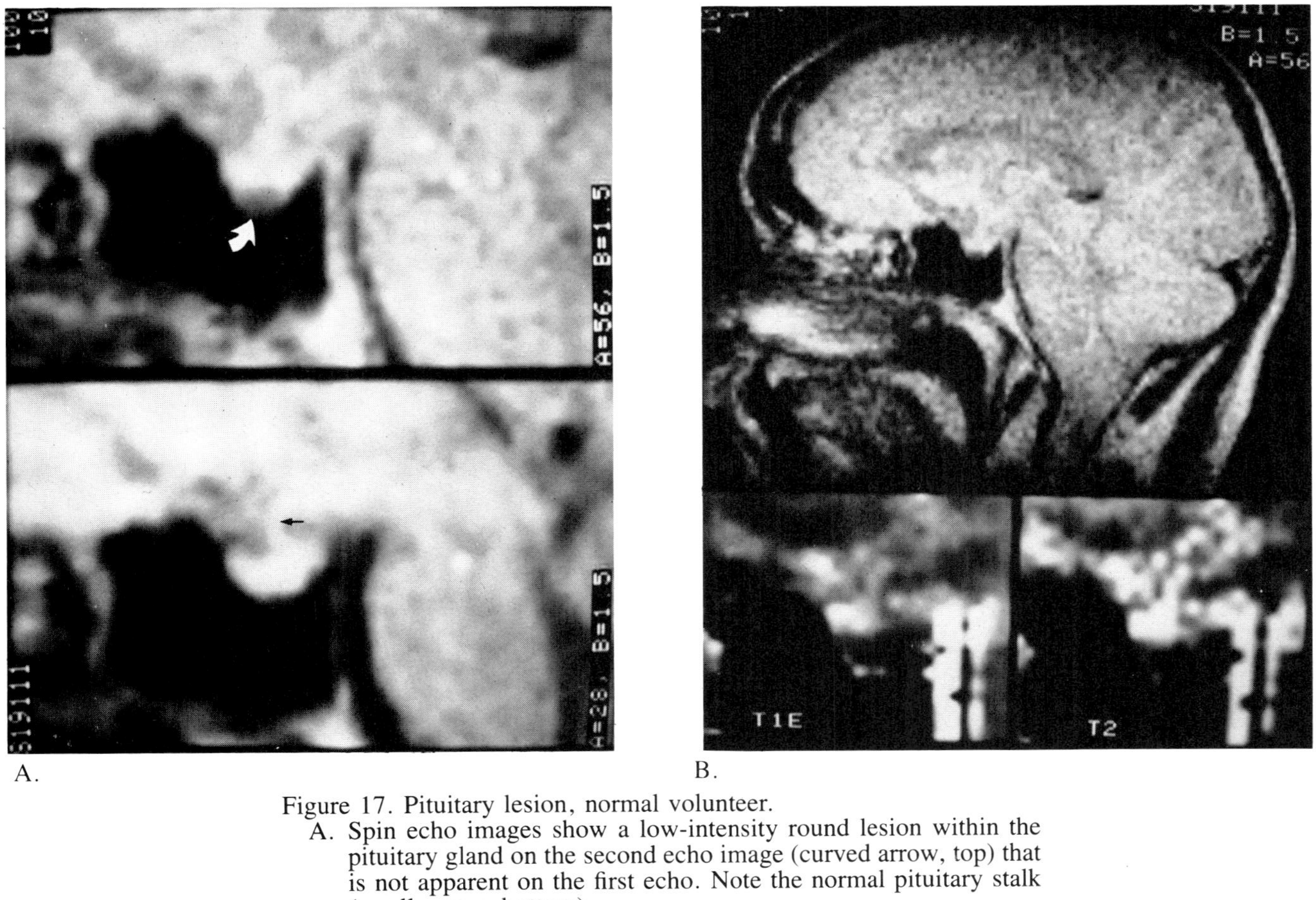

A. B.

Figure 17. Pituitary lesion, normal volunteer.
A. Spin echo images show a low-intensity round lesion within the pituitary gland on the second echo image (curved arrow, top) that is not apparent on the first echo. Note the normal pituitary stalk (small arrow, bottom).
B. The T_1 and T_2 calculated images (lower insets) indicate short relaxation times (dark pixels) in this presumably benign lesion (the volunteer had no known hormonal or other abnormalities). Full brain image with TR 1.5 sec shown on top.

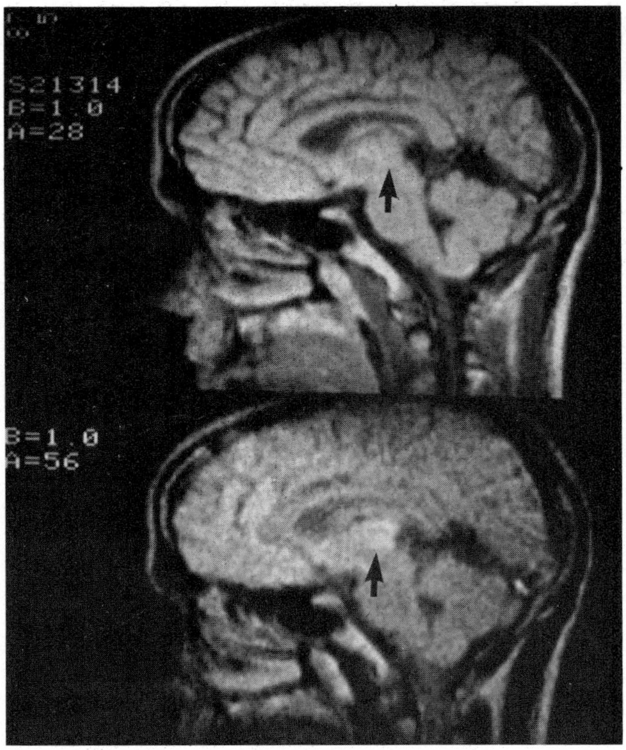

A.

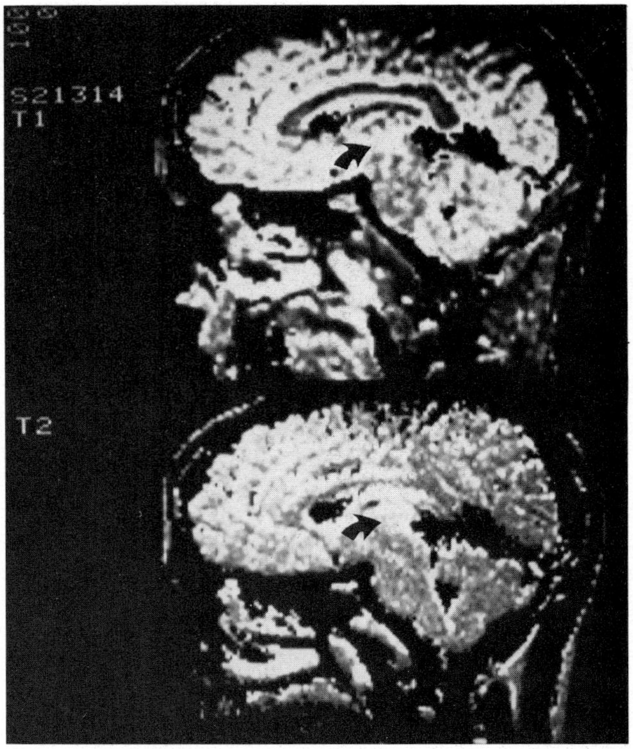

B.

Figure 18. Germinoma.

A. Young male with abnormal mass in the region of the posterior third ventricle (arrows), seen to better advantage on the second echo (bottom) than the first echo image (top).

B. Calculated T_1 (top) and T_2 images help verify the slight prolongation of both relaxation times by showing bright pixels in the tumor focus (curved arrows).

(Reproduced, with permission, from Brant-Zawadzki M, Davis PL, Crooks LE, et al: NMR demonstration of cerebral abnormalities: Comparison with CT. AJNR 4:117, 1983. © 1983, American Roentgen Ray Society.)

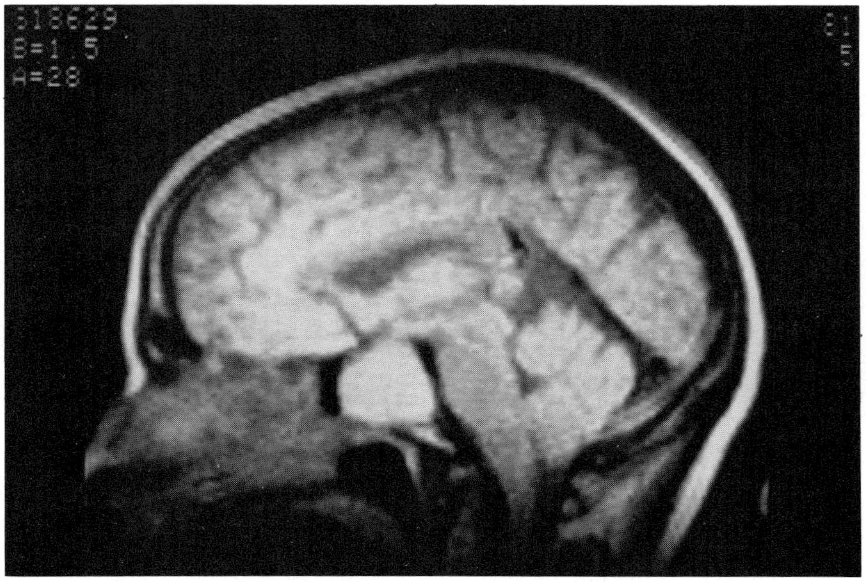

A.

Figure 19. Pituitary adenoma.

A. A large tumor of the pituitary gland is seen in this patient whose contrast allergy prevented a CT study.

B.

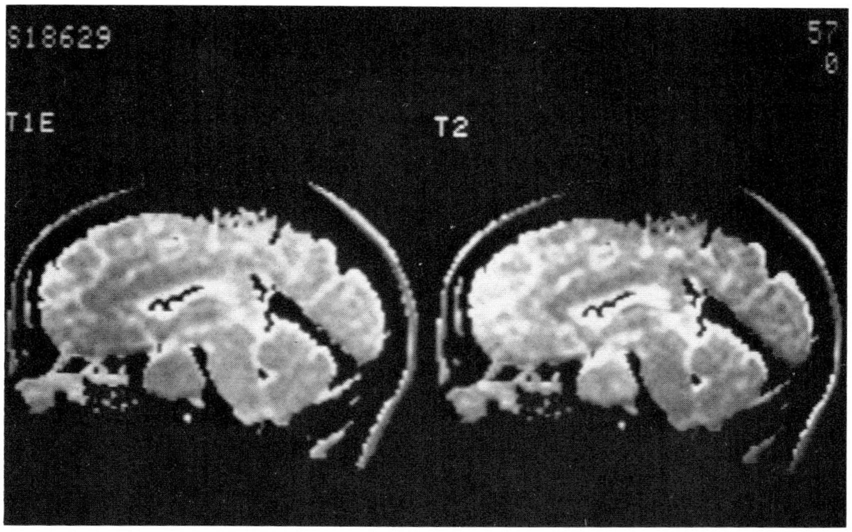

B. Relaxation value calculations and resulting analogue images show slight T_1 shortening (dark pixels) and T_2 prolongation (bright pixels) when compared with the adjacent brainstem.

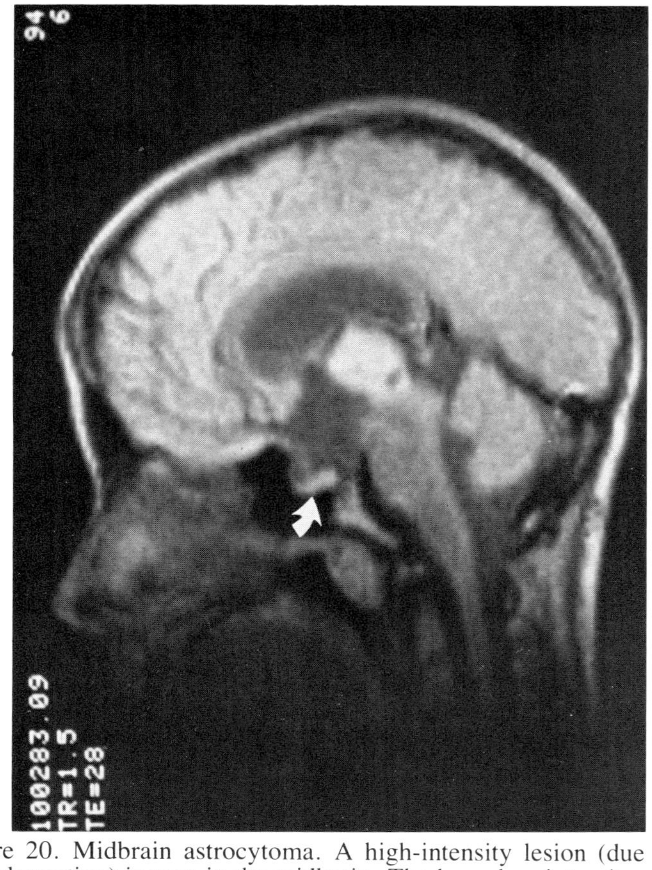

Figure 20. Midbrain astrocytoma. A high-intensity lesion (due to T_2 prolongation) is seen in the midbrain. The large low intensity region located anteriorly is the obstructed and dilated third ventricle that exhibited long T_1 relaxation. Note the pituitary in the expanded sella (arrow).

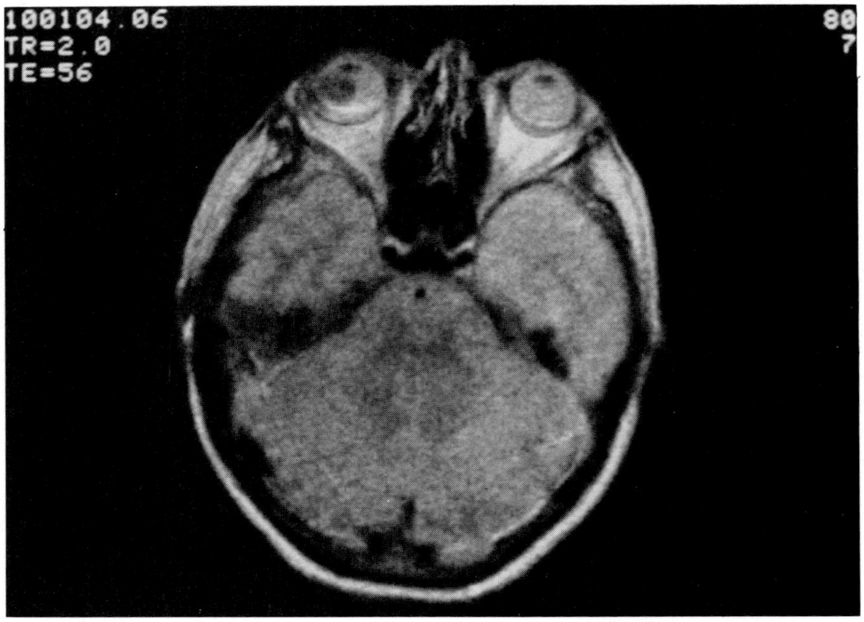

Figure 21. Choroidal melanoma. This malignancy shows a low intensity relative to the vitreous and brain, and exhibited a short T_2 relaxation time on calculated images (see text).

Infarction

The pathophysiology of infarction was discussed partly in the preceding section on cerebral edema, since the earliest manifestation of cerebral ischemia, at least at the molecular level, is the accumulation of excess water. As stated, NMR is exceedingly sensitive to the development of cerebral edema and, consequently, infarction. Our early experience bears this out.

Both early and subacute infarction were visualized by intensity changes that correlate well with increased water content. Edema in the gray matter was seen in early infarct (Fig. 3), and white matter intensity change compatible with edema was noted in subacute infarction (Fig. 22). Our experience was limited, and the sensitivity of NMR to hyperacute infarcts (those within the first 24 hr) was not studied, but should be at least as good as and probably better than that of CT. Early changes in brain water occur within the first 6 hr of infarction and have been detected with NMR in an experimental model of edema produced by focal cryogenic injury inflicted 3 hr prior to imaging. An experimental model of infarction in the primate showed the lesion only 90 min after its production.[17] Reversible changes of low perfusion and relative anoxia may be demonstrable, if not with proton imaging, then perhaps with imaging such nuclei as sodium or with spectroscopy of phosphorus. This ability would have a tremendous impact on the management of patients with cerebrovascular disease.

Trauma

Our experience in evaluating patients with trauma to the brain and spinal cord was hampered by the logistical problems of imaging acutely traumatized subjects on life-support equipment in a non-hospital-based laboratory. However, one patient with extra-axial hematomas was studied after surgery for a parenchymal hematoma. The separation of the epidural from the subdural component of the hematoma was clear, as was the postsurgical abnormality of the brain parenchyma (Fig. 23). Others have reliably shown traumatic hematoma with NMR,[7] and the probability of detecting subtle contusion is high.

Infection

Patients examined with focal neurologic signs resulting from infectious encephalomyelitis have shown focal intensity abnormalities that are nonspecific. These abnormalities probably represent the edema or demyelinization, or both, caused by the inflammatory process or its consequences on brain vasculature (Fig. 24).

Brain abscess has been seen both in a healing capsule stage and in the cerebritis stage in an ex-

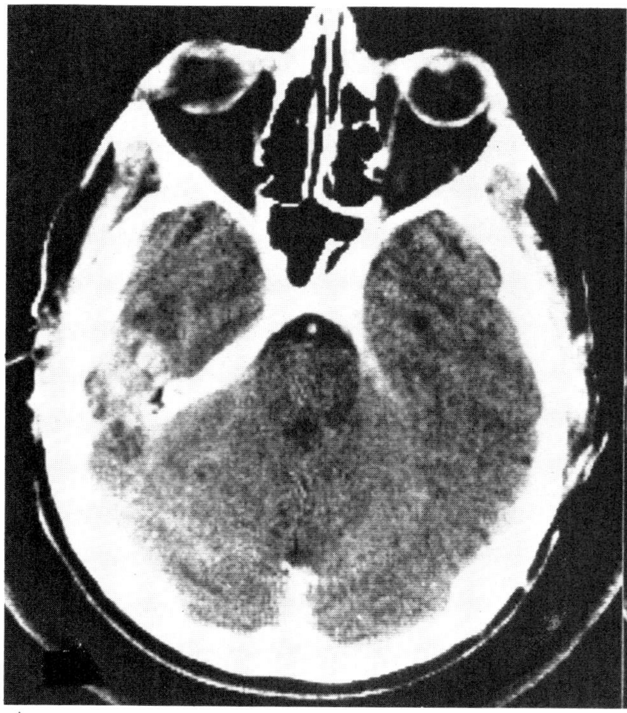

A.

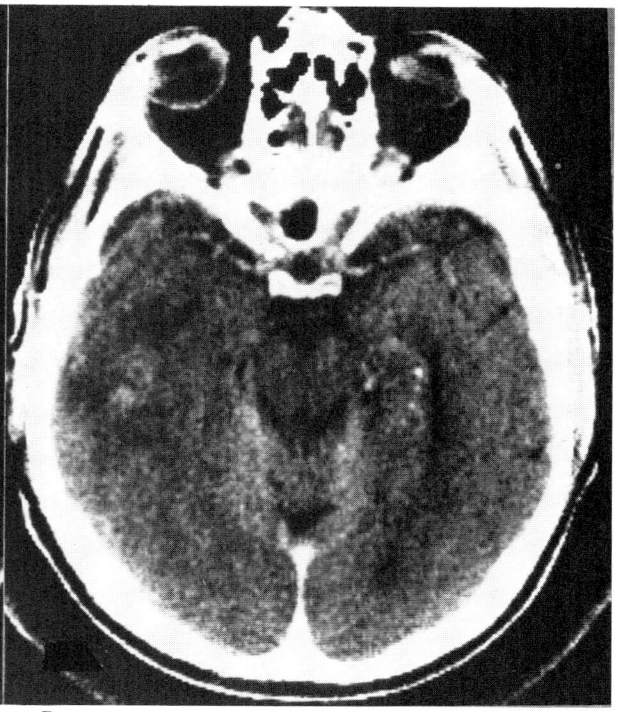

B.

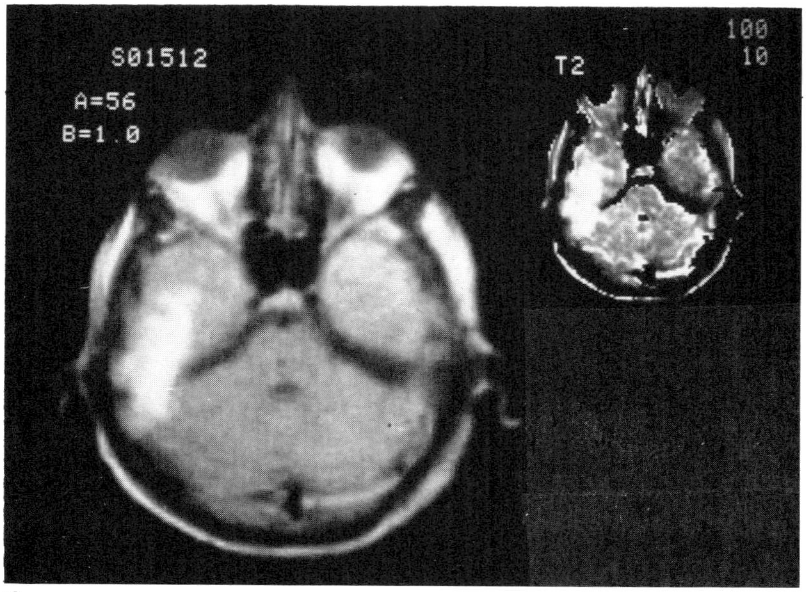

C.

Figure 22. Subacute infarction, right temporal lobe.

A and B. Two contiguous axial CT sections at the level of the temporal lobe show an area of inhomogeneous enhancement and low attenuation in the right temporal region. The patient had suffered an infarction in the distribution of the right middle cerebral artery 2 weeks prior to this study.

C. Spin echo NMR image shows a focus of high intensity, indicating edema in the right temporal lobe. Note that the inset shows a calculated T_2 image that verifies the prolongation of T_2 relaxation within the infarction.

(Reproduced, with permission, from Crooks LE, Mills CM, Davis PL, et al: Visualization of cerebral and vascular abnormalities by NMR imaging: The effects of imaging parameters on contrast. Radiology 144:843, 1982.)

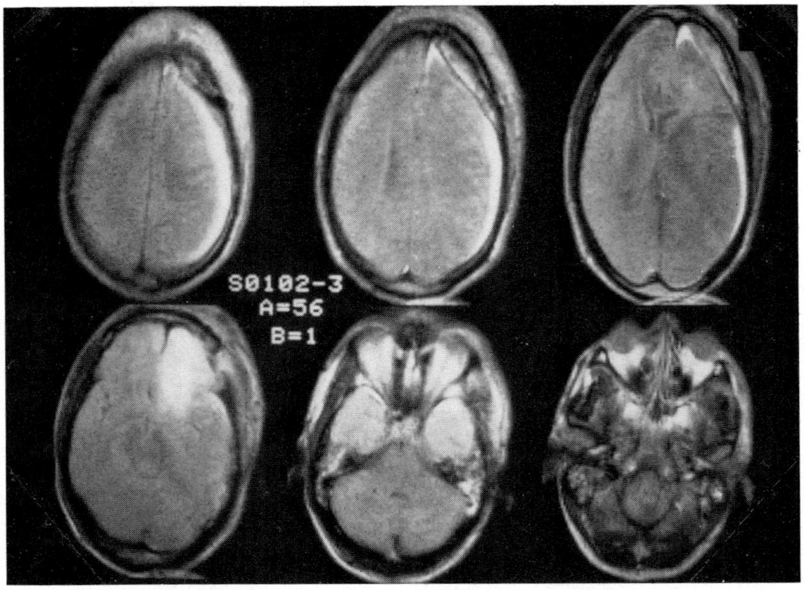

Figure 23. Trauma. Status after removal of left frontal hematoma. These spin echo NMR sections show a subdural hematoma on the upper sections, partially invaginated by an epidural hematoma (upper right images). Note the low intensity of the dura separating the epidural and the subdural compartments. The bed of the evacuated hematoma is seen in the lower left image.

(Reproduced, with permission, from Crooks LE, Mills CM, Davis PL, et al: Visualization of cerebral and vascular abnormalities by NMR imaging: The effects of imaging parameters on contrast. Radiology 144:843, 1982.)

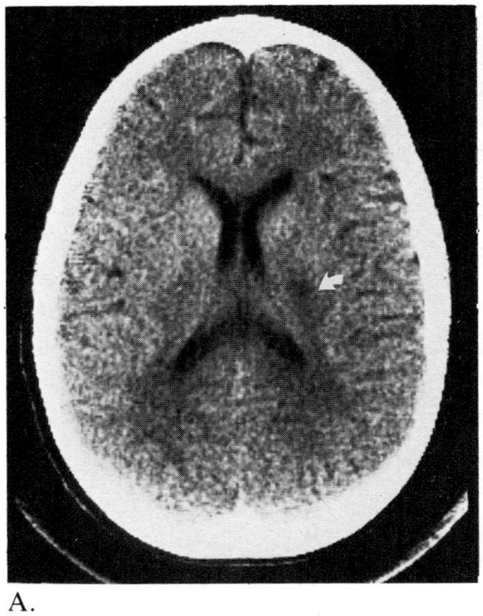

A.

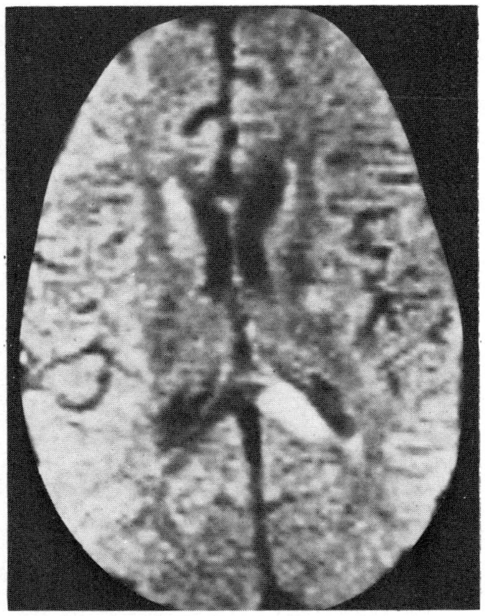

B.

Figure 24. Postinfectious encephalitis.

A. Axial CT section in this young patient with right hemiparesis shows a focus of low attenuation in the left internal capsule (arrow).

B. Spin echo NMR image at the same level shows a focus of high intensity in the left internal capsule, but a more obvious abnormality in the left major forceps of the corpus callosum adjacent to the left trigone. The latter lesion was not apparent on the CT study.

(Reproduced, with permission, from Brant-Zawadzki M, Davis PL, Crooks LE, et al: NMR demonstration of cerebral abnormalities: Comparison with CT. AJNR 4:117, 1983. © 1983, American Roentgen Ray Society.)

perimental model studied in our laboratory.[18] Nuclear magnetic resonance is able to detect central liquefaction in cerebritis; thus, eventually we will be able to guide aspiration of the necrotic center through a burr hole. This procedure is vital to both the determination of the causative organism and decompression of the mass effect associated with larger abscesses.

Vascular Lesions

Because rapidly flowing protons have little chance to fully participate in the RF pulse and magnetic gradient sequence during imaging of their particular plane, blood vessels characteristically contain little or no intensity, since the NMR signal is very low within the lumen of such structures. Slow flow, however, introduces interesting phenomena into the NMR signal, phenomena that may eventually permit the quantitation of flow velocities. Such capability would have a great impact on the evaluation of cerebrovascular disease.

An understanding of the protons in stationary and slowly flowing blood is necessary to appreciate the signal intensity variations seen with slowed flow. The signal intensity of stationary protons in an occluded vessel (or of extravascular protons) is dependent on the amount of magnetization achieved by them between repeated RF perturbations, an amount of magnetization smaller than would accrue in the absence of RF perturbation. The signal intensity of these stationary repeatedly perturbed protons can be arbitrarily defined as 100%.

Slowly flowing blood yields a higher intensity signal than stationary blood. This effect is the result of the addition of fully magnetized protons into the image plane, protons from a region in the body that did not experience RF perturbation. Such fully magnetized protons produce a larger signal when first excited than the partially demagnetized protons that experience the full RF pulse sequence while remaining within the imaged section. Consequently, the NMR intensity of slowly flowing protons will exceed 100%. This augmentation of NMR signal intensity is referred to as paradoxical enhancement.

As mentioned, nuclei in blood flowing at a high velocity will pass through the imaged volume during the interval required for RF and gradient pulse

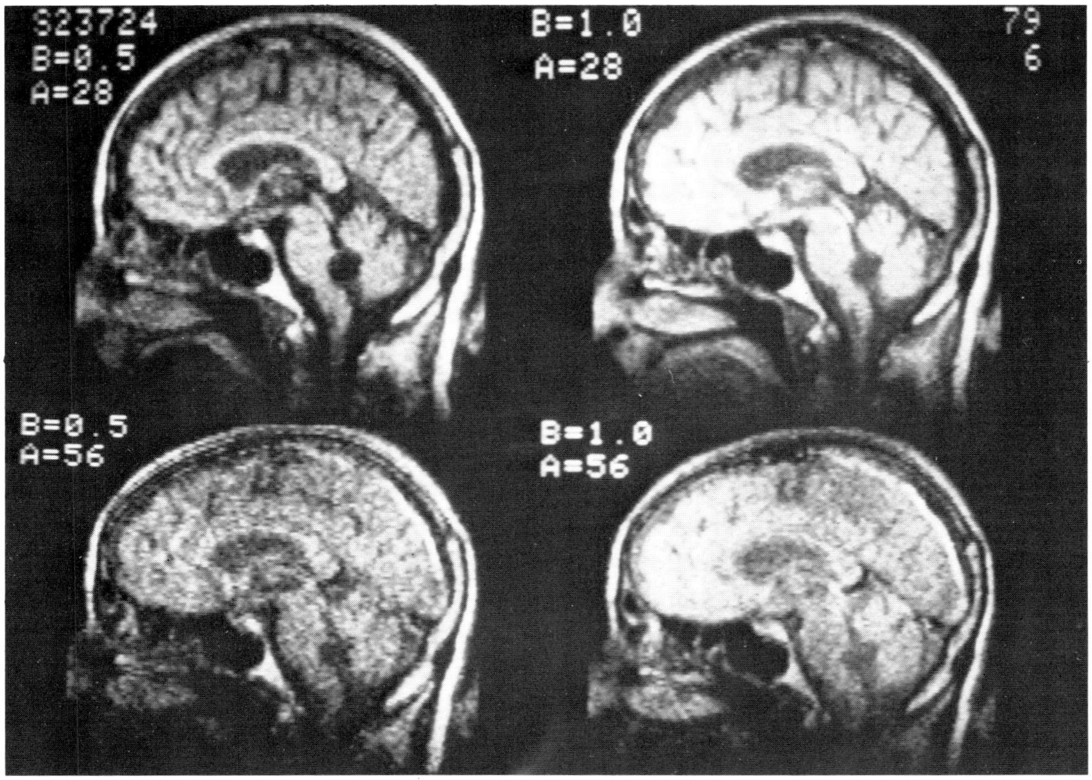

Figure 25. Flow in sagittal sinus; enhancement with second echo. Sagittal spin echo images show that the superior sagittal sinus is black and inseparable from the inner table of the skull with a 28-msec echo delay. The second echo image at 56 msec shows the increase in signal intensity of the sinus and the CSF.

administration. These protons will, therefore, yield no detectable signal. The velocity above which moving nuclei do not yield signal is designated the threshold velocity.

The signal intensity of flowing blood is affected not only by its velocity, but also by the instrument parameters. Images obtained with one pulse sequence interval and two echo delays will show, for any one velocity, a signal of higher intensity in the second echo as compared with the first echo image (Fig. 25).

Pathologic alterations in blood flow patterns may be characterized by the change of signal intensity of blood with different instrument parameters. Slowly flowing blood is distinguished from stationary blood by the signal intensity variation between the first and second echo delays. A high-intensity signal is seen with vascular occlusion in the first echo image. The second echo will show a decrease in signal intensity. However, if there is not occlusion, the slow flow is present, the signal intensity will be equal or higher for the second echo image (Fig. 26).

Nuclear magnetic resonance imaging defines vascular structures in the sagittal, axial, and coronal planes. The advantage of multiplanar imaging is the definition of the entire cross-sectional surface of a vessel in one of the three section planes (Fig. 27). The correct diagnosis of pathologic atherosclerotic narrowing is difficult unless the vessel is completely within the imaged section. Plaque produces a positive intensity signal where there would normally be no detectable signal.

Arteriovenous malformations are ideally suited to NMR imaging because of the contrast afforded by flowing blood. The rapid blood flow in the arteries supplying and veins draining the malformation result in a characteristic collection of low-intensity serpiginous structures (Fig. 28). Cryptic ateriovenous malformations are often composed of vessels so small they cannot be seen at angiography (hence the term "cryptic"). Their clinical course is one of chronic seizures, punctuated by episodes of hemorrhage that may be devastating or subclinical. The latter event can produce a very perplexing CT appearance, since CT has trouble distinguishing the low attenuation of old clot from that of edema (Fig. 29A). Nuclear magnetic resonance, on the other hand, easily distinguishes clot from edema by virtue of the former's short T_1 relaxation characteristics (Fig. 29B).

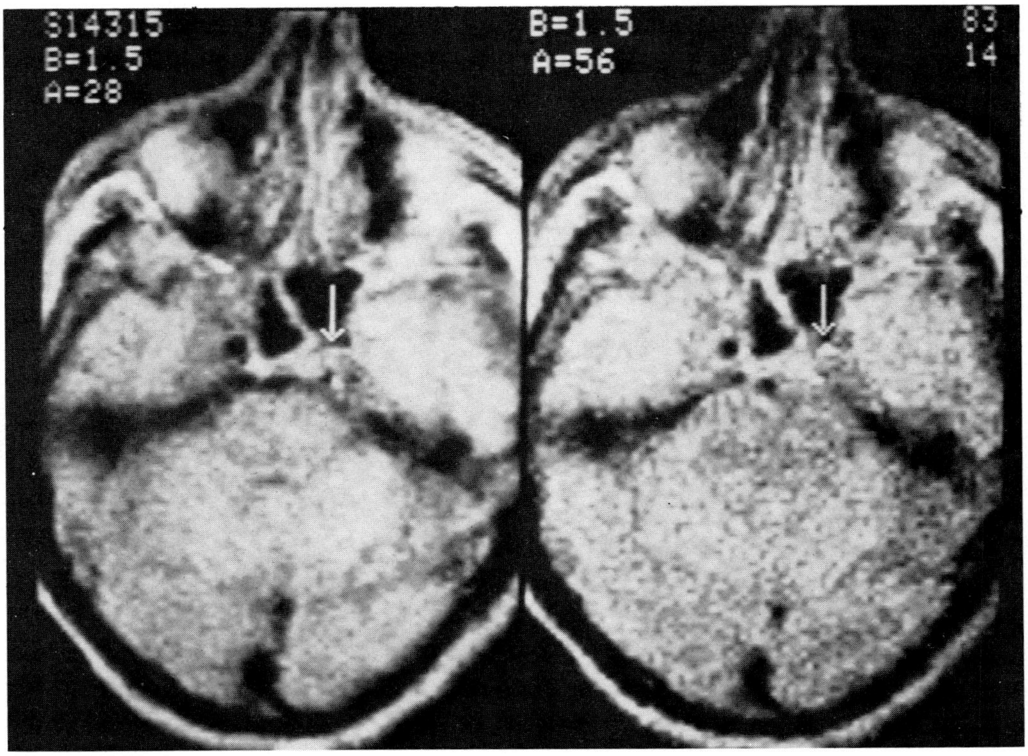

Figure 26. Occlusive disease, intracranial carotid artery. The right internal carotid and basilar arteries are normal. Paradoxical enhancement is seen in the left internal carotid artery as a result of slow flow (arrows).

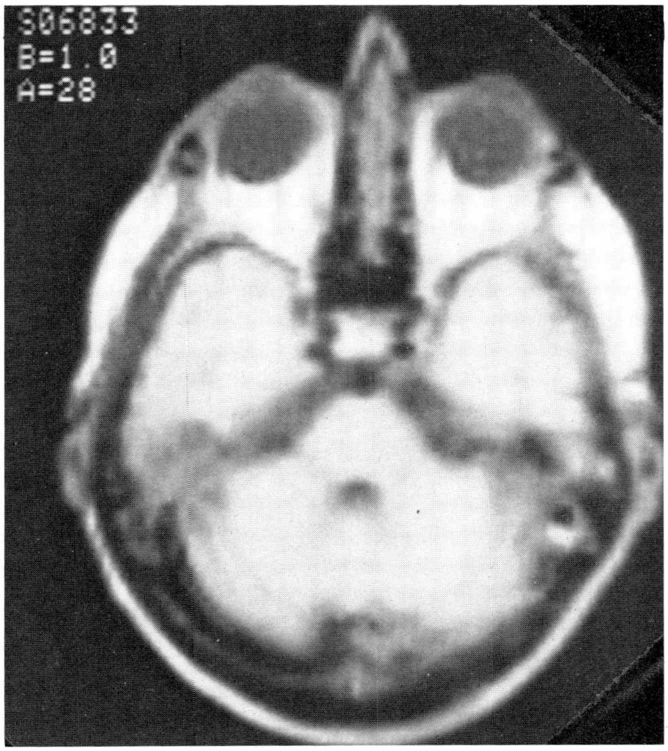

A.

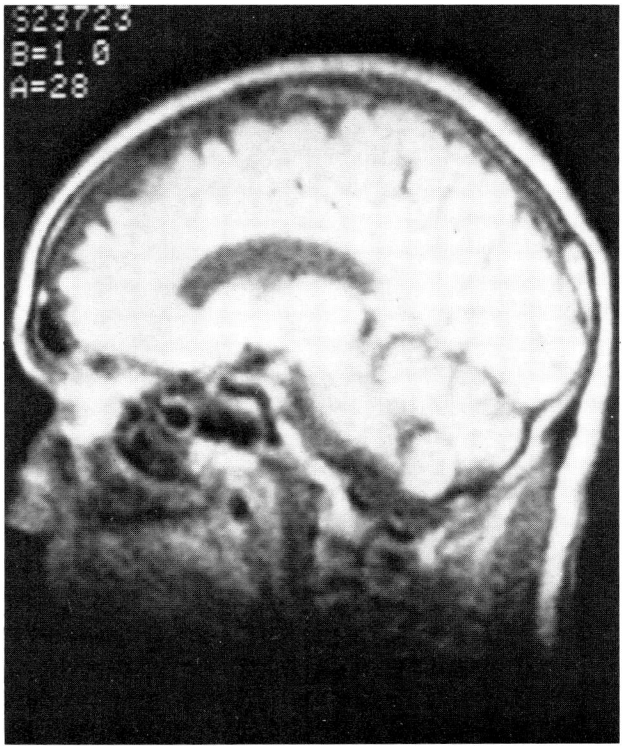

B.

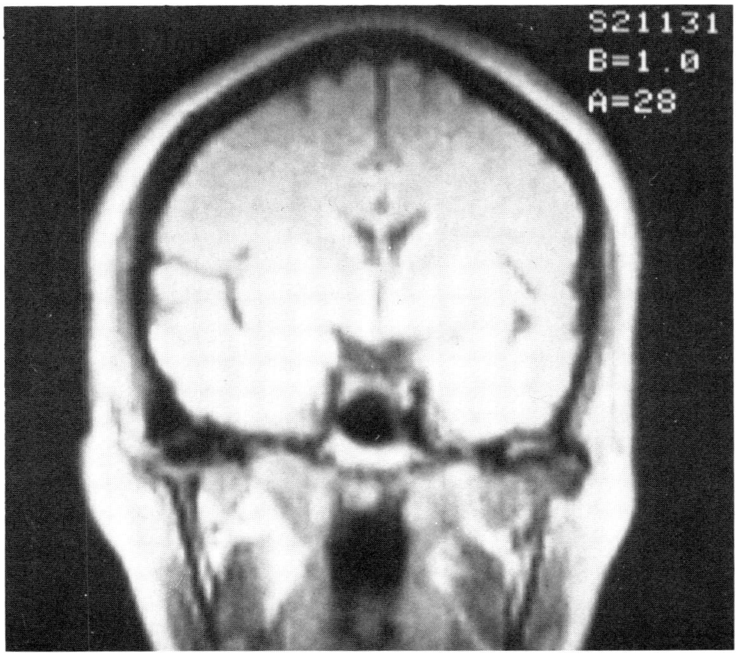

C.

Figure 27. Normal internal carotid artery and cavernous sinus. The internal carotid artery has no signal because of rapid flow, while the cavernous sinus has higher intensity secondary to slower venous flow.
A. Axial spin echo image.
B. Sagittal spin echo image.
C. Coronal spin echo image.

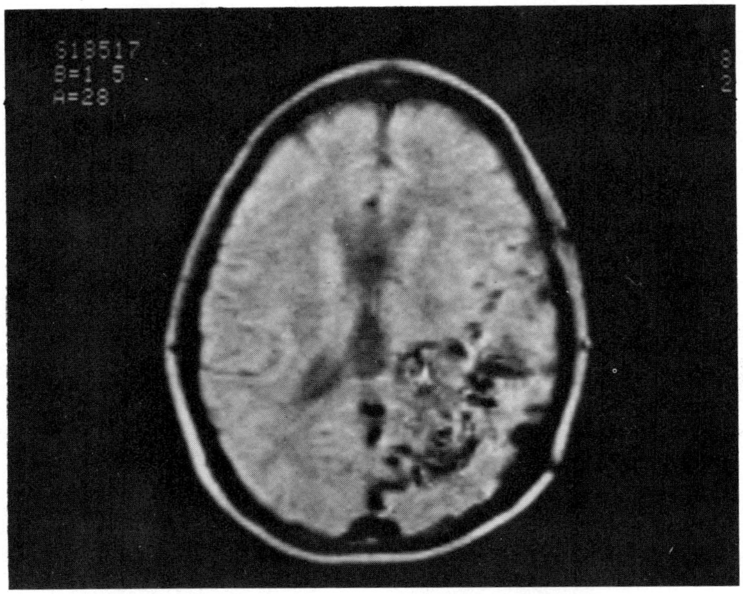

Figure 28. Arteriovenous malformation. Multiple, serpiginous structures of low intensity are seen in the left temporo-occipital region in this patient with a known arteriovenous malformation.
(Reproduced, with permission, from Brant-Zawadzki M, Davis PL, Crooks LE, et al: NMR demonstration of cerebral abnormalities: Comparison with CT. AJNR 4:117, 1983. © 1983, American Roentgen Ray Society.)

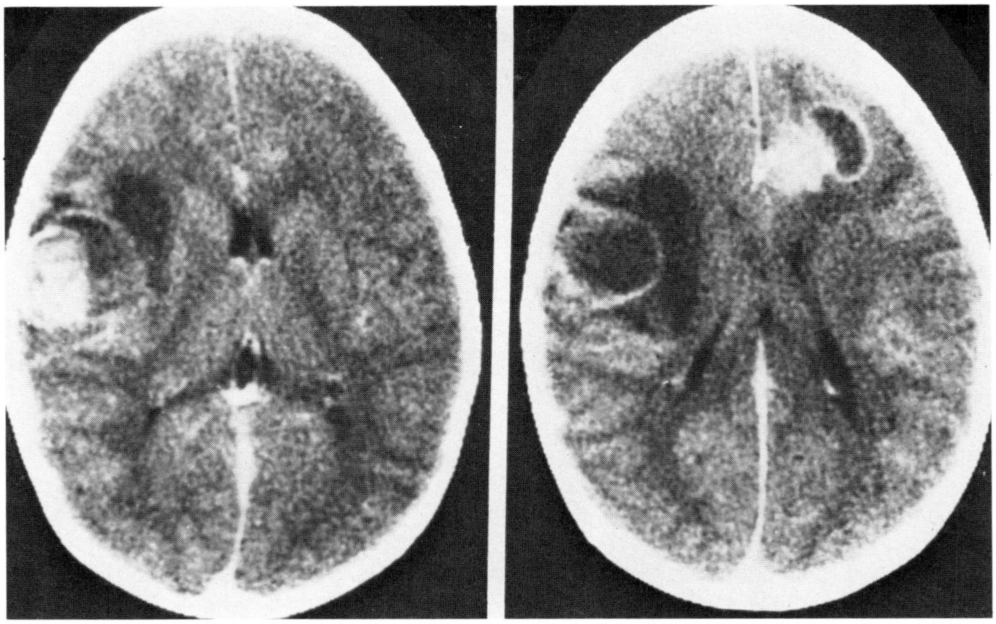

A.

Figure 29. Cryptic arteriovenous malformations.
A. Contiguous axial CT sections depict bilateral lesions simulating neoplasms.

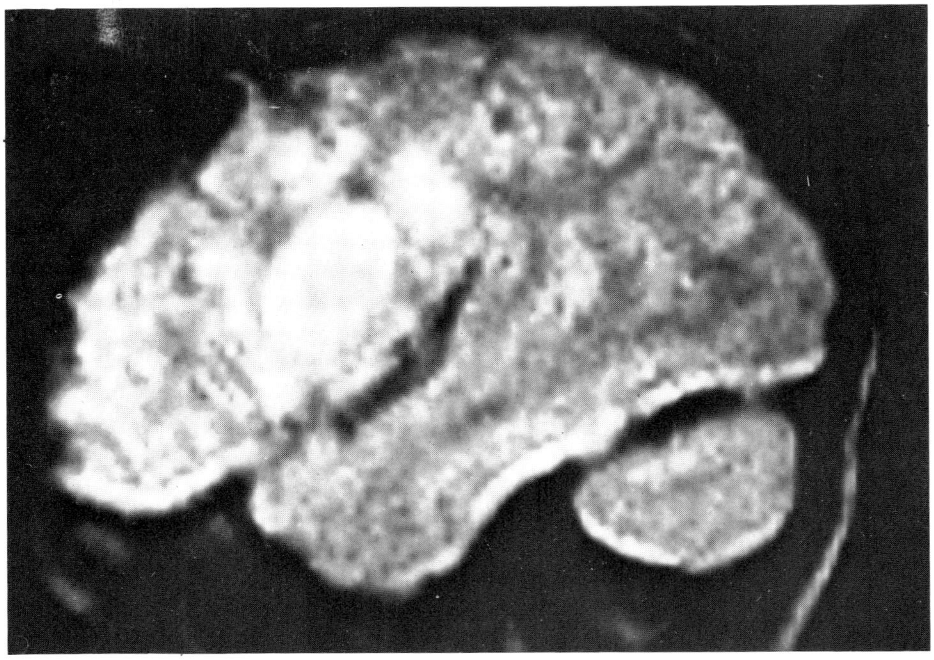

B.

B. Sagittal NMR section through right-sided lesion shows an oval lesion of high intensity (which exhibited shortened T_1 relaxation consistent with clot) surrounded by a focal area of lesser high intensity (with long T_2 characteristics) consistent with edema. The oval area high intensity corresponds to the CT low-density focus that showed peripheral enhancement (A, right). The NMR appearance and the long-standing clinical symptomatology suggested cryptic arteriovenous malformation, which was proved at surgery (see text).

(Reproduced, with permission, from Brant-Zawadzki M, Davis PL, Crooks LE, et al: NMR demonstration of cerebral abnormalities: Comparison with CT. AJNR 4:117, 1983. © 1983, American Roentgen Ray Society.)

Future Relation of Nuclear Magnetic Resonance to Computed Tomographic Imaging

The preceding discussion should suggest the tremendous potential of NMR in clinical application to diagnosis of CNS disease. Certainly the role of NMR will be at least complementary to CT, and in a number of entities it may become the primary diagnostic modality, supplanting CT. Computed tomography will remain necessary to detect subtle bone disruption caused by trauma, or early erosion of bone caused by other processes. Extremely rapid, millisecond CT scanners may also be essential for perfusion studies. Intravascular contrast agents used with CT also affect its specificity. Agents for NMR contrast enhancement are only beginning to be developed, and some questions as to their role remain. However, it is clear that the major factor in further advances in the specificity of proton NMR imaging is the accumulation of experience with this modality. The early clinical material certainly illustrates that such experience is forthcoming, and the tremendous clinical demand for this imaging modality is already beginning to be felt. Hospitals that can afford only one computer-assisted tomographic machine should choose x-ray CT. If they can accommodate two such units, one must be an NMR imager.

REFERENCES

1. Hawkes RC, Holland GN, Moore WS, et al: Nuclear magnetic resonance (NMR) tomography of the brain: A preliminary clinical assessment with demonstration of pathology. J Comput Assist Tomogr 4:429, 1980.
2. Doyle FH, Gore JC, Pennock JM: Imaging of the brain by nuclear magnetic resonance. Lancet 2:53, 1981.

3. Young IR, Burl M, Clarke GJ, et al: Magnetic resonance properties of hydrogen: Imaging the posterior fossa. AJR 137:895, 1981.

4. Young IR, Bailes DR, Burl M, et al: Initial clinical evaluation of a whole body nuclear magnetic resonance (NMR) tomograph. J Comput Assist Tomogr 6:1, 1982.

5. Alfidi, RJ, Haaga JR, El Yousef SJ, et al: Preliminary experimental results in humans and animals with a superconducting, whole-body nuclear magnetic resonance scanner. Radiology 143:175, 1982.

6. Crooks LE, Mills CM, Davis PL, et al: Visualization of cerebral and vascular abnormalities by NMR imaging: The effects of imaging parameters on contrast. Radiology 144:843, 1982.

7. Bydder GM, Steiner RE, Young IR, et al: Clinical NMR imaging of the brain: 140 cases. AJNR 3:459, 1982.

8. Brant-Zawadzki M, Davis PL, Crooks LE, et al: NMR demonstration of cerebral abnormalities: Comparison with CT. AJNR 4:117, 1983.

9. Brooks RA, DiChiro G, Keller MR: Explanation of cerebral white-gray contrast in computed tomography. J Comput Assist Tomogr 4:489, 1980.

10. Melamed E, Lavy S, Bentin S, et al: Reduction in regional cerebral blood flow during normal aging in man. Stroke 11:31, 1980.

11. Crooks L, Sheldon P, Kaufman L, et al: Quantification of obstructions in vessels by nuclear magnetic resonance (NMR). IEEE Trans Nucl Sci NS-29:1181, 1982.

12. Kaufman L, Crooks LE, Sheldon PE, et al: Evaluation of NMR imaging for detection and quantification of obstructions in vessels. Invest Radiol 17:554, 1982.

13. Fishman RA: Brain edema. N Engl J Med 293:706, 1975.

14. James AE, Jr, Flor WJ, Novak GR, et al: The ultrastructural basis of periventricular edema: Preliminary studies. Radiology 135:747, 1980.

15. Singer JR, Crooks LE: Some magnetic studies of normal and leukemic blood. J Clin Eng 3:237, 1978.

16. Vinuela FV, Fox AJ, Debrun GM, et al: New perspectives in computed tomography of multiple sclerosis. AJNR 3:277, 1982.

17. Spetzler RF, Zabramski JM, Kaufman B, et al: Acute NMR changes during MCA occlusion: A preliminary study in primates. Stroke 14:185, 1983.

18. Brant-Zawadzki M, Enzmann DR, Placone RC, et al: NMR and CT correlation of experimental brain abscess. AJNR 4:250, 1983.

9

Nasopharynx

WILLIAM P. DILLON, M.D.
JACK DEGROOT, M.D., Ph.D.

Nuclear magnetic resonance offers several advantages over CT in the evaluation of the nasopharynx and surrounding compartments. Although both imaging modalities allow direct cross-sectional assessment of superficial and deep nasopharyngeal anatomy, NMR is superior to CT in the depiction of anatomic detail, has direct multiplanar imaging capability, and allows greater tissue characterization of both normal and pathologic anatomy without the hazards of intravenous contrast or ionizing radiation.[1–8] An additional advantage of NMR is the inherent high contrast between blood vessels and surrounding tissue.[9,10] This facilitates assessment of deep cervical and retropharyngeal nodes in the normal and pathologic nasopharynx without intravenous contrast.

In this chapter we will emphasize the normal NMR appearance of superficial and deep nasopharyngeal anatomy, including the infratemporal fossa and carotid sheath structures. Our initial experience with spin echo imaging of nasopharyngeal pathology will be presented to demonstrate the potential clinical use of this new imaging modality.

Imaging Techniques and Parameters

The precise and delicate anatomy of the nasopharynx requires imaging with both high spatial and contrast resolution. Computed tomography has overcome limitations in spatial resolution through advances in target reconstruction algorithms, but the contrast resolution of CT is still determined only by the inherent electron density differences of tissues. Nuclear magnetic resonance contrast resolution is superior to that of CT because of its dependence on several tissue characteristics: hydrogen proton density, T_1 and T_2 relaxation times, and blood flow. Spatial resolution on NMR and CT is primarily a function of pixel size, slice thickness, and signal-to-noise ratio. Our prototype imager (3.5-kGauss superconducting) allows 7-mm sections spaced 12 mm apart with matrix display capability of 128 × 256 or 256 × 256 pixels, resulting in a 1.6 mm^2 or 0.7 mm^2 pixel size, respectively. The price paid for smaller pixel size is increased imaging time and decreased signal-to-noise ratio per pixel. Although imaging time is relatively fixed, the signal may be increased by increasing the pulse sequence repetition rate (TR interval). Lengthening TR to 2.0 sec allows more time for perturbed protons to realign with the magnetic field; therefore, greater signal is emitted per pulse interval. With experience we have found that optimal imaging parameters for the nasopharynx include a 15-slice axial sequence, a TR of 2.0 sec, and a display matrix of 256 × 256 pixels (image time of 34 min). This sequence provides spatial detail comparable to our high-resolution CT scanning techniques (5-mm sections, GE 8800). Our NMR studies also include a 5-slice sequence (axial or coronal) at a TR interval of 0.5 sec (126 × 256 pixel, image time of 4.3 min). Pulse interval (T) for each section imaged will primarily reflect the influence of T_1 relaxation on the tissue NMR signal. Signals are sampled at 28 msec and 56 msec (TE) after each 90° radiofrequency pulse. Variation in signal intensity between the 28 and 56 msec echo for any given TR interval will primarily reflect the effect of T_2 relaxation on tissue signal.

In our technique we have developed a 55-cm elliptical neck coil surrounding the upper torso and neck. This permits centering of the antenna at selected sites in the neck or nasopharynx without shoulders encumbering the coil position, as is the case with smaller-diameter head coils.

A.

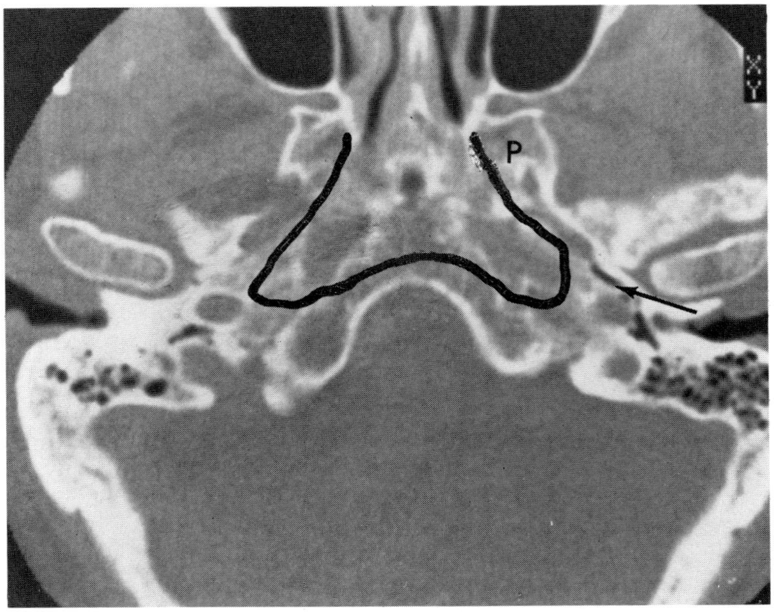

B.

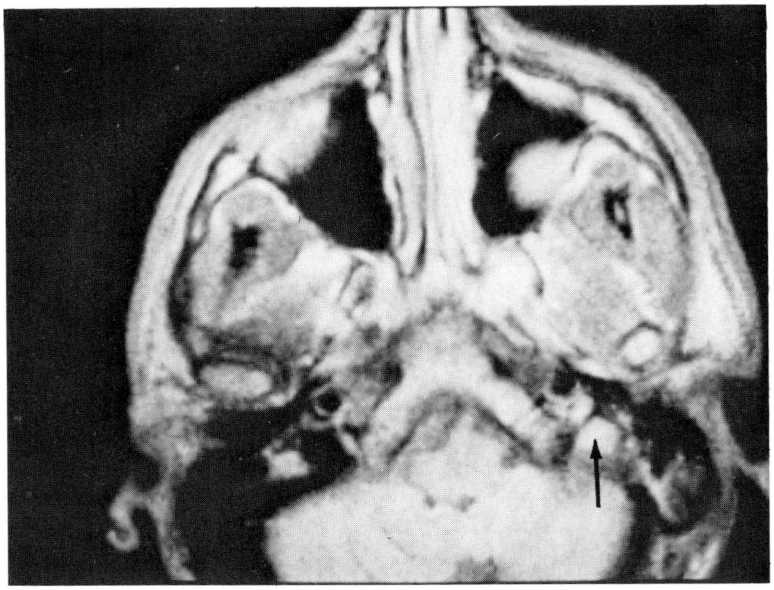

Figure 1.

A. High-resolution axial CT scan of skull base details the osseous roof of the nasopharynx. Black line indicates the attachment of the pharyngobasilar fascia. Note that the pterygoid plates (P), the mandibular condyles, and the clivus all have marrow cavities surrounded by dense rim of cortical bone. The bony detail of the basilar foramina are better depicted by CT than NMR at the present time. (arrow, eustachian tube)

B. Spin echo image (TR of 2.0 sec, TE of 28 msec) of the skull base. Marrow cavities are of high intensity because of fat, slow flowing blood, or both. Cortical bone is of low intensity because of lack of hydrogen, but is not imaged if adjacent to air. The mastoid trabeculae are not visualized because of lack of contrast with surrounding air cells. The internal carotid artery is of low intensity because of its quickly flowing blood while the slowly flowing or turbulent blood within the internal jugular vein (arrow) returns a high-intensity signal.

Normal Anatomy

Superficial Landmarks

The nasopharynx functions as an airway for nasal respiration, but also participates in deglutition. Its lumen is held patent by the tough, pharyngobasilar fascia, which provides a foundation for the mucosa and the intrinsic musculature of the nasopharynx. The attachment of the pharynx and its fascia to the skull base is rhomboidal, the most posterior attachment sweeping horizontally across the prevertebral musculature anterior to the carotid canal and foramen magnum and meeting in the midline at the pharyngeal tubercle. The lateral attachments of the fascia lie opposite to the sphenoidal spines, then form lateral recesses as it sweeps anterior to its attachment on the medial pterygoid plate (Fig. 1). The pharyngeal mucosa and lymphoid tissue rest on the medial surface of this fascia, while laterally the pharyngeal constrictors and palatal musculature (mucosa, adenoids, and muscle) border this fascial plane. These superficial tissues are displayed with poor contrast resolution on CT because of their similar electron density (Fig. 2A). As a result, normal variations in superficial contours may be difficult to differentiate from early superficial submucosal carcinomas by CT.[1–3] The spin echo axial images (TR of 2.0 sec, TE of 28 msec) optimally depict the mucosal surfaces and adenoidal tissue as high-intensity structures with prolonged T_1 and T_2 relaxation times; the prolonged times are probably caused by increase in water content. The spin echo technique provides exceptional contrast for delineation of surrounding intrinsic, prevertebral, and palatal musculature, including the tensor and levator veli palatini muscles (Figs. 2B, 3).

The osseous and cartilaginous structures provide structural integrity to the nasopharyngeal soft tissues and are convenient landmarks on CT. Although they are not as readily apparent on NMR, bone and cartilage are visualized as low-intensity structures because of their paucity of hydrogen nuclei (Figs. 1–4). Marrow-containing bones such as the mandible, clivus, vertebral bodies, and occasionally the pterygoid plates and styloid processes will have focal central high intensity because of the short T_1 relaxation time of fat or slowly flowing blood within bone marrow (Fig. 2).

It may sometimes be impossible to image certain osseous or cartilaginous structures on NMR, particularly if they are surrounded by air. Neither air nor dense bone returns an NMR signal. If air and bone are adjacent to each other, no contrast will exist. This is evident in the normal aerated maxillary sinus, since the presence of the posterior bony antral wall may only be inferred by its position just anterior to the pad of retroantral fat (Fig. 2B). This principle also applies to the cartilaginous torus tubarius, which likewise is not imaged when adjacent to the aerated nasopharynx. If soft tissue (tumor, infection, fluid, or unerupted teeth) exists within the maxillary sinus, or if prominent adenoids are present in the nasopharynx, the higher intensity from these tissues will provide the necessary contrast for delineation of these low-intensity osseous or cartilaginous structures (Figs. 2B, 3).

Deep Compartments

Fascial sheaths separate the deep soft tissues of the nasopharynx into four compartments: the parapharyngeal space, the infratemporal fossa, the carotid sheath, and the retropharyngeal space. Because it has a more prolonged T_1 relaxation time than muscle or fat, fibrous connective tissue has low intensity on images acquired at TR of 2.0 sec. While inseparable by CT, the actual fascial tissue is occasionally distinguished on NMR from surrounding musculature by virtue of differing relaxation times. These fascial structures are detected as fine low-intensity envelopes marginating certain muscle groups, such as the lateral pterygoid muscles (Fig. 5A) or the intrinsic pharyngeal muscles (Fig. 5B).

Lateral to the nasopharyngeal musculature lies a compartment containing fibrous and fatty tissue, the parapharyngeal space. It separates the muscles within the infratemporal fossa from the nasopharyngeal constrictors and allows mobility of these muscles during mastication and swallowing.[3] Computed tomography demonstrates the parapharyngeal space as a region of low attenuation separating pterygoid from pharyngeal musculature. The fatty nature of the content of this space is confirmed with NMR because this region has a short T_1 relaxation time and is of high intensity on all TR intervals. Blood flowing within pharyngeal veins and branches of the external carotid artery may occasionally be seen as focal areas of low intensity within this space (Fig. 4B).

The infratemporal fossa is a compartment lateral to the parapharyngeal space and is bounded laterally by the zygomatic arch. It contains the pterygoid, temporalis, and masseter muscles surrounding the mandible. Contrast provided by the high-intensity signal of fat separates these muscle bundles to greater advantage than CT does (Fig. 2). As it curves posterior to the mandibular ramus, the deep lobe of the parotid gland also enters the infratemporal fossa. A prominent high-intensity fat pad is present between the muscles

A.

B.

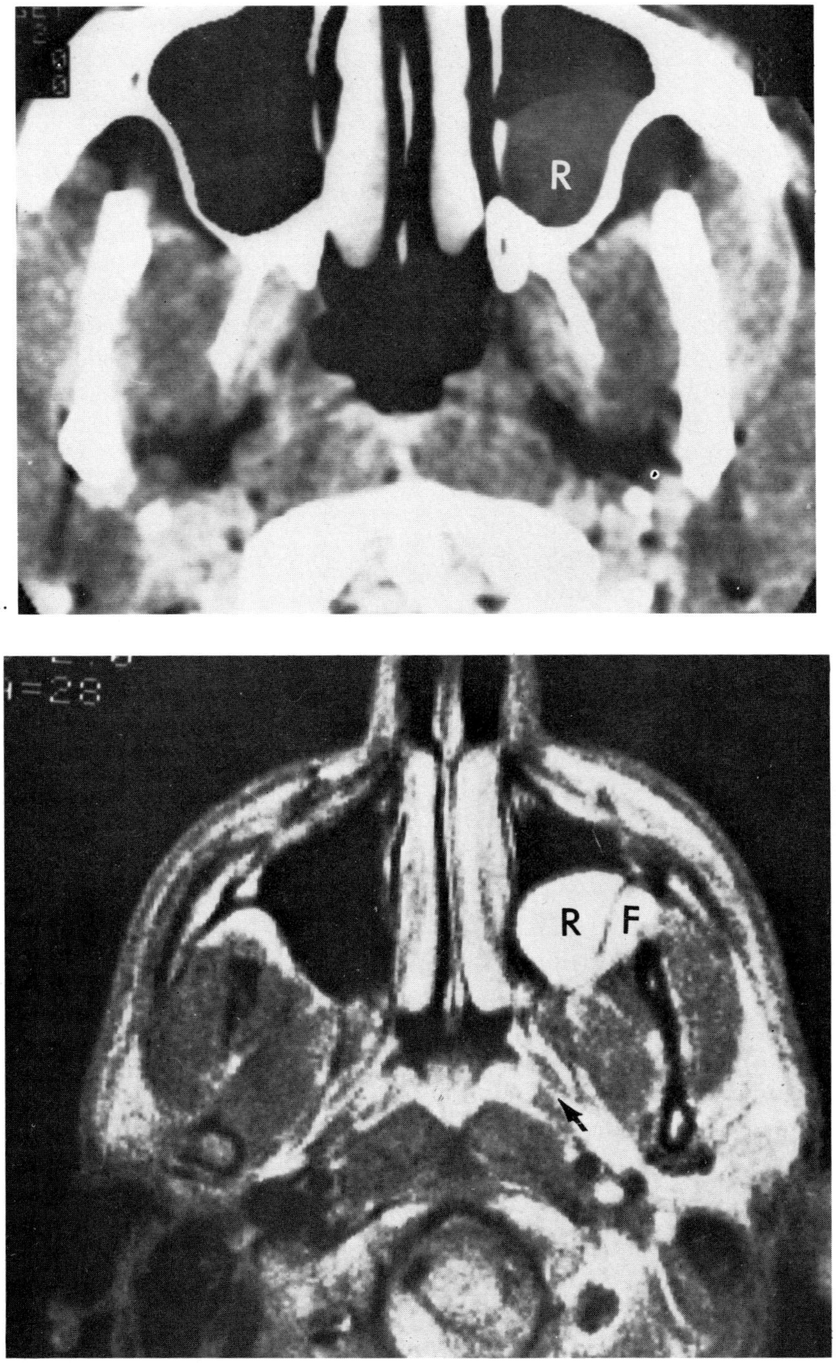

Figure 2.

A. Axial CT scan through normal mid-nasopharynx. Symmetrical low-attenuation parapharyngeal fat planes separate intrinsic musculature of nasopharynx from infratemporal fossa. Little soft-tissue contrast exists between mucosal surfaces and muscle. Bolus intravenous contrast defines the internal carotid and jugular vein from surrounding parenchyma. There is an incidental left maxillary sinus retention cyst (R).

B. Spin echo axial image (TR of 2.0 sec, TE of 28 msec) through upper nasopharynx of same patient as A. Mucosal surfaces are of high intensity on TR of 2.0-sec images, providing contrast for delineation of prevertebral, tensor, and levator palatini musculature (arrow). High-intensity fat (F) separates the infratemporal fossa musculature. Retention cyst (R) provides contrast necessary for visualization of posterior maxillary sinus wall.

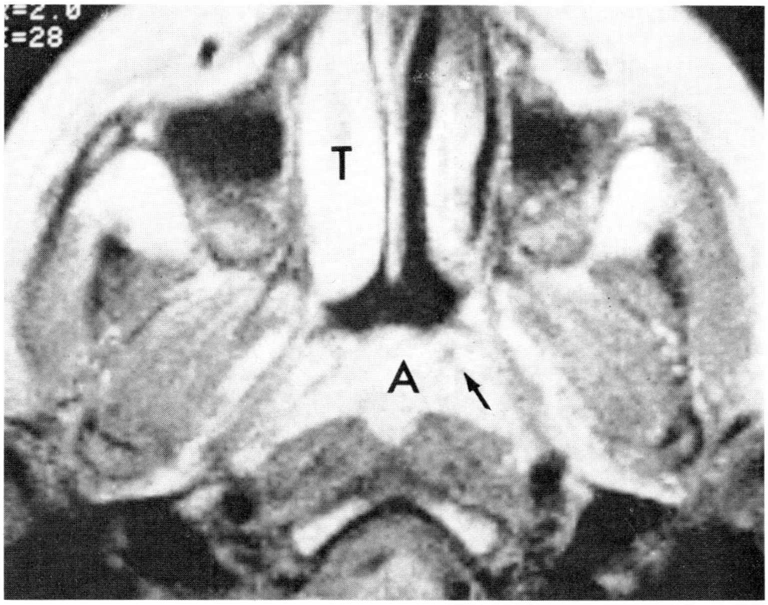

Figure 3. Axial spin echo image (TR of 2.0, TE of 28 msec) in a young child with prominent adenoidal tissue (A). Mucosa and adenoids are of high intensity. Curved low-intensity torus is visualized (arrow). Note edematous high-intensity right nasal turbinate (T).

and the maxillary antrum, and provides contrast for delineation of maxillary sinus disease (Fig. 2B).

The carotid sheath structures (internal carotid artery, internal jugular vein, cranial nerves IX–XII, and sympathetic fibers) lie posterolateral to the retropharyngeal and parapharyngeal space. By virtue of the characteristics of blood flow,[8] the carotid and jugular vessels have no NMR intensity and provide contrast for surrounding parenchymal structures such as lymph nodes and muscles (Figs. 4B, 6). Occasionally, slow or turbulent blood flow within these vessels, especially in a small internal jugular vein, will produce a high-intensity signal that increases on the 56-msec echo (Fig. 2B). The bony styloid process descends from the skull base just lateral to the internal jugular vein. It may be confused with these vessels because it has either a low-intensity signal or, occasionally, a high-intensity signal if marrow is present within its lumen (Fig. 2B).

The retropharyngeal space, a potential compartment bounded by the pharyngobasilar fascia anteriorly and the prevertebral fascia posteriorly, is not normally visualized with either CT or NMR. The prevertebral musculature, the longus colli and the longus and rectus capitis, lies between the nasopharynx and spine. The longus colli lies medially and extends as high as the C2 vertebral body. The longus capitis extends from the C2 to the skull base, coursing medial to the rectus capitis anterior and rectus capitis lateralis (Fig. 3). The muscle groups are of low intensity on spin echo images and may be separated from one another by their enveloping lower-intensity connective tissue and intervening high-intensity fat. Large nasopharyngeal neoplasms will exert mass effect on the prevertebral muscles, a concept better shown with NMR than CT (Fig. 5).

Pathology

In our relatively early experience, pathologic processes such as nasopharyngeal squamous carcinoma, lymphoma, denervation atrophy of muscles, and adenoidal hyperplasia were all imaged with NMR as well as or better than with CT. Compared with CT, the advantages of greater contrast resolution and

A.

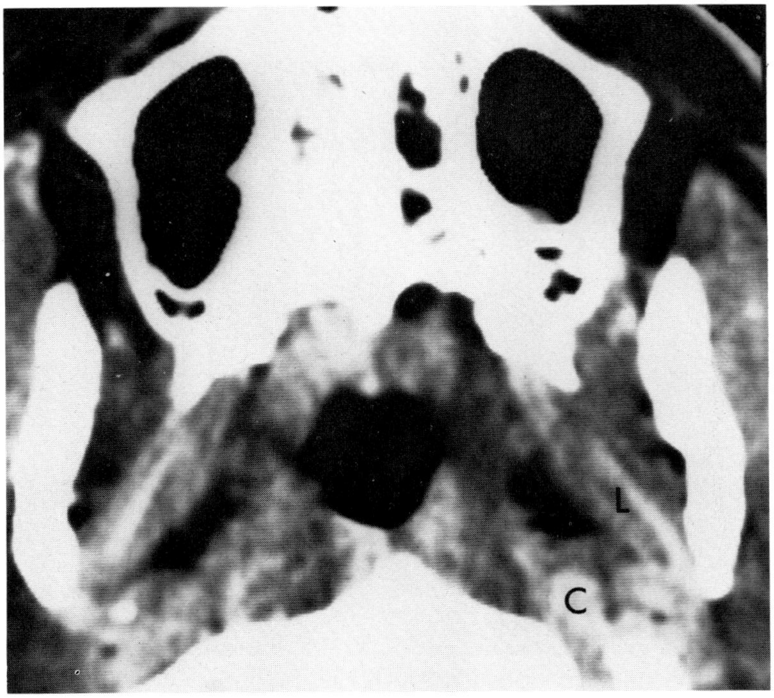

B.

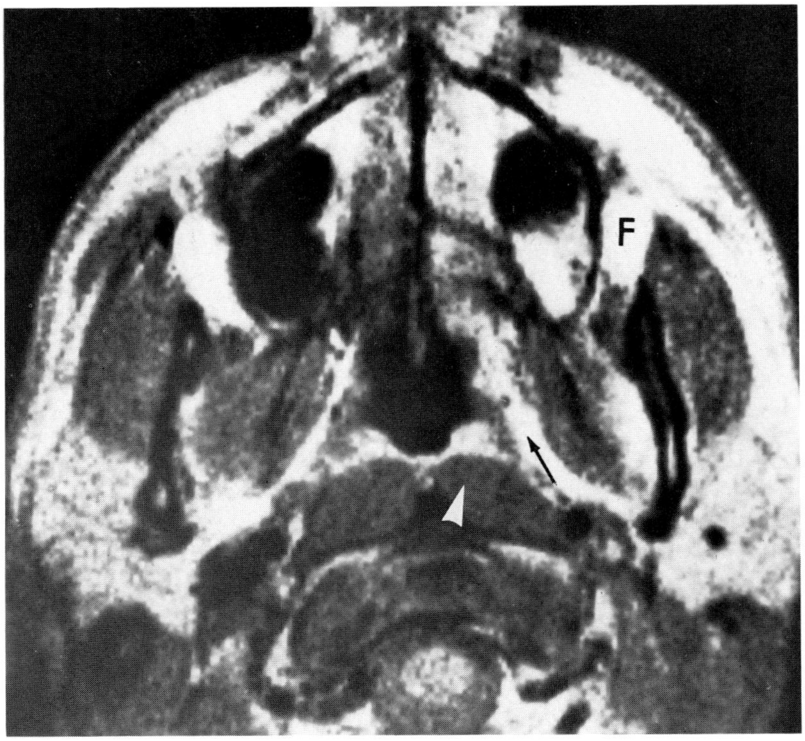

Figure 4.

A. Normal axial CT scan of nasopharynx at the level of the hard palate and Passavant's muscular sphincter. Styloid musculature effaces the carotid sheath vessels. Prevertebral musculature is poorly differentiated from nasopharyngeal soft tissues. (C, internal carotid; L, lateral pterygoid muscle)

B. Nuclear magnetic resonance image of same patient as in 4A (TR of 2.0 sec, TE of 28 msec). Pharyngeal veins within parapharyngeal space (arrow) return low-intensity signal. Deep lobe of parotid gland hooks around mandibular ramus to efface the carotid sheath and parapharyngeal space. (F, retroantral fat; arrowhead, longus colli)

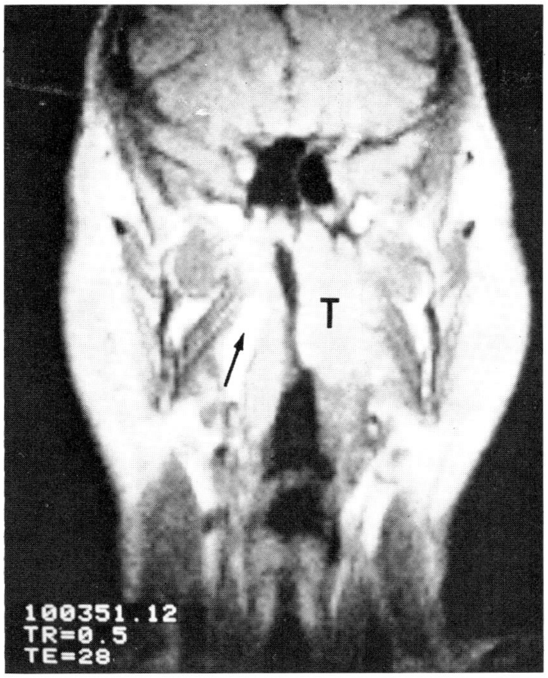

A.

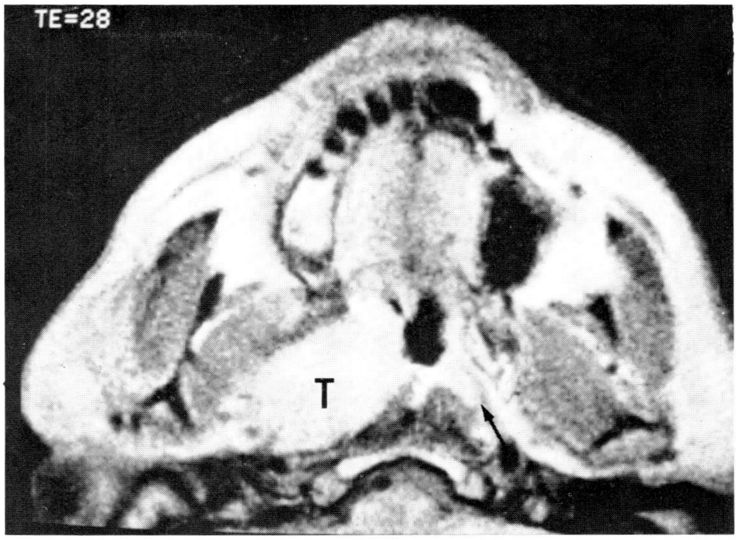

B.

Figure 5.
A. Coronal spin echo image (TR of 0.5 sec, TE of 28 msec). Patient's right is viewer's right. The patient had an infiltrating squamous carcinoma (T) invading the parapharyngeal space (arrow) on the right. Normal anatomy for comparison on the right. Note the low-intensity connective tissue enveloping the pterygoid musculature. The tumor extends to, but not through, the skull base.
B. Same patient as in A (TR of 2.0 sec, TE of 28 msec). Infiltrating squamous carcinoma (T) invades parapharyngeal space and displaces medial pterygoid and prevertebral muscle on right. Note low-intensity connective tissue surrounding left pterygoid muscle. The tumor has increased in intensity, with a longer TR interval reflecting prolongation of T_1 relaxation time. (arrow, parapharyngeal space on the left)

multiparametric image acquisition allow NMR to better define the margins of neoplastic disease from surrounding normal structures in the nasopharynx (Figs. 5–7). Although this was not surgically proven, two patients were felt to have retropharyngeal adenopathy imaged more clearly with NMR than CT (Figs. 6, 7). These early results would indicate that NMR may become the preferred imaging modality for evaluation of nodal disease in the head and neck, and has important implications for surgical and radiotherapy planning in such patients. Histologic specificity may not be possible with current NMR imaging techniques, but this remains to be evaluated.

Artifacts induced by dental fillings and swallowing have been the bane of CT imaging in the oropharynx and nasopharynx. Nuclear magnetic resonance appears to overcome at least some of these problems because it is less sensitive to small intermittent motion artifacts and nonferromagnetic dental appliances and fillings.[11] Pathology obscured by artifacts on CT is easily imaged with NMR (Fig. 8). Ferromagnetic dental bridgeworks and root canal posts, however, will induce large areas of image distortion on NMR images. Patients with such dental work will probably be better evaluated by CT.

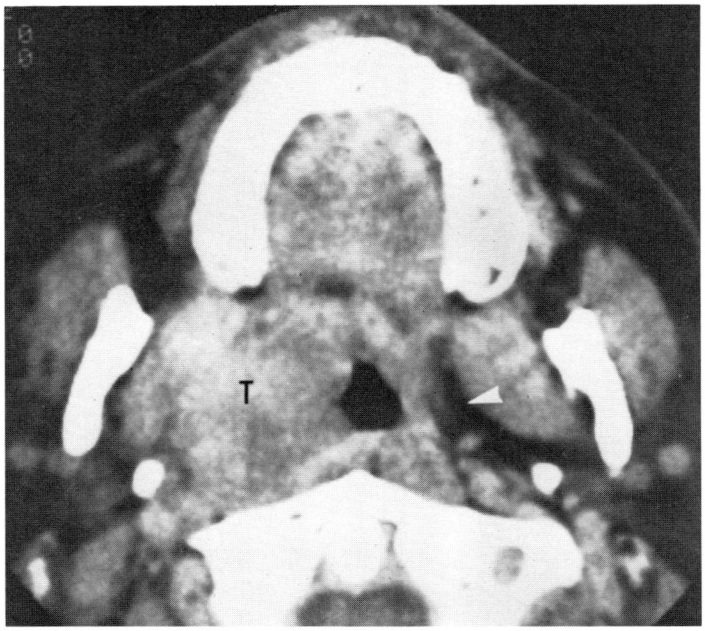

A.

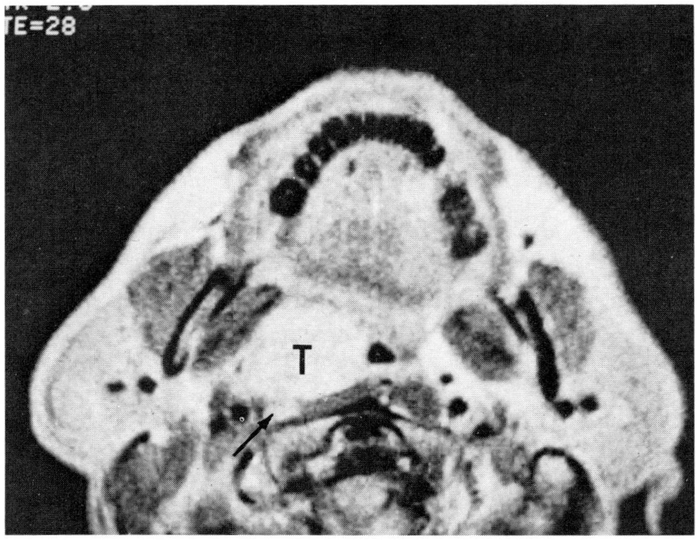

B.

Figure 6.

A. Axial CT demonstrates right nasopharyngeal soft-tissue mass (T) invading parapharyngeal fat. Contrast between tumor and prevertebral and pterygoid musculature is suboptimal. The carotid sheath is effaced on the right. (arrowhead, parapharyngeal fat)

B. Axial spin echo image depicts tumor (T) displacing prevertebral and medial pterygoid muscle and invading the soft palate. A focal high-intensity structure (arrow) is in anatomic position of the medial retropharyngeal nodal chain and probably represents adenopathy (patient received only radiation therapy).

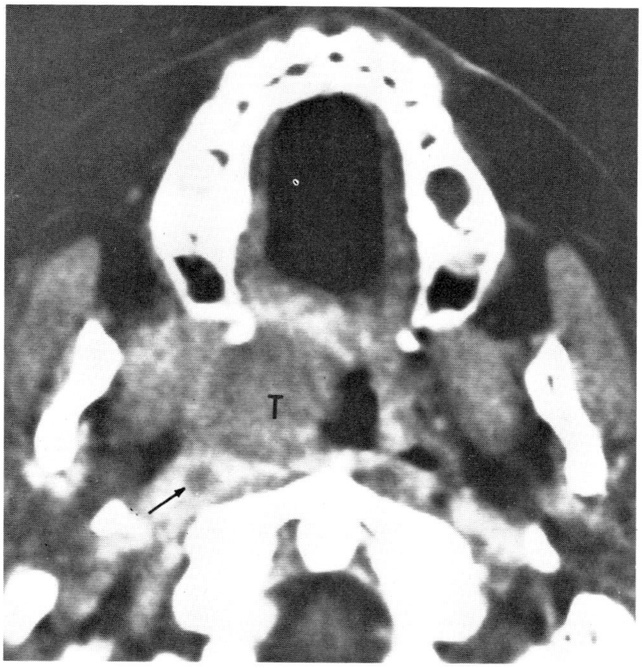

A.

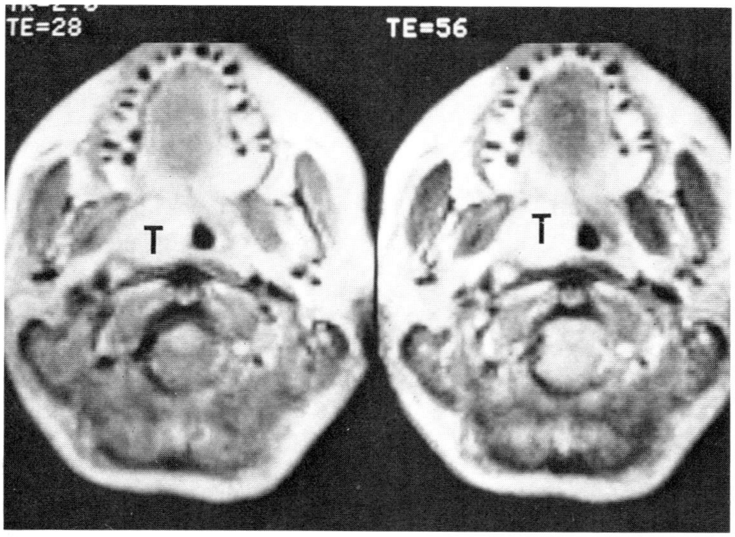

B.

Figure 7.

A. A CT scan of 60-year-old woman with primary histocytic lymphoma of the nasopharynx (T). A localized mass obliterates the parapharyngeal space. A low density mass near the right carotid sheath is suspicious for adenopathy (arrow).

B. First (left) and second (right) echoes (TR of 2.0 sec, TE of 28, 56 msec) from a spin echo sequence. The high-intensity tumor and retropharyngeal node are easily differentiated from surrounding soft tissues and vessels. Little change in the intensity between the first and second echo reflects a paucity of T_2 relaxation time effect on image intensity.

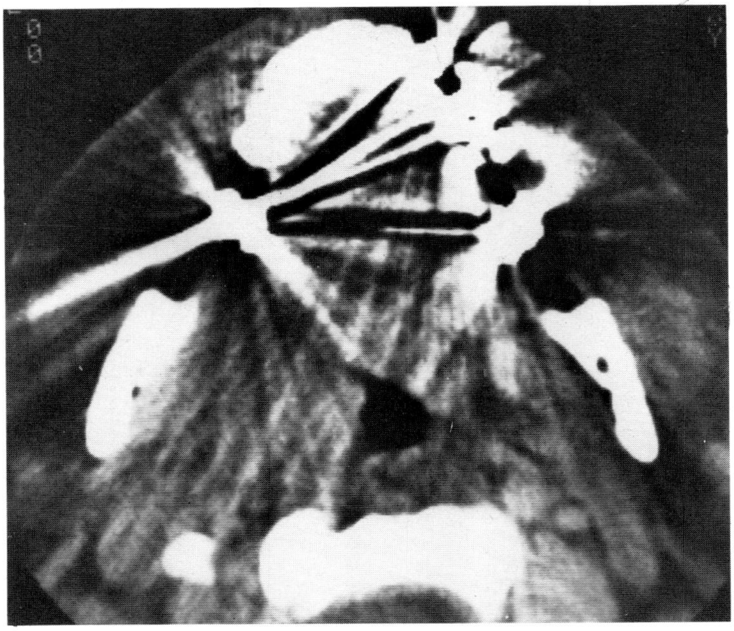

A.

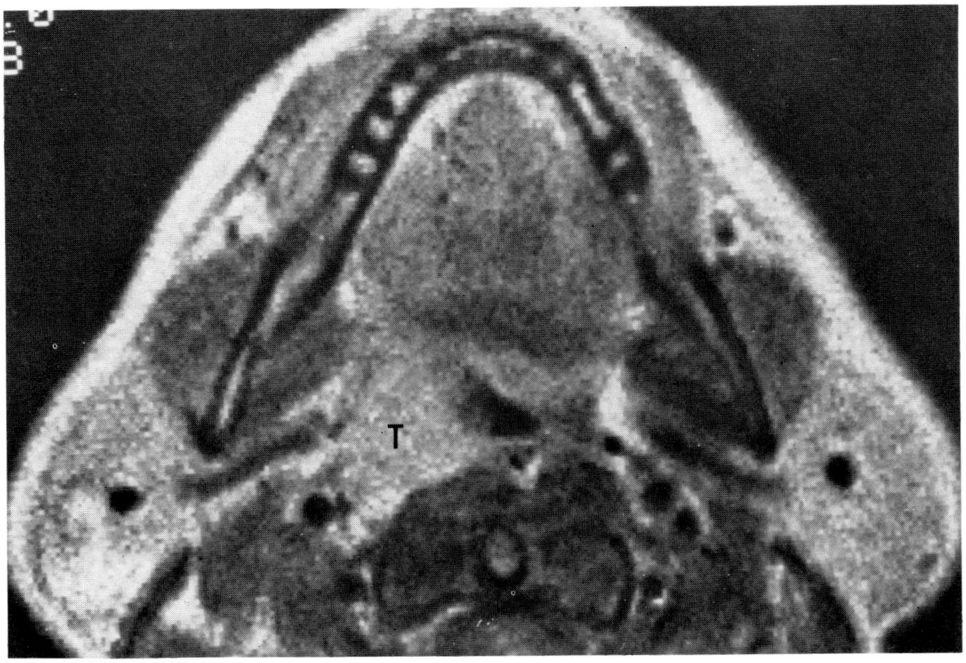

B.

Figure 8.
A. Axial CT image demonstrates marked dental filling artifact, which obscures pathologic condition in right oropharynx.
B. Same patient as in A. Nasopharyngeal carcinoma (T) extending to right glossopharyngeal sulcus is depicted without obscuring artifact from nonferromagnetic dental fillings.

REFERENCES

1. Mancuso AA, Bohman L, Hanafee W, et al: Computed tomography of the nasopharynx: Normal and variants of normal. Radiology 137:113, 1980.
2. Silver JA, Mawad ME, Hilal SK, et al: Computed tomography of the nasopharynx and related spaces. Radiology 147:725, 1983.
3. Mancuso AA, Hanafee WN: *Computed Tomography of the Head and Neck*. Baltimore: Williams and Wilkins, 1982.
4. Bydden GM, Steiner RE, Young IR, et al: Clinical NMR imaging of the brain: 140 cases. AJR 139:215, 1982.
5. Brant-Zawadzki M, Davis P, Crooks L, et al: NMR demonstration of cerebral abnormalities: Comparison with CT. AJR 140:847, 1983.
6. Moon K, Genant H, Helms C, et al: Musculoskeletal applications of nuclear magnetic resonance. Radiology 147:161, 1983.
7. Alfidi RJ, Haaga JR, El Yousef SJ, et al: Preliminary experimental results in humans and animals with a superconducting, whole-body nuclear magnetic resonance scanner. Radiology 143:175, 1982.
8. Crooks L, Arakawa M, Hoenninger J, et al: Nuclear magnetic resonance whole-body imager operating at 3.5 kGauss. Radiology 143:169, 1982.
9. Crooks L, Ortendahl D, Kaufman L, et al: Clinical efficiency of nuclear magnetic resonance imaging. Radiology 146:123, 1983.
10. Crooks L, Mills C, Davis P, et al: Visualization of cerebral and vascular abnormalities by NMR imaging: The effects of imaging parameters on contrast. Radiology 144:843, 1982.
11. New PFJ, Rosen BR, Brady TJ, et al: Potential hazards and artifacts of ferromagnetic and nonferromagnetic surgical and dental materials and devices in nuclear magnetic resonance imaging. Radiology 147:139, 1983.

10

Neck

DAVID D. STARK, M.D.
GORDON GAMSU, M.D.
ALBERT A. MOSS, M.D.

Nuclear magnetic resonance imaging is particularly suitable for demonstrating the anatomy of the neck. As a cross-sectional imaging modality, NMR has advantages similar to those of radiographic CT and ultrasound. Computed tomography has proved valuable in the diagnosis of laryngeal lesions, providing information about deep laryngeal structures that is unavailable by conventional radiographic techniques or endoscopy.[1] Similarly, CT and ultrasound can be helpful in the evaluation of thyroid and parathyroid lesions.[2-5] However, CT requires the administration of intravenous contrast material to allow identification of vessels, and image quality is often degraded by motion-related streak artifacts. Ultrasound suffers from limited acoustic beam penetration and cannot be used to evaluate the larynx or superior mediastinum.

Although the spatial resolution of NMR is currently a limiting factor, these inherent limitations of CT and ultrasound do not affect NMR. Nuclear magnetic resonance offers excellent soft-tissue contrast without the need for intravenous contrast material, is resistant to motion-related artifacts, and can be used to examine the entire neck and thoracic inlet. This chapter reviews our initial experience with NMR imaging of the normal and pathologic anatomy of the neck and thoracic inlet.

Technique for Nuclear Magnetic Resonance Imaging of the Neck

The NMR spin echo and inversion recovery images referred to in this chapter were obtained by using a 0.35-T imaging system that has been described in detail.[6,7] The spin echo sequence repetition rate (TR) can be varied from 0.5 to 2.0 sec, and the echo delay (TE) can be either 28 or 56 msec. The inversion recovery sequence TR can be either 1.0 or 1.8 sec, and the inversion delay (TI) is either 420 or 278 msec, respectively. A specially designed elliptical coil with a minor axis of 30 cm and a major axis of 58 cm is used for neck examinations. Standard images with 128 y-axis lines and 256 x-axis lines are obtained by using a resolution element of 1.7 × 1.7 × 7 mm. Standard images have a 22- × 44-cm field of view. High-resolution images with 256 y-axis lines and 256 x-axis lines are obtained by using a resolution element of 0.8 × 0.8 × 7 mm. High-resolution images have a 20- × 20-cm field of view. Multiple anatomic sections are imposed simultaneously; however, the 7-mm sections are separated from each other by a gap of 4 mm.

The normal cervical and thoracic inlet structures usually seen on CT scans are readily identified on axial spin echo images. The spin echo technique with TR of 2.0 sec and TE of 28 msec has the greatest signal-to-noise ratio, and the high-resolution technique has better spatial resolution than the standard technique (Fig. 1). Although the spatial resolution of NMR is slightly inferior to high-resolution CT, the CT images are frequently degraded by streak artifacts caused by swallowing or respiratory motion and/or x-ray beam hardening. Unlike CT, NMR is only minimally affected by the relatively limited movements that occur during NMR imaging sequences that vary from 4.5 to 17 min.

Normal Anatomy

The neck can be defined as the region of the body that lies between the plane demarcated by the lower margin of the mandible and the superior nuchal line of the occipital bone, and the plane demarcated by the suprasternal notch and first thoracic vertebral body.[8]

Fascial planes divide the neck into several compartments. For the interpretation of axial images, it is useful to define three anatomic compartments: visceral, posterior, and lateral.[9] The visceral compartment of the neck is the most anterior and contains the structures of the aerodigestive tract, including the larynx, trachea, and esophagus. The thyroid gland also lies within the visceral compartment. The infrahyoid strap muscles and sternocleidomastoid muscles form the anterior and lateral boundaries, respectively, of the visceral compartment.

The structures of the visceral compartment of the neck are well demonstrated by NMR when using morphological criteria analogous to those used with CT. Near the level of the hyoid bone (Figs. 1A, 2), the epiglottis, valleculae, and hypopharynx can be identified. The epiglottis and pharyngeal constrictor muscles have an intermediate intensity similar to that of skeletal muscle (Fig. 1A) and can be differentiated from the fatty tissues of the pre-epiglottic space and aryepiglottic folds, which have a high intensity, approaching the intensity of subcutaneous fat (Fig. 1B). At the level of the laryngeal vestibule the paralaryngeal fatty tissues merge with the false vocal cords (Fig. 1C). The true vocal cords are seen as symmetric structures of intermediate intensity at the level of the arytenoids (Fig. 1D). The intensity of the true vocal cords is lower than that of the false vocal cords. Presumably this is because the vocalis muscle reduces the amount of paralaryngeal fat within the true vocal cords.

The pyriform sinuses are well seen as symmetric elliptical structures with the intensity of air lateral to the laryngeal vestibule and medial to the thyroid cartilage (Fig. 1B). The cartilaginous portions of the laryngeal skeleton (thyroid cartilage, arytenoids, and cricoid) (Figs. 1C–1F) are vaguely delineated as intermediate-intensity linear structures that are indistinguishable from fascial planes. However, calcified cartilage has a distinctive low intensity, and the fatty medullary cavities of the thyroid cartilage, arytenoids, and cricoid are seen as high-intensity regions (Figs. 1C–1F, 2). The normal high-intensity medullary cavity of the cricoid is separated from the lumen of the subglottic larynx by a 1- to 2-mm intermediate-intensity zone. Cartilaginous tracheal rings cannot be identified on either axial or sagittal images.

The collapsed muscular hypopharynx anterior to the spine has a lower intensity than that of the fatty paralaryngeal tissues (Figs. 1C, 1D). The esophagus (Figs. 1G, 1H) is identified as a cylindrical structure of intermediate intensity immediately posterior to the trachea below the level of the cricoid cartilage (Fig. 1F). At the thoracic inlet the esophagus deviates to the left of the midline. The lumen of the esophagus is readily identified when distended with air or fluid (Figs. 1G, 1H).

The normal thyroid gland is readily differentiated from the sternothyroid muscles because of its greater intensity (Fig. 1F). The thyroid is seen on two to four contiguous sections as a homogeneous and symmetric wedge-shaped structure anterior and lateral to the trachea. The intensity of the thyroid gland increases relative to muscle on images obtained with long TR and TE because the thyroid has a slightly longer T_1 and markedly longer T_2 than muscle.

The posterior compartment of the neck includes the cervical vertebrae, posterior extensor muscles, and anterior flexor muscles, including the scalenes and the longus colli muscles. These muscle groups have an intermediate intensity and are separated by adipose tissue of high intensity (Figs. 1A–1F). The fascial tissue investing major muscle groups (Fig. 1D) and the fibrous tissue of the nuchal ligament (Figs. 1A, 1B) have a similar low intensity. Dense cortical bone of the vertebral bodies, posterior elements, and ribs also has a low intensity. However, bone marrow within the central cancellous bone has a high intensity, approaching the intensity of fat (Fig. 1C). Cerebrospinal fluid is seen as a low-intensity region surrounding the spinal cord, which has an intermediate intensity (Fig. 1G).

The lateral compartments of the neck contain the carotid sheaths. The cervical carotid arteries and jugular veins are easily distinguished from adjacent fat, muscle, and lymph nodes because of the low intensity of flowing blood, which is black on the NMR image gray scale (Fig. 1F). Because of the superb contrast resolution and absence of streak artifacts, NMR is superior to CT for delineation of vessels at the thoracic inlet. The major cervical vessels can be traced to the innominate arteries or veins in all patients (Figs. 1G, 1H).

Normal-size cervical lymph nodes are rarely seen with NMR because of the limited spatial resolution and volume averaging with surrounding fat. In one patient, a 1.0- × 0.5-cm node appeared as an intermediate-intensity structure in the thoracic inlet

A.

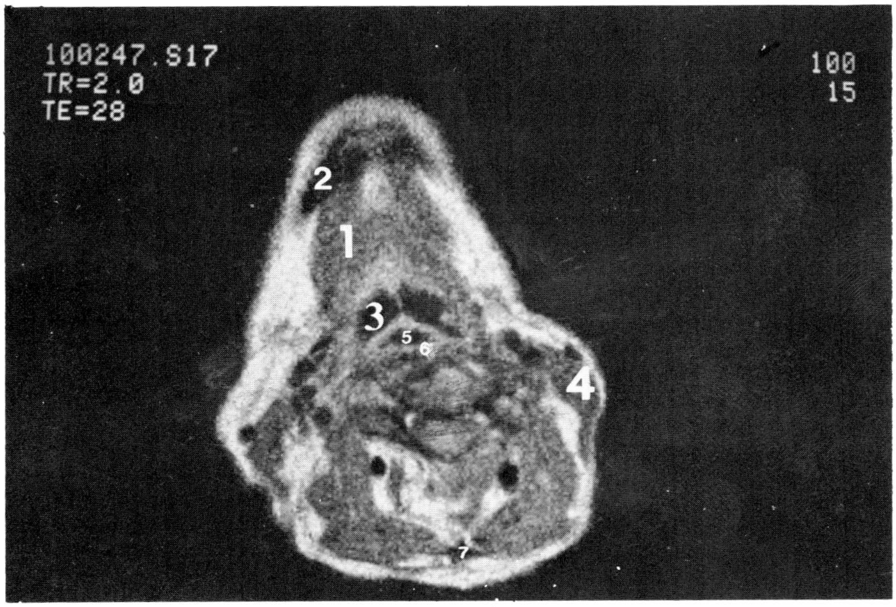

B.

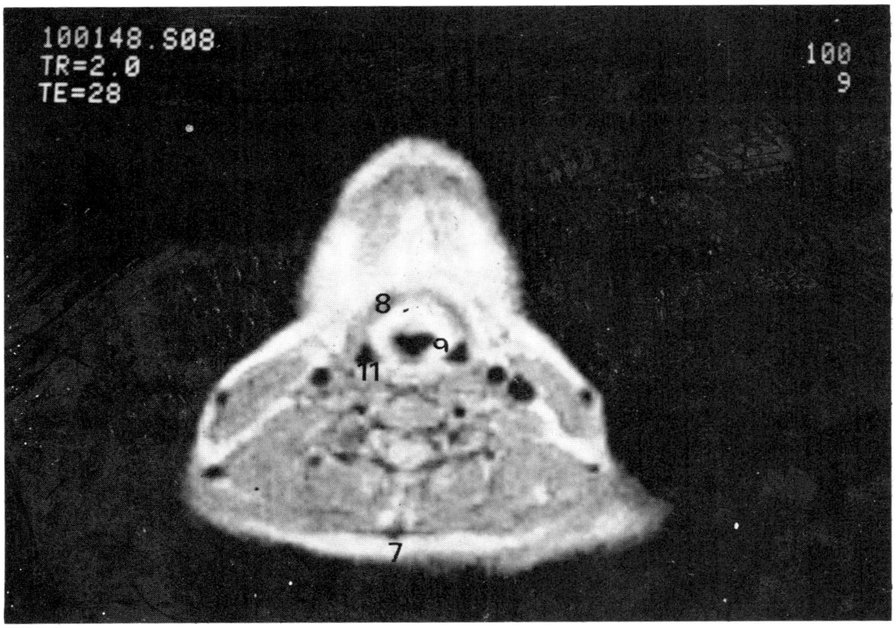

Figure 1. Normal axial anatomy displayed by using high-resolution spin echo technique (TR of 1.5 or 2.0 sec, TE of 28 msec).

A. Axial high-resolution NMR image at the cranial border of the neck. The myelohyoid muscle (1) is seen inserting into the mandible (2). The valleculae (3) are of the same intensity as air and are separated by the median glossoepiglottic fold. The epiglottis and hypopharyngeal constrictor muscles (6) have an intermediate intensity, slightly greater than that of the sternocleidomastoid (4) or other skeletal muscle groups, presumably because of the increased amount of fat within the hypopharyngeal tissues. The nuchal ligament (7) is seen at the midline posteriorly as a low-intensity structure contiguous with the interspinous ligament and fascia of the trapezius muscle. (5, hypopharynx)

B. The infrahyoid strap muscles (8) are separated by subcutaneous fat in the midline. The fatty tissues of the pre-epiglottic space have a high intensity and can be seen extending into the aryepiglottic folds (9) medial to the pyriform sinuses (11), which have the same low intensity as air. (7, nuchal ligament)

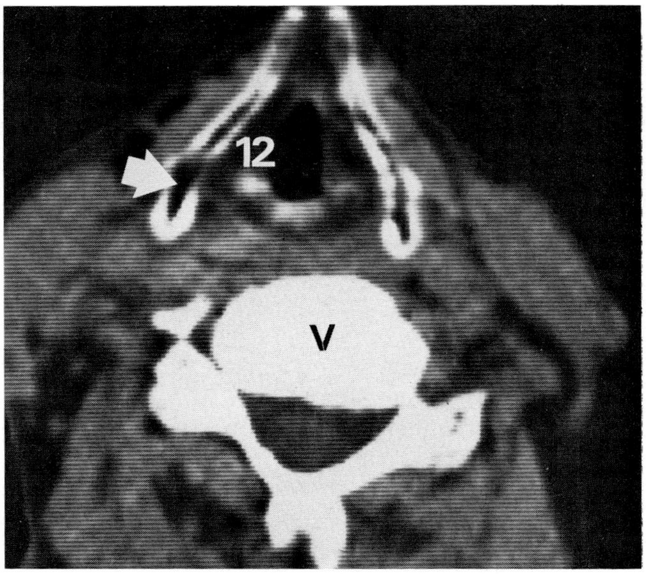

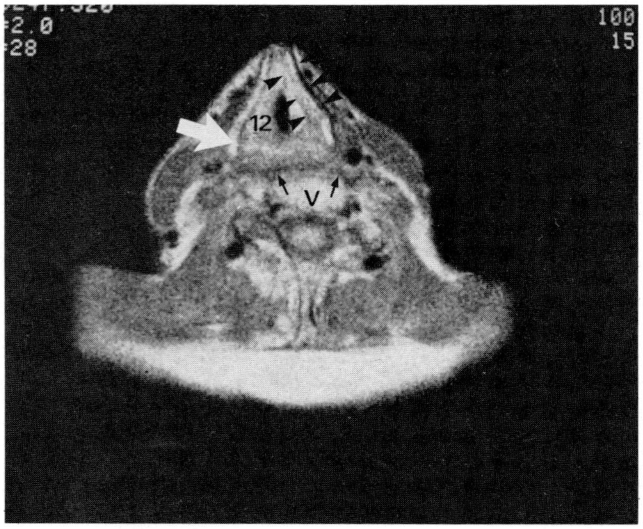

C. On the left, a CT scan demonstrates calcification of the thyroid cartilage surrounding the fatty low-density medullary cavity (arrow). At the level of the laryngeal vestibule the fatty paralaryngeal connective tissues (12) merge with the false vocal cords. The cranial margins of the calcified arytenoids are seen within the posterior portion of the false vocal cords. (V, vertebral body)

On the right, an NMR image demonstrates the calcified margin of the thyroid cartilage as a low-intensity margin (arrowheads) surrounding the high-intensity medullary cavity (arrow). The paralaryngeal fatty tissues (12) have a lower intensity than that of subcutaneous fat, presumably because of the presence of fibrous tissue and muscle. Marrow within the cancellous bone of a vertebral body (V) has a high intensity, surrounded by low-intensity compact bone of the cortex (small arrows).

D. 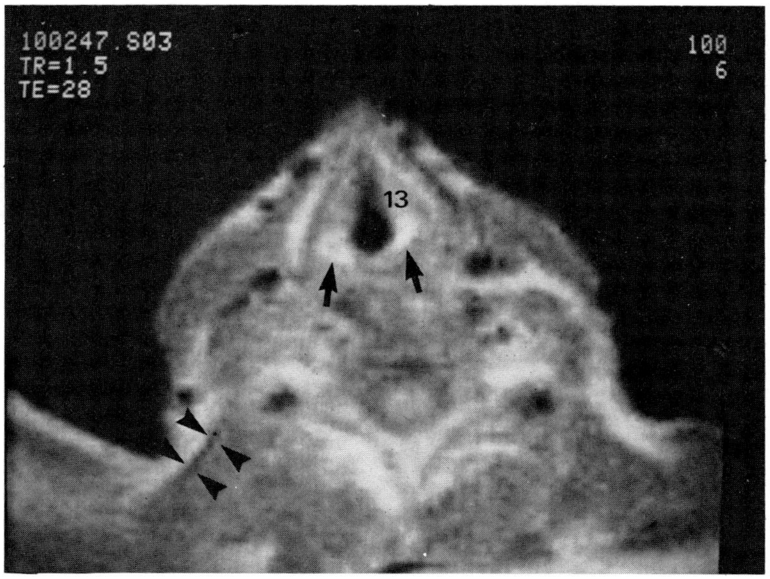

D. Standard-resolution NMR image. The true vocal cords (13) have a lower intensity than the false vocal cords (compare with C), presumably because of replacement of paralaryngeal fatty tissue by the vocalis muscle. The fatty medullary cavities of the arytenoids (arrows) have a high intensity. Low-intensity fascial tissue is shown investing an extensor muscle group (arrowheads).

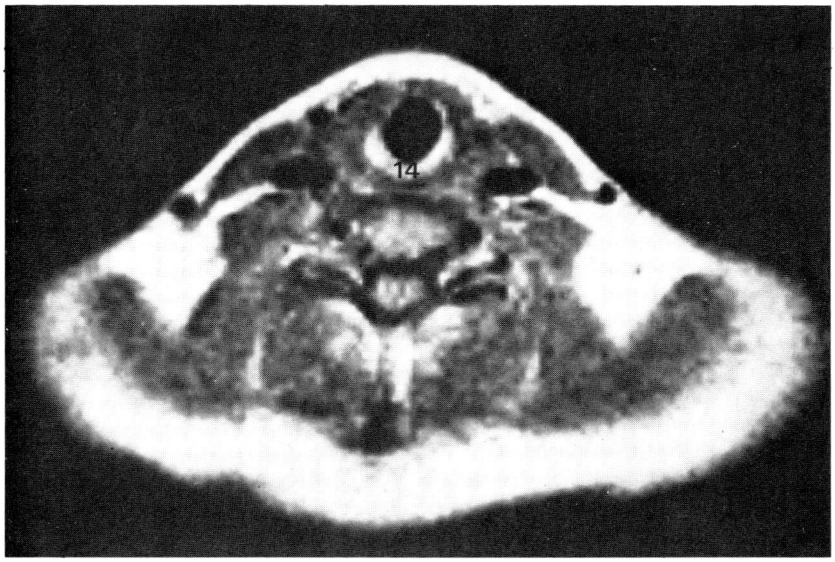

E.

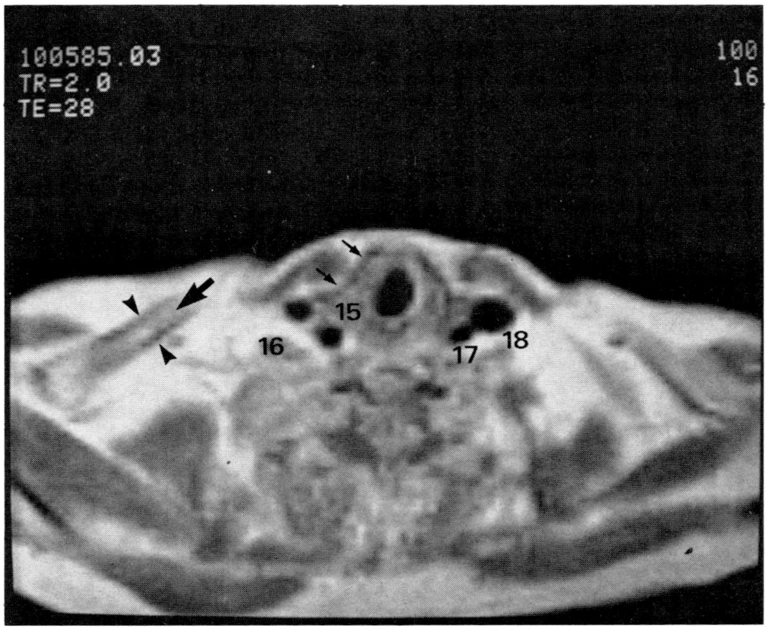

F.

E. The medullary cavity of the cricoid cartilage (14) is seen as a high-intensity structure separated from the lumen of the subglottic hypopharynx by an intermediate-intensity stripe less than 2 mm in width.

F. The thyroid gland (15) is located immediately posterior to the sternothyroid muscle (small arrows) and has a higher intensity than muscle on this spin echo image obtained using TR of 2 sec and TE of 28 msec. Bone marrow within the medullary cavity of the clavicle (large arrow) is surrounded by low-intensity cortical bone (arrowheads). (16, anterior scalene muscle; 17, carotid artery; 18, jugular vein)

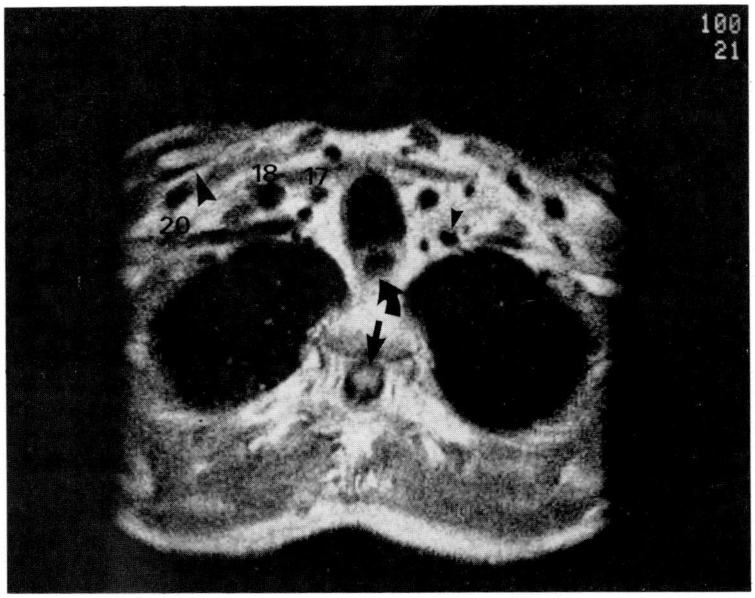

G.

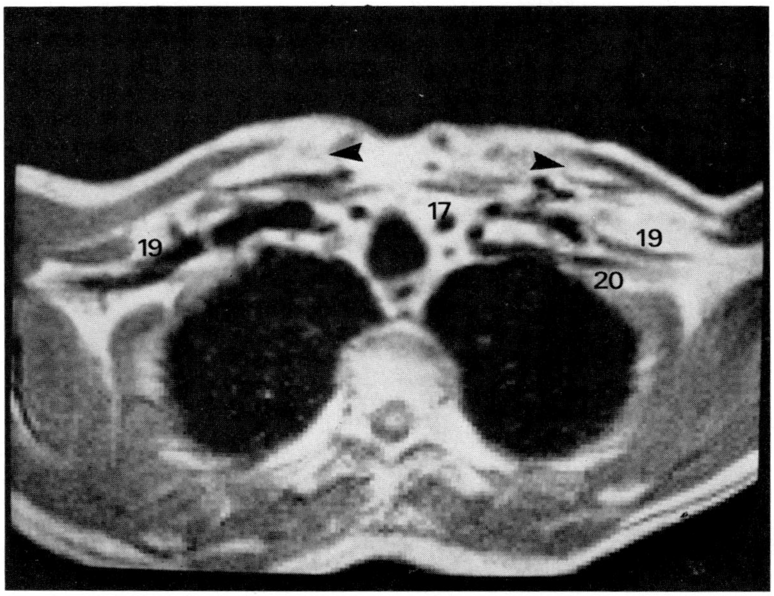

H.

G. High-resolution image at the level of the thoracic inlet. The muscular wall of the esophagus (curved arrow) can be differentiated from the gas-filled lumen. All of the vascular structures are well seen (17, carotid artery; 18, jugular vein; 20, subclavian artery). Vertebral artery (small arrowhead), right clavicle (large arrowhead), and thoracic spinal cord (small arrow) are surrounded by low-intensity cerebrospinal fluid.

H. Standard-resolution NMR image 1 cm caudal to F. The subclavian veins (19) are shown entering the thorax anterior to the subclavian arteries (20). (arrowheads, clavicles; 17, carotid artery)

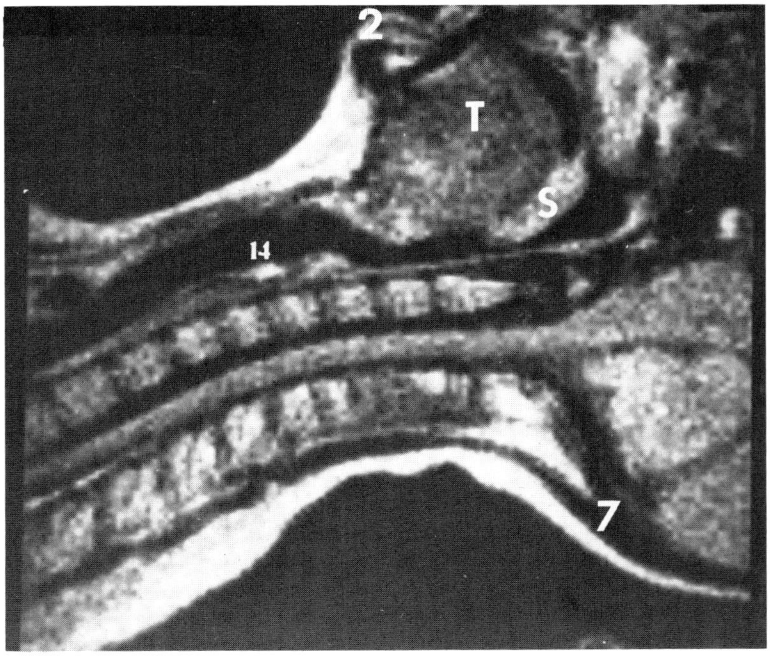

Figure 2. Midline sagittal section shows the mandible (2) or a rim of low-intensity cortical bone surrounding high-intensity bone marrow; tongue (T), soft palate (S), and other aerodigestive tract structures. (7, nuchal liagment; 14, cricoid cartilage)

adjacent to the brachiocephalic artery. The node has a long T_1 and long T_2, similar to those of the thyroid gland, and similarly increased in intensity relative to muscle at long TR or long TE.

Pathology

Thyroid Gland

Palpable thyroid nodules are usually evaluated by radionuclide scintigraphy, percutaneous needle biopsy, or both. Computed tomography and ultrasound are supplemental tests that can influence the management of thyroid nodules by demonstrating cystic areas or by detecting multiple nonpalpable nodules,[2] findings that reduce the likelihood that the lesion is malignant. Computed tomography is also useful in assessing cervical and mediastinal extension of malignant and benign thyroid neoplasms.[3,4] Unfortunately, CT is often limited by streak artifacts caused by swallowing or respiratory motion and x-ray beam hardening. Furthermore, large amounts of iodinated contrast material must be injected to identify the complex vascular structures of the neck and thoracic inlet. Ultrasound is superior to CT for visualization of the cervical thyroid gland;[2] however, the lower pole of the thyroid gland often lies within the thoracic inlet, where bone and air limit acoustic transmission. Nuclear magnetic resonance has the combined advantages of CT and ultrasound in that both the neck and thoracic inlet can be examined without streak artifacts and vascular structures can be separated from soft-tissue masses without the infusion of intravenous contrast material.

Nuclear magnetic resonance can differentiate thyroid nodules from normal thyroid tissue with spin echo technique with long TR and TE. Solid thyroid nodules increase in intensity relative to normal thyroid tissue on images obtained with long. TR and TE because of the longer T_1 and T_2 relaxation times of the nodules. Colloid cysts have a longer T_1 and T_2 than solid tissues do (Fig. 3). This difference in relaxation times can be correlated with the water content of the cyst fluid, which is greater than the water content of solid thyroid nodules. The protein content of colloid cyst fluid can be greater than the protein content of thyroid nodules. Spin echo images can identify fibrous tissue planes and fibrous septations within thyroid cysts (Fig. 3C). Fibrous septations are seen as low-intensity structures similar to fascial tissue.

Intrathoracic extension of a goiter is identified by NMR using criteria analogous to those used with CT.[4] However, NMR images are free of artifacts, and NMR demonstrates the patency of displaced vascular structures. The inversion recovery techniques are less useful than the spin echo techniques for

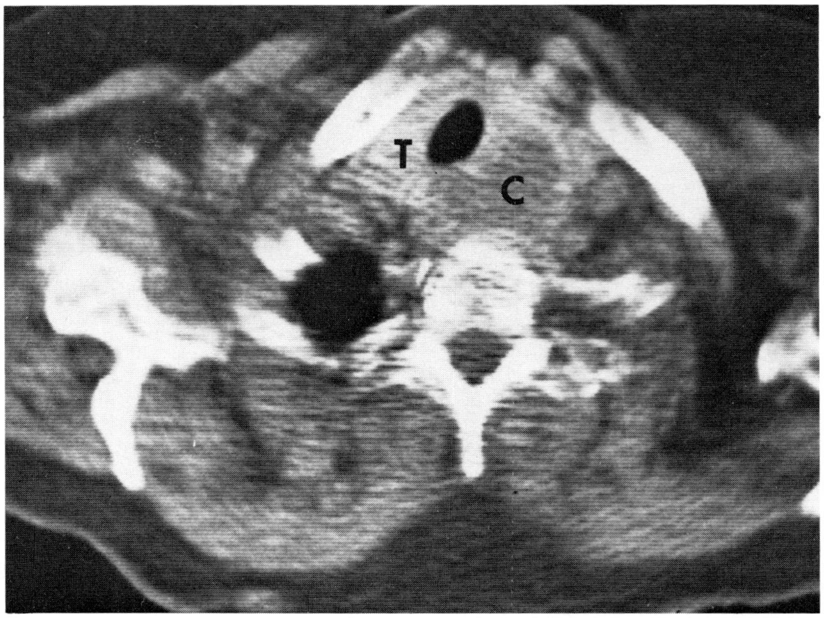

A.

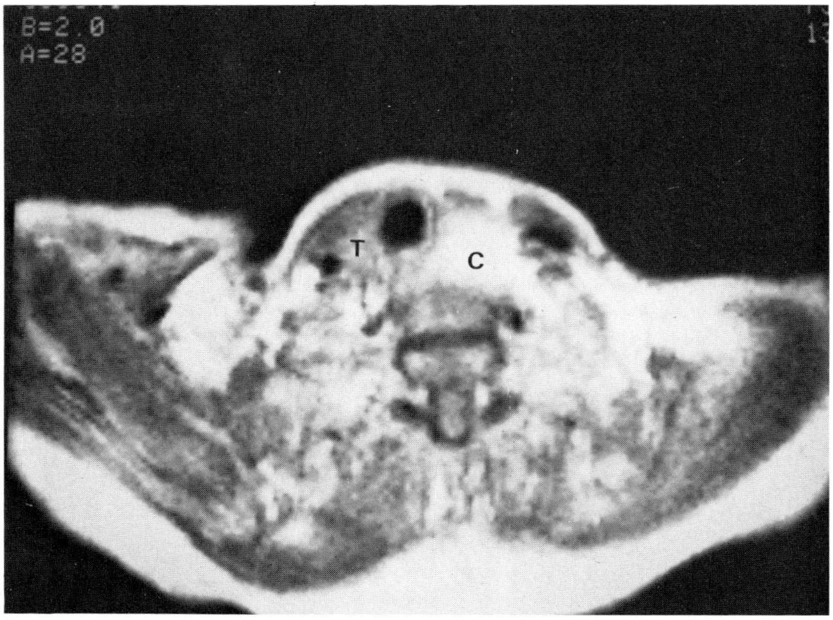

B.

Figure 3.

A. A CT scan demonstrates a cystic mass (C) arising from the thyroid gland (T).

B. Standard-resolution spin echo NMR image obtained using TR of 2.0 sec and TE of 28 msec demonstrates the cystic mass (C) as a high-intensity structure relative to the thyroid (T) gland because the cyst fluid has a long T_1 and T_2.

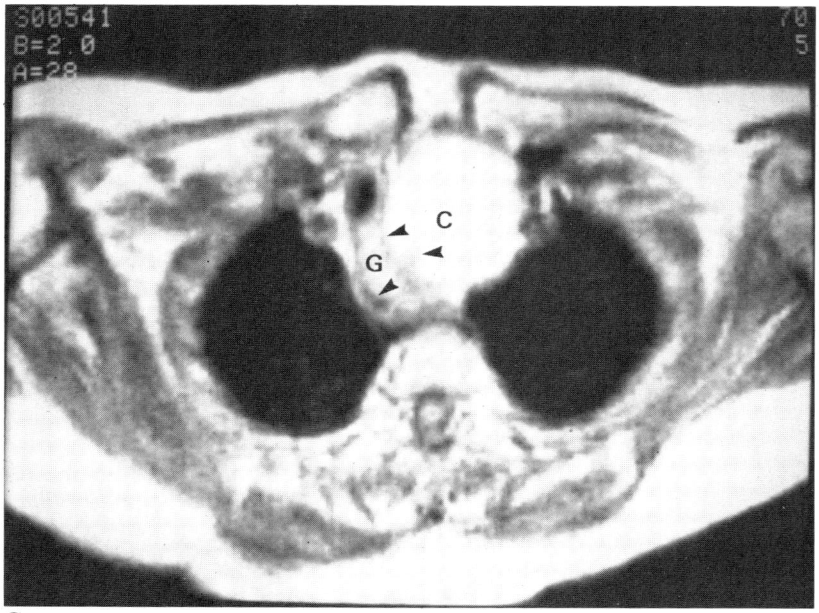

C.

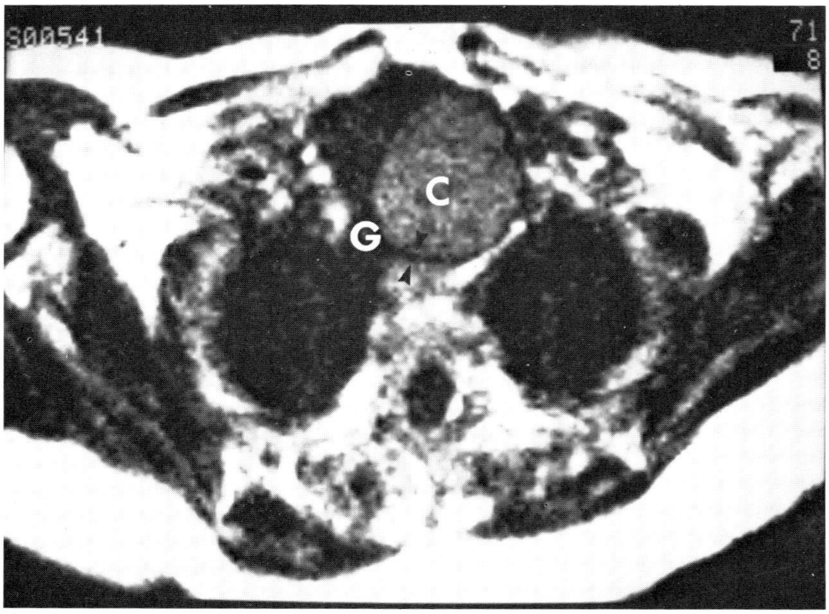

D.

C. Caudal to B, NMR differentiates the cystic mass (C) from the solid goiter (G). Fibrous tissue planes and the fibrous septation within the cyst have a relatively low intensity (arrowheads).

D. Inversion recovery image (TR of 1.8 sec and TE of 278 msec) at the same level as C identifies the cyst (C) as intermediate-intensity structure surrounded by a low-intensity capsule (arrowheads). However, the solid portion of the goiter (G) cannot be differentiated from blood vessels, lung, or muscle.

differentiating thyroid tumors from muscle and thyroid gland because all of these tissues have long T_1 relaxation times (Fig. 3D). Furthermore, because of their low signal intensity on the inversion recovery images, these tissues are difficult to distinguish from vascular structures.

Parathyroid Glands

Preoperative localization of parathyroid tumors is indicated when the surgeon wishes to perform a unilateral dissection or whenever the patient has undergone previous neck surgery. Although both ultrasound and CT are useful localizing tests, accuracy rates have been limited to 60%–80% because of the small size (an average of 1.5 cm) of parathyroid tumors. The resolving power of ultrasound and CT is not limited by spatial resolution, but rather by technical limitations such as acoustic beam penetration, radiographic streak artifacts, and poor tissue contrast.[5]

Normal parathyroid glands cannot be identified by either NMR or CT. Computed tomography can identify parathyroid tumors as small as 0.6 cm; however, to date only tumors larger than 1 cm in size have been identified by NMR. Although the spatial resolution of NMR is slightly inferior to that of CT, the absence of streak artifact and ease of identification of blood vessels are advantages. Spin echo technique readily differentiates parathyroid tumors from adjacent vessels, muscle, fat, and thyroid gland (Fig. 4). Parathyroid tumors have a long T_1 and can be differentiated from fat, which has a short T_1, by using spin echo techniques with short TR (Fig. 4C) or inversion recovery techniques. Parathyroid tumors also have a long T_2 and can be differentiated from the thyroid gland, which has a shorter T_1 and T_2, by using spin echo techniques with long TR and TE (Fig. 4C). Parathyroid adenomas have a slightly longer relaxation time than hyperplastic parathyroid glands, but this difference has not been useful in individual cases. To date, parathyroid adenomas, hyperplastic glands, and parathyroid carcinoma have not been discriminated by intensity or T_1 or T_2 measurements.

Lymph Nodes

Pathologic lymph nodes are detected by NMR by using morphological criteria analogous to those used with CT (Fig. 5). Intensity measurements, T_1, and T_2 appear to be similar for normal lymph nodes (less than 1 cm in size) and pathologic lymph nodes (greater than 1.5 cm in size).

Enlarged lymph nodes due to Hodgkin's disease, non-Hodgkin's lymphoma, and active tuberculosis all have a similar intensity, T_1, and T_2. The T_1 and T_2 are slightly longer than those of thyroid nodules.

A.

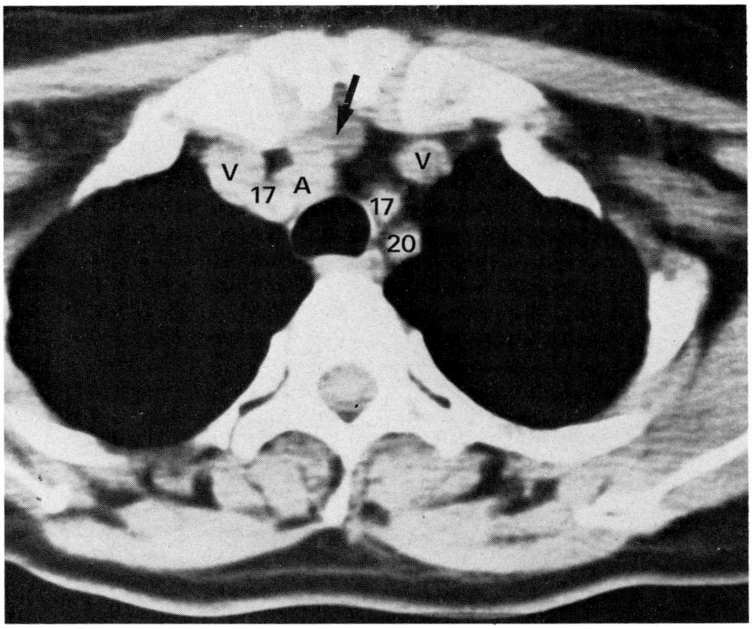

Figure 4.

A. A CT scan at the level of the thoracic inlet fails to differentiate a parathyroid adenoma (arrow) from adjacent sternothyroid muscle or vascular structures. Brachiocephalic veins (V) and artery (A) are seen. Calcifications are seen at the origin of the subclavian artery (20). (17, carotid artery)

B.

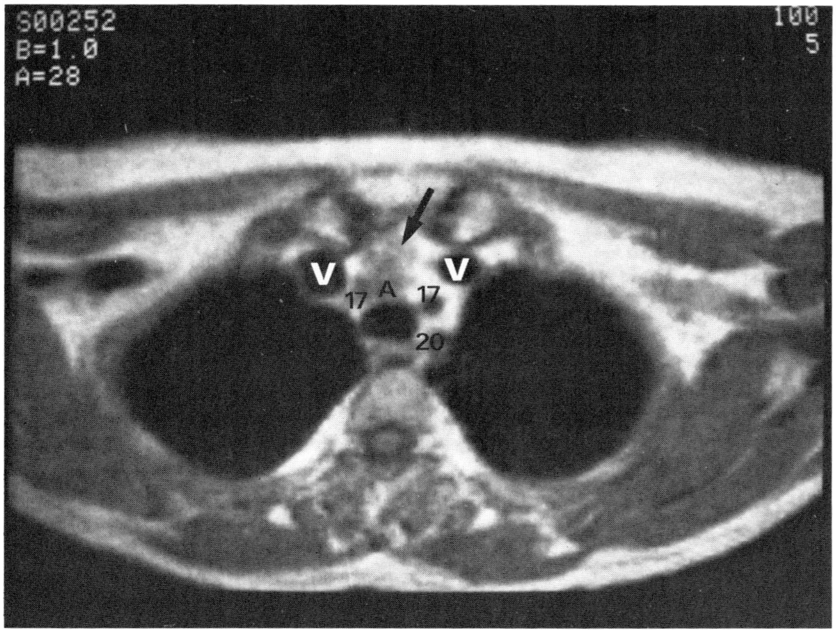

C.

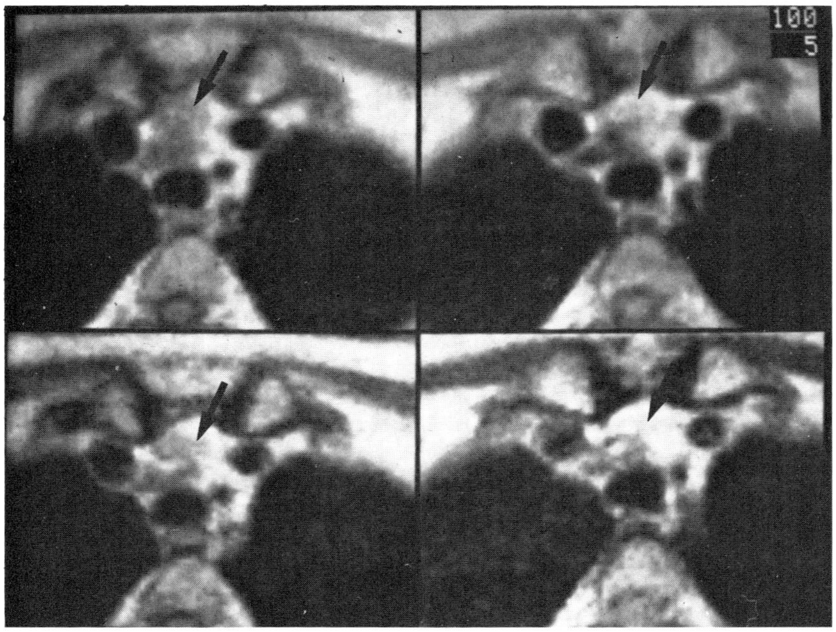

B. An NMR image differentiates the parathyroid adenoma (arrow) from adjacent fat and vessels. The brachiocephalic artery (A) has increased intensity, presumably because of turbulent blood flow at the origin of the subclavian artery (20). The calcification is not detected by NMR. (17, carotid artery; V, brachiocephalic veins)

C. Top left, TR is 1.0 sec and TE is 28 msec; top right, TR is 2.0 sec, TE is 28 msec; bottom left, TR is 1.0 sec, TE is 56 msec; bottom right, TR is 2.0 sec, TE is 56 msec. The parathyroid adenoma (arrow) increases in intensity relative to fat at long TR and TE, indicating that the adenoma has a long T_1 and T_2.

A.

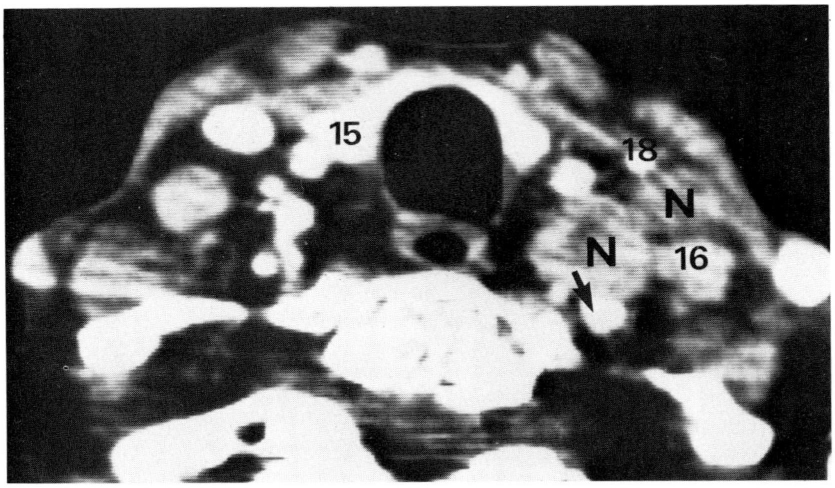

B.

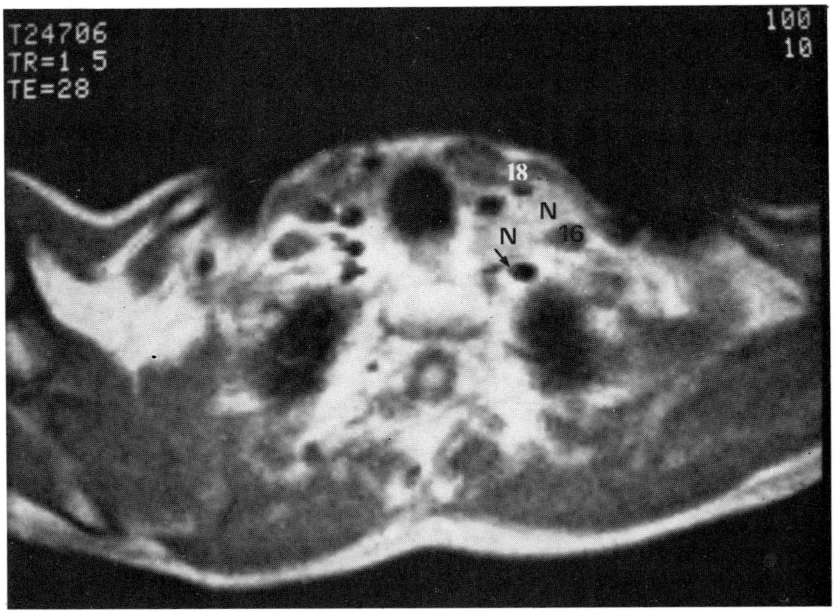

Figure 5.

A. A CT scan following administration of iodinated contrast material demonstrates pathologically enlarged lymph nodes (N) displacing the internal jugular vein (18) and vertebral artery (arrow). The anterior scalene muscle (16) cannot be differentiated from the lymph nodes. (15, thyroid gland)

B. An NMR spin echo image obtained using TR of 1.5 sec and TE of 28 msec. Lymph nodes (N) can be differentiated from adjacent fat, vessels, and muscle. T_1 contrast is reduced because of the long TR. Fat is the brightest tissue because it has a short T_1 and is fully remagnetized between pulse repetitions (TR); additionally, fat has a long T_2 and maintains most of its intensity at TE of 28 msec when the signal is obtained. The lymph node has a long T_2, but also has a long T_1 and, therefore, is not fully remagnetized between pulses and has a slightly lower intensity than fat does. Thyroid tissue is less intense than the lymph nodes and muscle is less intense than the thyroid because of shorter T_2 relaxation times. (16, anterior scalene muscle; 18, jugular vein)

Visualization of tissue planes separating lymph nodes from the thyroid gland, muscles, and vessels aids in their identification when using morphological criteria analogous to those used with CT. Fatty tissue planes are best detected by using spin echo techniques with a short TR or inversion recovery technique to maximize contrast between fat that has a short T_1 and pathologic tissues that have a long T_1. Muscle has a long T_1 similar to that of lymph nodes, and therefore both muscle and lymph nodes have low intensity on spin echo images with short TR and on inversion recovery images. However, on spin echo images with a long TR or a long TE, lymph nodes have a greater intensity than muscle, with its shorter T_2 (Fig. 5B).

A major advantage of NMR over CT is differentiation of lymph nodes and muscles from tortuous vessels in the neck and thoracic inlet. Because of blood flow, vascular structures are usually seen as very low-intensity structures. However, when blood flow is abnormally slow or turbulent, particularly in venous structures, increased signal intensity can be identified within the vascular lumen (Fig. 4B). Blood vessels with focal regions of increased intraluminal signal intensity can be differentiated from other soft-tissue structures by examining adjacent sections to demonstrate continuity with other blood vessels.

Overview

Imaging techniques that accentuate contrast (intensity differences) between tissues with different T_1 relaxation times are useful in detecting lesions surrounded by adipose tissue or in delineating fatty tissue planes. Therefore, spin echo technique with short TR or inversion recovery techniques are useful initial imaging procedures because of the excellent contrast between fat (which has a high intensity as a result of its short T_1) and pathologic tissues (which have a low intensity because of their long T_1). Imaging techniques that accentuate contrast between tissues with different T_2 relaxation times are useful in differentiating various normal and pathologic tissues. Spin echo techniques with long TR allow full recovery of magnetization between pulses and reduce image contrast due to T_1 differences. The use of a long TE results in low signal intensity from tissues with a short T_2 (muscle, thyroid) and a high signal intensity from tissues with a long T_2 (fat, tumors). Therefore, spin echo technique with long TR and long TE accentuates intensity differences between tissues with different T_2 relaxation times. Fascial planes and scar tissue have a paucity of NMR-observable protons and, therefore, have a low intensity that does not change significantly with variations in imaging technique.

Tissue characterization does not yet allow differentiation of benign and malignant processes. T_1 and T_2 values overlap for a variety of thyroid, parathyroid, and lymph node lesions. This finding is consistent with observations in other organ systems that indicate that NMR is not yet tissue specific.[10] It appears that a high tissue-water content correlates best with a long T_1 and T_2 in both neoplastic and inflammatory conditions.[11,12] Although tumors with rapid growth rates tend to have a high water content, current techniques show that there is considerable overlap between benign and malignant neoplasms and between neoplastic and inflammatory lesions. Cystic structures are readily detected by NMR because of their high water content and long T_1 and T_2.

Our preliminary experience indicates that NMR has several potential advantages over CT and ultrasound for imaging normal and pathologic structures in the neck. With improvements in spatial resolution and further experience with tissue characterization, it appears likely that NMR imaging of the neck will become a clinically significant diagnostic modality.

REFERENCES

1. Gamsu G: Computed tomography of the larynx and pyriform sinuses. In *Computed Tomography of the Body*, Moss AA, Gamsu G, Genant H (Eds). Philadelphia: W. B. Saunders, 1983 (in press).
2. Stark DD, Clark OH, Gooding GAW, et al: Localization of thyroid lesions by high resolution ultrasound and high resolution CT in patients with hyperparathyroidism. Surgery (in press).
3. Mancuso AA, Hanafee WW: *Computed Tomography of the Head and Neck*, pp 92-111. Baltimore: Williams and Wilkins, 1982.
4. Glazer GM, Axel L, Moss AA: CT diagnosis of mediastinal thyroid. AJR 138:495, 1982.
5. Stark DD, Gooding GAW, Moss AA, et al: Parathyroid imaging: Comparison of high resolution CT and high resolution ultrasound. AJR (in press).
6. Crooks LE, Arakawa M, Hoenninger J, et al: Nuclear magnetic resonance whole-body imager operating at 3.5 kGauss. Radiology 143:169, 1982.
7. Crooks LE, Ortendahl DA, Kaufman L, et al: Clinical efficiency of nuclear magnetic resonance imaging. Radiology 146:123, 1983.
8. Snell RS: *Clinical Anatomy for Medical Students*, p 619. Boston: Little, Brown, 1973.

9. Mancuso AA: Computed tomography of the neck. In *Computed Tomography of the Body,* Moss AA, Gamsu G, Genant H (Eds). Philadelphia, W. B. Saunders, 1983 (in press).

10. Moss AA, Goldberg HI, Stark DD, et al: Nuclear magnetic resonance imaging of hepatic tumors. (Submitted for publication.)

11. Herfkens RJ, Davis PL, Crooks LE, et al: Nuclear magnetic resonance imaging of the abnormal live rat and correlations with tissue characteristics. Radiology 141:211, 1981.

12. Stark DD, Bass NM, Moss AA, et al: NMR imaging of experimentally induced liver disease. Radiology (in press).

11

Chest

W. RICHARD WEBB, M.D.
GORDON GAMSU, M.D.

A clinical evaluation of NMR in the diagnosis of diseases of the lung, pleura, hila, and mediastinum has been under way at the University of California, San Francisco since mid-1982. Although our experience to date is preliminary, some conclusions can be drawn regarding the utility of NMR in the diagnosis of chest diseases.

Image Acquisition

Imaging techniques that have been used in the NMR evaluation of the chest include inversion recovery, saturation recovery, and spin echo.[1,2] At the University of California, San Francisco, we have relied primarily on spin-echo images because of their excellent resolution and ability to differentiate tissues on the basis of both T_1 and T_2 relaxation times.[2–5]

The imager used in these studies was a 3.5-kGauss (0.35-T) superconducting magnet that has a hydrogen resonant frequency of 15 MHz. A 55-cm body coil gives a resolution element of 1.7, 2.1, or 2.5 mm, depending on patient size.[2,3]

Generally, at least two imaging sequences are performed. The repetition times (TR) vary from 0.5 to 2.0 sec. Repetition times of 0.5 and 1.5 sec are the most commonly used. In each sequence, images are obtained with echo delay times (TE) of 28 and 56 msec.

The imaging sequences with a TR of 1.5 and 2.0 sec provide the best anatomic detail[2] and allow the simultaneous production of images at 15 adjacent levels. This permits the evaluation of a large portion of the chest and a determination of the extent of pathologic lesions. The imaging sequence of five slices with a TR of 0.5 sec is usually performed in the area of greatest interest and allows the discrimination of tissues on the basis of T_1 differences.[2,4,5] The images obtained at each level with different TE values (28 or 56 msec) allow a determination of T_2 differences.

Coronal and sagittal NMR images (Fig. 1) are sometimes useful in defining the extent of a pathologic process that is more difficult to evaluate in the transaxial plane, and they have a spatial resolution superior

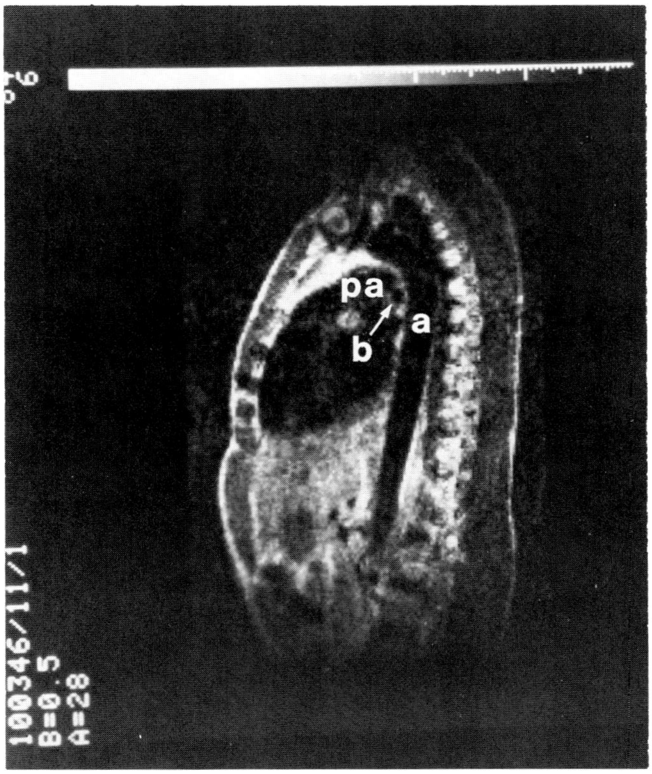

Figure 1. Normal sagittal image. A direct sagittal image in a normal patient shows the main pulmonary artery (pa), left main bronchus (b), and descending aorta (a). The origin of the left subclavian artery from the superior margin of the aortic arch is also visible.

to that of reformatted CT. Also, lesions involving vessels lying primarily in the coronal or sagittal planes are very well demonstrated.

Although we obtain coronal or sagittal images in one-third to one-half of our cases, transaxial images are always performed; they offer the advantage that they are more readily comparable with axial CT images. However, one word of caution is necessary. Because CT is usually performed at a total lung capacity (full inspiration) and NMR is performed at a lower lung volume during quiet breathing, CT and NMR scans may not be entirely comparable on a centimeter-to-centimeter basis. For example, on NMR the lung hila are usually several centimeters more cephalad relative to the suprasternal notch than they appear on CT scans.

Normal Findings

In normal subjects, most mediastinal structures visible on CT are also visible using NMR, but their resolution is somewhat less, probably as a result of the combination of slightly larger pixel size and the respiratory motion inherent in the much longer scan times.[2] Although electrocardiographic gating of the NMR image can enhance the resolution of hilar and mediastinal structures (Fig. 2), our experience with this technique is limited. Respiratory gating may offer some advantage in improving the resolution of the lungs and hila.

The aortic arch and its large branches are always seen on spin echo images because their blood-filled lumina produce no signal, clearly distinguishing them from the intense signal of mediastinal fat (Fig. 3). Similarly, the subclavian veins, brachiocephalic veins, and vena cava are always visible. The azygos vein or arch is seen in most patients. In areas where a vessel contacts lung, the vessel wall is visible as a white stripe slightly less intense than fat. Within the upper mediastinum, at the cervicothoracic junction, small vessels are more easily seen with NMR than with CT.

The trachea, main bronchi, and bronchus intermedius are always seen, but cannot be distinguished from vessels, except by location. Only some segmental bronchi are clearly seen.

As with the systemic vasculature, no signal is recorded from the lumina of pulmonary arteries or veins. The main pulmonary artery, left and right pulmonary arteries, right interlobar pulmonary artery, right anterior trunk (right upper lobe artery), and descending left pulmonary artery are all usually seen, as are large pulmonary veins (Fig. 3). Small hilar vessels and bronchi are often visible but can be difficult to identify. In approximately half of patients, small collections of fat are visible in the hila.

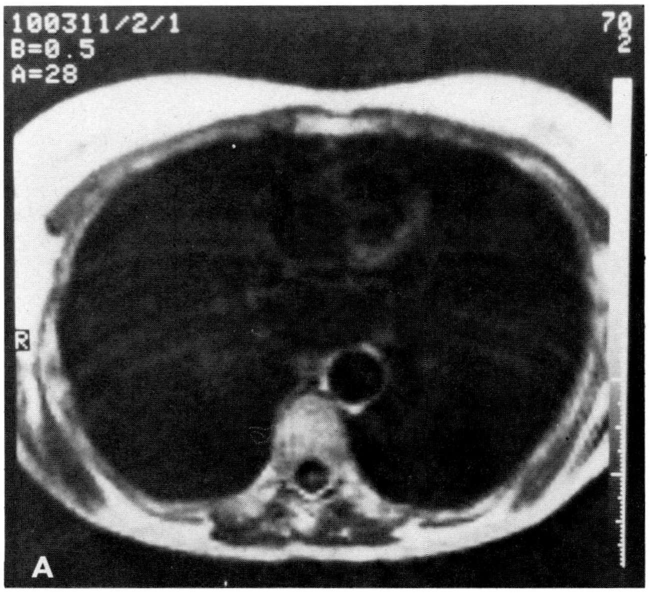

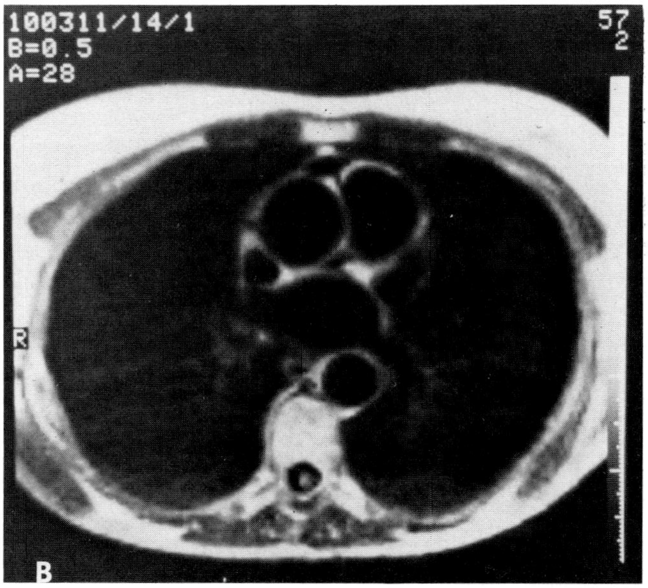

Figure 2. Electrocardiographic gating. In addition to improving the resolution of cardiac structures, electrocardiographic gating greatly improves visualization of the pulmonary hila. A nongated spin echo image (A) in a normal patient (TR = 0.5 sec, TE = 28 msec) shows poor resolution of the inferior pulmonary hila and left atrium. In a gated image at the same level (B), performed with the same imager parameters, the hilar structures are much more clearly seen.

In several patients whom we have studied, normal-sized lymph nodes (less than 1 cm in diameter) on CT in the mediastinum that were noted were not

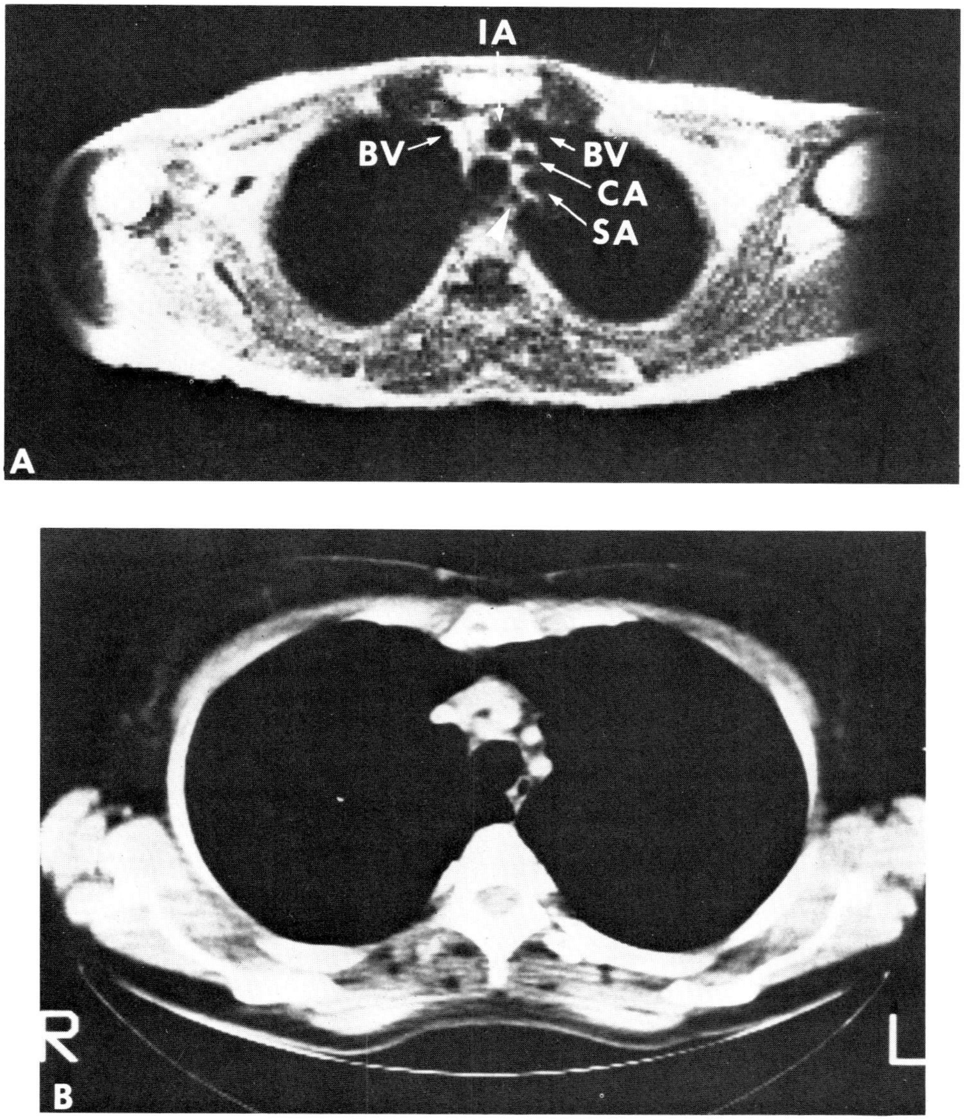

Figure 3. Normal mediastinal anatomy.
A. Spin echo NMR image (TR = 1.5 sec, TE = 28 msec) at the level of the manubrium clearly shows the left subclavian (SA), left carotid (CA), and innominate arteries (IA), sharply contrasted with adjacent mediastinal fat. The right and left brachiocephalic veins (BV) are not as sharply defined because no fat outlines their lateral margins. The esophagus (arrowhead) along the left lateral tracheal wall is visible because of luminal air.
B. A contrast-enhanced CT scan at the same level as A also demonstrates each of these structures. Note that the left brachiocephalic vein, anterior to the innominate artery, is in a different relative position than on the NMR image. This probably reflects the somewhat different lung volumes at which CT and NMR are performed.

visible on NMR scans. Because mediastinal vessels of this diameter are clearly seen by using NMR, the invisibility of normal nodes probably reflects the fact that their intensity is more nearly the same as that of surrounding mediastinal fat than is the very low intensity signal from flowing blood. The esophagus is usually distinguishable from mediastinal fat only when it contains air. In some patients, a low-intensity convex area, representing the anterior vertebral cortex or an osteophyte, can be confused with air-filled

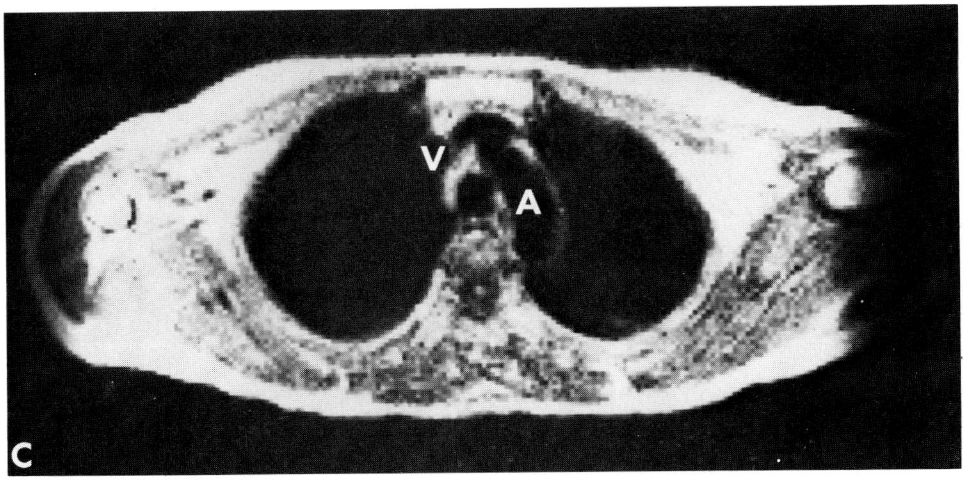

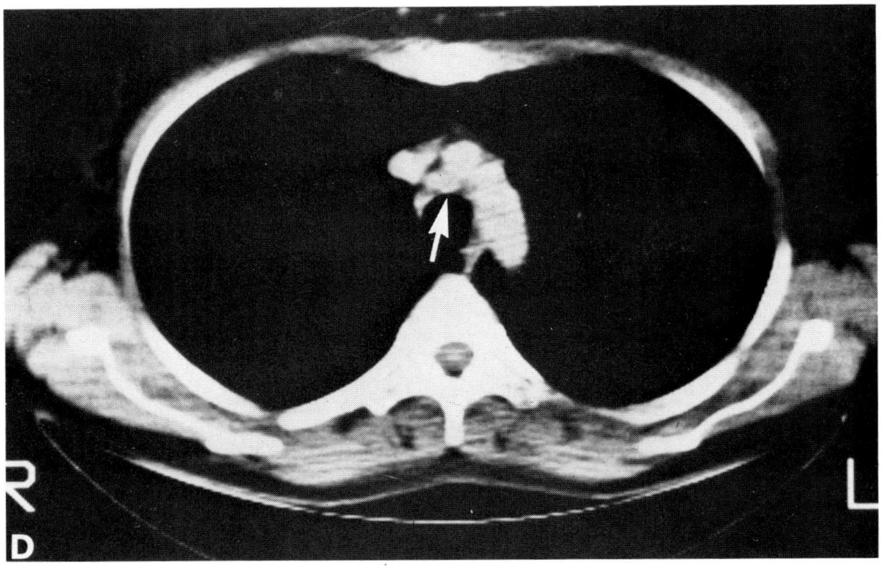

C. Nuclear magnetic resonance image at the level of the upper aortic arch (A) also shows the left brachiocephalic vein and the superior vena cava (V). An arcuate lucency behind the trachea represents anterior vertebral cortex. The esophagus lies between the trachea and aorta.

D. A CT scan at the same level as C shows a normal-sized (1-cm) pretracheal lymph node (arrow). This is not visible on the spin echo NMR image.

esophagus, but its characteristic shape and location usually allow their differentiation.

Because of the very low proton density of normal air-filled lung, little NMR signal is obtained from it. Also, because pulmonary parenchymal vessels are usually invisible when the spin echo technique is used, this important clue to lung anatomy is lacking. Thus, on normal NMR images, pulmonary parenchymal architecture is generally not visible.

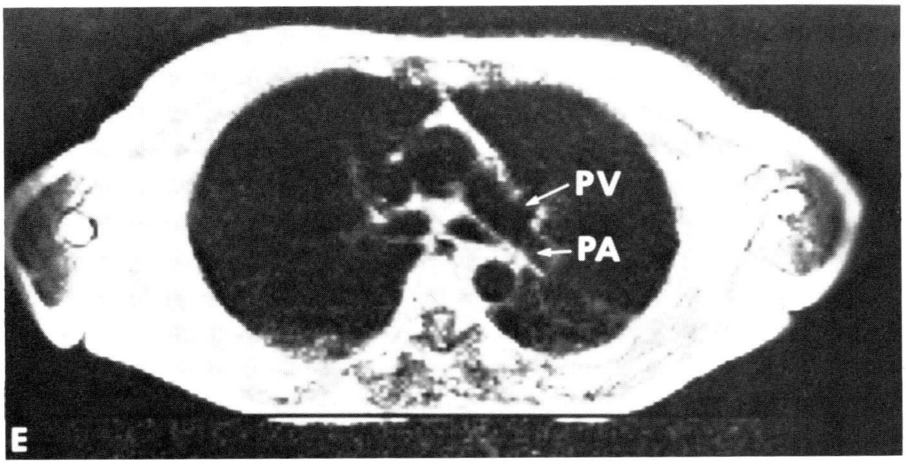

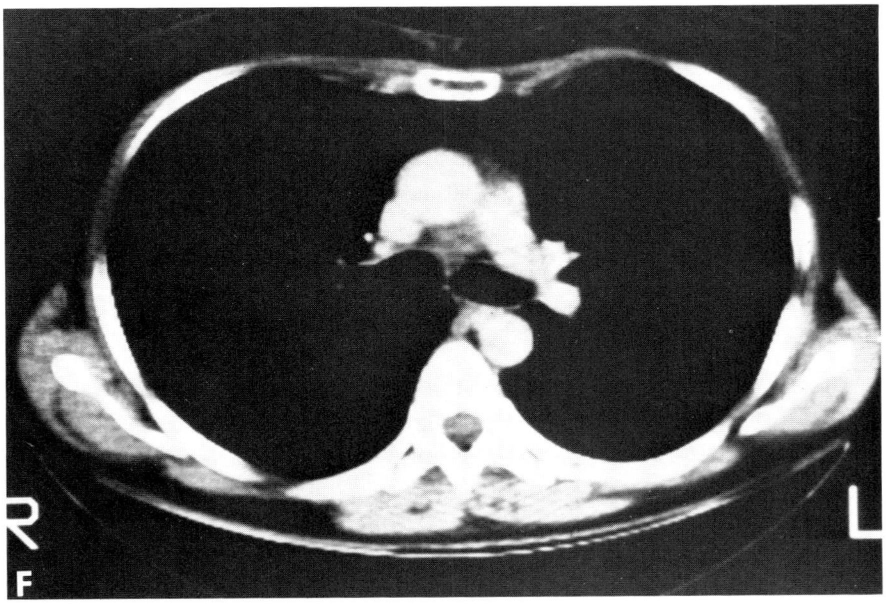

E. An NMR image slightly below the tracheal bifurcation shows the right and left main bronchi, superior vena cava, ascending and descending aorta, and air-filled esophagus. Within the left hilum, the walls of the left pulmonary artery (PA) and left superior pulmonary vein (PV) are visible, but little other tissue is seen.

F. A CT scan at the same level as E shows similar anatomy. Vessels within the left hilum are clearly defined because of contrast injection and closely parallel the NMR appearance.

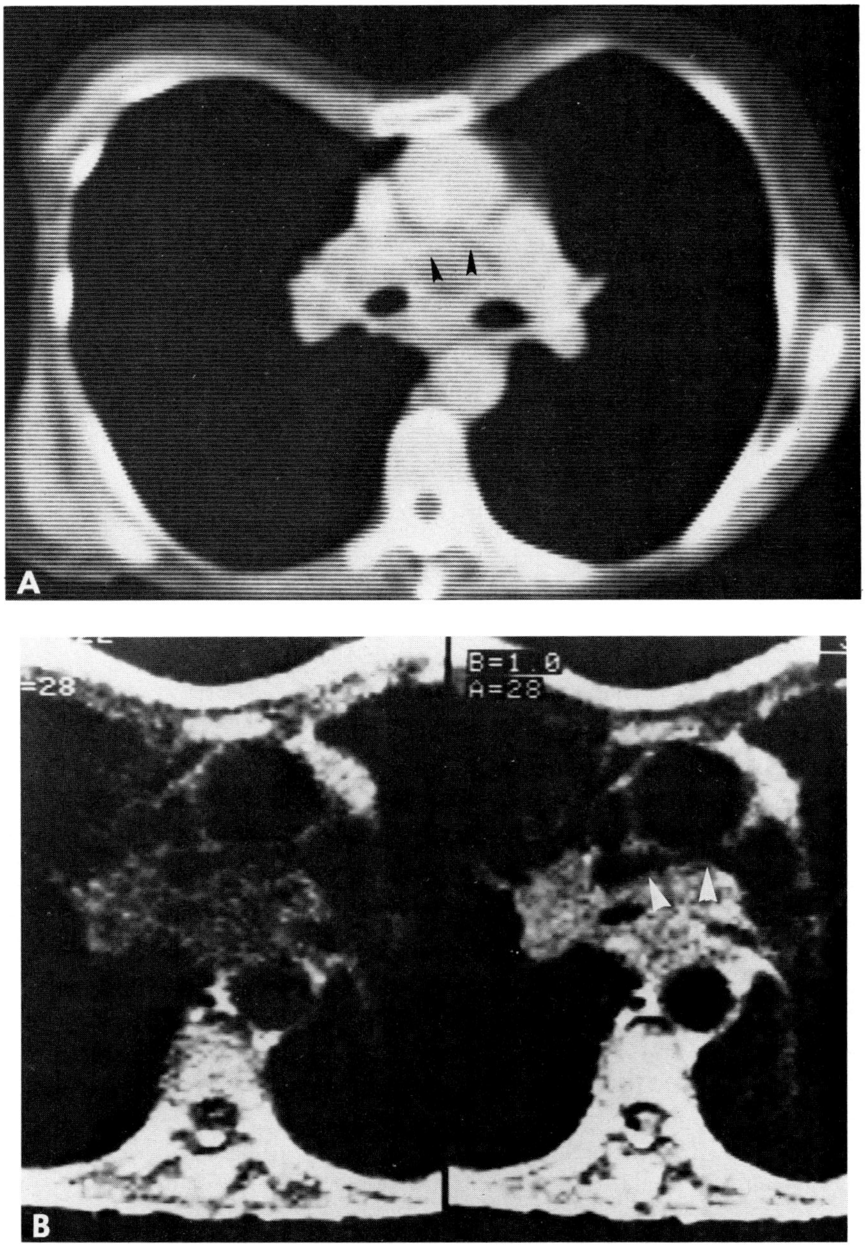

Figure 4. Hilar and mediastinal mass.

A. Contrast-enhanced CT scan in a patient with bronchogenic carcinoma involving the right hilum and adjacent subcarinal space. The mediastinal tumor mass is narrowing the right pulmonary artery (arrowheads). However, the right hilar mass cannot be clearly distinguished from the right pulmonary artery. Note that normal mediastinal fat surrounds the descending aorta and azygos vein.

B. Spin echo NMR image at the same level with a TE of 28 msec and TR values of 0.5 sec (left) and 1.0 sec (right). In the image with a TR of 1.0 sec, hilar and mediastinal tumor are only slightly less intense than normal mediastinal fat surrounding the aorta and azygos vein. However, with a shorter TR (0.5 sec), tumor decreases markedly in intensity relative to fat because of its longer T_1 value. T_1 and T_2 values of tumor measured greater than 2.0 sec and 50 msec respectively, while these values for mediastinal fat were 444 msec and 37 msec. Note that compression of the right pulmonary artery (arrowheads) by tumor and the right hilar mass are more easily seen on the NMR image.

Tissue Characterization

Studies of thoracic tissue characterization by NMR have concentrated on determining T_1 and T_2 values for different tissues. Although, the physiologic bases of T_1 and T_2 differences are not well understood, it appears that T_1 and T_2 values correlate best with the water content of tissues.[4,5] One exception is the long T_2 value of fat; it is not related to water content.[5]

In general, the T_1 and T_2 values of different normal tissues fall into distinct ranges, as do the T_1 and T_2 values of certain pathologic conditions such as sterile or pyogenic abscess and hematoma. However, tumors have a wide range of T_1 and T_2 values and tend to overlap with those of normal and other abnormal tissues.[5]

In the diagnosis of intrathoracic neoplasms, and particularly bronchogenic carcinoma, the detection of mediastinal tumor is of primary importance because it often affects the choice of therapy. Differentiation of tumor from normal mediastinal vessels is easily accomplished because of the absence of signal from moving blood.[1,2] Differentiation of tumor from normal mediastinal fat can usually be accomplished by taking advantage of their differences in T_1.[2] Although the T_2 values of tumors and mediastinal fat overlap, their T_1 values are usually quite different, with fat having a much shorter T_1 (Figs. 4, 5).

Thus, if we first examine the spin echo image of a patient with a mediastinal mass obtained using a TR value of 1.0 or 1.5 sec (a relatively T_1-independent image), both tumor and mediastinal fat will produce an intense signal and can be difficult to distinguish from each other. However, if this image is compared with one obtained with a TR of 0.5 sec, differences will become apparent. Although the signal from fat remains intense because of its short T_1, the signal from tumor, having a longer T_1 value, decreases in intensity (Fig. 4). This difference can also be appreciated on inversion recovery images in which contrast is primarily T_1-dependent.[4] The signal from fat is intense on inversion recovery images, while that from tumor is very low in intensity.

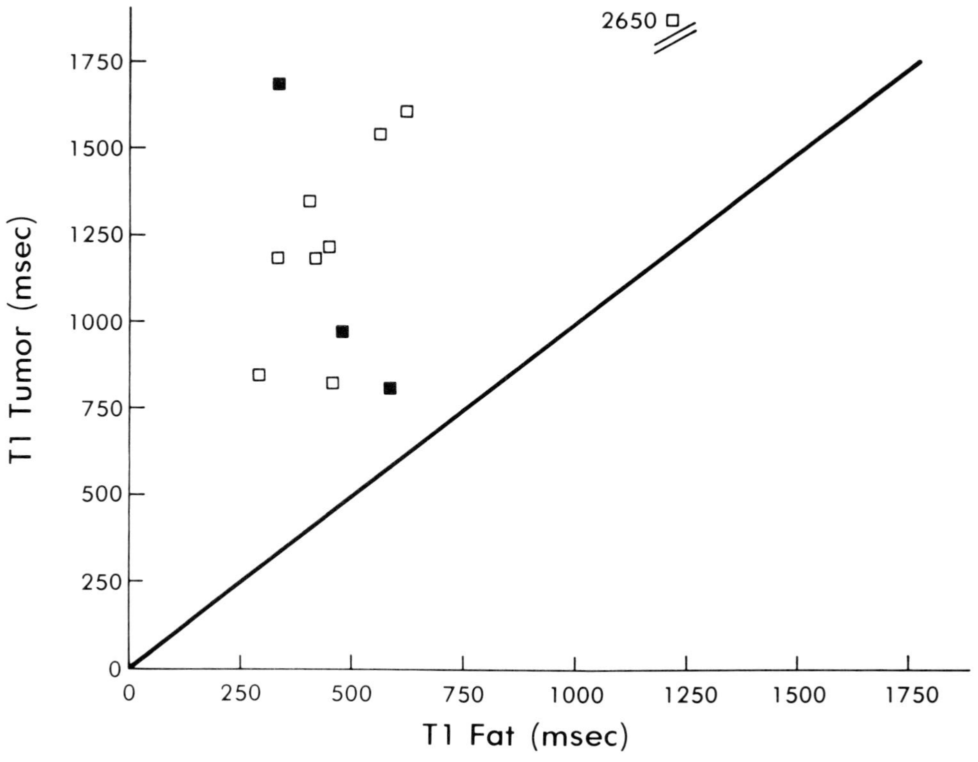

Figure 5. T_1 differences between fat and tumor. Average values of T_1 for masses in nine patients with bronchogenic carcinoma (open squares) and three patients with benign lesions (solid squares) are compared with T_1 values of mediastinal fat recorded from the same images. Note that in all instances the T_1 of the mass is greater than that of fat (i.e., all points fall above the line of equality). Also, note that T_1 values of benign and malignant lesions overlap.
(Reproduced, with permission, from Gamsu G, Webb WR, Sheldon P, et al: Nuclear magnetic resonance imaging of the thorax. Radiology 147:473, 1983.)

This difference in the T_1 value can be used in another way to help distinguish fat and tumor. If an image is calculated in which each pixel represents the T_1 value of tissue at that point (T_1 image), fat, because of its short T_1 value, will appear much different from tumor, with its longer T_1 value. In a T_1 image, mediastinal fat will appear gray or black, and mediastinal tumor will be white. However, in the interpretation of clinical NMR studies, comparing images with differing recovery times is usually sufficient to make a diagnosis. Unless T_1 differences are visibly detectable by comparing images, a T_1 image will not usually be helpful in identifying an area of abnormality.

Mediastinal Masses

Within the mediastinum, vascular structures are clearly demonstrated, and mediastinal masses representing enlarged, tortuous, or abnormal vessels are easily diagnosed. Also, the anatomic relationship between mediastinal masses and adjacent vessels is well demonstrated using NMR, as is compression of normal vessels by a mass (Fig. 4).

Because fat produces an intense NMR signal, the normal mediastinum is intense on spin echo NMR images. Normal-sized lymph nodes are not visible using current techniques (Figs. 3C, 3D).

The detection of enlarged mediastinal lymph nodes or mediastinal tumor masses by using NMR is based on alterations of normal anatomic contours, vascular displacement or compression, and differences in NMR intensity between tumor and mediastinal fat resulting primarily because of differences in T_1.[2] In patients having a mediastinal mass whom we have studied, NMR imaging allowed the diagnosis of the mass and its differentiation from normal mediastinal tissues, and clearly showed vascular compression when present (Figs. 4, 5). Although contrast-enhanced CT scans also showed these findings, vascular compression was more clearly demonstrated when NMR imaging was used. An advantage of CT at the present time is its better spatial resolution.

In patients with mediastinal masses resulting from bronchogenic carcinoma, lymphoma, or benign disease, the T_1 values of mediastinal fat and mass were significantly different (Figs. 4, 5). On the other hand, T_2 values showed significant overlap between normal and abnormal tissues.[2] In the one example of sarcoid that we have seen, and in several other patients with benign lesions, the values of T_1 and T_2 did not differ significantly from those of patients with malignant tumors.

In many patients whom we have studied, mediastinal masses appeared rather homogeneous on NMR images regardless of the TE and TR values used. However, in at least one patient a mediastinal mass had local areas of increased intensity on images with a TE of 56 msec, reflecting a long T_2. This correlated with low density (12 H) measured on CT, and suggested liquefaction or necrosis. A long T_2 is common with fluids.

Hilar Masses

The normal pulmonary hilum consists primarily of bronchi and pulmonary arteries and veins. In the interpretation of hilar CT, the diagnosis of a hilar mass requires the differentiation of normal vasculature from abnormal soft tissue. In some locations, this differentiation can be made on anatomic grounds alone, but in other areas, mass and vessels are difficult to distinguish unless the vessels are opacified with a large bolus of contrast agent. In a significant number of cases, however, the degree of opacification achieved following intravenous contrast infusion is insufficient to allow a confident differentiation of mass and vessel on CT. This problem can be avoided by using NMR.

Because rapidly flowing blood results in little NMR signal, only the walls of pulmonary arteries and veins are visible on NMR. Thus, NMR images of the normal hilum show only the walls of bronchi and hilar vessels (Fig. 3), and hilar masses should be easy to detect (Figs. 4, 6). For the most part, this is true. However, at several levels, small collections of fat are present within the normal hilum, and these can cause some confusion. One common site of fat deposition within the right hilum is at the level of the upper bronchus intermedius, at the point at which the ascending and descending branches of the right pulmonary artery diverge.

In patients whom we have studied with hilar mass resulting from bronchogenic carcinoma or sarcoidosis, NMR allowed a confident diagnosis of hilar mass in each, and clearly showed the relationship of mass to normal vessels and bronchi.[2] When compared with CT scans performed following the injection of a large bolus of contrast medium, NMR and CT findings were comparable (Fig. 6). Nuclear magnetic resonance was superior to CT in defining the mass lesion when the bolus injection was suboptimal or when no contrast was given. Moreover, in several cases, hilar NMR demonstrated lymph nodes measuring approximately 0.5 cm in diameter that could not be appreciated on CT (Fig. 7). Obstruction or

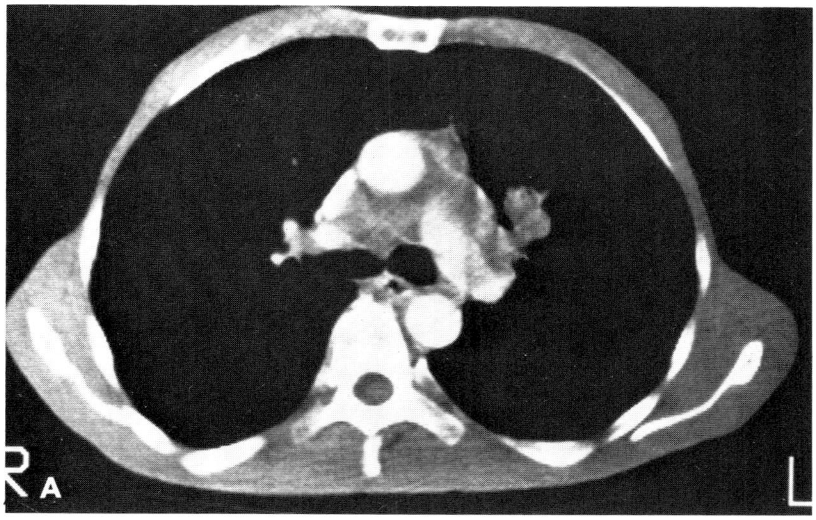

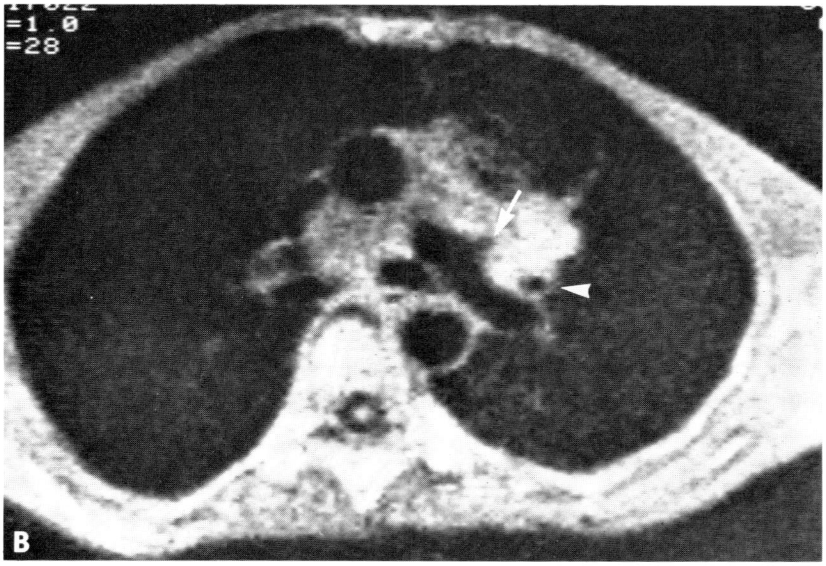

Figure 6. Hilar and mediastinal mass.

A. A CT scan performed during the injection of a large contrast bolus shows a bronchogenic carcinoma involving the left hilum, lateral to the left pulmonary artery, and adjacent mediastinum. Tumor in the pretracheal space is displacing the superior vena cava anteriorly.

B. Spin echo NMR image at the same level (TR 1.0, TE 28 msec) clearly shows the left anterior hilar mass and correlates closely with the CT image. The left apical-posterior segmental bronchus (arrowhead) and a left superior pulmonary vein branch (arrow) are best seen using NMR. Displacement of the superior vena cava is also visible.

(Reproduced, with permission, from Gamsu G, Webb WR, Sheldon P, et al: Nuclear magnetic resonance imaging of the thorax. Radiology 147:473, 1983.)

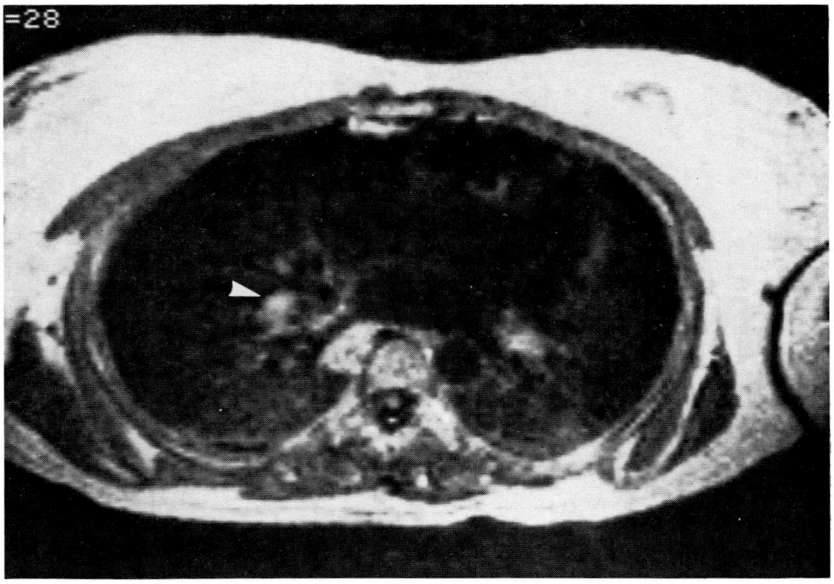

Figure 7. Hilar node. Spin echo NMR image with TR = 1.0 sec and TE = 28 msec in a patient with sarcoidosis shows a small right hilar lymph node (arrowhead), easily distinguished from bronchi and pulmonary vessels. A lymph node of this size would be difficult to detect using CT. Lymph node enlargement in the azygoesophageal recess and left hilum are also visible. Also note that the cardiac silhouette is very poorly defined because of cardiac motion.
(Reproduced, with permission, from Gamsu G, Webb WR, Sheldon P, et al: Nuclear magnetic resonance imaging of the thorax. Radiology 147:473, 1983.)

narrowing of large bronchi is diagnosed as well with NMR as CT, but the resolution of small bronchi is superior with CT. Nuclear magnetic resonance does not clearly distinguish a hilar mass from adjacent pulmonary consolidation resulting from bronchial obstruction, but this is not always possible with CT either.

Lung Nodules

On spin echo images little morphological detail is visible within the pulmonary parenchyma because of the very low proton density of normal air-filled lung. Thus, areas of intense signal within the lung are abnormal, representing mass, consolidation, or other pathologic processes. Because of respiratory motion, one would expect the resolving power of NMR to be less than that of CT in detecting lung nodules, and, in fact, NMR appears less able than rapid CT scanning to resolve small (less than 1 cm) lung nodules. Also, because NMR slices at adjacent levels are separated by 5 mm with the University of California, San Francisco imager at the present time, it is possible that small nodules located between scans could be missed. One additional difficulty with NMR is that calcification is invisible. Thus, calcification within a lung nodule—evidence of a benign lesion on CT or plain radiographs—would be missed by using NMR alone.

However, NMR has several potential advantages. First, because of its ability to discriminate between tissues on the basis of T_1 and T_2, it can allow the detection of fluid within lung masses, a finding of potential diagnostic importance. In addition, because blood vessels containing flowing blood result in little NMR signal, differentiation of nodules and normal (or dilated) vessels should be made more easily than with CT (Fig. 8). Whether or not NMR has the ability to discriminate between benign and malignant nodules is unknown at the present time.

Fluid Collections

Fluids generally have long T_1 and T_2 values and thus give their most intense signal on images having a TE of 56 msec and a long TR (1.5 and 2.0 sec) (Figs. 9, 10). The actual intensity, however, depends on the fluid's composition. On spin echo NMR with

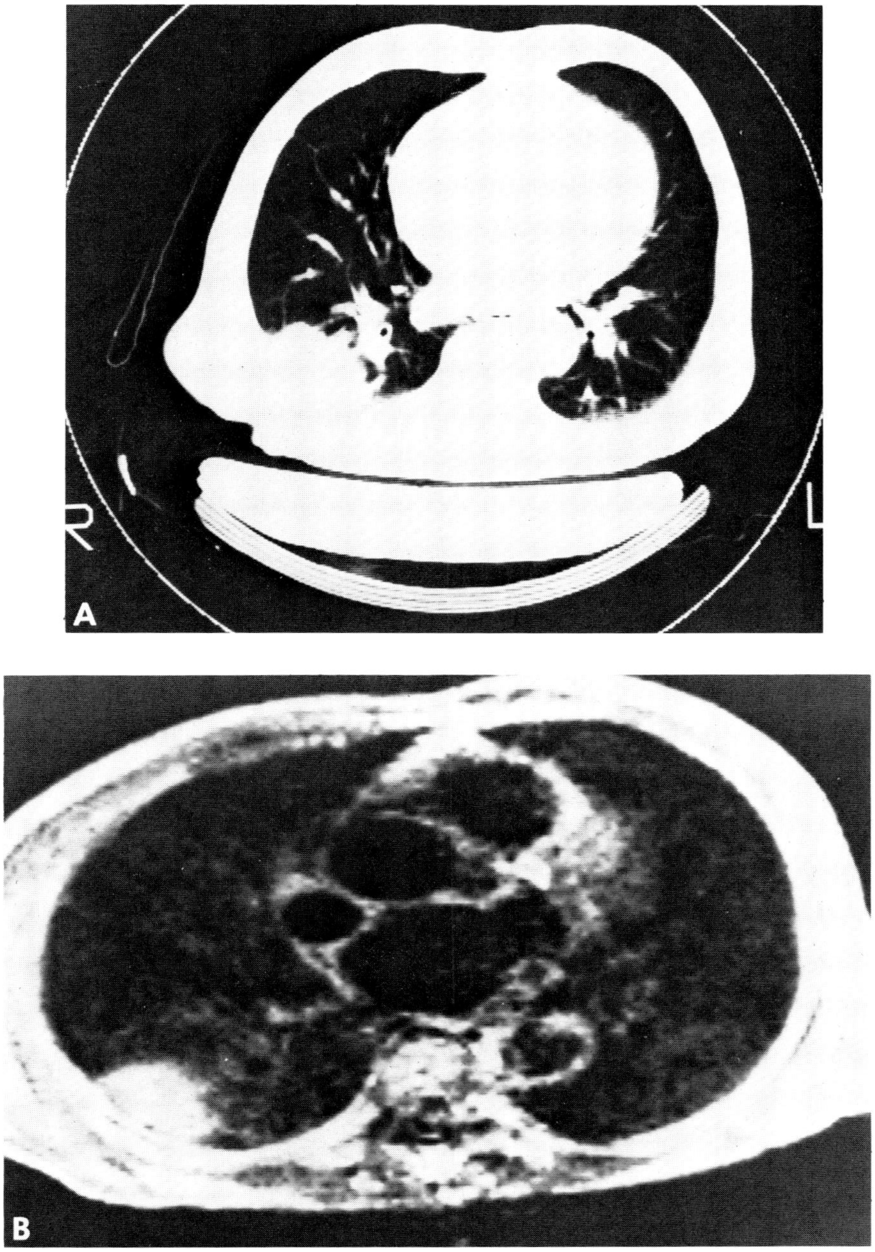

Figure 8. Lung mass.

A. A CT scan in a patient with a bronchogenic carcinoma shows a peripheral lung mass. Adjacent densities could reflect vessels or additional nodules. Also notice some fullness of the posterior right hilum.

B. Spin echo NMR shows the lung mass, but no evidence of additional nodule or hilar mass. The densities seen on CT are, therefore, vascular in nature.

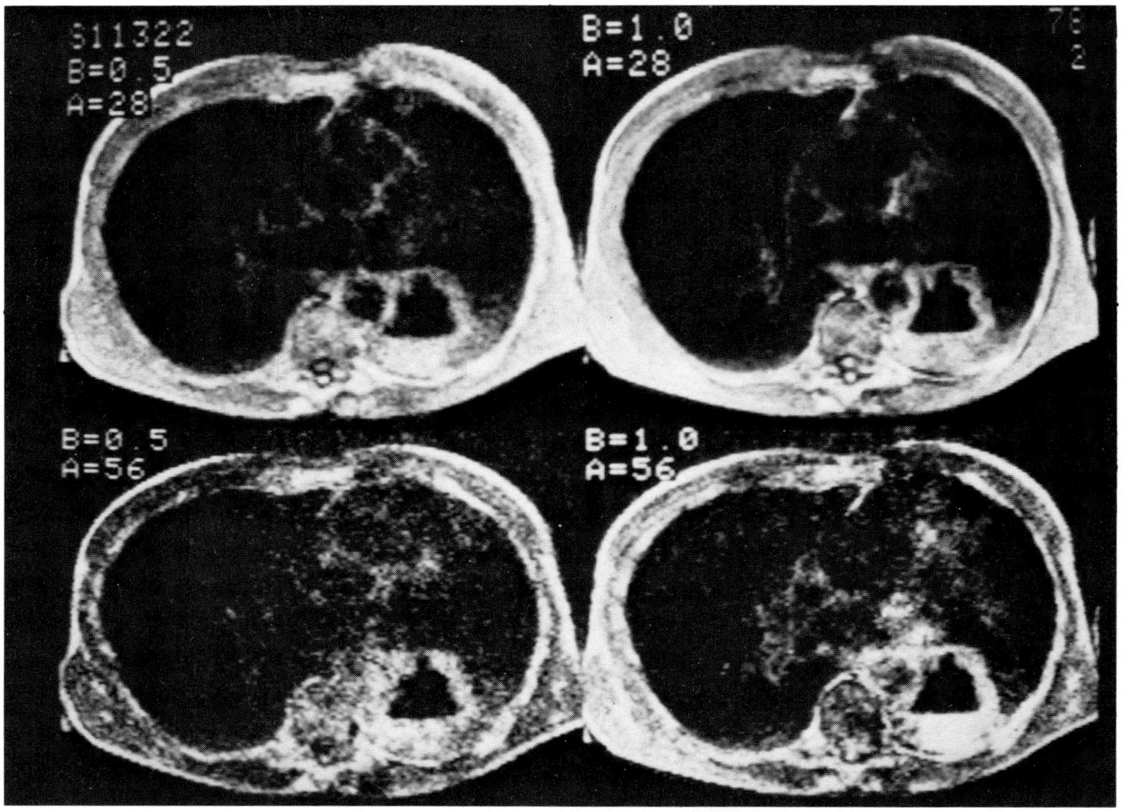

Figure 9. Fluid-containing bronchogenic carcinoma. Spin echo NMR images at the same level with TR values of 0.5 and 1.0 sec and TE values of 28 and 56 msec. Fluid within the mass is most intense in the image with long TR and TE values.
(Reproduced, with permission, from Gamsu G, Webb WR, Sheldon P, et al: Nuclear magnetic resonance imaging of the thorax. Radiology 147:473, 1983.)

a TE of 28 msec and a TR of 0.5 sec, pleural fluid may be invisible.

In patients with fluid-filled empyema and cavitary lung mass (Fig. 9) whom we have studied, fluid could be distinguished from the wall of the lesion on the basis of NMR intensity. The lesion's walls were best seen on images with TR of 0.5 sec and TE of 28 msec, with fluid being less intense using these imaging parameters.

Other Areas of Potential Interest

Because of its ability to detect hydrogen nuclei and measure proton density, NMR should allow quantitation of lung water, and could prove useful in the management of diseases characterized by pulmonary edema or hemorrhage.

Nuclear magnetic resonance may also be helpful in the diagnosis of pulmonary embolism. A clot within large pulmonary vessels should be visible, and turbulence or slow flow distal to an obstructing clot should result in an NMR signal.

General Considerations

The extent to which NMR will prove useful in the diagnosis of chest diseases cannot yet be stated with certainty. Its primary advantage over CT is its ability to distinguish vessels from nonvascular structures without the injection of contrast agents. This ability is of particular value in diagnosing hilar masses. Another potential advantage of NMR may be in its ability to directly image in nonaxial planes (coronal, sagittal). In these planes, the resolution of NMR is

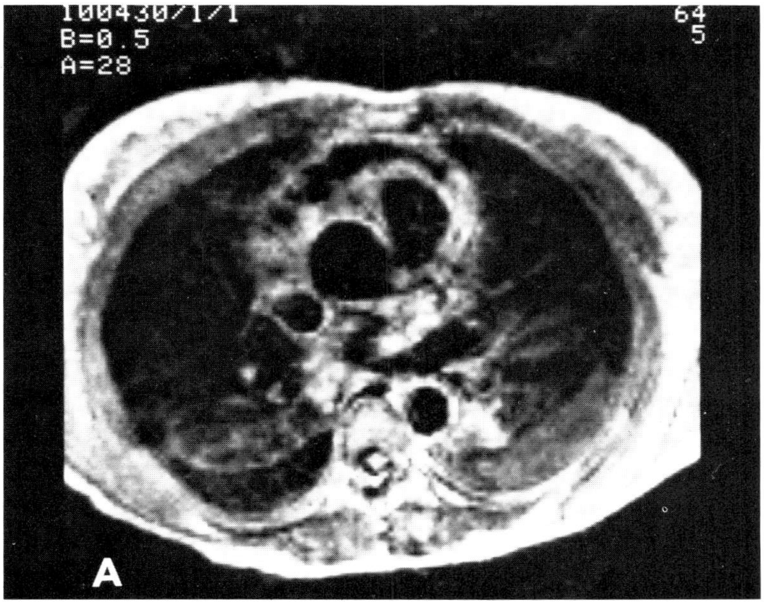

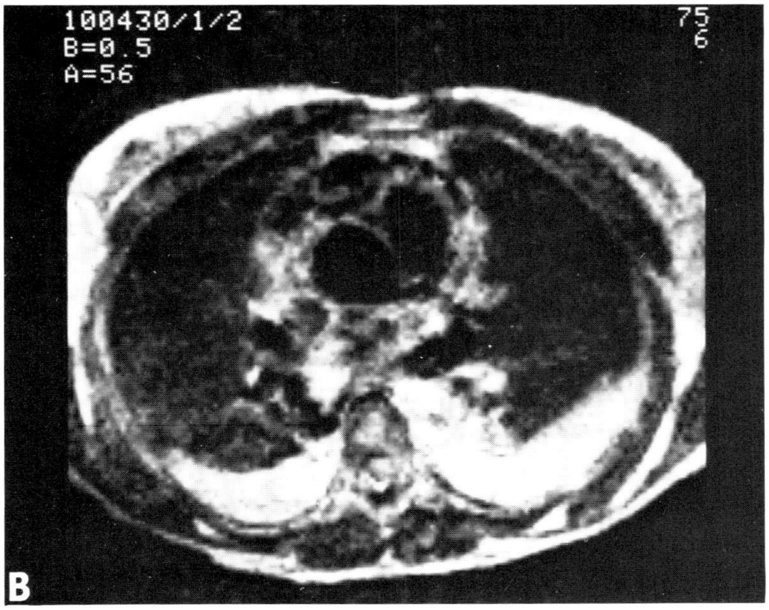

Figure 10. Bilateral pleural effusions. Spin echo images with a TR of 0.5 sec and TE of 28 msec (A) and 56 msec (B) show typical findings of pleural effusion.

A. With a short TE, the right pleural effusion is invisible and the left effusion is only faintly seen.

B. With a longer TE, the fluid collections are intense because of their long T_2 values.

(Courtesy of C. Higgins, M.D.)

superior to reformatted CT. On the other hand, the resolving power of transaxial NMR in the thorax is less than that of transaxial CT, and small normal or abnormal structures (such as lung nodules or mediastinal lymph nodes) may not be clearly seen.

REFERENCES

1. Alfidi JR, Haaga JR, El Yousef SJ, et al: Preliminary experimental results in humans and animals with a superconducting, whole-body, nuclear magnetic resonance scanner. Radiology 143:175, 1982.
2. Gamsu G, Webb WR, Sheldon P, et al: Nuclear magnetic resonance imaging of the thorax. Radiology 147:473, 1983.
3. Crooks C, Arakawa M, Hoeninger J, et al: NMR whole body imager operating at 3.5 kGauss. Radiology 143:169, 1982.
4. Crooks LE, Mills CM, Davis PL, et al: Visualization of cerebral and vascular abnormalities by NMR imaging: The effects of imaging parameters on contrast. Radiology 144:843, 1982.
5. Herfkens RJ, Davis PL, Crooks LE, et al: Nuclear magnetic resonance imaging of the abnormal live rat and correlations with tissue characteristics. Radiology 141:211, 1981.

12

Cardiovascular System

CHARLES B. HIGGINS, M.D.
PETER LANZER, M.D.
ROBERT HERFKENS, M.D.
ELIAS BOTVINICK, M.D.
MARTIN J. LIPTON, M.D.

Nuclear magnetic resonance imaging is a completely noninvasive method for visualization of the cardiovascular system. It does not involve ionizing radiation and, unlike echocardiography, it does not depend upon the inconsistencies of operator-manipulated transducers or body habitus to obtain diagnostic images. Unlike radiographic imaging techniques, no contrast medium is needed to distinguish blood vessels and cardiac chambers from the blood pool. There is considerable natural contrast between motional blood within the cardiac chambers and blood vessels and the walls of these structures since nuclei of blood flowing at normal velocities generally produce much less NMR signal than the vessel wall and surrounding tissues do. An added advantage of NMR imaging is that it can provide at least an estimate of flow velocity and may detect sites of slow or turbulent flow by specific changes in the luminal signal.[1–3] Nuclear magnetic resonance signals from flowing blood can be accentuated by variations in imaging parameters.[1–3] Thus, NMR imaging seems capable of defining both anatomic and physiologic abnormalities of the cardiovascular system.

Cardiac imaging has two important limitations. The acquisition of the NMR signals needed for reconstruction of images requires several minutes during which extensive cardiac motion occurs. Since motion of the compositional nuclei of the heart result in considerable loss of signal intensity, nongated images of the heart are generally not sufficiently distinct for diagnostic purposes. Moreover, the reconstructed image is an average image that corresponds neither to end-diastole nor end-systole. From such images precise quantitation of cardiac dimensions or magnetic relaxation properties of the cardiac walls is not possible. Because of these constraints, the full potential of NMR imaging of the heart requires gated acquisition of NMR signals.

With adequate gating of NMR data acquisition, myocardial pathology should be discriminated better than previously achieved by radiographic methods. While soft-tissue contrast attainable with CT resides at the less than 5% level, soft-tissue contrast achievable with NMR imaging lies in the range of 20%–250%.[4] Through the estimation of hydrogen density, as well as the magnetic relaxation rates of tissues (referred to as T_1 and T_2 values), it seems to be possible to attain tissue characterization.[5] This latter capability would be particularly important for the diagnosis and investigation of myocardial diseases.

Effect of Blood Flow on Nuclear Magnetic Resonance Images

The NMR signal from hydrogen nuclei in blood is dependent on flow. The signal intensity of flowing blood is inversely proportional to the fraction of hydrogen nuclei that are moving and to their velocity.

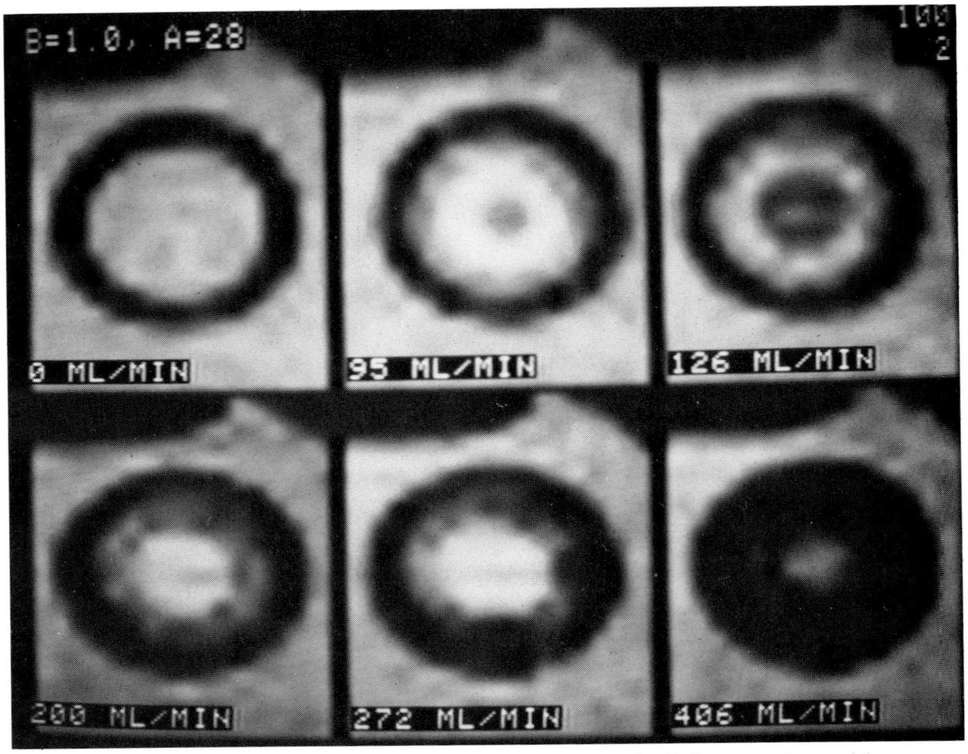

Figure 1. Examples of flow signals produced in 9.6-mm tubes with blood flowing at various velocities. Note the circumferential lucency separating the laminated flow signal from the wall of the tube. As the flow increases from zero to 406 ml/min in a 9.6-mm diameter tube, the average intensity of the signal from moving blood first increases and then decreases to nearly zero. The distribution of signal intensity is not uniform; rather, it takes on the pattern of concentric rings. The walls of the plexiglass tube are dark since this material produces no NMR signal.

If the hydrogen nuclei flow through the volume being imaged in the time that it takes to perform one data acquisition sequence (approximately 50 msec), their signal is lost. However, under some imaging conditions flowing blood can actually produce high signal intensity (a phenomenon called "paradoxical enhancement"). Concepts regarding the NMR imaging of flowing blood have been derived from in vitro experiments in which blood was pumped at varying velocities through tubes placed in the NMR imager.[1-3] The pattern of signals generated from blood flowing at various velocities is shown in Figure 1. The signal intensity from blood is related to the velocity of blood flow (Fig. 2). For any velocity below the cut-off velocity (velocity at which signal intensity of flowing blood approaches background levels), the signal at sequential cross-sectional levels varies in a characteristic manner (Figs. 2, 3). With the NMR imaging parameters used in the current study, blood flow at velocities greater than 10 to 15 cm/sec produces a signal intensity approaching the background level, that is, there is little or no signal intensity within the vascular lumina.[1,3] However, even at normal flow velocities flow signals are frequently observed on images produced by using a longer echo delay (TE) time (TE of 56 msec) (Figs. 3, 4). The intensity of the signal generated within the lumina of blood vessels increased with longer echo delay (56 msec, or second spin echo). In multisection images strong signal intensity from blood may be observed at the initial cross-sectional level into which blood is flowing (Figs. 2, 4).

Paradoxical enhancement is the phenomenon produced when flowing blood at the entrance section (transverse) of an image volume generates NMR signal intensity that is actually greater than stationary blood. When blood flows into a region of the subject being excited by radio-frequency (RF) pulses, the

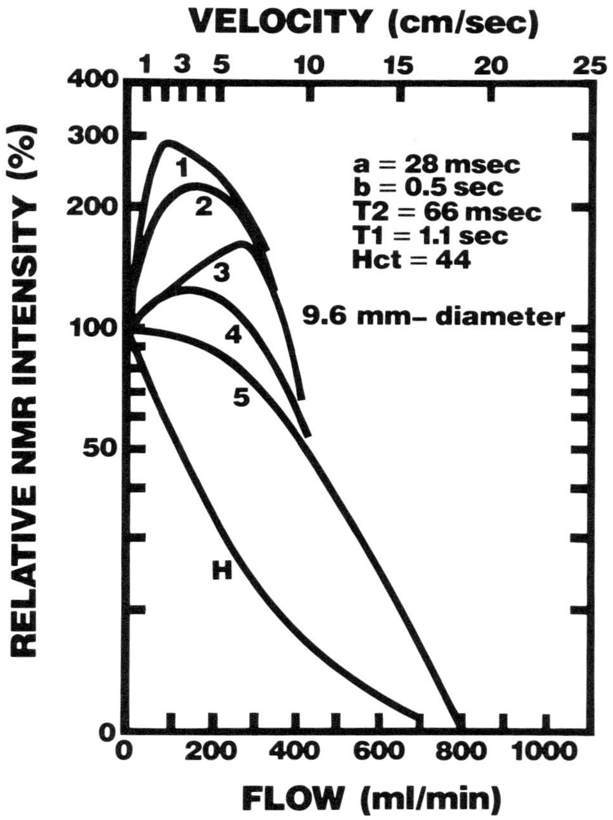

Figure 2. Nuclear magnetic resonance signal intensity versus flow velocity (rate) of flowing blood in 9.6-mm plexiglass tubes. Signal intensity is expressed as a percentage of that produced by stationary blood. Curves are displayed for multiple (five) cross-sectional levels within the imaged volume. The section at which blood flow enters the imaged volume (1) shows the greatest relative signal intensity. Note that at the entrance section (1) and near the entrance section (2–4) the flow signal at lower velocities is actually higher than that generated from stationary blood. (H, horizontal flow, or flow parallel to the plane of section) (Reproduced, with permission of the American Heart Association, from Kaufman L, Crooks LE, Sheldon PE, et al: The potential impact of nuclear magnetic resonance imaging on cardiovascular diagnosis. Circulation 67:251, 1983.)

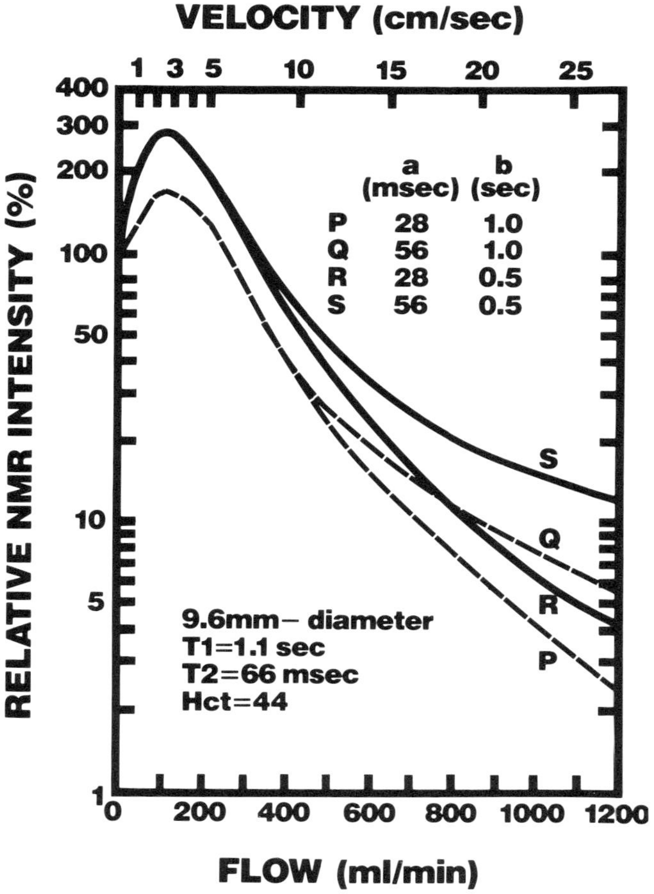

Figure 3. Signal intensity of flowing blood is plotted against velocity and rate of flow for a single cross-sectional level with four different imaging parameters. Intensity is expressed as a percentage of the signal generated from stationary blood. The imaging parameters for each curve are shown in the legend. (TR, b; TE, a) (Reproduced, with permission of the American Heart Association, from Kaufman L, Crooks LE, Sheldon PE, et al: The potential impact of nuclear magnetic resonance imaging on cardiovascular diagnosis. Circulation 67:251, 1983.)

nuclei flowing into the entrance level have not previously undergone excitation. Consequently, they become fully magnetized at this first cross-sectional level (entrance image) and they produce comparatively large signal intensities. Not only is the signal intensity even greater than that expected from stationary blood, but it is also considerably greater than the signal from muscle. This phenomenon is recognized as a bright, frequently laminated or solid high-intensity signal within the lumen (Figs. 1, 4). These bright areas within the lumen are frequently separated from the vessel wall by a concentric black ring (signifying decreased or no intensity). In general, paradoxical enhancement and other flow artifacts are most prominent with longer echo delay times (TE of 56 msec) of an imaging sequence. Within the abdomen, the intensity of this flow signal is most prominent at the most cranial level of the imaged volume for arteries and the most caudal level for veins, corresponding to the direction of flow for these vessels (Fig. 4). Paradoxical enhancement is greater at lower flow velocities and consequently tends to be most prominent for venous structures.

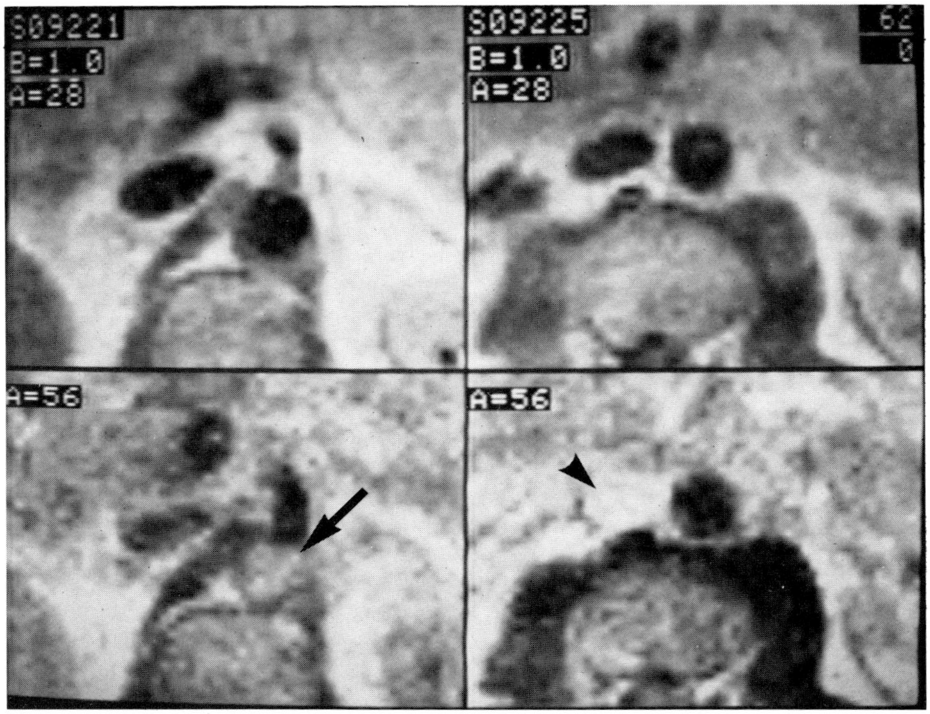

Figure 4. Cross-sectional magnified NMR images of the aorta and inferior vena cava in an asymptomatic patient reveal a circular contour of the aorta, thin wall, and smooth endocardial surface. Images of the most cranial cross-sectional level of a 5-cm imaged volume are shown in the left-hand panels, and images of the most caudal cross-sectional level in the right-hand panels. Images using a longer echo delay (A of 56 msec) at the entrance level of the imaged volume show an intraluminal flow signal (arrow) in the aortic lumen. Note circumferential separation between the intraluminal flow signal and the vessel wall. Intraluminal flow signal also is present in the vena cava (arrowhead) at the most caudal level of the imaged volume.

(Reproduced, with permission, from Herfkens RJ, Higgins CB, Hricak H, et al: Nuclear magnetic resonance imaging of the cardiovascular system: Normal and pathologic findings. Radiology 147:749, 1983.)

Gated Cardiac Nuclear Magnetic Resonance Imaging

Obtaining an electrocardiographic signal for synchronization during NMR imaging presents unique problems. With a standard electrocardiographic monitoring circuit, the electromagnetic fields within the NMR imager induce signal in electrocardiographic leads, and these leads also radiate signals that are detected by the NMR antenna and produce noise. In spite of this limitation, adequate cardiac images have been achieved using three gating techniques: pressure cuff plethysmography, laser Doppler flowmeter transducer, and electronically isolated electrocardiographic leads.

In cuff plethysmography, a cuff placed around the subject's arm detects changes in limb dimension with arterial pulsation. The output of this signal is used to initiate the NMR data acquisition sequence. A cardiac image produced by this technique displays sharp definition of the cardiac wall and internal cardiac morphology, including the moderator band of the right ventricle and papillary muscles of the left ventricle (Fig. 5).

The laser Doppler flowmeter transducer detects a signal that is generated in relation to changes in capillary blood volume and velocity during systole. Laser light is directed to a portion of the body that has a rich superficial capillary bed (such as the lip or ear lobe); the light is reflected by tissue and carried out of the imager through the fiberoptic light cables. Since all of the electrical and electronic components of the gating circuit are outside the RF shield of the

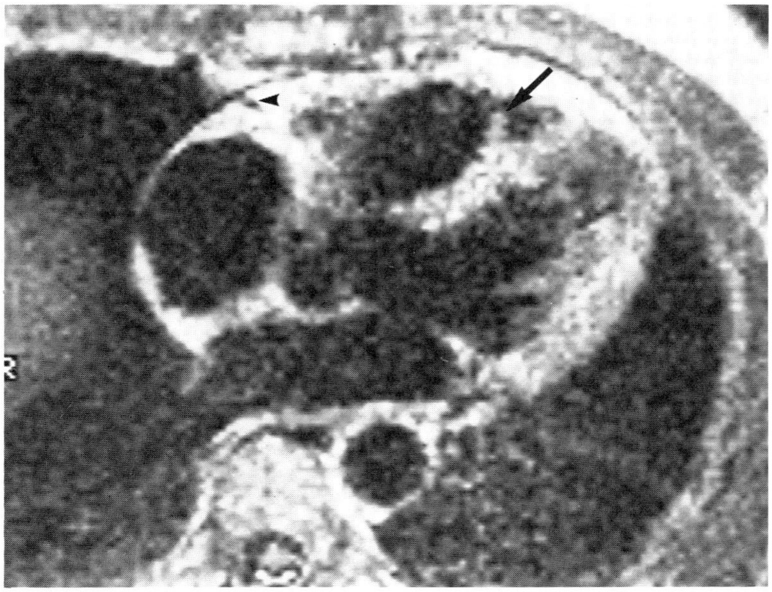

Figure 5. Gated NMR image at the midventricular level of a normal volunteer shows all four cardiac chambers. Internal topography of the ventricles is demonstrated, including the moderator band (arrow) of the right ventricle and a papillary muscle of the left ventricle. Note the right coronary artery in the right atrioventricular groove (arrowhead). (Reproduced, with permission, from Herfkens RJ, Higgins CB, Hricak H, et al: Nuclear magnetic resonance imaging of the cardiovascular system: Normal and pathologic findings. Radiology 147:749, 1983.)

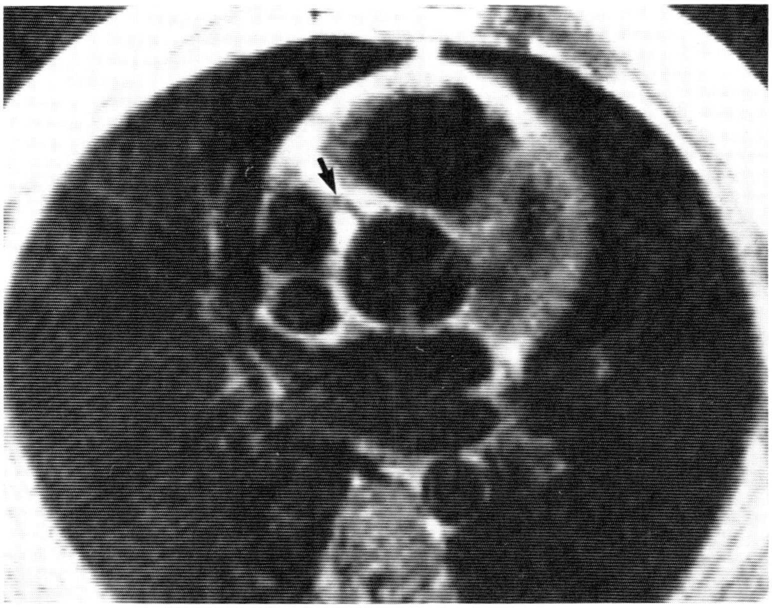

Figure 6. Gated NMR image at the level of the aortic valve of a normal volunteer. The image clearly discriminates the wall of the cardiac chambers and great vessels. At this level, portions of a tricuspid aortic valve are discernible. The right ventricular outflow tract, the left atrium, and the junction of the superior vena cava and right atrium are also visualized. Note the right coronary artery arising from the right sinus of Valsalva (arrow). Discrimination of the vessel is accentuated by the adjacent high intensity produced by the fat at the base of the aorta.

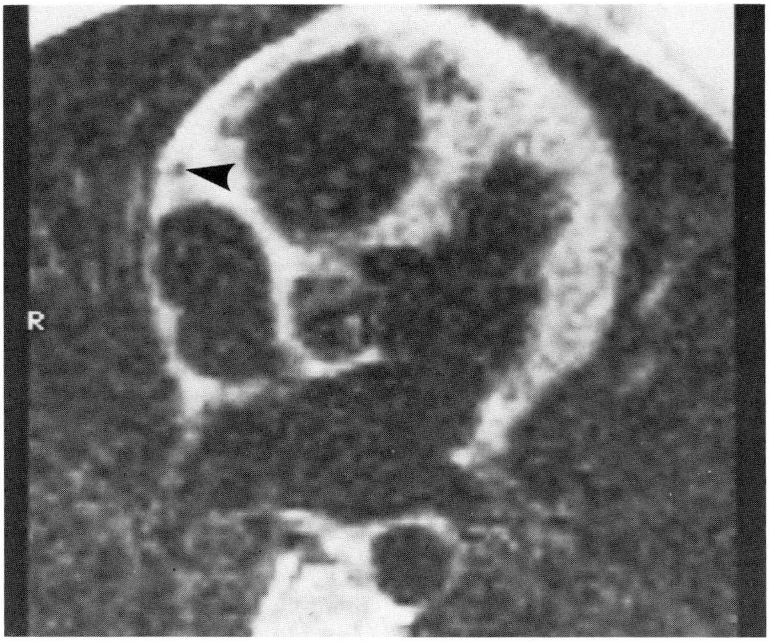

Figure 7. Gated NMR image at the midventricular level of a normal volunteer. There is clear discrimination of the interface between the ventricular wall and flowing blood within the cardiac chambers. The thickness of the septum and free wall of both ventricles is clearly defined. Note the right coronary artery (arrowhead) lying within and contrasted by the high-intensity fat within the right atrioventricular groove.

NMR imager, there is no noise introduced. Excellent cardiac images have been produced with this technique. These images have clearly displayed all four cardiac chambers, atrial and ventricular septa, and cardiac valves (Figs. 6, 7). Even at this early stage in the development of NMR imaging of the heart, the proximal portion of the coronary arteries can be clearly discerned on images by using this gating technique (Figs. 6, 7).

The first two techniques share a common disadvantage; there is delay caused by the distance between the heart and the sampling site. Consequently, it has not been possible to easily and predictably obtain images during the first third of the cardiac cycle. For this reason, and to more precisely define end-systole and end-diastole, the preferable technique is electrocardiographic gating. To minimize introduction of noise to NMR images, the electrocardiographic gating circuit uses low-resistance electrodes and short leads. The electrical signals from the leads are converted into light pulses that are transmitted out of the RF enclosure over fiberoptic light guides. Again, precise definition of internal cardiac anatomy has been attained on NMR images by using this gating technique (Figs. 8, 9).

Cardiac Morphology: Normal and Pathologic

The cross-sectional imaging format is the one generally used for cardiac NMR studies; it displays cardiac anatomy in a form similar to that done with CT and sector scan echocardiography. In contradistinction to CT, but similar to echocardiography, the walls of the heart and vascular structures are clearly distinguished without the need for contrast medium.

In most patients, cardiac motion causes loss of NMR signal; consequently cardiac structures are poorly defined on nongated images. Occasionally, however, the cardiac chambers are sufficiently defined for anatomic diagnosis on nongated images (Fig. 10). These nongated NMR images display defined regional contour abnormalities and mural thrombus of the left ventricle that result from transmural myocardial infarction (Fig. 10). Corresponding to the site of transmural infarction, stasis of blood can be distinguished from normally flowing blood on NMR images (Figs. 8, 10).

Gated NMR images of the heart have displayed the internal architecture of the heart (Figs. 5–9). The natural sharp interface in contrast between moving blood in the cardiac chamber and the cardiac walls

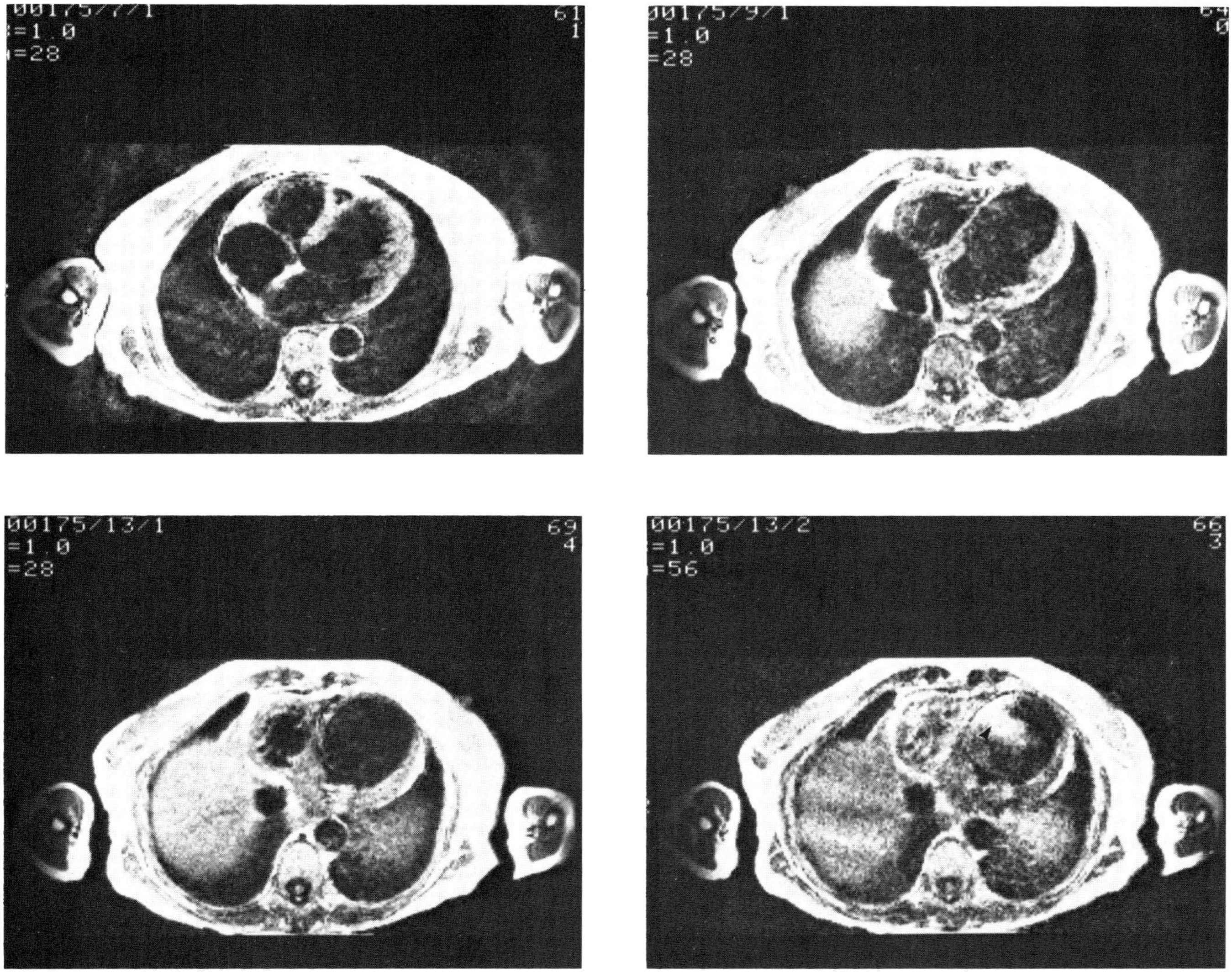

Figure 8. Series of gated NMR images extending from the middle of the left ventricle to the apex of a patient with an anteroapical aneurysm of the left ventricle. Note the severe thinning of the septum and anteroapical segment. Intraluminal signal (arrowhead) is evident in the region of the aneurysm images with the longer echo delay (56 msec) (right lower panel). This finding is consistent with stasis of blood in this region of the left ventricle. The most cranial image (left upper panel) shows normal wall thickness at all regions except the anterior region of the septum.

results in clear discrimination of the ventricular and atrial septa. Consequently, it should be possible to measure wall thickness at all sites within the ventricles. Nuclear magnetic resonance cross-sectional images at the level of the aortic valve have demonstrated coaptation of the cusps of a normal tricuspid aortic valve (Fig. 7). Even at this early time in the development of NMR imaging, the spatial resolution of the imager, coupled with the current gating system, has permitted visualization of the coronary arteries (Figs. 5–7). The origin and proximal portions of the coronary arteries have been displayed on NMR axial images at the base of the heart (Fig. 6), and the right coronary artery has been visualized within the atrioventricular groove on axial images through the middle of the ventricles (Fig. 7). Resolution of the internal topography of the ventricles has been sufficient to depict the moderator band of the right ventricle (Fig. 5), the papillary muscles (Fig. 6), junction of the chorda tendineae cordis and papillary muscle (Fig. 5), leaflets of the atrioventricular valves (Fig. 5), and cusps of the aortic valve (Fig. 7).

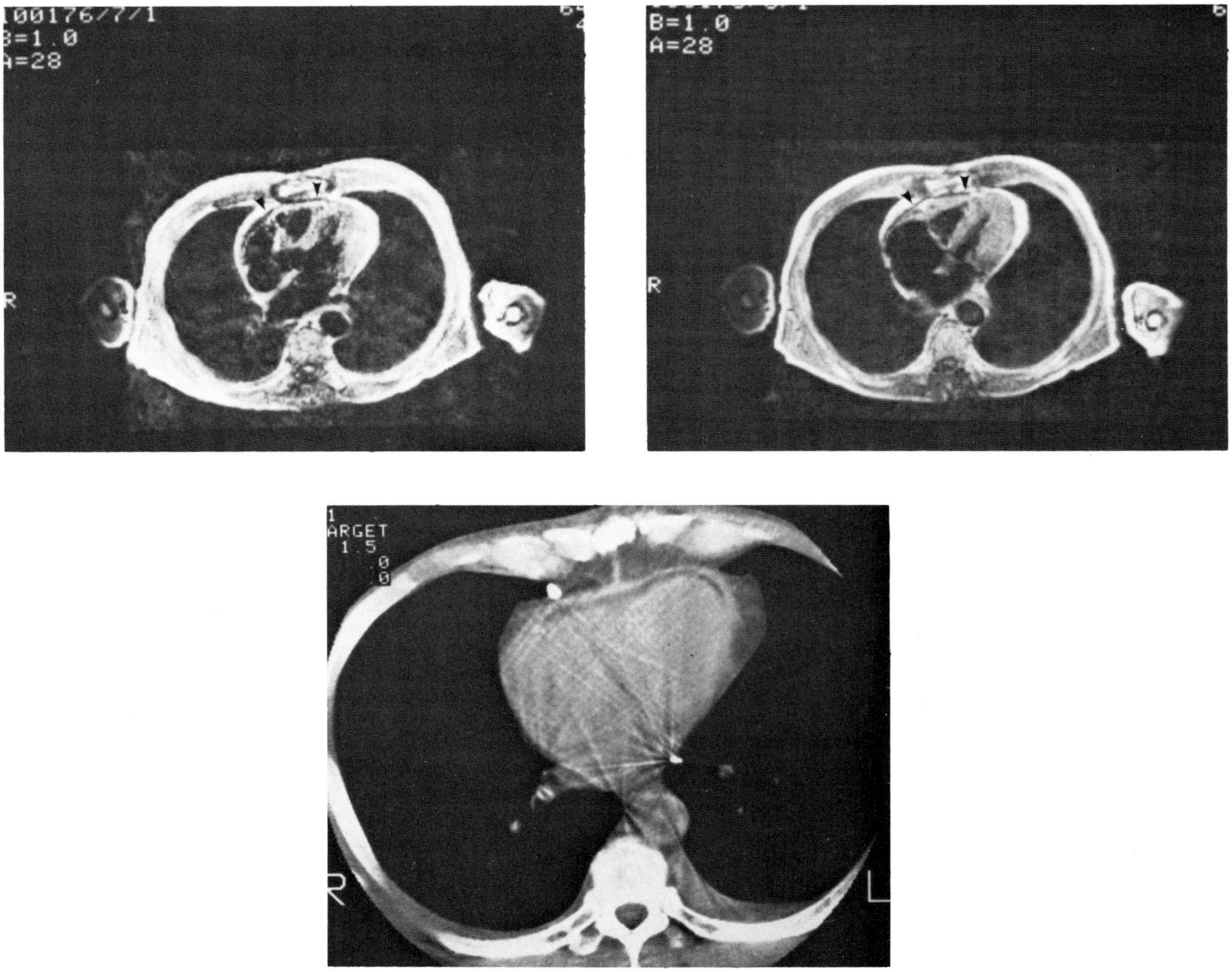

Figure 9. Transverse gated NMR images of the left ventricle (top left and right) and nongated CT images (bottom) display thickened pericardium, large right atrium, and small ventricles. Notice that the pericardium in constrictive pericarditis has low NMR signal intensity, indicating that it is composed of fibrosis. In this patient who developed constrictive pericarditis after coronary artery bypass graft surgery, the NMR image shows greatest pericardial thickening around the anterior surface of the heart (arrowheads). The large amount of epicardial fat produces intense NMR signal.

Gated NMR images have shown the extreme wall-thinning associated with transmural infarction and aneurysm formation (Fig. 8). The images have also displayed the marked increase in wall thickness of the right and left ventricle (Fig. 9). It is interesting to note the change in intensity of the signal within the chamber of the left ventricle in the area of the aneurysm on the image produced with the second spin echo (TE of 56 msec) as compared with those generated using the first spin echo (TE of 28 msec) (Fig. 8). Presumably this finding indicates static blood within the aneurysm rather than mural thrombosis, which would produce signal within the chamber using either spin echo sequence.

Pericardial Diseases

The normal pericardium is only 1–2 mm thick and the visceral and parietal layers are closely applied to each other. The normal pericardium is composed primarily of fibrous tissue, which produces little NMR signal. Epicardial and subepicardial fat produce an area of high NMR signal intensity at the external

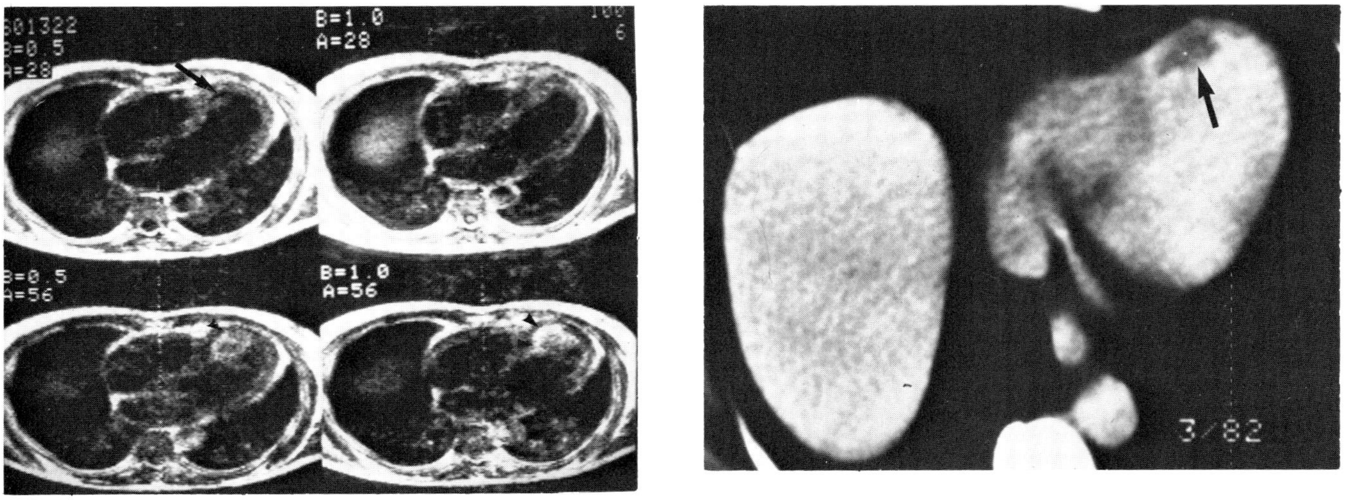

Figure 10. Nuclear magnetic resonance images (left) at different imaging parameters through the same cross-sectional level of the left ventricle in a patient with postinfarctional aneurysm of the anteroapical segment of the left ventricle. The NMR images show good delineation of the left ventricular walls, presumably because of decreased myocardial motion. In the area of the infarction the myocardial wall is thin and displays localized dilatation (arrow). In addition, the intense NMR signal within the lumen at this site (arrowheads) indicates the presence of mural thrombus and stagnant flow (lower panels). The intraluminal flow abnormality is displayed best on the images produced with the second spin echo (A of 56 msec) (lower panels). The corresponding CT scan demonstrates the mural thrombus (arrow).
(Reproduced, with permission, from Herfkens RJ, Higgins CB, Hricak H, et al: Nuclear magnetic resonance imaging of the cardiovascular system: Normal and pathologic findings. Radiology 147:749, 1983.)

margin of the heart. The normal pericardium itself is recognized as a thin lucent line surrounding portions of the circumference of the heart where it is highlighted by adjacent fat (Fig. 5). In some patients the normal pericardium is not distinguished on gated NMR images. Increase in thickness of the dark rim around the myocardium has been observed in the presence of thickened pericardium in a patient with constrictive pericarditis (Fig. 9). The low signal intensity of the thickened pericardium suggests that it is composed mostly of fibrous tissue, as would be expected in chronic constrictive pericarditis.

In the few clinical cases examined to date, gated NMR images have defined several inflammatory pericardial processes. Figure 11 shows a moderate pericardial effusion and an abundant inflammatory exudate of the pericardium. In contradistinction to the normal pericardium, the inflamed pericardium produced strong NMR signal intensity. Moreover, the signal intensity of the pericardium is greater on the image with a longer echo delay time (28 msec versus 56 msec), which is consistent with the long T_2 relaxation time observed with edematous tissues. Figure 12 displays a small pericardial effusion and thickened pericardium in a patient with tuberculous pericarditis. Again, the inflammatory process of the pericardium causes relatively high signal intensity from the anatomically thickened pericardium. Gated

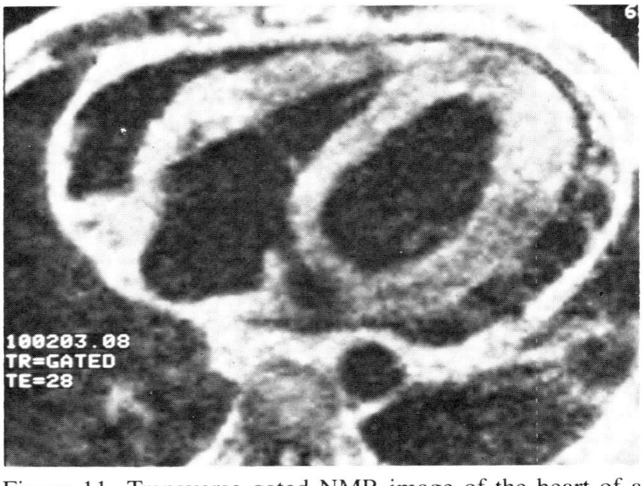

Figure 11. Transverse gated NMR image of the heart of a patient with uremic pericarditis. Image shows shaggy, thickened pericardium with high signal intensity. The thickened edematous visceral and parietal layers are separated by the pericardial effusion.

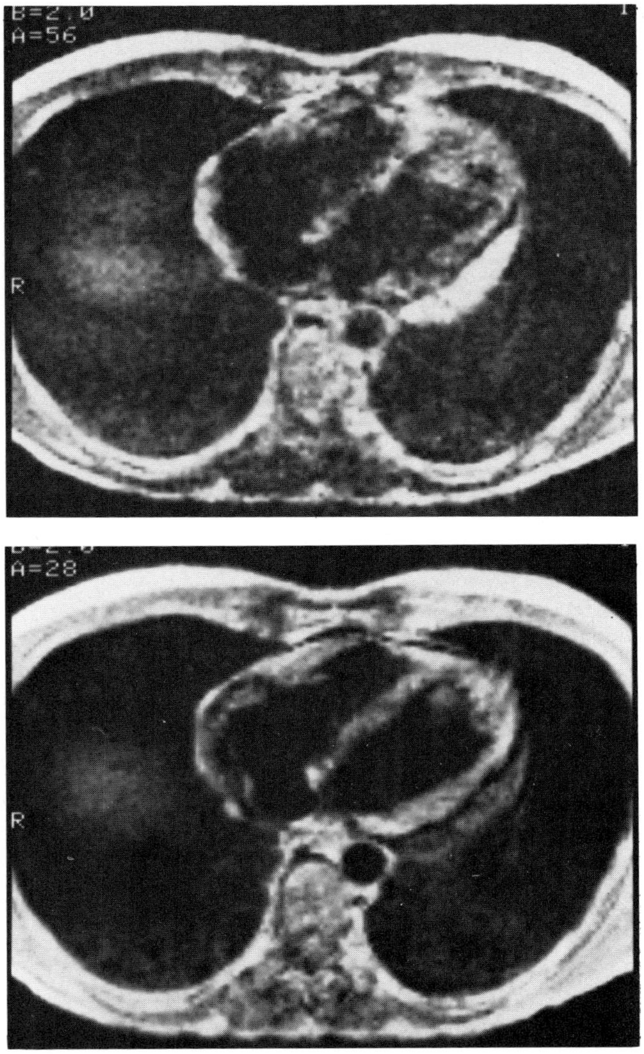

Figure 12. Transverse gated images of the heart of a patient with chronic infectious (tuberculous) pericarditis. The irregularly thickened visceral and parietal pericardial layers are separated by a lucent space, that represents the pericardial effusion. Note that the NMR intensity of the thickened posterior pericardium increases on the image using the long echo delay time (56 msec). The finding suggests inflamed edematous pericardium. As opposed to the fibrosis characteristic of constrictive pericarditis, inflammatory reaction in the pericardium results in relatively high NMR signal intensity.

NMR images of the heart in a patient with constrictive pericarditis revealed small-volume right ventricle, left ventricle, and left atrium, and thickened pericardium (Fig. 9). The low signal intensity of the pericardium suggests that there is no active inflammatory process, and probably a relatively low water content of the pericardium. The myocardial wall of both ventricles was also considerably thickened in this patient.

Our experience to date with NMR imaging of the normal and diseased pericardium indicates that NMR provides at least two important diagnostic concepts. First, it gives anatomic information (thickened pericardium, effusion, etc.). Second, the images, by virtue of the intensity of tissue, give some clue to the nature of the pathologic process. This information is supplemented by measurement of T_1 and T_2 relaxation rates of the specific tissue.

Characterization of Normal and Infarcted Myocardium

Initial reports suggested that infarcted myocardium could not be distinguished from normal myocardium without a paramagnetic contrast medium to mark myocardial perfusion,[6] even though other studies using tissue samples found a significant increase in the T_1 relaxation time of ischemically damaged myocardium.[7] The NMR technique used in the imaging study (steady-state free precession) was apparently not sensitive to the differences in T_2 relaxation rates that exist between normal and ischemically damaged muscle. A recent study in our laboratory[8] used the spin echo technique to image excised canine hearts with 24-hr-old myocardial infarcts within 1 hr after extirpation. In each heart the area of infarction had increased signal intensity in comparison with normal myocardium (Fig. 13).

Myocardium, like other tissues, can be characterized by NMR imaging according to its hydrogen density and magnetic relaxation times. When considering the animals with experimental myocardial infarctions as a group, the mean value for T_1 relaxation times of infarcted myocardium (728.3 ± 94.8 msec) was not significantly higher than the value for normal myocardium (650.2 ± 87.4 msec) (Fig. 14). There was a wide range of T_1 values for infarcted and normal myocardium, and there was overlap of individual T_1 values between infarcted and normal myocardium. However, the T_1 value of the infarcted myocardium was higher than that of normal myocardium for each animal. The percent differences in T_1 values of infarcted as compared with normal myocardium, when each dog was used as its own control, are depicted in Figure 15.

The mean value for T_2 relaxation time of infarcted myocardium (48.4 ± 2.4 msec) was significantly greater ($P < 0.01$) than that of normal myocardium (42.1 ± 1.2 msec) (Fig. 14). There was a narrow range of T_2 values for normal myocardium (40.2–43.5 msec), and no overlap of individual values between the infarcted and normal groups existed. The percent differences in T_2 values of infarcted as compared with normal myocardium, when each dog was used as its own control, are depicted in Figure 15.

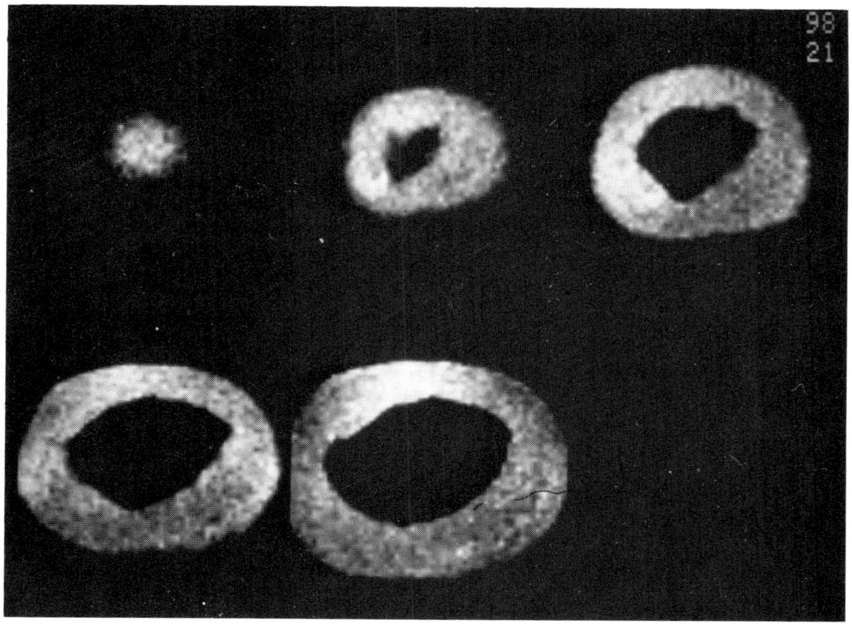

Figure 13. Sequential cross-sectional NMR images extending from the apex to midportion of the left ventricle in a dog at 24 hr after ligation of the left anterior descending coronary artery. The high-intensity area (increased brightness) involves almost the entire cross-sectional area of the caudal sections (upper panels), but only the anterior region on two cranial sections. Images are viewed from above.

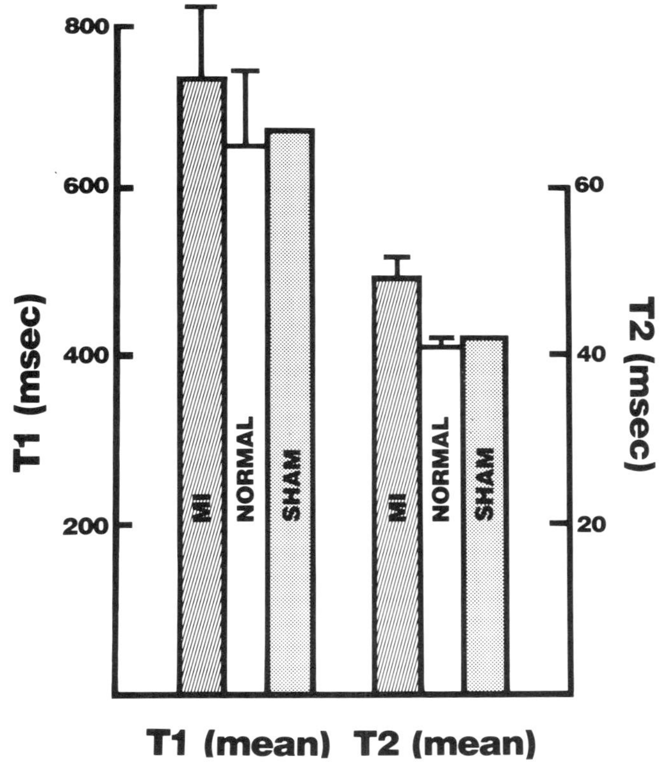

Figure 14. Block diagram shows the mean values ± standard deviation for T_1 and T_2 values for infarcted myocardium (MI) and normal myocardium (normal) of dogs with 24-hr old myocardial infarctions. The value for sham animals is also shown.
(Reproduced, with permission, from the American Journal of Cardiology.)

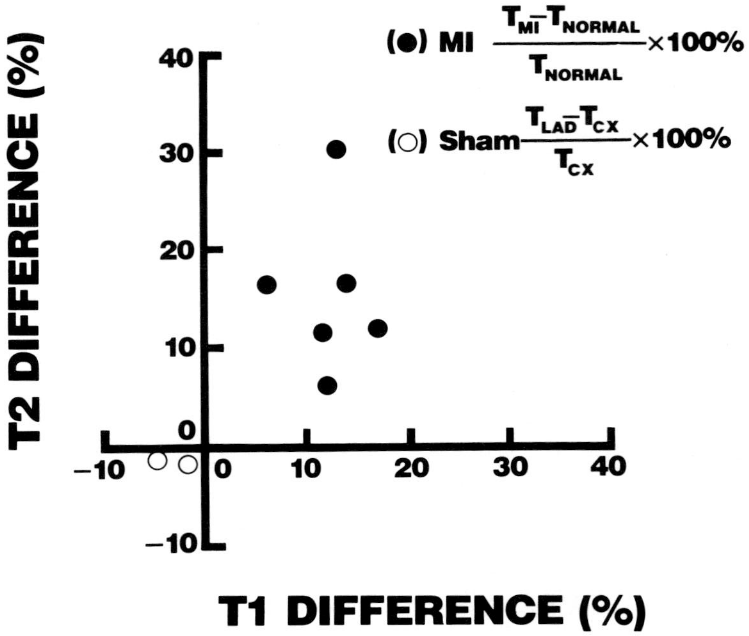

Figure 15. The percent of differences in T_1 and T_2 values between normal and infarcted myocardium is plotted for each of the six dogs with infarcts. Likewise, the percent of differences for relaxation times of the left anterior descending and circumflex distribution is plotted for each of the two sham dogs.
(Reproduced, with permission, from the American Journal of Cardiology.)

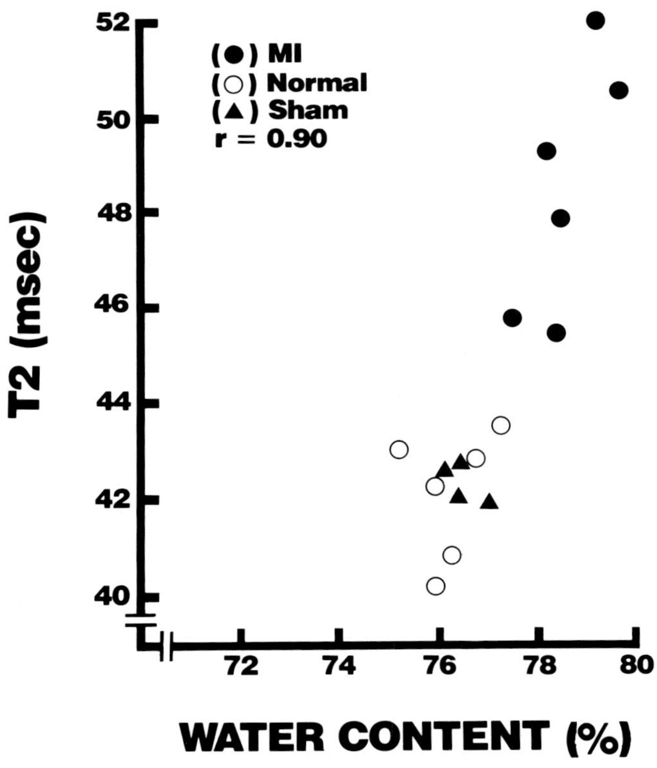

Figure 16. The T_2 relaxation times are plotted against the percent water content.
(Reproduced, with permission, from the American Journal of Cardiology.)

There was no significant relationship between T_1 relaxation time and the water content of myocardium ($r = 0.47$, $P = NS$). On the other hand, there was a good correlation ($r = 0.90$, $P < 0.01$) for the linear relationship between T_2 values and percent water content of myocardial samples (Fig. 16).

Recently, gated NMR images performed in vivo on intact dogs with one-week old myocardial infarctions have also shown that the infarcted myocardium can be distinguished from normal myocardium without the use of contrast medium. The infarcted myocardium has high signal intensity as compared with normal myocardium on NMR images produced with the spin echo technique. One can optimistically conjecture that we will soon be able to assess the effects of therapeutic intervention by defining T_1 and T_2 values using NMR imaging of ischemic processes. Indeed, with proton NMR imaging, tissue characterization must be shown to have clinical efficacy since the anatomic information provided by NMR imaging is similar to that achieved by CT and ultrasonography. A distinct advantage of NMR imaging in ischemic heart disease and other cardiovascular diseases is the ability to define internal cardiac topography without contrast medium. This is particularly important in acute myocardial ischemia

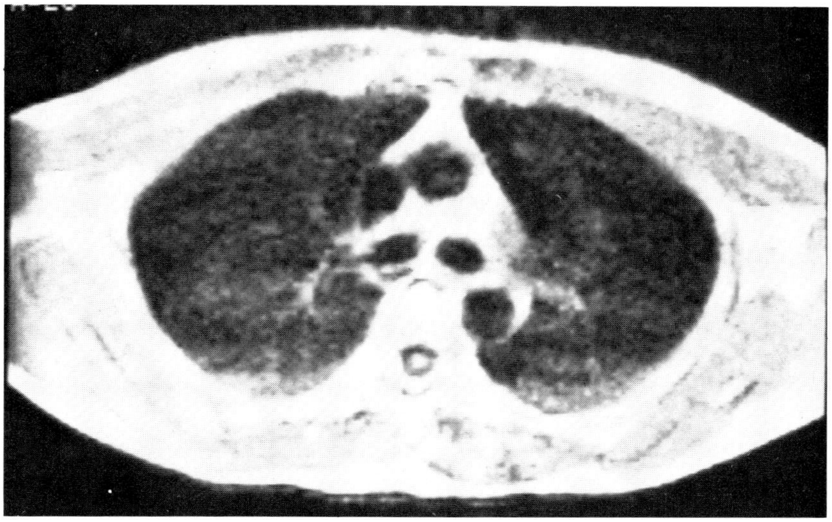

Figure 17. Nongated NMR image of the mediastinum just below tracheal bifurcation. There is clear discrimination of the walls of the superior vena cava, ascending and descending, as well as the bronchi and aorta.

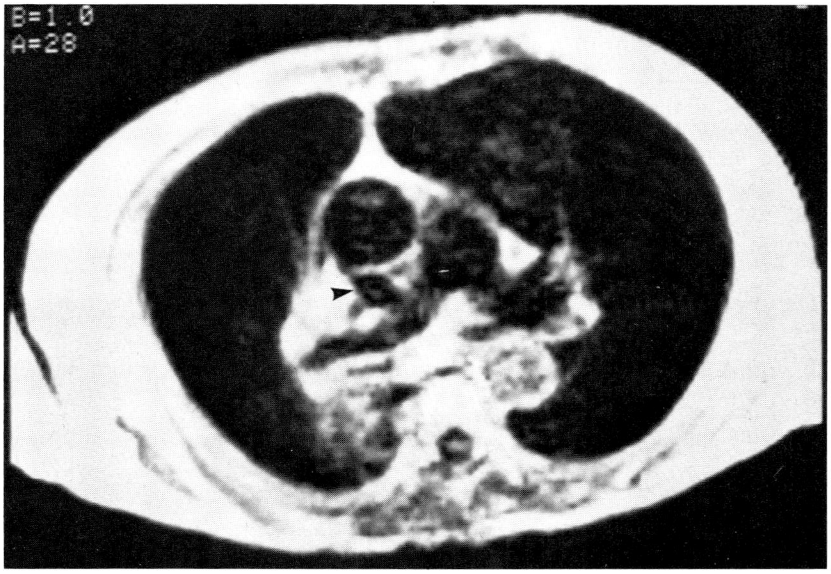

Figure 18. Nuclear magnetic resonance image of the thorax in a patient with hilar involvement from bronchogenic carcinoma. The solid tumor in the right hilum is clearly distinguished from the hilar vessels. The right bronchus is encased by tumor, and the superior vena cava is displaced medially (arrowhead).

and infarction since contrast media in current use have potentially deleterious hemodynamic and electrophysiologic effects.

Thoracic Aortic Abnormalities

Recent reports have shown that mediastinal vessels are well defined on nongated NMR images (Fig. 17).[9,10] Since vessels have little or no signal within the lumen, the superior vena cava, ascending and descending aorta, and pulmonary vessels are easily distinguished from solid structures such as mediastinal or hilar tumors and lymph nodes (Fig. 18).

The substantial natural contrast between flowing blood and the vessel wall permits clear demonstration of abnormalities of the aortic wall such as intimal flaps of aortic dissections displaced into the aortic lumen (Figs. 19, 20), localized dilation of the lumen and distortion of the contour of the wall of the aorta

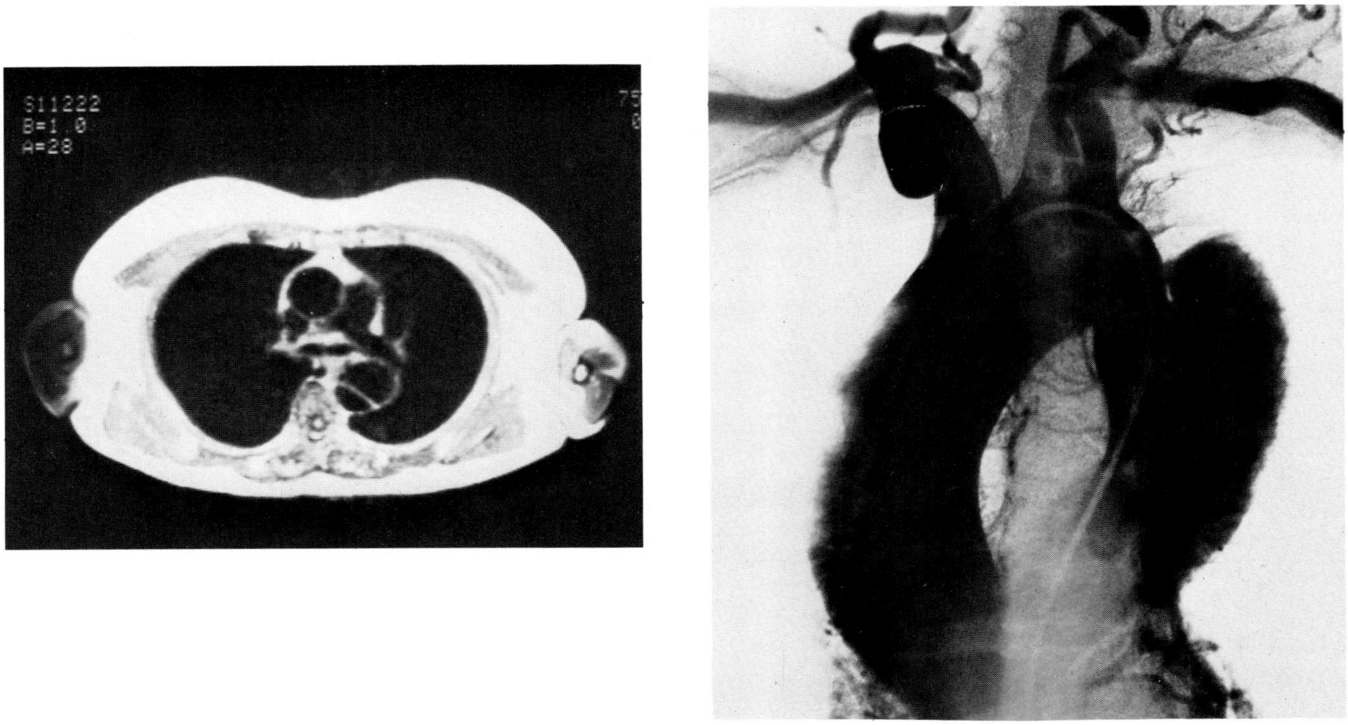

Figure 19. Nuclear magnetic resonance image (left) and angiogram (right) of the descending aorta in a patient with type B aortic dissection. The NMR image shows the intimal flap and the false and true lumina of the aorta.

(Reproduced, with permission, from Herfkens RJ, Higgins CB, Hricak H, et al: Nuclear magnetic resonance imaging of the cardiovascular system: Normal and pathologic findings. Radiology 147:749, 1983.)

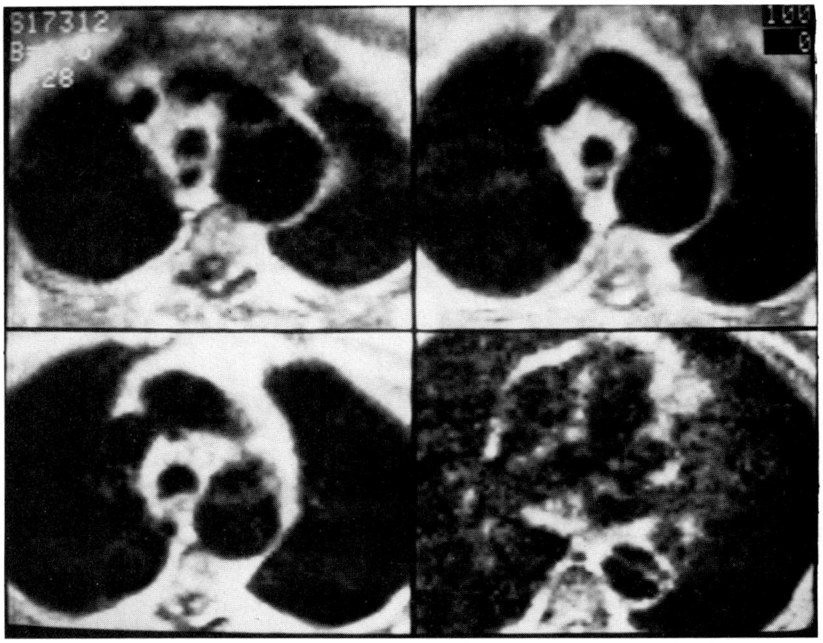

Figure 20. Sequential cross-sectional images in a patient with an aneurysm of the aortic arch and a type B dissection of the descending aorta. Images at the level of the arch display localized dilatation of the aortic arch. Image through the descending aorta shows the intimal flap.

(Reproduced, with permission, from Herfkens RJ, Higgins CB, Hricak H, et al: Nuclear magnetic resonance imaging of the cardiovascular system: Normal and pathologic findings. Radiology 147:749, 1983.)

caused by aneurysms (Fig. 20), and mural thrombus and sluggish flow within the false channel of aortic dissection.

In several patients, saphenous vein bypass grafts to the left anterior descending coronary artery have been identified on NMR images.[9] The absence of signal within the lumen of the graft suggests patency of the graft with blood flow at a velocity greater than 10–15 cm/sec.[1,3] The sensitivity and specificity of NMR imaging for defining graft patency and normal blood flow in grafts have not yet been determined.

Abdominal Vasculature

Blood vessels within the abdomen are easily distinguished from viscera and other surrounding tissues without contrast medium (Fig. 21).[11,12] The origin of the celiac axis and superior mesenteric artery can be visualized in both cross-sectional (Fig. 21) and sagittal images (Fig. 22). The renal arteries, which are generally not identified on standard contrast-enhanced CT scans, are usually demonstrated on NMR images (Fig. 21). The relative size and position of the superior mesenteric artery and vein are also shown (Fig. 21).

Because the aorta and inferior vena cava contain little or no intraluminal signal on images using the first spin echo, these retroperitoneal vessels have been readily distinguished from retroperitoneal masses (Fig. 23). The effect of retroperitoneal processes impinging upon the infrarenal segment of the vena cava were noted to cause a considerable reduction in the diameter of the inferior vena cava as compared with the diameter of the portal vein (Fig. 24).

The venous structures within the liver (Figs. 25, 26), in the periportal area (Fig. 25), and within the pancreas are well demonstrated (Fig. 25). In fact, body and tail of the pancreas can be definitely identified on NMR images by the clear visualization of the transversely oriented splenic vein (Fig. 25). Likewise, the superior mesenteric vein, splenic vein, and confluence of these veins to form the portal vein can be shown on sequential NMR images (Figs. 21, 25).

The normal relationships of venous structures within the liver and porta hepatis are particularly well shown (Figs. 26, 27). The course of the hepatic veins within the liver and their confluences with the inferior vena cava can be demonstrated on sequential cross-sectional images through the liver (Fig. 27). Because the inferior vena cava is well depicted on NMR images, the normal changes in shape of the inferior vena cava at infrarenal level, at the level of

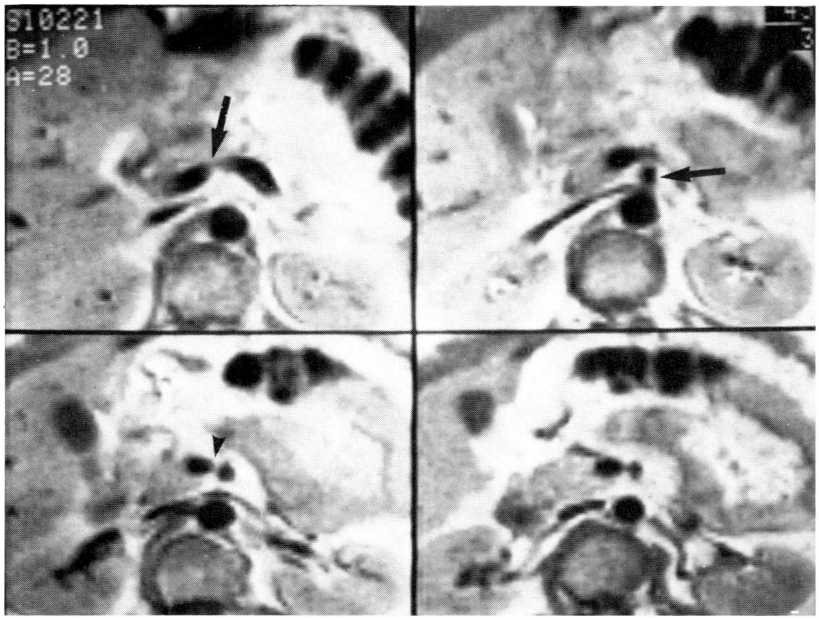

Figure 21. Nuclear magnetic resonance cross-sectional images display junction of splenic and mesenteric veins (upper left, arrow), origin of superior mesenteric artery (SMA) from aorta (upper right, arrow), and left renal vein coursing beneath the SMA (upper panel). More caudal image (below) shows the SMA and superior mesenteric vein (arrowhead) coursing together, and origin of left renal artery from the aorta.

(Reproduced, with permission, from Higgins CB, Goldberg H, Hricak H, et al: Nuclear magnetic imaging of vasculature of abdominal viscera: Normal and pathologic features. AJR 140:1217, 1983.)

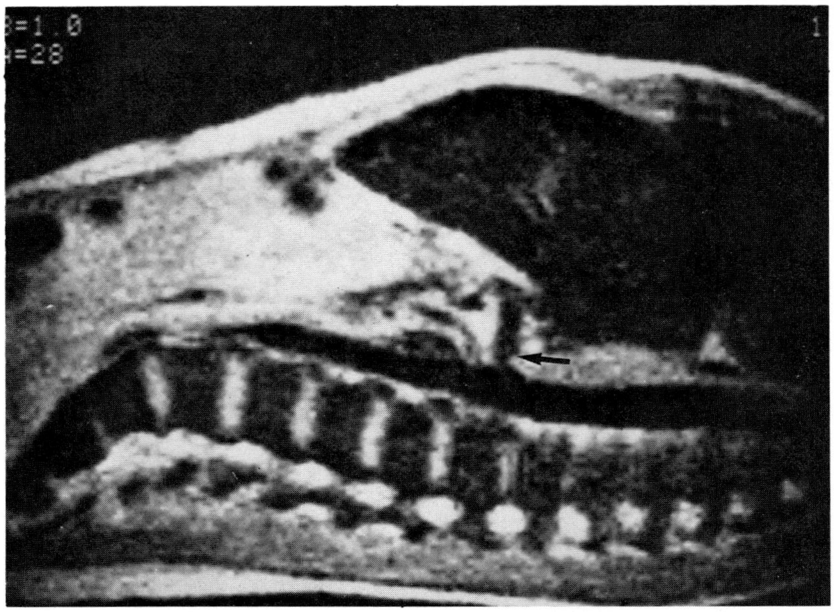

Figure 22. Sagittal image through the abdominal aorta in a child with hemochromatosis due to thalassemia major. Origin of the celiac artery from the aorta is shown (arrow). Note the low NMR signal intensity of the liver as a consequence of its high content of iron.
(Reproduced, with permission, from Higgins CB, Goldberg H, Hricak H, et al: Nuclear magnetic imaging of vasculature of abdominal viscera: Normal and pathologic features. AJR 140:1217, 1983.)

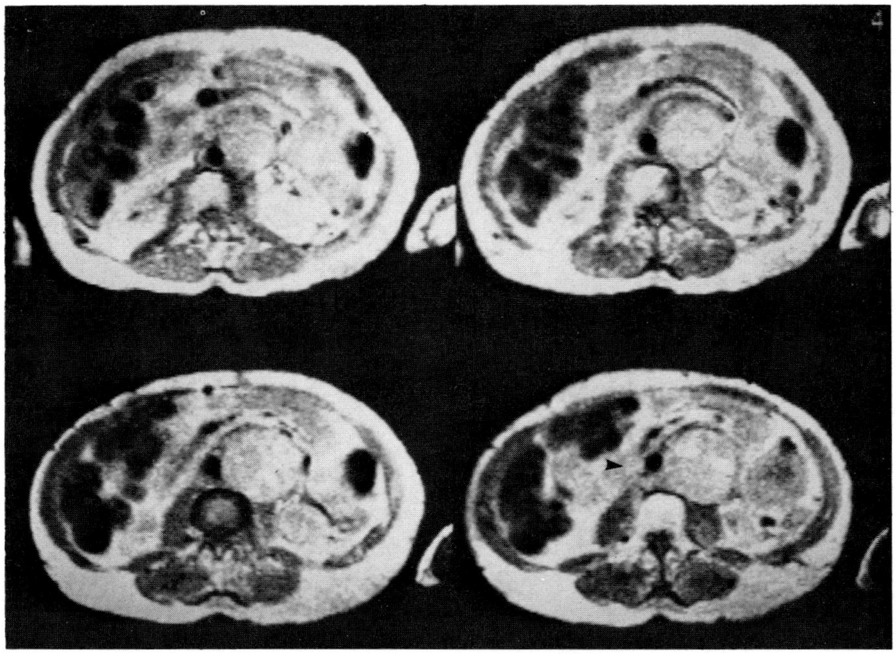

Figure 23. Solid retroperitoneal mass (apudoma) can be distinguished from the adjacent aorta because of absence of intraluminal signal. The aorta is displaced and distorted in contour but is not invaded by the tumor. The presence of signal filling the inferior vena cava (arrowhead) at all levels suggests very sluggish flow in the inferior vena cava.
(Reproduced, with permission, from Higgins CB, Goldberg H, Hricak H, et al: Nuclear magnetic imaging of vasculature of abdominal viscera: Normal and pathologic features. AJR 140:1217, 1983.)

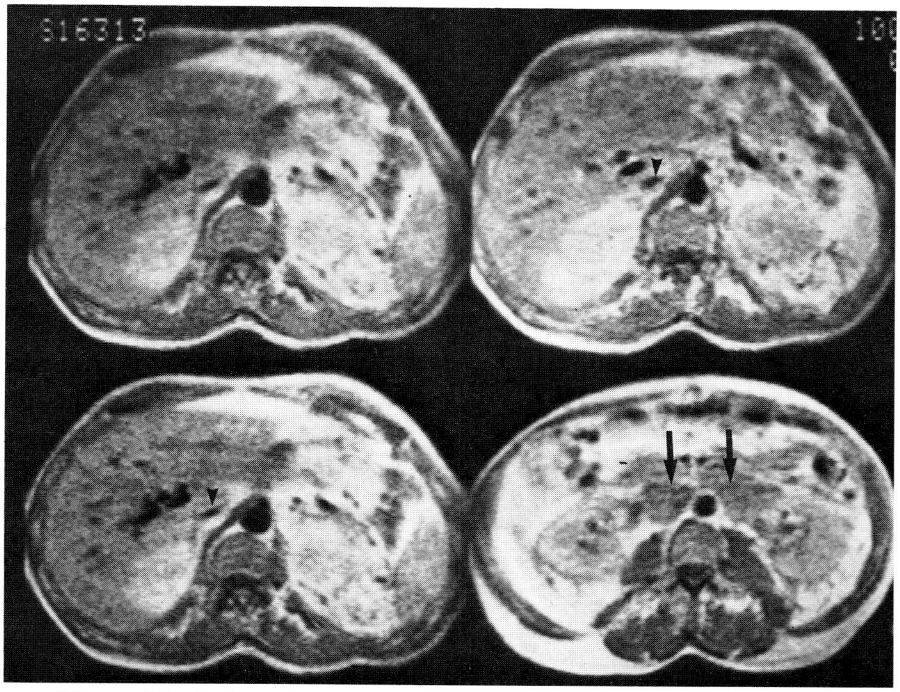

Figure 24. Series of NMR cross-sectional images of a patient with retroperitoneal fibrosis. Two large plaques of retroperitoneal fibrosis (arrows) are shown on most caudal image. On this image the IVC is severely narrowed (arrowhead). On more cranial images the disparity in the diameter of the narrow inferior vena cava and portal venous system is evident.

(Reproduced, with permission, from Higgins CB, Goldberg H, Hricak H, et al: Nuclear magnetic imaging of vasculature of abdominal viscera: Normal and pathologic features. AJR 140:1217, 1983.)

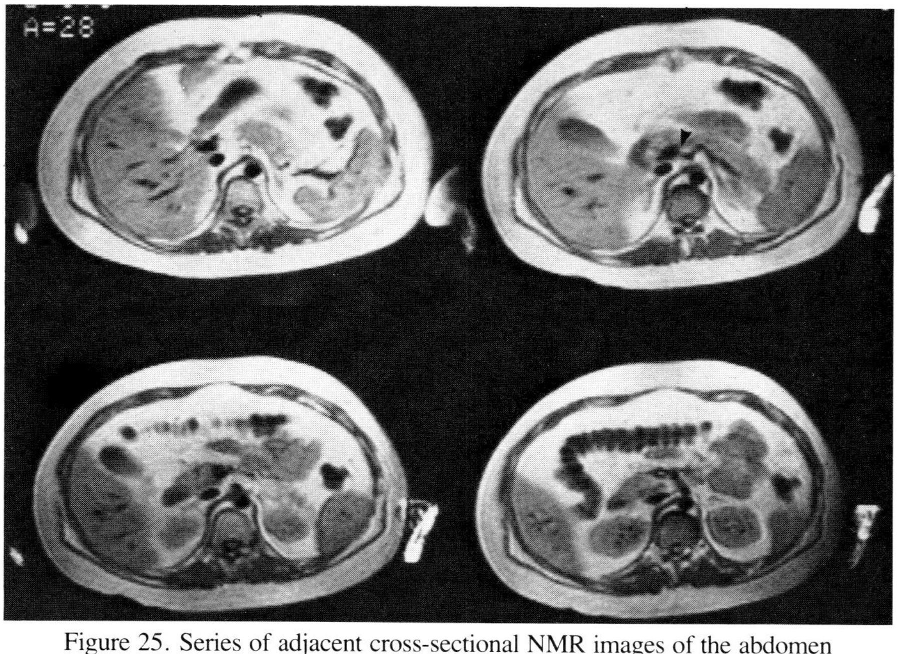

Figure 25. Series of adjacent cross-sectional NMR images of the abdomen of a normal volunteer reveal the formation of the splenic vein in the hilum of the spleen, midsplenic vein in contiguity with the pancreas, and junction with the portal vein (arrowhead). Pancreas can be definitely identified by the splenic vein coursing in contiguity with it (right upper panel).

(Reproduced, with permission, from Higgins CB, Goldberg H, Hricak H, et al: Nuclear magnetic imaging of vasculature of abdominal viscera: Normal and pathologic features. AJR 140:1217, 1983.)

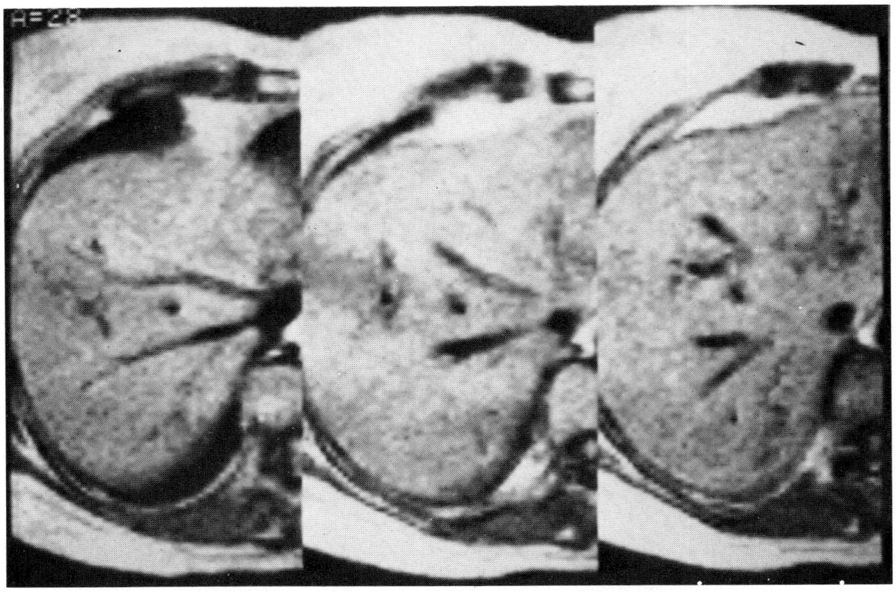

Figure 26. Series of adjacent NMR images through the liver show the hepatic veins coursing from the periphery of the liver on caudal image (right) toward their confluence with the inferior vena cava on the most cranial image (left).

(Reproduced, with permission, from Higgins CB, Goldberg H, Hricak H, et al: Nuclear magnetic imaging of vasculature of abdominal viscera: Normal and pathologic features. AJR 140:1217, 1983.)

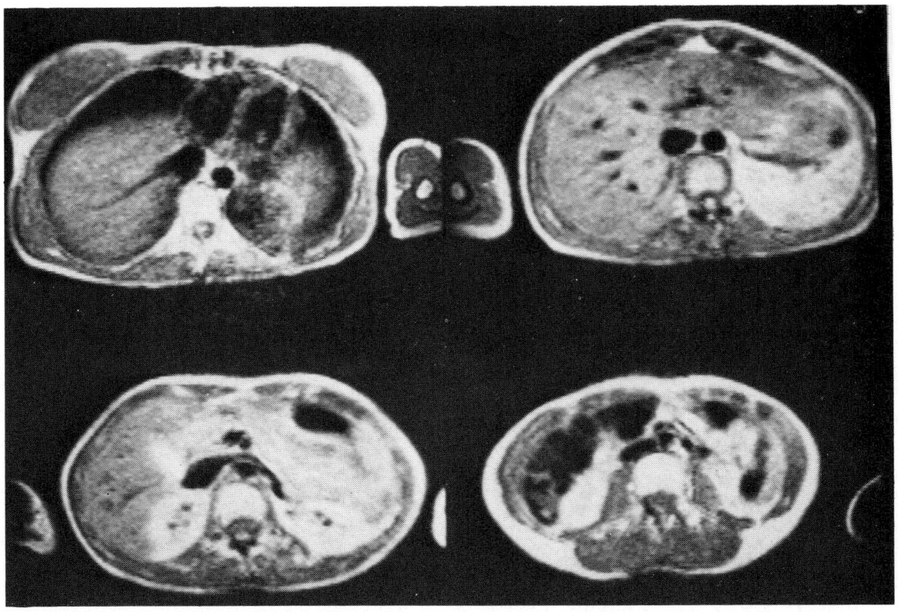

Figure 27. Series of NMR images of the abdomen in a normal volunteer display variations in the shape of the vena cava below the renal veins, at junction of renal veins, within the liver, and at the junction with the right atrium. Prosthetic breast implants are shown on the image in the left upper panel.

(Reproduced, with permission, from Higgins CB, Goldberg H, Hricak H, et al: Nuclear magnetic imaging of vasculature of abdominal viscera: Normal and pathologic features. AJR 140:1217, 1983.)

renal veins, within the liver, and at the junction with the right atrium can be appreciated (Fig. 27).

Identification of masses within the pancreas was facilitated by the clear delineation and displacement of the splenic vein coursing adjacent to the pancreas (Fig. 28). The absence of intraluminal signal in the splenic vessel should indicate patency of this vessel (Fig. 28). This conjecture, however, has not yet been confirmed by correlative angiographic and NMR studies in a group of patients with pancreatic abnormalities.

The effect of hepatic masses on intrahepatic veins can be precisely assessed on NMR images (Figs. 29, 30). Intrahepatic masses were shown to cause displacement of intrahepatic veins, the portal vein, and the inferior vena cava. Distortion of the shape of the inferior vena cava by hepatic masses, particularly those in the caudate lobe, were demonstrated (Fig. 30). An intrahepatic tumor (hepatoma) extending with the inferior vena cava as far as the junction with the right atrium was also demonstrated (Fig. 29).

In patients with portal hypertension, this diagnosis is evident by the marked dilation of the portal venous system (Fig. 31). Similarly, venous collaterals and splenic congestion are demonstrated on NMR images.

Atherosclerosis

Atherosclerotic disease involves the walls of arteries with a variety of lesions: from a nonprotrusion fatty streak to complex lesions characterized by the accumulation of lipid, smooth muscle, fibroblasts, and, occasionally, calcification, which eventually projects into and narrows the vascular lumina. Ex vivo NMR imaging of iliac arteries displayed atherosclerotic plaques that correlated with postmortem examination.[1,3] With a knowledge of the anticipated appearance of atherosclerotic vascular lesions, it has been possible to identify atherosclerotic lesion of the aorta and the iliac and femoral arteries on NMR images.[13]

In young patients and volunteers without known vascular disease, the aortic and ileofemoral vessels had a thin wall and a uniform thickness around their circumference. The endothelial surface was smooth and devoid of lesions projecting into the lumen (Figs. 21, 22). In several of these individuals, sagittal images of the abdominal aorta (Fig. 22) demonstrated thin anterior and posterior walls without evidence of localized thickening or endothelial irregularity.

Nuclear magnetic resonance images in a number of patients have shown eccentric thickening of the aorta (Fig. 32) and also depicted discrete lesions

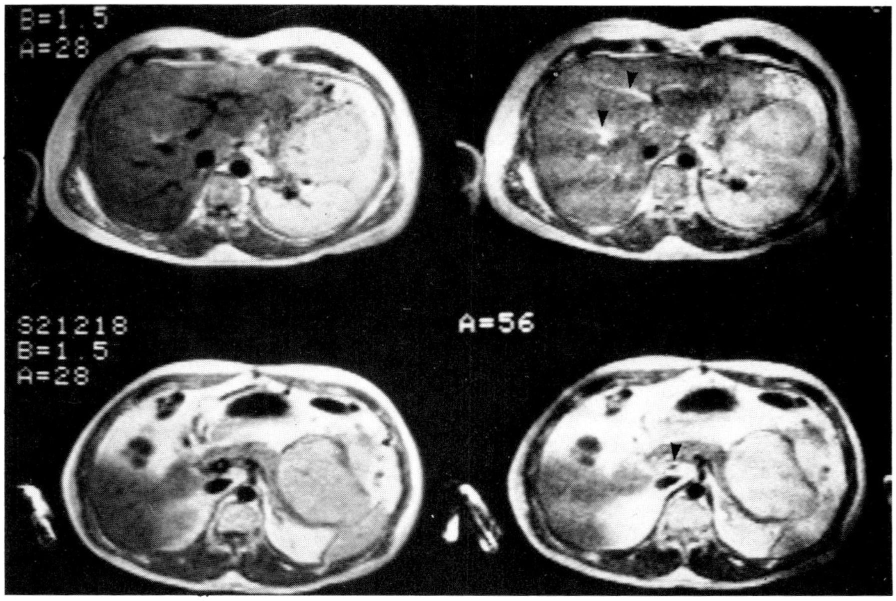

Figure 28. Nuclear magnetic resonance images of the upper abdomen in a 49-year-old man with a tumor of the tail of pancreas (apudoma). Images produced with first (1.5 sec/28 msec, right) and second (1.5 sec, 56 msec, left) spin echoes are shown. High-intensity signal (paradoxical enhancement) is seen in the portal venous radicles (arrowheads) and superior mesenteric vein (arrowhead), but not in the inferior vena cava.

(Reproduced, with permission, from Higgins CB, Goldberg H, Hricak H, et al: Nuclear magnetic imaging of vasculature of abdominal viscera: Normal and pathologic features. AJR 140:1217, 1983.)

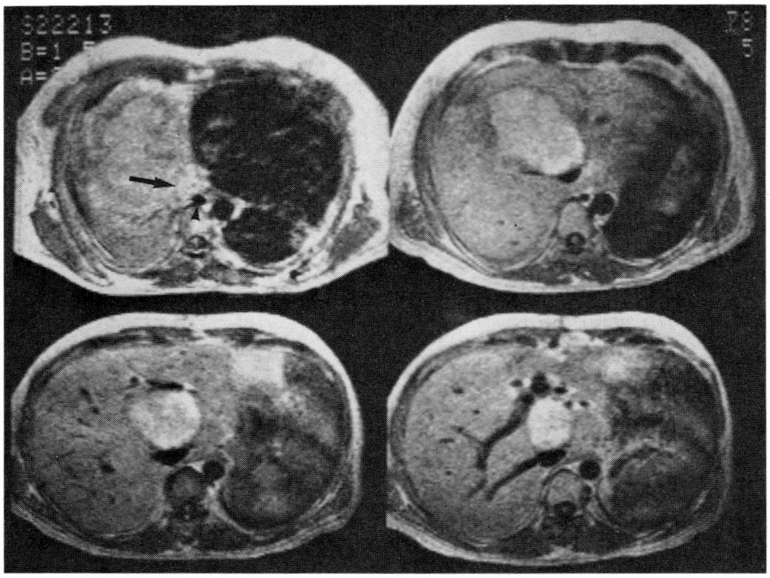

Figure 29. Series of NMR images through the liver show an extensive hepatoma that has high signal intensity. The most cranial images show that the high-intensity tumor has extended into the inferior vena cava (arrow) and advanced to the level of the inferior vena cava-right atrial junction. There is marked distortion and displacement of the inferior vena cava and portal vein. Both of these vessels are markedly dilated on the two caudal images. Because of obstruction of the intrahepatic inferior vena cava by tumor, the azygos vein is markedly dilated (arrowheads).

(Reproduced, with permission, from Higgins CB, Goldberg H, Hricak H, et al: Nuclear magnetic imaging of vasculature of abdominal viscera: Normal and pathologic features. AJR 140:1217, 1983.)

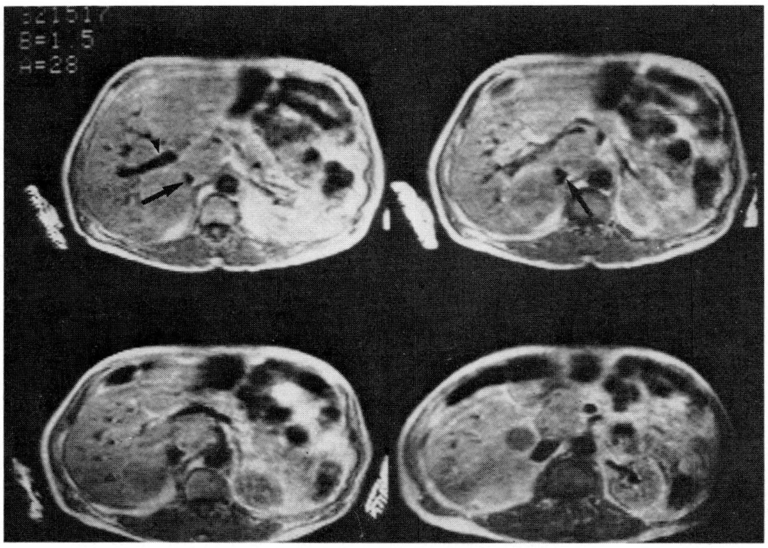

Figure 30. Nuclear magnetic resonance cross-sectional images of a patient with hepatic metastasis from colonic carcinoma. Multiple low-intensity regions in the liver correspond to metastatic lesions also seen on CT scan (not shown here). The inferior vena cava (arrow) is distorted by these metastases, particularly by one in the caudate lobe. Portal venous system is mildly dilated (arrowhead).

(Reproduced, with permission, from Higgins CB, Goldberg H, Hricak H, et al: Nuclear magnetic imaging of vasculature of abdominal viscera: Normal and pathologic features. AJR 140:1217, 1983.)

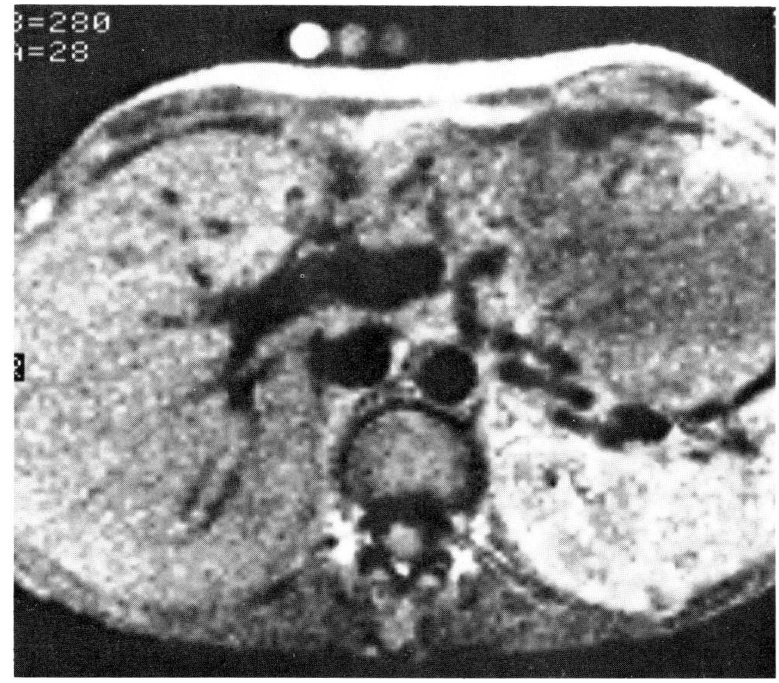

Figure 31. Nuclear magnetic resonance images show a markedly dilated portal vein in a patient with portal hypertension. The portal vein and inferior vena cava are normally of nearly similar size. Venous collaterals emanating from the hilum of the spleen are also well visualized.

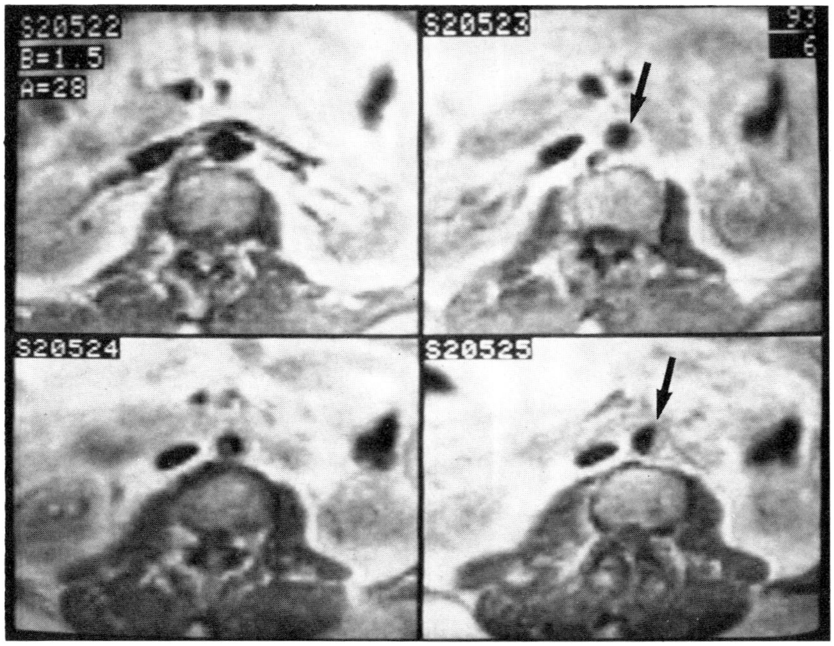

Figure 32. Multilevel cross-sectional NMR images of the infrarenal abdominal aorta shows eccentric narrowing (arrows) of the aorta and localized increase in wall thickness at multiple levels. Magnified views at one level show the eccentric narrowing produced by an atherosclerotic plaque. Note that this plaque has an intensity greater than muscle and near that of retroperitoneal fat.

(Reproduced, with permission, from Herfkens R, Davis PL, Crooks LE, et al: NMR imaging of atherosclerotic disease. Radiology 148:161, 1983.)

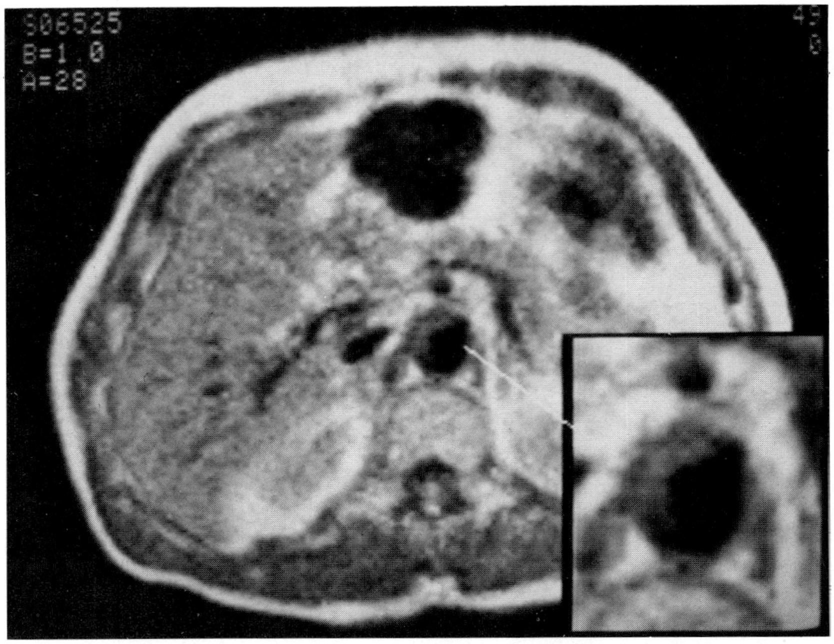

Figure 33. Large irregular atherosclerotic plaque protruding into the abdominal aorta. Inset contains a magnified view of the aorta.
(Reproduced, with permission, from Herfkens R, Davis PL, Crooks LE, et al: NMR imaging of atherosclerotic disease. Radiology 148:161, 1983.)

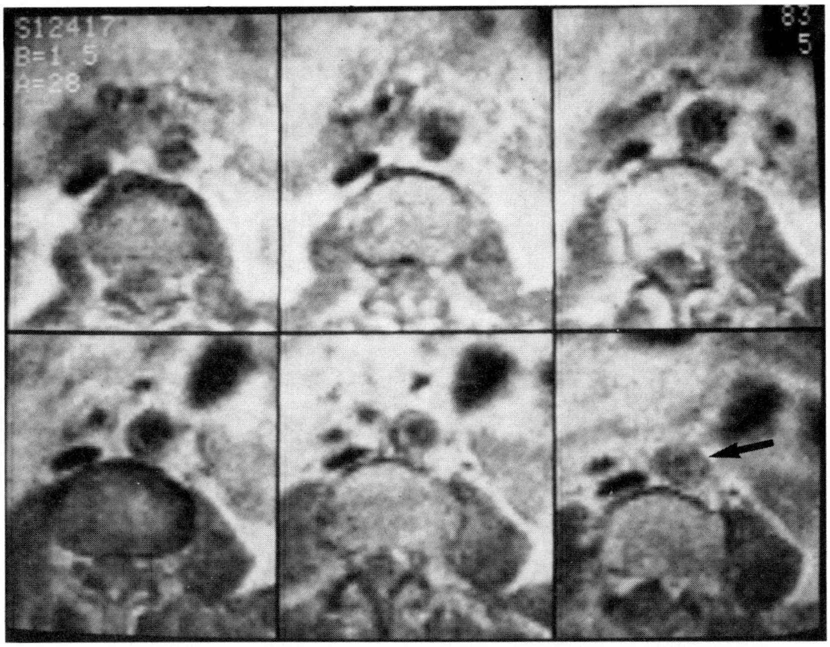

Figure 34. Sequential NMR images of abdominal aorta below the origin of the renal arteries reveal concentric narrowing of the infrarenal aorta. At the most caudal level of the aortic lumen is almost completely filled with atherosclerotic debris (arrow).
(Reproduced, with permission, from Herfkens R, Davis PL, Crooks LE, et al: NMR imaging of atherosclerotic disease. Radiology 148:161, 1983.)

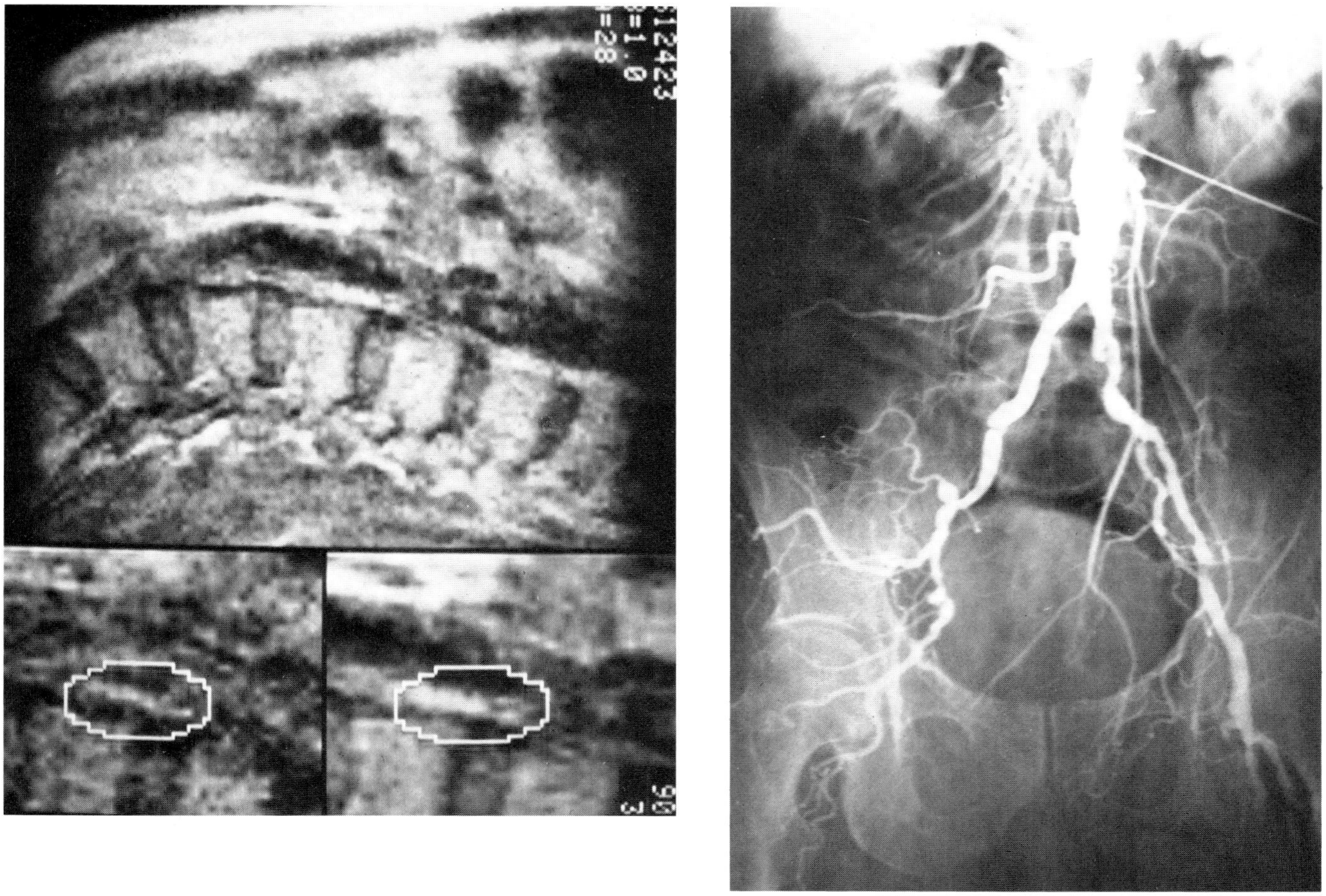

Figure 35. Sagittal NMR image (left) of the abdominal aorta reveals generalized thickening and high-intensity areas at several sites in the posterior wall. Inset shows magnified views of the high-intensity areas, which presumably represent lipid-laden atherosclerotic plaques (within white outlines). Aortogram (right) confirms the severe and diffuse atherosclerotic disease of the aorta and iliac arteries.
(Reproduced, with permission, from Herfkens R, Davis PL, Crooks LE, et al: NMR imaging of atherosclerotic disease. Radiology 148:161, 1983.)

protruding into the infrarenal aorta in some (Fig. 33). In some patients multilevel images also demonstrated concentric narrowing of the distal aorta (Fig. 34). Atherosclerotic vascular disease detected on NMR images has been confirmed by angiography.

Cross-sectional NMR images at levels between the renal arteries and the aortic bifurcation revealed substantial thickening of the aortic wall and marked narrowing of the aortic lumen in a man in whom aortography showed diffuse atherosclerosis of the aorta and iliac arteries (Fig. 34). Sagittal NMR images through the abdominal aorta of a patient with angiographic evidence of severe aortic atherosclerosis revealed irregular thickening of the posterior aortic wall (Fig. 35). At two sites of prominent mural thickening there was increase in intensity of the NMR signal. Presumably the greater intensity of signal in these regions represents lipid-laden atherosclerotic plaques. A cross-table lateral view of the abdominal aorta was not available in this patient with severely symptomatic peripheral vascular disease. The frontal aortogram revealed severe diffuse aortoiliac atherosclerosis (Fig. 35).

Calcification, such as in cortical bone, produces low NMR signal intensity. Calcification of the aortic wall was represented on the NMR images as an arc or circumferential rim of decreased intensity (Fig. 36). A CT scan at the same level as the NMR image confirmed the presence of calcification within the aortic wall.

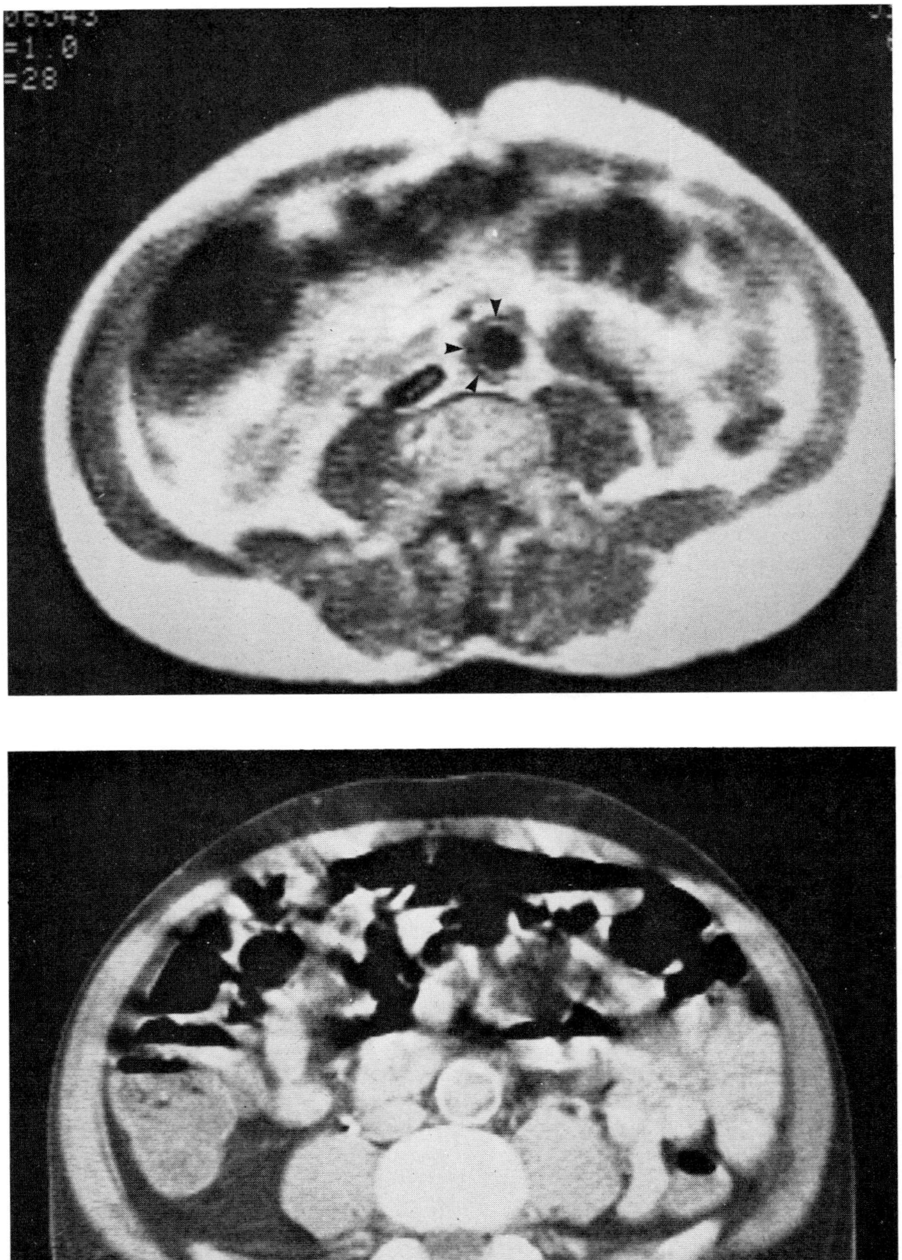

Figure 36. NMR image of the infrarenal aorta shows irregular thickening of the vessel wall and a lucent arc in the right side of the aortic wall (arrowheads). This represents mural calcification; calcification produces little or no NMR signal. A CT scan at the same level confirms the presence of aortic mural calcification. Calcification seems more extensive on the CT scan.

(Reproduced, with permission, from Herfkens R, Davis PL, Crooks LE, et al: NMR imaging of atherosclerotic disease. Radiology 148:161, 1983.)

Future Directions

There is currently too little clinical experience with NMR imaging to predict with any reliance the eventual role of this technique in cardiovascular diagnosis. However, it offers three intriguing insights into the assessment of cardiovascular disease processes and their response to therapeutic interventions: direct tissue characterization, noninvasive regional blood flow measurements, and chemical and metabolic imaging.

The magnetic relaxation times (T_1 and T_2 times) of protons in different tissues are currently the basis for tissue characterization. These relaxation times and their diagnostic implications are poorly understood at present. Nevertheless, both normal and pathologic tissues have been shown to have combinations of T_1 and T_2 values that fall into a characteristic range for a specific tissue.[5,14,15] These specific relaxation times for various tissues have been estimated in experimental animals and in man. Various disease processes, including ischemia, have been found to alter the magnetic relaxation times of tissues and organs.[5,14,16]

Besides the qualitative identification of abnormal flow patterns discussed in this chapter, it should be possible to noninvasively quantitate blood flow using NMR imaging. A number of techniques have been suggested for quantitating flow with NMR, including measurement of transit time of excited nuclei[17] and the generation of "flow graphs" that plot the number of nuclei flowing within various velocity intervals.[18]

Perhaps the most promising aspect of NMR imaging is the potential for metabolic imaging. This capability exploits another parameter of NMR, the ability to measure chemical shifts. The chemical nuclei (other than hydrogen) that seem most attractive for in vivo chemical studies because of a compromise among their NMR sensitivity, abundance in biological systems, and biochemical interests are Na-23, C-13, and P-31. Most of the current work on NMR spectroscopy of the myocardium in vivo has involved P-31. These P-31 NMR spectroscopic studies have provided a method for monitoring (1) shifts in the relative concentration of high-energy phosphate compounds and inorganic phosphate, (2) intracellular pH during myocardial ischemia, and (3) interventions intended to attenuate the deleterious myocardial effects of ischemia.[19,20]

Research into the use of NMR imaging for the diagnosis and assessment of severity of cardiovascular disease has just recently begun. Our knowledge of NMR imaging at this time will undoubtedly be shown in the next decade to be quite primitive. The observations presented in this chapter are gleaned from our initial experience.

REFERENCES

1. Crooks LE, Sheldon P, Kaufman L, et al: Quantification of obstructions in vessels by nuclear magnetic resonance (NMR). IEEE Trans Nucl Sci NS-29:1181, 1982.
2. Kaufman L, Crooks LE, Sheldon PE, et al: The potential impact of nuclear magnetic resonance imaging on cardiovascular diagnosis. Circulation 67:251, 1983.
3. Kaufman L, Crooks, LE, Sheldon PE, et al: Evaluation of NMR imaging for detection and quantification of obstruction in vessels. Invest Radiol 17:554, 1983.
4. Hansen G, Crooks LE, Davis PL, et al: In vivo imaging of the rat with nuclear magnetic resonance. Radiology 136:695, 1980.
5. Davis PL, Kaufman L, Crooks LE, et al: NMR characteristics of normal and abnormal rat tissue. In *Nuclear Magnetic Resonance Imaging in Medicine,* Kaufman L, Crooks LE, Margulis AR (Eds), p. 71. New York: Igaku-Shoin, 1981.
6. Brady TJ, Goldman MR, Pykett IL, et al: Proton nuclear magnetic resonance imaging of regionally ischemic canine hearts: Effects of paramagnetic proton signal enhancement. Radiology 143:169, 1982.
7. Williams ES, Kaplan JI, Thatcher F, et al: Prolongation of proton spin lattice relaxation times in regionally ischemic tissue from dog hearts. J Nucl Med 21:449, 1980.
8. Higgins CB, Herfkens R, Lipton MJ, et al: Nuclear magnetic resonance imaging of acute myocardial infarctions: Alterations in magnetic relaxation times. Am J Cardiol (in press).
9. Herfkens RJ, Higgins CB, Hricak H, et al: Nuclear magnetic resonance imaging of the cardiovascular system: Normal and pathologic findings. Radiology 147:749, 1983.
10. Alfidi RJ, Haaga JR, El Yousef SJ, et al: Preliminary experimental results in humans and animals with superconducting, whole body, nuclear magnetic resonance scanner. Radiology 143:175, 1982.
11. Hawkes RC, Holland GN, Moore WS, et al: Nuclear magnetic resonance (NMR) tomography of the normal abdomen. J Comput Tomogr 5:613, 1981.
12. Higgins CB, Goldberg H, Hricak H, et al: Nuclear magnetic imaging of vasculature of abdominal viscera: Normal and pathologic features. AJR 140:1217, 1983.
13. Herfkens R, Davis PL, Crooks LE, et al: NMR imaging of atherosclerotic disease. Radiology 148:161, 1983.
14. Herfkens R, Davis PL, Crooks LE, et al: Nuclear magnetic resonance imaging of the abnormal rat and correlation with tissue characteristics. Radiology 141:211, 1981.

15. Hutchison JMS, Smith FW: Human NMR imaging. In *Nuclear Magnetic Resonance Imaging in Medicine,* Kaufman L, Crooks LE, Margulis AR (Eds), p. 101. New York: Igaku-Shoin, 1981.

16. Herfkens RJ, Lipton MJ, Crooks LE, et al: NMR imaging of infarcted muscle: A rat model. Radiology 147:761, 1983.

17. Battocletti JH, Halback RE, Salles-Cunha SX, et al: NMR blood flowmeter: Theory and history. Med Phys 8:435, 1980.

18. Singer JR: Blood flow measurements by NMR. In *Nuclear Magnetic Resonance in Medicine,* Kaufman L, Crooks LE, Margulis AR (Eds), p. 11. New York: Igaku-Shoin, 1981.

19. Jacobus WE, Taylor GJ, Hollis DP, et al: Phosphorus nuclear magnetic resonance of perfused working rat hearts. Nature 265:756, 1977.

20. Nunnally RL, Bottomley PA: ^{31}P NMR studies of myocardial ischemia and its response to drug therapies. J Comput Assist Tomogr 5:296, 1981.

13

Liver, Gallbladder, Alimentary Tube, Spleen, Peritoneal Cavity, and Pancreas

ALBERT A. MOSS, M.D.
DAVID D. STARK, M.D.
HENRY I. GOLDBERG, M.D.
ALEXANDER R. MARGULIS, M.D.

Nuclear magnetic resonance imaging of the abdomen can accurately display both normal anatomy and a variety of pathologic conditions of the organs.[1-4] The liver, pancreas, and spleen are readily demonstrated as structures with an intermediate NMR signal that are surrounded by fat, which has a very intense signal. The gallbladder signal varies from very intense to low, depending on the concentration of bile. The gastrointestinal tract can be well delineated when it contains air or is surrounded by fat, but is less well differentiated from adjacent solid organs when it is fluid-filled. Optimal imaging of the gastrointestinal tract requires administration of oral or rectal contrast material or both. Air and positive paramagnetic contrast materials are being investigated.

Liver and Gallbladder

Normal

The normal liver and its contents are well delineated by spin echo, inversion recovery, and saturation recovery NMR imaging techniques.[4-7] The normal liver appears as an organ of homogeneous moderate intensity in the right upper abdomen (Fig. 1).

Normal hepatic vascular structures— the hepatic veins, portal veins, and intrahepatic portion of the inferior vena cava—are clearly shown without contrast medium administration as low-intensity structures because of the rapid flow of blood within the hepatic vascular system (Fig. 2). The biliary system can be differentiated from the vascular system by varying imaging techniques. Periportal fat permits the anatomy of the porta hepatis to be clearly displayed without administration of biliary or vascular contrast agents.

The normal liver, as measured on our system, has an average T_1 relaxation time (440 msec) that is eight to 10 times the T_2 relaxation time (48 msec). While the reported ratio of T_1 and T_2 relaxation times obtained on NMR systems having different magnetic field strengths and employing different NMR data acquisition techniques has been similar, individual calculated T_1 and T_2 times have varied greatly. Thus T_1 and T_2 relaxation times as calculated by imaging systems should not be considered as "absolute" values, and times obtained on one system cannot be quantitatively compared with times calculated on a different NMR system.

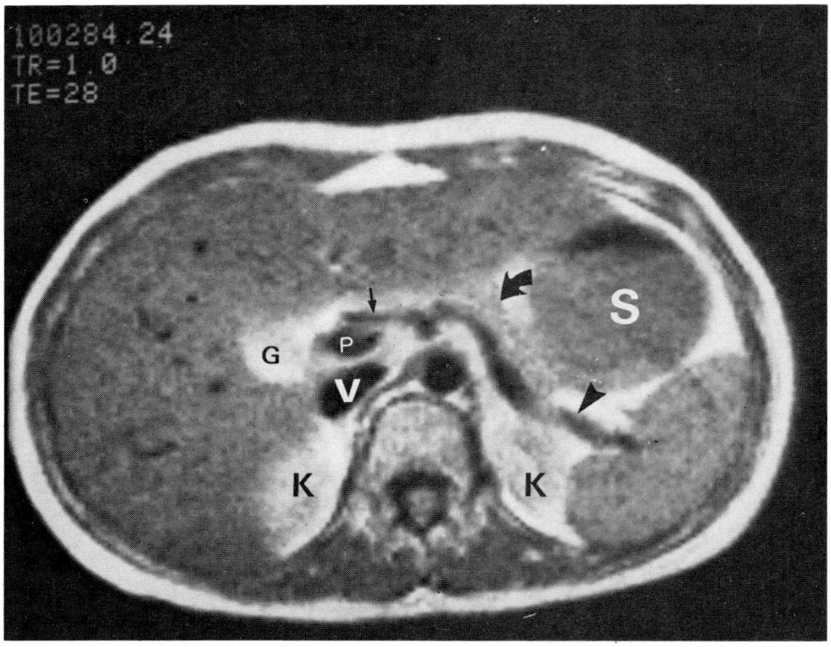

Figure 1. Normal upper abdomen: spin echo image. The liver and spleen are seen as organs having homogeneous moderate intensity. The intensity of the liver is less than that of the kidneys (K) and greater than that of the paraspinous muscles. (S, fluid-filled stomach; V, inferior vena cava; P, portal vein; hepatic artery [arrow]; splenic artery [arrowhead]; pancreas [curved arrow]; G, gallbladder)

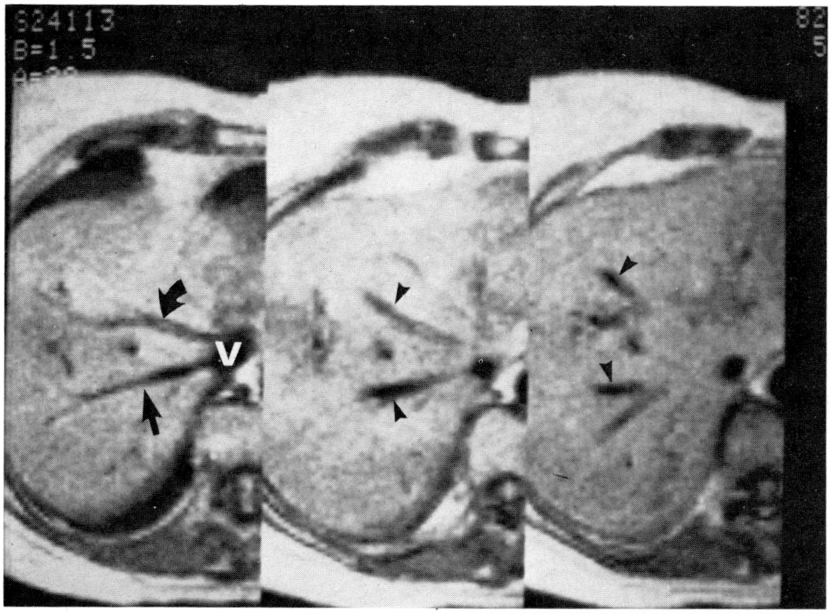

Figure 2. Normal hepatic veins: three spin echo images. The right (arrow) and middle hepatic vein (curved arrow) are seen entering the inferior vena cava (V). At more caudal levels the hepatic veins (arrowheads) are demonstrated within the liver parenchyma. The vascular structures are patent because the rapid flow of blood provides a signal of very low intensity.

Technique

In hepatic NMR imaging, the spatial and contrast resolution of the hepatic image vary greatly, depending on the data acquisition technique employed to obtain the image. Selection of the optimal NMR technique is both critical and difficult because an infinite number of techniques are theoretically possible. The NMR timing parameters TR and TE can be varied independently over an enormous range, and hepatic images can be obtained using spin echo, inversion recovery, partial saturation recovery, free induction decay imaging techniques, or some combination of these. Hepatic NMR images shown in this chapter were obtained using spin echo and inversion recovery techniques. Spin echo images were obtained with TR intervals of 0.5, 1.0, 1.5 and/or 2.0 sec and TE intervals of 28 and 56 msec.

Inversion recovery images had a TR interval of either 1.0 or 1.8 sec and an acquisition delay recovery time (TI) of either 420 or 277 msec. T_1 and T_2 values were calculated from the spin echo images. Imaging time was 4.5 or 8.5 min for five sections and 13 min for 15 sections.

Direct coronal and sagittal NMR images can be obtained and have proved useful in defining the relationship of the liver to the diaphragm and in evaluating lesions involving the inferior vena cava or right atrium.

The ideal method of NMR imaging the liver has yet to be determined. However, while it is premature to endorse any particular technique, certain trends are evident. Most reported tumors have prolongation of both T_1 and T_2 values, although the degree of prolongation varies from patient to patient and even within various lesions in the same patient.[4,8] Therefore, it appears prudent to image the liver with techniques that are both relatively more T_1-dependent (inversion recovery) and T_2-dependent (spin echo). Spin echo techniques with long TR and TE times and inversion recovery techniques with long TR and short TI intervals appear to most frequently demonstrate hepatic mass lesions clearly.[4,8] Further investigation is required to further refine hepatic NMR imaging techniques.

Primary Hepatic Tumors

Hepatoma

The intensity of hepatomas as compared with normal liver varies, depending on the NMR imaging parameters employed to obtain the image. Hepatocellular carcinomas have prolonged T_1 values[8] and thus are usually best visualized with spin echo techniques using a long interval between excitations (TR of 1.5 sec). Hepatomas usually have a higher intensity than normal liver on spin echo images with long intervals between excitations; when shorter TR intervals are employed, however, distinction of tumor from surrounding liver becomes less apparent (Fig. 3). On inversion recovery images, hepatomas usually appear as masses having a lower intensity than normal liver. Inversion recovery images obtained with long TR and short TI intervals enhance contrast between normal liver and hepatomas. Because most hepatomas have prolonged T_2 values, contrast between hepatoma

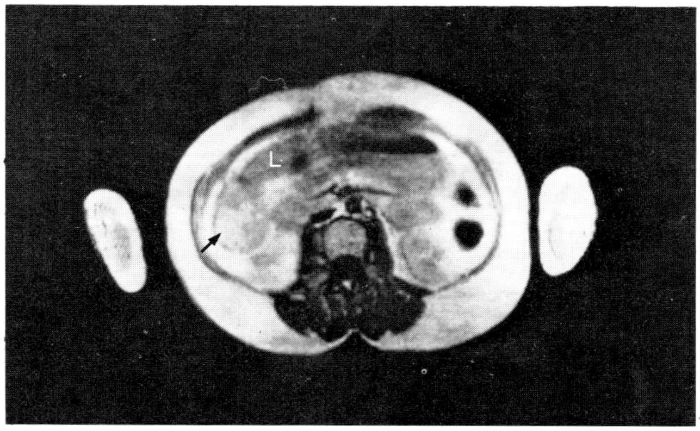

A.

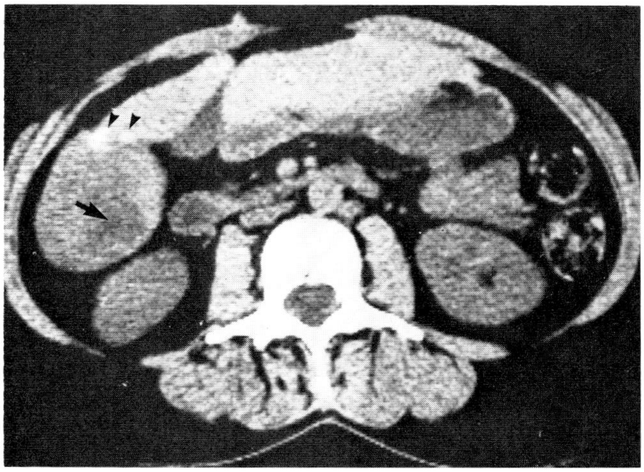
B.

Figure 3. Hepatoma.
A. Spin echo image demonstrates pedunculated hepatoma arising from the right hepatic lobe. The tumor has an irregular central area of increased intensity (arrow) as compared with normal liver (L).
B. A CT scan at same level reveals the hepatoma to have a necrotic center (arrow). Foci of calcification (arrowheads) are seen that were not detected on the NMR image.

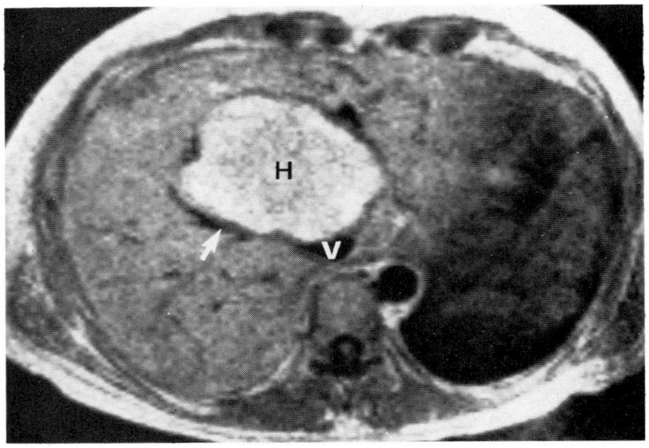

A.

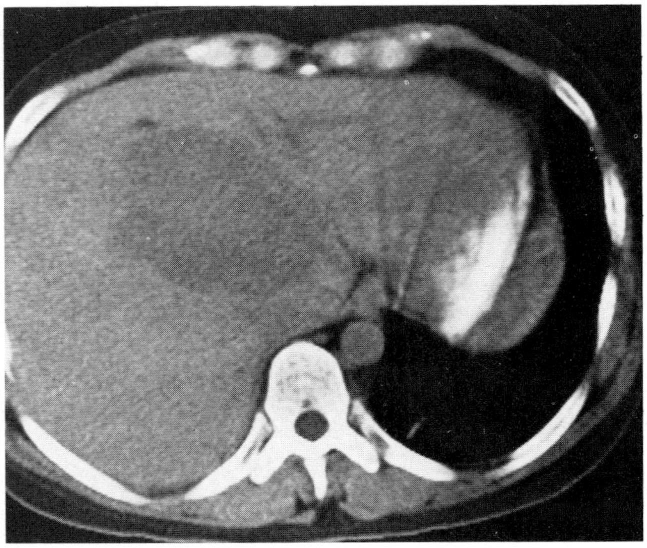

B.

Figure 4. Hepatoma.
A. Spin echo image clearly shows a centrally positioned hepatoma (H) as a mass of high intensity. The hepatoma is displacing but not obstructing the inferior vena cava (V) and portal vein (arrow). The internal structure and surrounding capsule are well delineated.
B. A CT scan at the same level. The internal architecture of the liver capsule and relationship to hepatic vascular structures are poorly delineated.

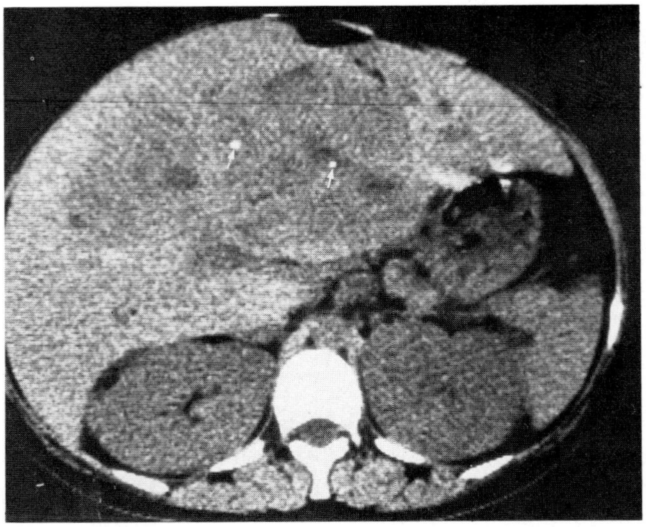

A.

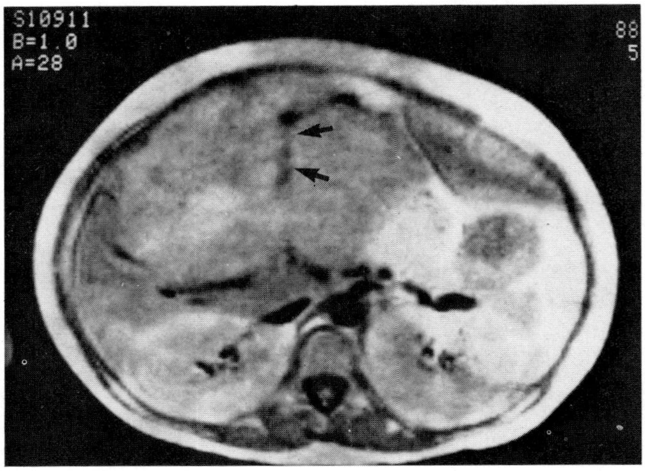

B.

Figure 5. Hepatic adenoma.
A. A CT scan demonstrates a large hepatic mass containing two foci of calcification (arrows). The tumor is not homogeneous, and a surrounding capsule is not seen.
B. Spin echo image reveals a bilobed adenoma that has a greater intensity than the normal liver. The internal septations are clearly shown as bands (arrows) having a low intensity. Calcifications are not seen.

and liver is usually increased on spin echo images obtained with an echo delay (TE) of 56 msec as compared with those obtained using a TE of 28 msec.

Nuclear magnetic resonance imaging has been of great value in demonstrating internal structure within hepatomas, displaying the margin of tumors, and in depicting the relationship of hepatic vascularity to the malignancy (Fig. 4). Both T_1 and T_2 relaxation times can vary among different regions of a solitary hepatoma, and among multifocal hepatomas in individual patients. Calculated T_1 times have tended to vary to a greater extent than T_2 relaxation times.[8]

The NMR diagnosis of solitary or multifocal hepatomas is usually straightforward because differences in intensity and in the T_1 and T_2 relaxation time values of tumor and liver are usually readily apparent. Infiltrating hepatomas that replace almost the entire liver are more difficult to diagnose because the tumor-liver interface is frequently ill defined. An accurate diagnosis can be made by noting that there is loss of the normal portal and hepatic vascular structures in the tumor mass and by comparing tumor T_1 and T_2 times with standards for normal liver.

Adenoma

There has been limited experience using NMR to evaluate benign hepatic adenomas. The NMR appearance did not differ from that of hepatoma in two reported cases,[8] and the calculated T_1 and T_2 relaxation times fell within the range encountered in hepatomas and metastatic tumors. Nuclear magnetic resonance depicted the internal architecture and surrounding capsule of adenomas better than CT did, but did not demonstrate foci of calcification as well (Fig. 5).

Hemangioma

A hepatic hemangioma, because of the very slow flow through the lesion, appeared as a mass having a very high intensity (Fig. 6) on a spin echo image obtained with a long (2.0-sec) TR interval. The NMR intensity of hemangiomas will probably vary depending on the size, degree of necrosis, and rate of blood flow to and through the parenchyma of the hemangioma.

Metastatic Tumors

A variety of hepatic metastases have been studied by NMR imaging using a variety of techniques.[2,4-8] Young et al.[2] reported that inversion recovery images have greater contrast than partial saturation recovery images and provide the best tumor-liver contrast difference. Alfidi and coworkers, comparing saturation recovery, inversion recovery, and calculated T_1 images, felt that inversion recovery and T_1 images were superior to saturation recovery images.[9] Comparison of spin echo and inversion recovery techniques by Moss et al.[8] revealed that some tumors were better demonstrated by spin echo techniques, whereas inversion recovery techniques displayed other tumors better.

On spin echo images, hepatic metastases usually appear as masses having a greater intensity than normal liver (Fig. 7), but some of them demonstrate a lower intensity on some spin echo images (Fig. 8). As in primary hepatic tumors, metastatic deposits are usually best demonstrated on spin echo images with long TR and TE intervals. On images obtained with short TR and TE intervals, lesions can appear isointense with normal liver and be undetectable on a particular spin echo image (Fig. 7). When this occurs, imaging with longer TR and TE times usually clearly demonstrates the metastatic focus.

Metastatic tumors typically are shown as masses of low intensity on inversion recovery NMR images (Figs. 8, 9), but tumors also can appear as hyperintense lesions. Inversion recovery images using different TR and TI parameters alter tumor-liver contrast, with slightly improved contrast on images obtained using long TR and short TI intervals. The center of a metastatic focus often has a longer T_1 time than the periphery of the lesion does. Thus metastatic deposits may appear as a mass that has a low-intensity peripheral rim surrounding a center of higher intensity (Fig. 10). This is analogous to the CT appearance of many metastatic lesions after a bolus injection of contrast material.

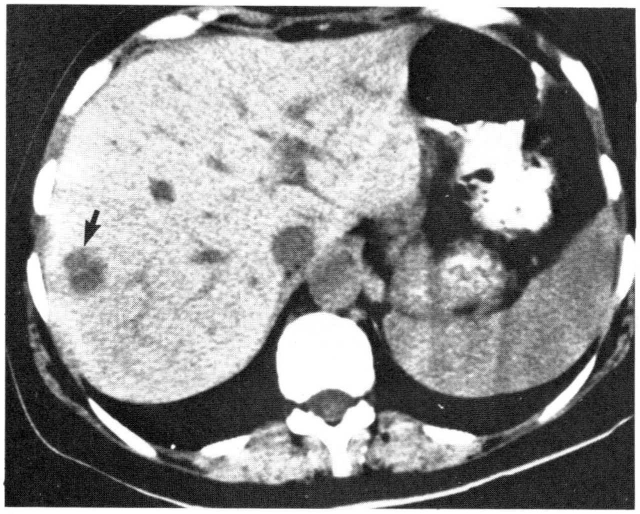

A.

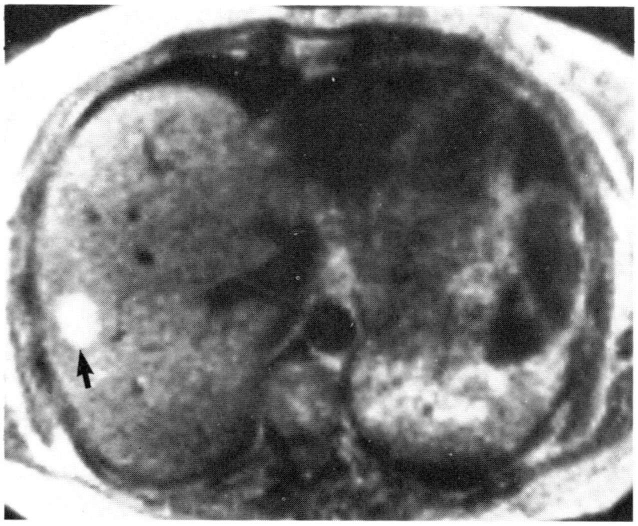

B.

Figure 6. Hemangioma of the liver.
A. A CT scan reveals a 2-cm hemangioma (arrow) in the right lobe of the liver.
B. Spin echo image demonstrates the hemangioma (arrow) as a high-intensity mass presumably because of the slow blood flow through the hemangioma.

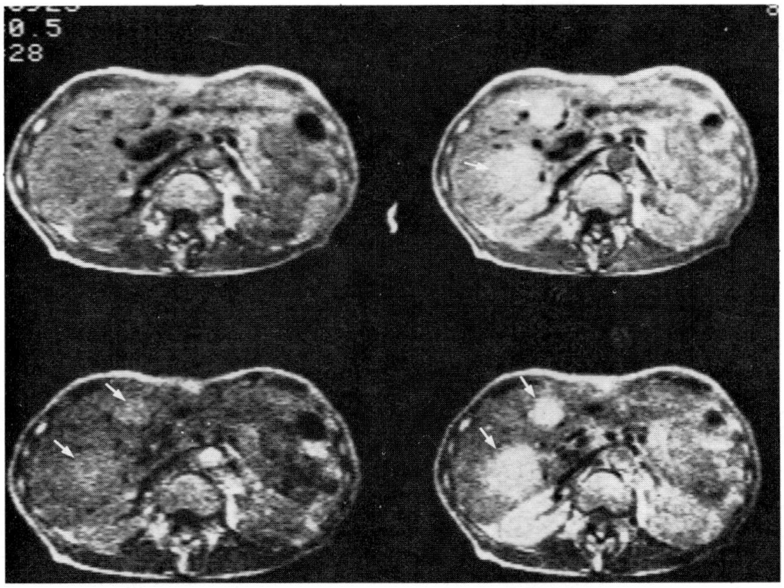

Figure 7. Carcinoid metastatic to liver. Spin-echo images: upper left, 0.5 sec and 28 msec; upper right, 0.5 sec and 56 msec; lower left, 1.0 sec and 28 msec; lower right, 1.0 sec and 56 msec. The metastatic foci (arrows) appear as masses having a higher intensity than normal liver. The metastases are best demonstrated on the 1.0-sec/56-msec image and are undetectable on the 0.5-sec/28-msec image.

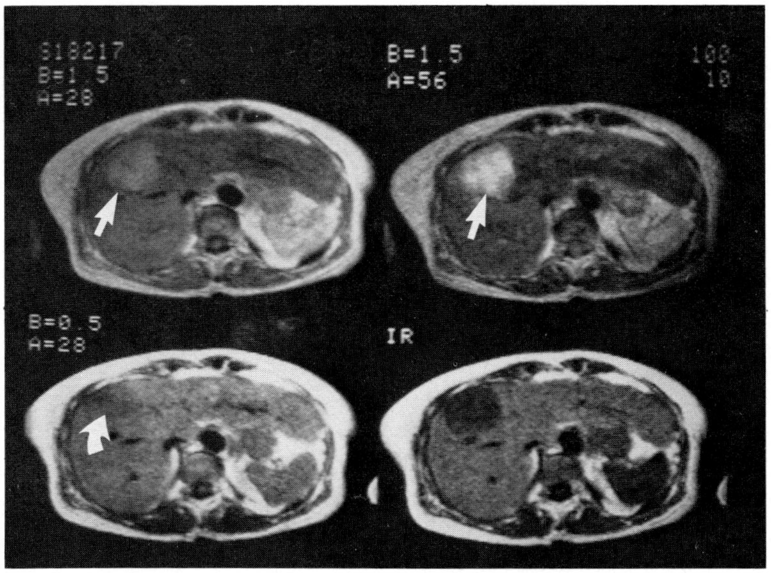

Figure 8. Hepatic metastasis. Spin echo images: upper left, 1.5 sec and 28 msec; upper right, 1.5 sec and 56 msec; lower left, 0.5 sec and 28 msec. Inversion recovery image at lower right. The metastasis (arrows) is demonstrated as a region of increased intensity on the 1.5-sec/28-msec and 1.5-sec/56-msec spin echo images and as a mass of slightly lower intensity (curved arrow) on the 0.5-sec/28-msec image. The metastasis has a very low intensity on the inversion recovery image.

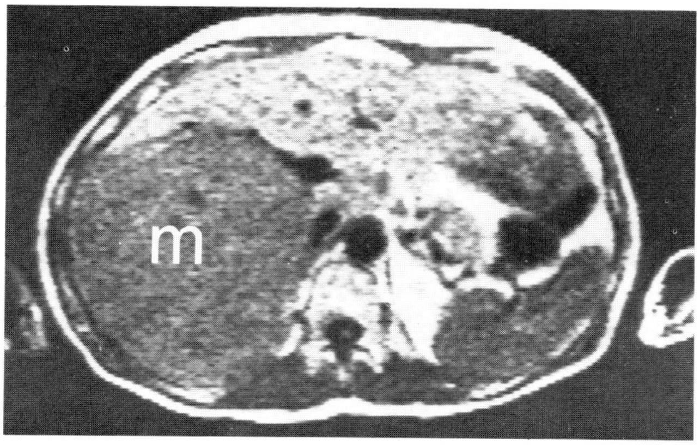

Figure 9. Colonic metastasis. Inversion recovery NMR image demonstrates a large metastatic focus (m) that has a lower intensity than normal liver.

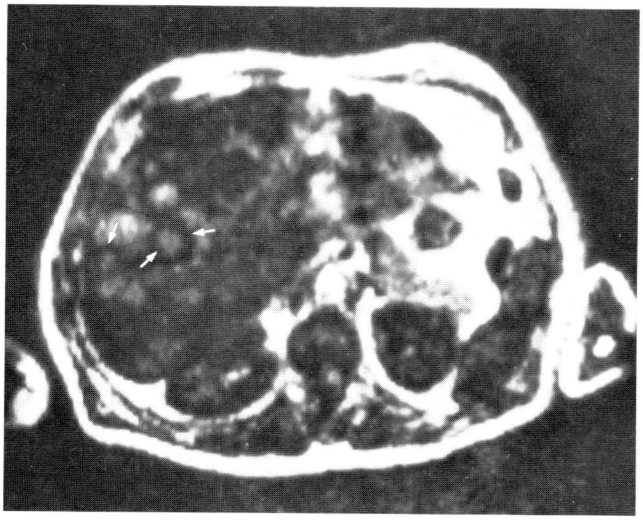

Figure 10. Hepatic metastasis. Inversion recovery image demonstrates multiple metastases, many of which have a low-intensity rim (arrows) around a mass of higher intensity.

The T_1 and T_2 relaxation times have been shown to vary between different tumors and similar tumors in different patients and even among tumors of the same cell type in a single patient. As in primary hepatic tumors, both the T_1 and T_2 values of metastatic tumors typically are longer than those of normal hepatic tissue; exceptions are found, however, where the T_2 relaxation time of a tumor is shorter than that of normal liver. The T_1 and T_2 relaxation times have been shown to vary slightly more in metastatic disease than in hepatomas.[8]

Although the average calculated T_1 and T_2 relaxation times of both primary and metastatic disease differ from those of normal liver, there is a large enough overlap between values that a single T_1 or T_2 calculation or combination of T_1 and T_2 times has not proved to be specific for a particular hepatic neoplasm. As compared with the relaxation times of normal livers, Moss et al. found tumors to have an average increase of 40% for T_1 and 21% for T_2, but found no consistent differences in relaxation times among various hepatic neoplasms.[8] The wide range of T_1 and T_2 relaxation times demonstrated in hepatic mass lesions probably reflects different degrees of cellularity, hemorrhage, blood flow, necrosis, and fat and water content within the tumors. Thus it appears unlikely that T_1 and T_2 measurements, either alone or in combination, will prove specific for particular hepatic tumors.

Comparison with Computed Tomography

At this time there is insufficient evidence to predict with assurance that NMR will be superior to CT in the detection and analysis of primary and metastatic tumors of the liver. Studies by Alfidi et al.[9] and Moss et al.[8] showed similar detection rates for the two technologies, although both compared early NMR images with mature CT technology. Computed tomography can best demonstrate calcified foci within tumors and assess tumor blood flow after intravenous bolus injections of contrast material. Surgical clips produce artifacts on CT scans, and direct coronal or sagittal images are not possible to obtain. Nuclear magnetic resonance displays the relationships of tumors to hepatic vasculature more clearly and permits more precise localization of tumors within the liver. Internal architecture of tumors is more clearly displayed by NMR, and in many cases surgical clips do not cause artifacts. An additional advantage of NMR is the ability to obtain direct sagittal and coronal images. Nuclear magnetic resonance also distinguishes blood vessels from the biliary ducts without contrast material injection.

Non-Neoplastic Liver Disease

A variety of metabolic, inflammatory, toxic, and autoimmune disease processes produce hepatic abnormalities. Imaging with NMR has proved capable of detecting and characterizing a variety of these hepatic diseases.[2,5–7,10,11] Moreover, quantitative analyses of NMR measurements have revealed significant correlations of NMR changes with histologic and biochemical changes.[12]

Hepatitis

Nuclear magnetic resonance has demonstrated prolongation of the T_1 and T_2 times in patients with proven hepatitis.[11] On spin echo images the involved liver appears more intense than normal liver and histologic sampling has shown that the most severe foci of hepatitis appear the most intensely (Fig. 11).

The abnormalities seen on NMR images of patients with hepatitis have been corroborated experimentally.[11] Hepatitis experimentally produced with carbon tetrachloride resulted in marked alterations in NMR signal intensity and T_1 and T_2 relaxation times. Relative to control livers, hepatic intensity, T_1 and T_2 relaxation times, and water and triglyceride content increased during the first 24 hr after carbon tetrachloride administration. At 48 hr, when hepatocellular necrosis occurred, liver intensity and T_2 relaxation time had increased, but water and triglyceride content remained unchanged. The maximum increase in T_1 relaxation time occurred just prior to hepatocellular necrosis, and reduction in T_1 time corresponded to hepatic cell destruction and the release of membrane constituents, mitochondrial debris, and ribosomal proteins into the cytoplasm and extracellular space. Measurements of T_2 relaxation time and hepatic intensity appear to be extremely sensitive in detecting the early changes associated with chemical hepatitis; both measurements become significantly elevated when even the earliest histologic alterations were demonstrated.[10] Thus, it appears that NMR will be capable of detecting and to some degree quantitating hepatocellular damage from hepatitis.

Steatosis

Nuclear magnetic resonance has not proved to be extremely sensitive to fatty infiltration of the liver. T_1 and T_2 relaxation times of the liver have been within the normal range in patients with obvious steatosis[10] (although hepatic intensity was slightly increased).

Nuclear magnetic resonance imaging of experimentally produced hepatic steatosis has tended to confirm the relative insensitivity of NMR to fat deposition within the liver. Massive increases in hepatic triglyceride content produced no statistical change in T_1 or T_2 relaxation times and only a small increase in hepatic intensity measurements.[10]

When experimentally induced hepatitis and steatosis are compared, important differences are revealed. In hepatitis, where both water content and triglyceride content increased, all three NMR measurements — intensity and T_1 and T_2—increased significantly. In steatosis, where water content did not increase but triglyceride content increased massively, there was only a slight increase in hepatic intensity. Thus NMR appears to be very sensitive to increases in water content but relatively insensitive to hepatic fat deposition. Although these changes have only been studied in experimental conditions, histologic and biochemical changes in humans are similar to changes produced experimentally in animals (Fig. 11).

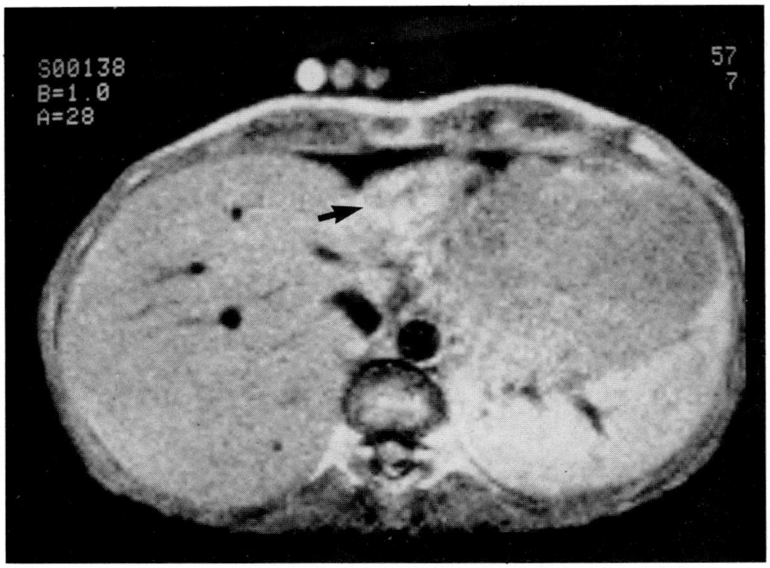

Figure 11. Hepatitis. Spin echo image demonstrates the lateral segment of the left hepatic lobe (arrow) to have a higher intensity than the rest of the liver. Biopsy revealed active hepatitis in lateral segment of left hepatic lobe and fatty infiltration of the rest of the liver. The intensity of the right hepatic lobe is within normal limits.

Hepatic Iron Overload

Excess iron deposited chronically in hepatic parenchymal cells is associated with hepatic injury, fibrosis, and, ultimately, cirrhosis.[10] Excessive total body iron stores can result from either a primary process (idiopathic hemochromatosis) or develop as a secondary disorder in patients with congenital or acquired anemia who receive repeated blood transfusions.

The diagnosis of either form of hemochromatosis is readily made by CT because scans demonstrate an increase in the density of the hepatic parenchyma.[12,13] By using dual-energy techniques, CT is capable of quantitating the amount of iron present within the liver and thus of noninvasively following the progress of therapy.[12]

Experimental results and experience in patients have shown that NMR can detect and quantify liver iron overload.[10] A dramatic reduction in hepatic NMR intensity is produced by iron overload (Fig. 12) in the range that is clinically important (1–8 mg/gm). Nuclear magnetic resonance has also been shown to be sensitive to the higher levels of iron overload found in some patients with chronic anemias.

Tissue iron is largely bound to ferritin and hemosiderin as ferric iron.[14] Ferric iron is detected by hydrogen NMR imaging not because it creates a signal, but because it is paramagnetic and decreases the T_1 and T_2 relaxation times of nearby hydrogen nuclei.[15] Decreases in NMR intensity show a linear relationship to liver iron levels, thus permitting quantification of hepatic iron content.

Cirrhosis

Experience with NMR imaging in patients with cirrhosis is still limited. Doyle et al.[5] reported a prolongation of the range of T_1 relaxation times from 210–270 to 280–450 msec. The prolongation was enough to increase the hepatic intensity on NMR images obtained with saturation recovery and inversion recovery techniques.[5] Thus it is possible that NMR will be capable of detecting cirrhosis on the basis of alterations in hepatic intensity and calculated T_1 and T_2 relaxation times. The increases in T_1 and hepatic intensity probably occur as a result of the elevated hepatic tissue-water content present in cirrhosis.

Miscellaneous Disorders

Nuclear magnetic resonance has demonstrated that the liver has a short T_1 relaxation time in a patient with primary biliary cirrhosis and excessive copper deposition in the liver, and a long T_1 time in biliary cirrhosis with normal copper levels.[5] Wilson's disease produced cirrhosis and a shrunken liver that had an increased intensity but a normal T_1 value.[5]

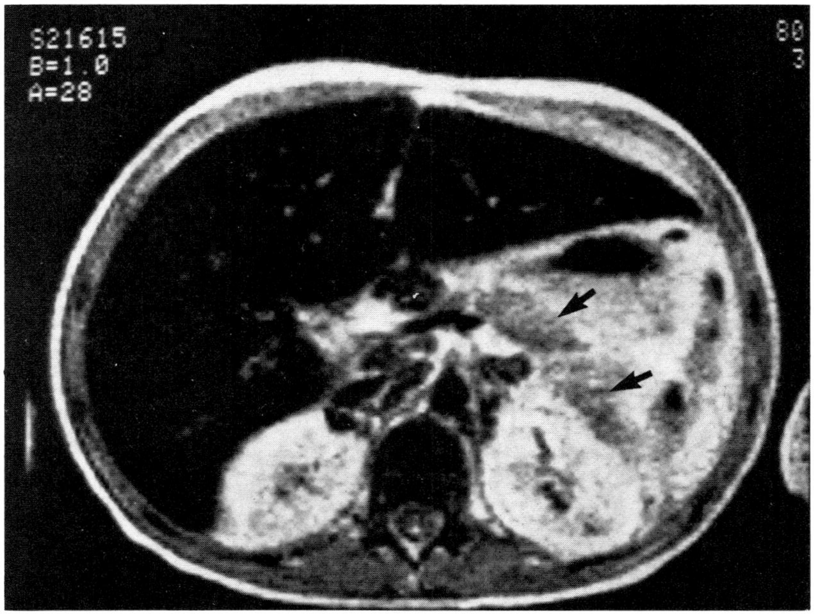

Figure 12. Hemochromatosis due to multiple transfusions. Spin echo image demonstrates the liver to have a very low intensity. The pancreas (arrows) also has a lower than normal intensity, reflecting excessive iron deposition within the pancreas.

Hepatic cysts, by virtue of their long T_1 and T_2 times, appear as low-intensity structures identical to renal cysts on spin echo images. Hemorrhage into a cyst changes the NMR appearance, giving the cyst a high intensity.

Dilated bile ducts appear as branching linear structures of decreased intensity. The degree and level of obstruction can be determined by observing changes on serial NMR images. Differentiation from portal and hepatic venous structures is easy because of the low signal intensity produced by the flowing blood in the vascular system and the increasing intensity on the second echo.

Gallbladder

Nuclear magnetic resonance has shown unique usefulness in the evaluation of the status of the gallbladder.[16] By using spin echo techniques, it is possible to ascertain whether the gallbladder is filled with concentrated bile and is therefore functioning or whether it is filled with dilute bile (Fig. 13). Hepatic bile is normally dilute and images generally with the intensity of water because of its low concentration of electrolytes and macromolecules of lipid materials. The normal gallbladder concentrates the bile by absorbing electrolytes and water and concentrating the bile approximately 10-fold.[17] Concentrated bile sends forth an intense NMR signal.

If the gallbladder is unable to concentrate bile, as in the case of chronic cholecystitis (Fig. 14), or is imaged shortly after it has been refilled with dilute hepatic bile following evacuation of concentrated bile, it will appear as a structure emitting a very low-intensity signal (Fig. 15). Thus, NMR is able to assess not only the ability of the gallbladder to concentrate bile, but also its ability to empty following the administration of a cholecystogogue. This permits direct evaluation of gallbladder function and of its ability to evacuate without having to resort to administration of a contrast medium such as Telepaque or Bilopaque. Gallbladder function can be studied without taking into consideration the absorption of contrast medium, the mechanism of excretion, or the metabolism of the particular contrast medium. There are no side effects, and there is no concern about the effect of iodine on the function of the thyroid gland. What one visualizes is the bile itself, its true composition as well as the results of normal or abnormal function, without any consideration of extraneous factors.

Gallstones are easily visualized in bile[16,18] because they emit no signal and therefore are seen as areas of low intensity (Fig. 16). The reason for the absence of signal is the lack of mobility of hydrogen protons in the gallstones.[19] This is true regardless of the nature of gallstones—whether composed of cholesterol, pigment, or calcium carbonate.[8] The hepatic duct (Fig. 17) is well seen in its extrahepatic portion as an area of low signal intensity perpendicular to the direction of the right portal vein.[16] It can be expected that in a distal section close to the duodenum

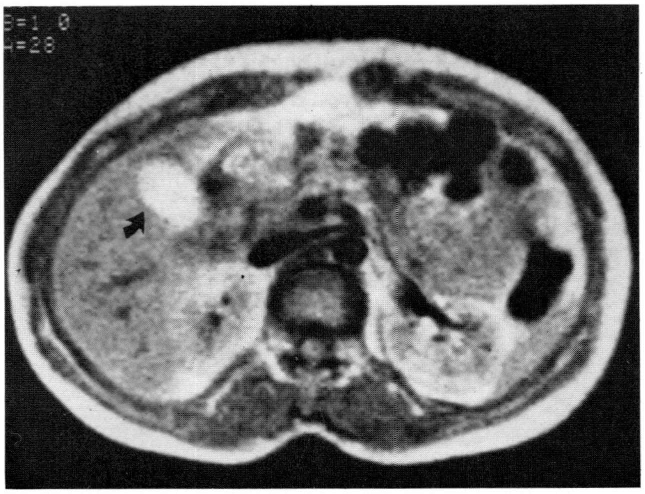

A.

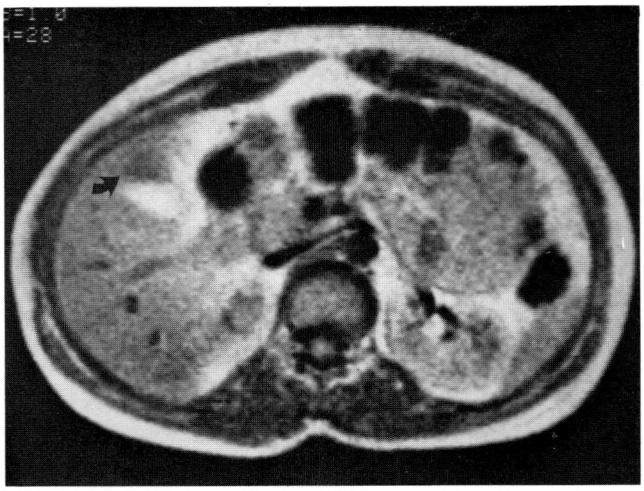

B.

Figure 13. Normal gallbladder.

A. Cross-section of image of the upper abdomen in a volunteer who has fasted for 20 hr. The gallbladder (arrow) is of high intensity, indicating the process of concentrated bile.

B. Image obtained 1 hr following stimulation with cholecystogogue. There is layering of dilute bile (arrow) above the residual concentrated gallbladder bile.

(Reproduced, with permission, from Hricak H, Filly RA, Margulis AR, et al: Work in progress: Nuclear magnetic resonance imaging of the gallbladder. Radiology 147:481, 1983.)

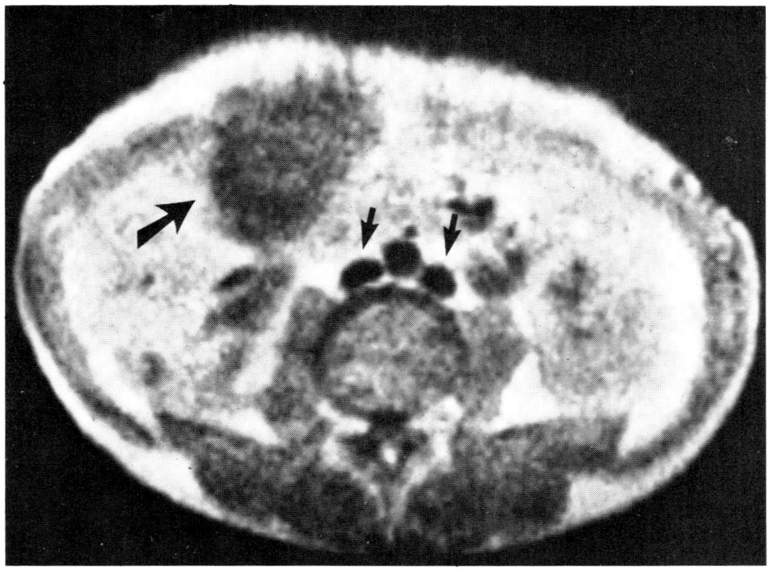

Figure 14. Large low-intensity gallbladder (large arrow) in a fasting patient. Other studies indicated that this patient had chronic cholecystitis. Notice a double inferior vena cava (small arrows).

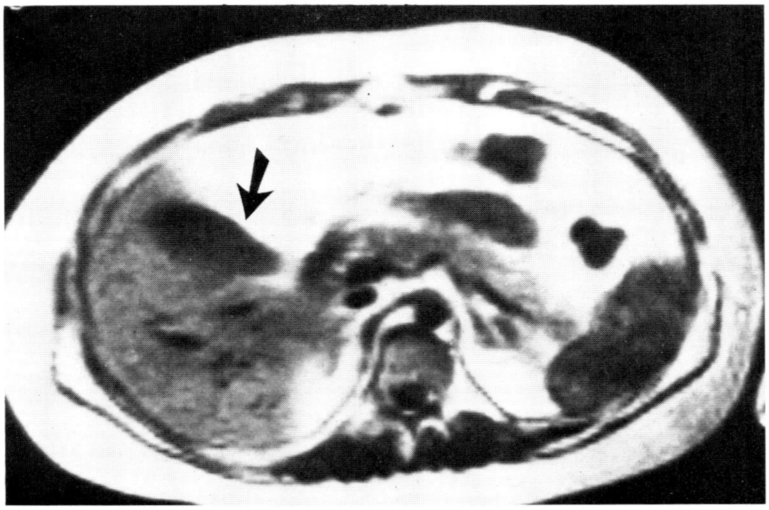

Figure 15. Image of abdomen in a patient obtained 2.5 hr after a full hospital breakfast containing large amounts of fat. The gallbladder, filled with dilute bile, emits a low-intensity signal (arrow).

the common bile duct will be seen as a rounded area of low intensity unless it is imaged during a period when the gallbladder evacuates concentrated bile after stimulation by cholecystogogue.

Experience with the examination of the gallbladder by NMR is not extensive. We have not yet studied hyperplastic cholecystoses or benign and malignant tumors of the gallbladder. It can be expected, however, because the gallbladder is readily imaged, that these conditions should be distinguished by NMR. Coronal as well as sagittal sections contribute further in the detection of intraluminal masses.

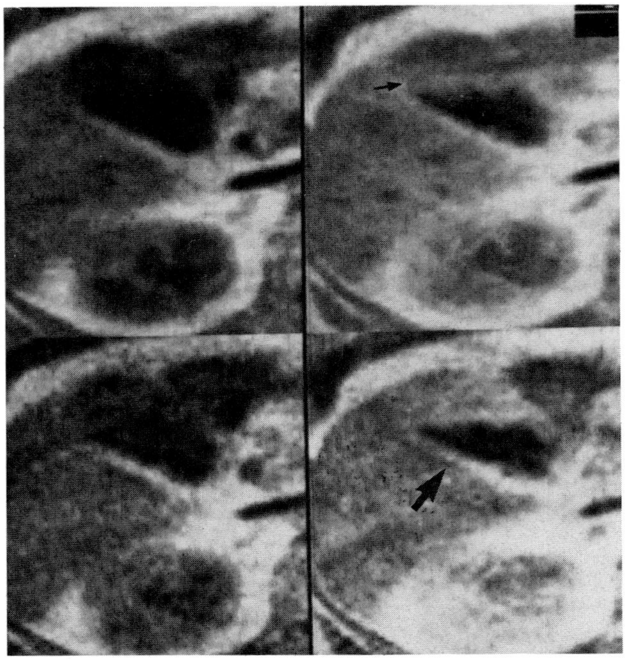

Figure 16. Four images obtained with spin echo technique showing a gallbladder full of stones. Image on the upper left was obtained with TR (pulse interval) of 0.5 sec, and the image on the upper right with TR of 1.0 sec. Both images in the upper row were obtained with a TE (echo delay) of 28 msec. Both images in the lower row were obtained with TE of 56 msec. The image in the left upper corner shows a gallbladder filled with dilute bile after fasting, indicating inability of the gallbladder to concentrate bile, which is consistent with chronic cholecystitis. Notice layering (small arrow) of the dilute bile on the image obtained with TR of 1.0 sec and TE of 56 msec.

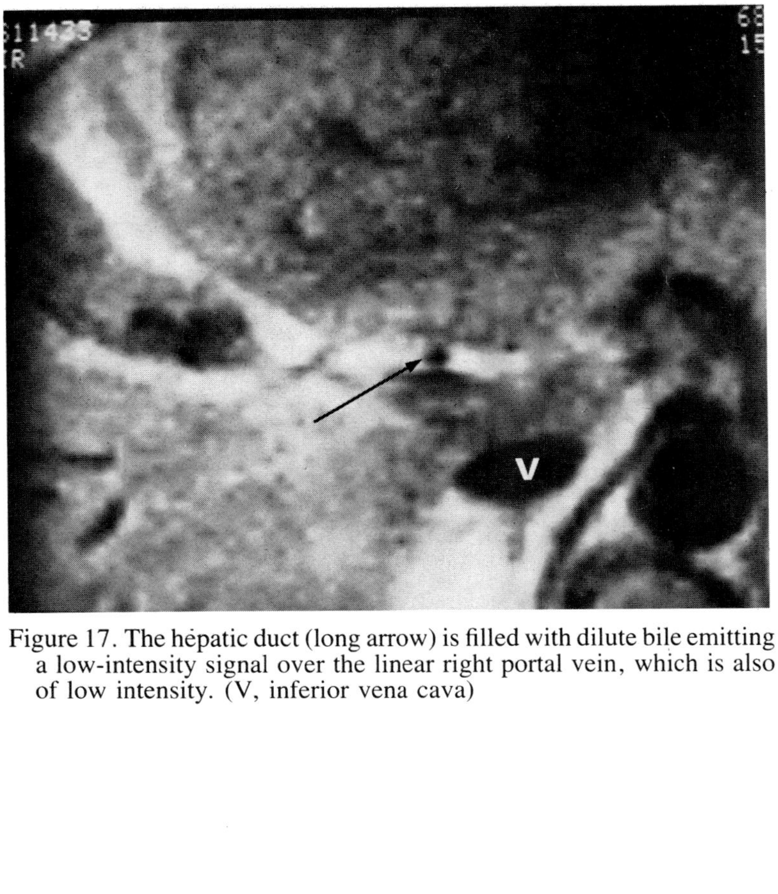

Figure 17. The hepatic duct (long arrow) is filled with dilute bile emitting a low-intensity signal over the linear right portal vein, which is also of low intensity. (V, inferior vena cava)

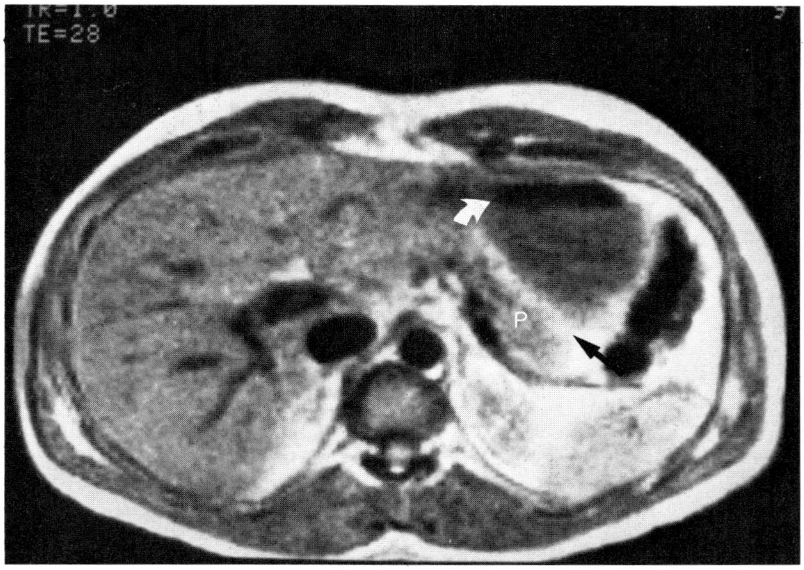

Figure 18. The stomach is filled with fluid and gas (curved arrow) and is seen located above the tail of the pancreas (P). There is a layer of fat between the two structures (arrow).

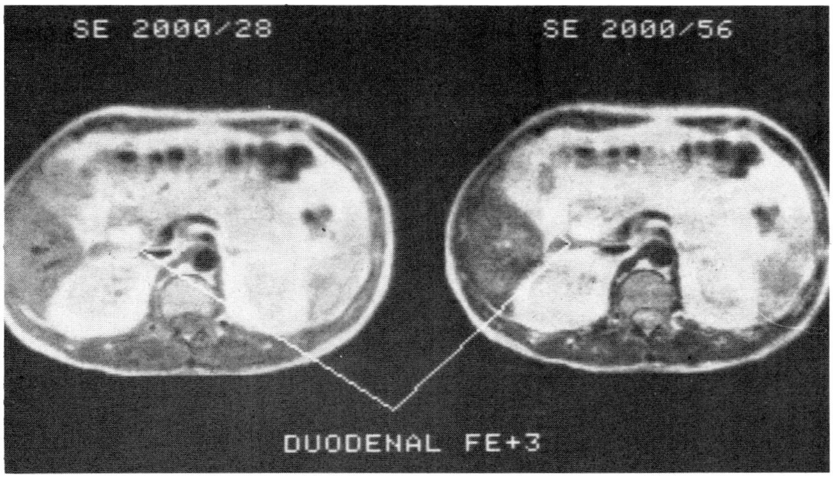

Figure 19. Normal duodenum following contrast ingestion. The duodenal lumen (arrow) has a high intensity produced by the $FeCl_3$ solution.

Alimentary Tube

The entire alimentary tube can be visualized by NMR when filled with fluid or gas (Fig. 18), although optimal demonstration of portions of the gastrointestinal tube will be dependent upon the development of oral NMR contrast material.

The esophagus is readily visualized by NMR in both axial and coronal sections when it contains air. Although experience has been limited, NMR has detected esophageal tumors[20] and demonstrated an ability to determine extraesophageal spread of disease. Further refinement in NMR technique should permit accurate staging of esophageal neoplasms as well as visualization of thickening of the wall of the esophagus in inflammatory conditions.

The upper portion of the stomach can be easily visualized (Fig. 18), and it is very likely that staging of neoplasms of the fundus and the body of the stomach will be possible by this method. Positive contrast media, such as those containing iron,[21] have been tried on volunteers and have permitted excellent visualization of the gastric lumen, wall of the stomach, and duodenum (Fig. 19). However, it is not clear whether negative contrast media such as air will prove to be superior to positive contrast in the evaluation of the upper gastrointestinal tract.

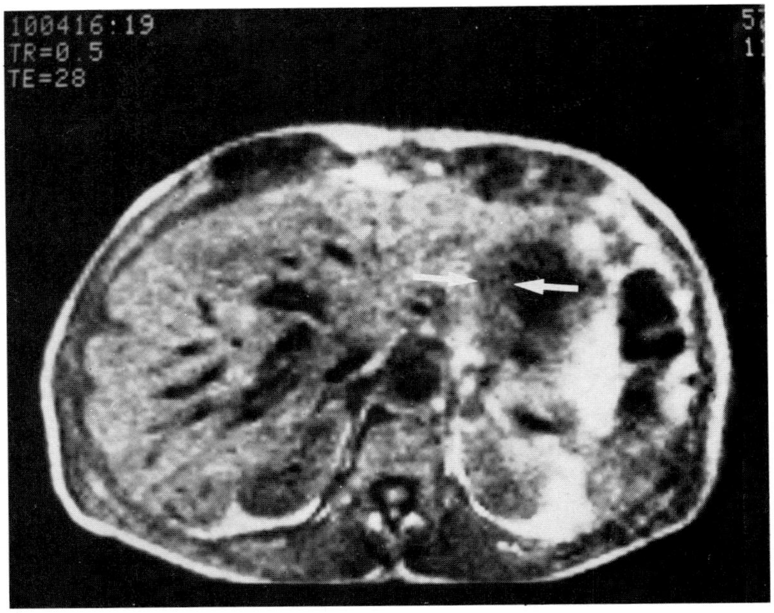

A.

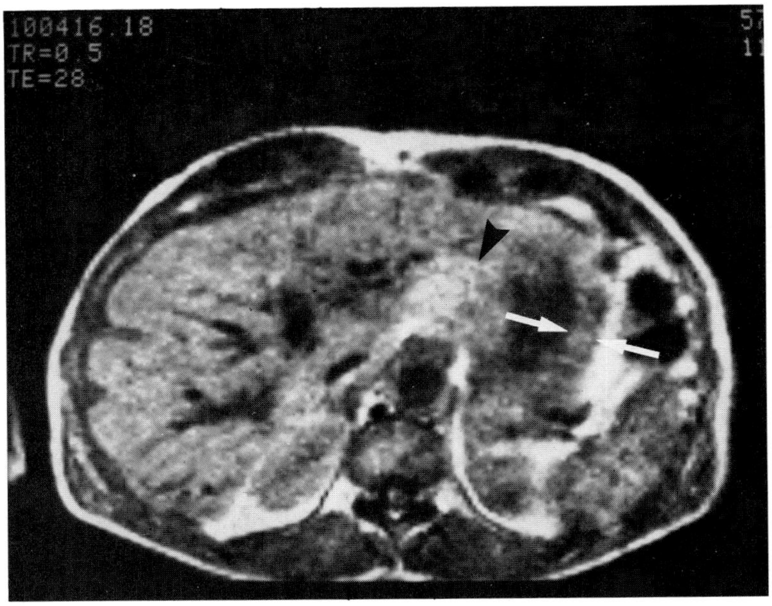

B.

Figure 20.
A. An NMR image of the upper abdomen showing a thickened wall of the stomach (between arrows) in a patient with extensive metastatic squamous carcinoma to the stomach.
B. At a slightly lower level the thickened wall of the stomach is again seen between white arrows. The arrowhead points to the large metastatic tumor.

The distal portion of the stomach is not as well visualized as the proximal stomach because of peristalsis. However, tumors and gastric wall thickening of the distal stomach have been identified (Fig. 20).

The duodenum can be readily examined, but the small bowel is not well seen because of peristaltic activity. It is very likely that glucagon or some drug with similar action will have to be used to paralyze peristalsis and thus permit detailed views of the small bowel. Excellent NMR images of the pelvic small bowel have been obtained when the small bowel was immobilized by adhesions.

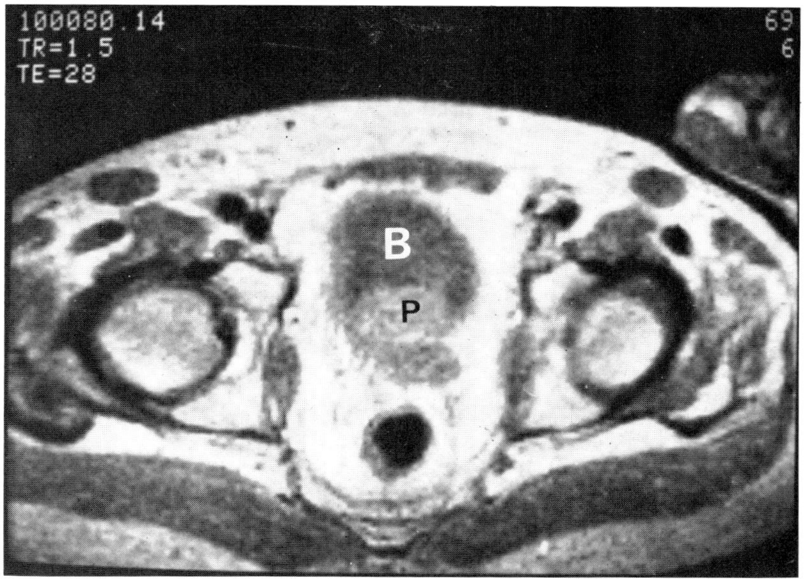

Figure 21. Normal rectum with fairly thin wall is seen in a patient with a large prostate (P) protruding into the urinary bladder (B).

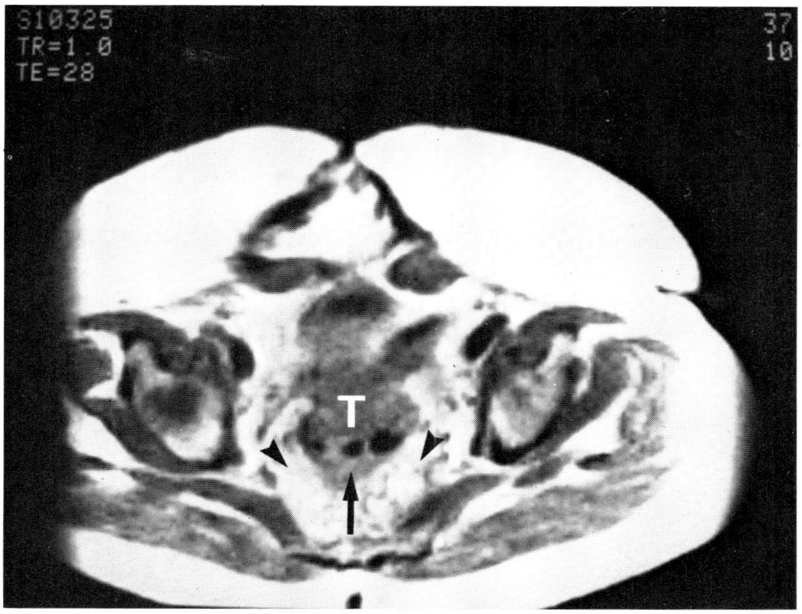

Figure 22. Large recurrent tumor of the upper rectum (arrow). Notice the infiltrated tumor emitting a signal of slightly higher intensity and appearing different from the subcutaneous fat (arrowheads).

The colon and, particularly, the rectum are clearly imaged when clean (Fig. 21) and filled with air. The mucosal surface of the large bowel will probably never be visualized with the detail and accuracy obtained with conventional or double-contrast barium studies. Nevertheless, it is very likely that NMR staging of colonic carcinoma, particularly of the lesions in the pelvis, will be possible (Fig. 22).[22] Based on experience obtained with CT,[23] it is very likely that fibrolipomatous infiltration of the mesentery as a result of Crohn's disease, abscesses, sinuses, and fistulae will be well suited to be studied by NMR.

Spleen

The normal spleen is clearly separable from the other structures in the left upper quadrant on proton NMR images because it is surrounded by retroperitoneal fat, fat in the transverse mesocolon, and mesentery of the small bowel (Fig. 23). A sharp margin is often demonstrated at the capsular border of the spleen both on spin echo and inversion recovery images (Figs. 1, 9, 23). This is seen as a thin black line outlining the organ, similar to that also seen around the liver and kidneys (Figs. 1, 23). The actual anatomic correlation is obscure; it does not appear to represent the capsule itself. The spleen appears somewhat more intense (less gray) than the liver and pancreas on spin echo images, and considerably more intense than the adjacent flank muscle, reflecting the greater intensity of signal from the normal spleen on proton NMR spin echo images as compared either with liver or pancreas.[7,24,25]

The visual difference in signal intensity from the spleen when compared with liver and pancreas is confirmed by differences in the calculated T_1 and T_2 relaxation times. In most cases, the T_2 relaxation time of the normal spleen is longer than that of the liver. In our experience with 11 normal patients, the spleen-to-liver T_2 ratio was 1.30 ± 0.33 SD, although in three cases the T_2 times were equal. In these same cases, the mean intraindividual spleen/liver ratio for T_1 relaxation times was 1.45 ± 0.29 SD.

The T_1 relaxation times of the normal spleen have been reported in the literature to be almost twice as long as those of normal liver,[24,25] and these results compare favorably with previous in vitro NMR spectroscopy data.[26,27] The spleen ordinarily contains approximately 76% water, while the liver contains 71%.[24] Apparently, a 4%–5% difference in water is a major factor responsible for the more intense signal from the spleen and prolonged T_1 and T_2 relaxation times in comparison with the liver. It is thought that the shorter T_1 and T_2 relaxation times in normal rat liver when compared with the spleen are due both to the lower liver water content and the higher fraction of macromolecular bound hepatic water protons.[26,27]

The homogeneous appearance of the normal spleen on images obtained with varying spin echo imaging sequences is in this respect is similar to the appearance seen on CT scans without contrast enhancement. The splenic arteries are identified as tubular structures without any detectable signal, an appearance similar to that of the portal vein (Fig. 23). The splenic artery is often seen ventral to the vein and slightly cephalad and may enter and exit the field of view because of its tortuosity. Occasionally the artery may be seen entering the splenic hilum at the same level at which the splenic vein exits the hilum. The splenic vein courses beneath the tail and body of the pancreas. It is separable from the overlying pancreas because of the difference in signal intensity between these two structures.

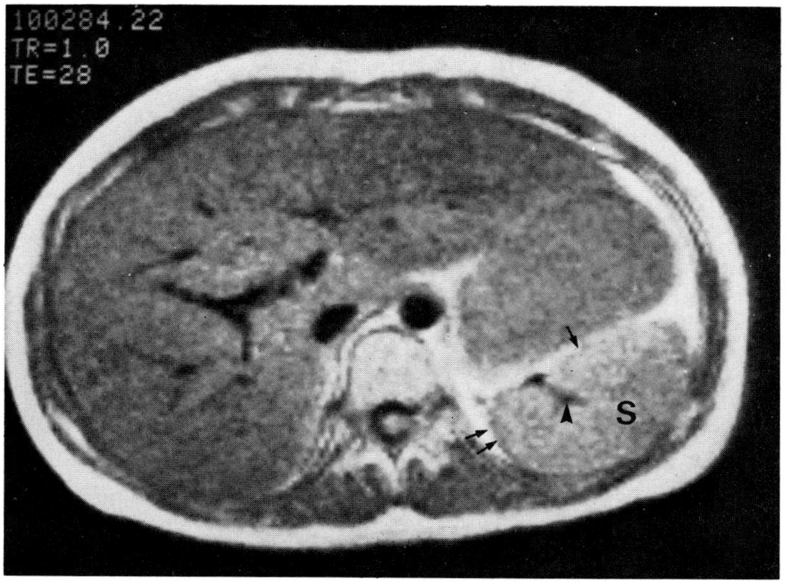

Figure 23. Normal spleen. The spleen (S) has a slightly higher intensity than the liver. A thin rim (arrows) of lower intensity surrounds the spleen. (arrowhead, splenic artery)

Splenic Abnormalities

Enlargement

The enlarged spleen is readily detectable by NMR scanning on sequences that include the top of the diaphragm and the tip of the spleen. In our experience, complete evaluation of splenic enlargement is best carried out on a sequence of 15 consecutive sections. On the sequences that contain only a few sections of the spleen, the true splenic size cannot be determined, but only the change in shape and the medial bulging of the enlarged spleen. A combination of transaxial scans and coronal views, which are easily obtained, provide complete information about the size and extent of splenic bulk.

Portal Hypertension

In portal hypertension with secondary splenomegaly, the intensity of the signal on spin echo images increases, and both T_1 and T_2 relaxation times have thus far been found to be prolonged. These findings probably are secondary to the increased water content associated with splenic congestion resulting from high portal pressures. When present, perisplenic and perigastric collateral vessels can be readily demonstrated by NMR imaging techniques.

Compression and obstruction in the splenic vein may also produce abnormalities similar to those seen with portal hypertension due to cirrhosis. In instances in which the splenic vein was compressed by tumor in the tail of the pancreas, the spleen was shown to have a more intense signal, increasing the spleen-to-liver intensity ratio above the normal level. In these cases there was also prolongation of the T_1 and T_2 relaxation times.

Hematomas

Hematomas of the spleen produced in dogs have been shown to be detected on NMR scans in the acute phase (6 hr) because the signal intensity from the hematoma differs considerably from that of the surrounding spleen (Fig. 24). Relatively fresh hematomas imaged with spin echo sequences demonstrate the hematoma as an area of increased intensity images obtained with short TR (0.5 sec) and TE (28 msec) times. The hematoma increases in intensity on images obtained with long TR (1.5 sec) and TE (56 msec) intervals, indicating that a fresh hematoma has long T_1 and T_2 relaxation times when compared with the surrounding spleen. In addition, the interphase of the hematoma with the splenic parenchyma is detectable as a thin rim of low intensity.

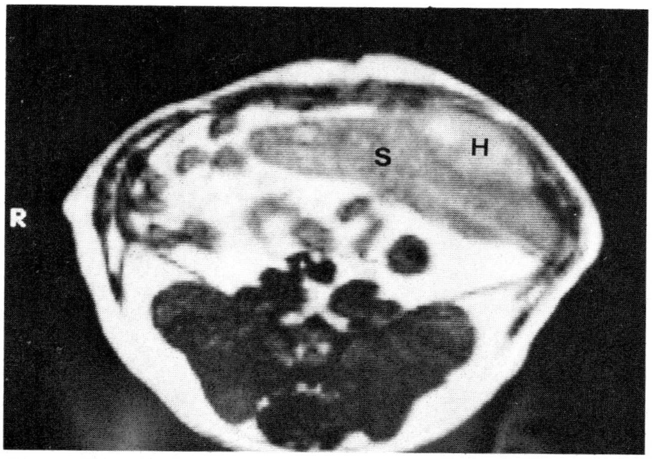

Figure 24. Acute splenic hematoma in a dog. The splenic subcapsular hematoma (H) is demonstrated as region having a high intensity as compared with normal spleen (S).

Miscellaneous Conditions

In one patient with adrenal pheochromocytoma, one portion of the spleen was noted to have a very intense signal with prolonged T_1 and T_2, while the remaining part of the spleen appeared normal (see Chap. 14). At surgery it was found that the pheochromocytoma was directly pressing on the spleen, and that splenic edema with some cellular infiltrate was present in the specimen.

Metastases from colon and melanoma can be detected within the splenic parenchyma because of the increased intensity and the long T_1 and T_2 relaxation times of the metastatic tumor in comparison with adjacent spleen.

Thus it appears that abnormalities that increase spleen volume, splenic blood flow, and water content, and which bring cellular infiltrate, all have the capability of altering the relaxation times of the involved splenic tissue sufficiently that these disease processes can be detected by proton NMR imaging.

Intraperitoneal and Retroperitoneal Fat

Adipose tissue is readily identified on proton NMR images because the high concentration of hydrogen in adipose tissue produces an intense signal. Adipose tissue is known to have not only a high hydrogen proton density, but also a longer T_2 than associated muscle, and a shorter T_1 than kidney, liver, or pancreas.[24] The signal intensity of fat is displayed as the brightest portion of the gray scale on NMR intensity images. Therefore, intraperitoneal and retroperitoneal adipose tissues are readily identified by virtue of their intense signal, and the intensity of fat provides a background for all the other intraperitoneal and retroperitoneal abdominal organs. The abdominal vasculature emanating from the superior mesenteric and celiac trunks is readily demonstrated coursing through the small bowel mesentery and mesocolon. Likewise, the stomach, small intestine, and colon are easily distinguishable from intraperitoneal and retroperitoneal fat—either when they are distended with air or water or when in a collapsed state—because of the difference in intensity between fat and the gastrointestinal tract.

Processes that alter, replace, or displace normal adipose tissue in the intraperitoneal or retroperitoneal location are easily detected on NMR images obtained with spin echo and inversion recovery techniques. The vast majority of abnormalities have a lower NMR intensity than fat and are therefore highlighted by surrounding intra-abdominal or retroperitoneal fat. Processes such as non-Hodgkins's lymphoma and metastasis produce obvious alterations in the usual homogeneous appearance of the mesenteric and omental fat, and result either in a well-marginated mass, such as is present in lymphoma (Fig. 25), or multiple defects of lesser intensity produced by metastatic carcinoma.

Retroperitoneal hematomas have been reported to be characterized by a low signal intensity compared with surrounding fat and appear as a mass displacing fat.[28] In rats, hematomas (produced by injecting uncoagulated blood into the subcutaneous fat) had a low-intensity signal on spin echo images obtained with short TR and TE intervals, but increased in intensity in comparison with muscle on images obtained with long echo delay and repetition rates.[29] The hematomas were found to have longer T_1 and T_2 times than normal liver or fat.[29]

Diffuse inflammation involving mesenteric fat, such as is seen in acute pancreatitis with phlegmon, can alter the normal appearance of abdominal and retroperitoneal fat, resulting in a diffuse, inhomogeneous decrease in signal intensity that is independent of variations in spin echo sequences. It appears that disease processes that either displace or infiltrate intra-abdominal or retroperitoneal fat are easily detected on proton NMR images because of the inherent contrast difference between the normal adipose tissue and any of the other disease processes likely to affect this tissue.

Pancreas

The normal pancreas has been one of the most difficult organs to image by NMR;[30] however, it is now possible to distinguish the pancreatic head, body, and tail in more than 60% of patients.[31] Respiratory motion and peristalsis create image unsharpness in the upper abdomen that limits spatial resolution, even for relatively immobile retroperitoneal structures. The pancreas appears as a homogeneous structure of moderate intensity, similar to bowel (Fig. 26). Retroperitoneal fat has a high intensity and serves to differentiate the ventral border of the pancreas from

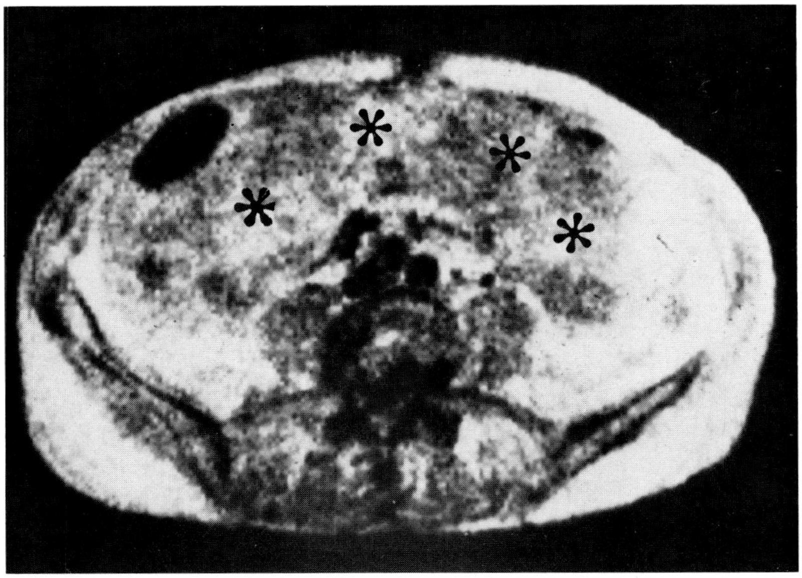

Figure 25. Mesenteric infiltration (*) by non-Hodgkin's lymphoma.

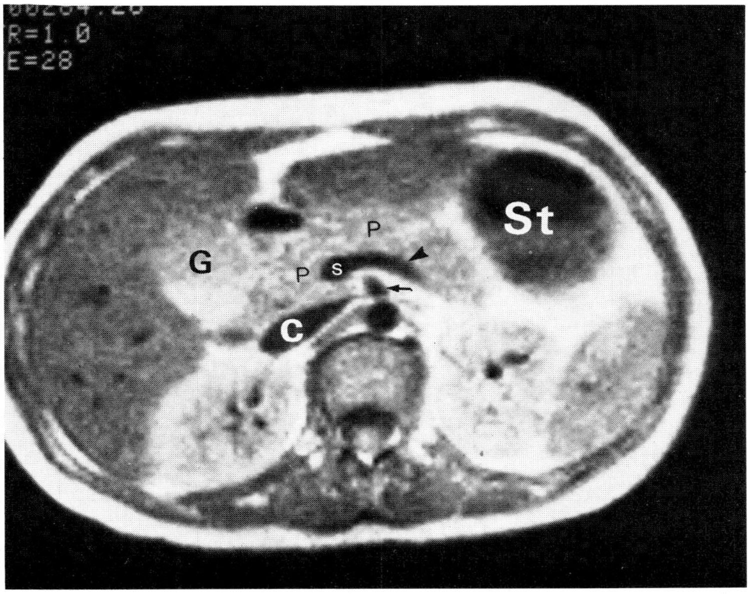

Figure 26. Normal pancreas. The pancreas (P) shown as a homogeneous organ of moderate intensity. (C, inferior vena cava; arrowhead, splenic vein; arrow, superior mesenteric artery; St, stomach; G, gallbladder with concentrated and fresh bile; S, superior mesenteric vein)

stomach or small intestine. Although gastrointestinal contrast materials such as air, water, or dilute solutions of paramagnetic compounds can be used to identify bowel, peristalsis and respiratory motion remain a problem. Furthermore, in patients with a paucity of retroperitoneal fat it is difficult to differentiate pancreas from bowel. Dorsal to the pancreas, however, the major vessels are reliable landmarks. The aorta, inferior vena cava, superior mesenteric artery and vein, and splenic vein can be identified in every patient.

The normal pancreas has an intensity similar to the liver on all spin echo and inversion recovery images. It follows that the calculated T_1 and T_2 relaxation times of pancreas and liver are similar. The intensity of the pancreas can be seen to be lower than the intensity of the kidneys and spleen by using spin echo techniques with long repetition rates (TR of 2.0 sec) or long echo delays (TE of 56 msec). It follows that the calculated relaxation times of the kidneys and spleen are relatively long. Inversion recovery images also demonstrate the short T_1 of normal pancreas relative to kidneys and spleen. With inversion recovery techniques (TR of 1.0 sec or 1.8 sec and TI of 420 msec or 277 msec, respectively), the pancreas is consistently greater in intensity than the kidneys and spleen.

Pancreatic Pathology

Neoplastic Disease

Adenocarcinoma of the pancreas larger than 3 cm in size has been detected by NMR using morphological criteria analogous to those of CT. Smaller lesions can be detected by NMR when there are cystic areas within the tumor or when the tumor has altered relaxation times. Adenocarcinomas generally have prolonged T_1 and T_2 times and can be differentiated from normal pancreatic tissue by selecting the appropriate imaging technique (Fig. 27).

T_1 differences can be accentuated by using spin echo techniques with short TR, and T_2 differences can be accentuated by using spin echo techniques with long TE. However, in some cases the contrast obtainable with existing spin echo techniques is insufficient to differentiate abnormal from normal pancreatic tissue. Further work on finding new techniques for enhancing the differences is needed.

Islet cell tumors have increased intensity relative to the normal pancreas on spin echo images because of T_2 prolongation (Fig. 28). This contrast can be accentuated by selecting a spin echo technique with a long echo delay (TE of 56 msec).

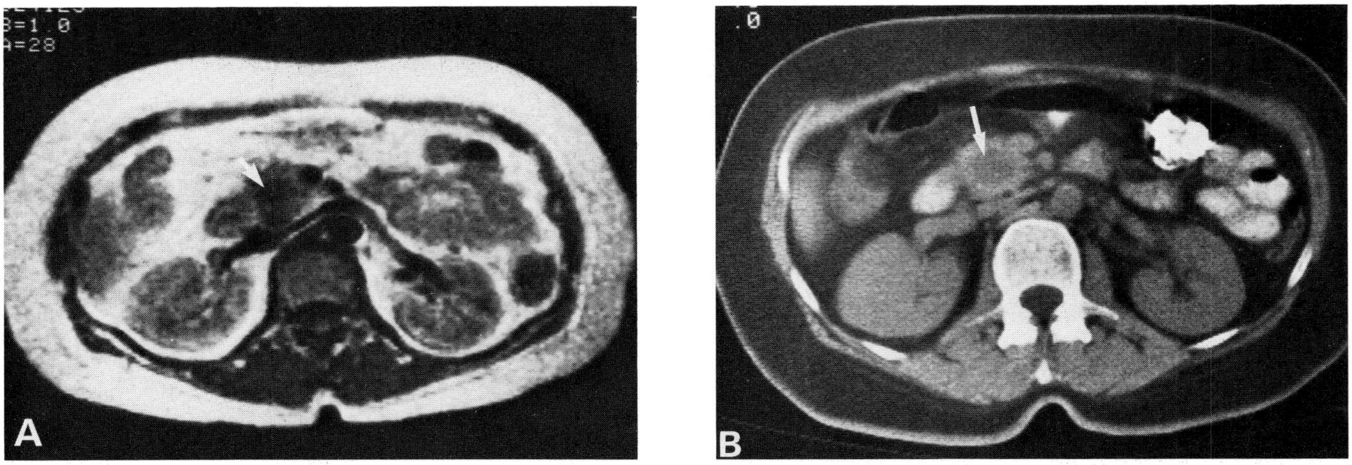

Figure 27. Pancreatic adenocarcinoma.
A. Spin echo image demonstrates a pancreatic carcinoma (arrow) as a rounded mass having a low signal intensity. The tumor was partially cystic.
B. A CT scan at same level reveals a 2-cm low-density mass (arrow) in the head of the pancreas.

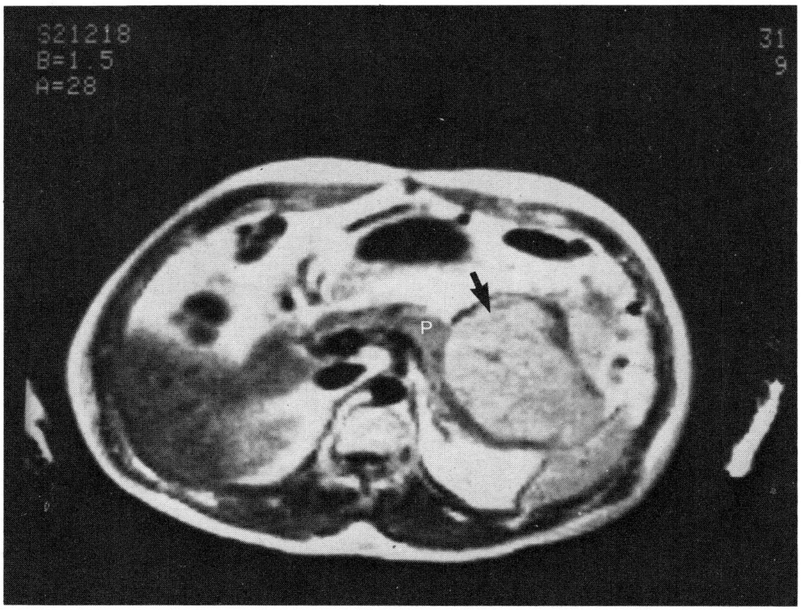

Figure 28. Pancreatic islet cell tumor shown on spin echo image as a mass (arrow) arising in the tail of the pancreas. The glucagonoma was encapsulated and had a higher intensity than the normal pancreas (P).

Extrapancreatic extension of tumors is readily identified by NMR. Tissue planes are sometimes more clearly seen with NMR than with CT, allowing identification of retroperitoneal invasion. Furthermore, vascular structures are always clearly delineated by NMR without administration of contrast material, and patency or occlusion by tumor is readily identified. Nuclear magnetic resonance can detect liver metastases using either spin echo or inversion recovery techniques, and biliary obstruction is detected because fluid within the dilated common bile duct can be distinguished from surrounding vessels and liver.

A.
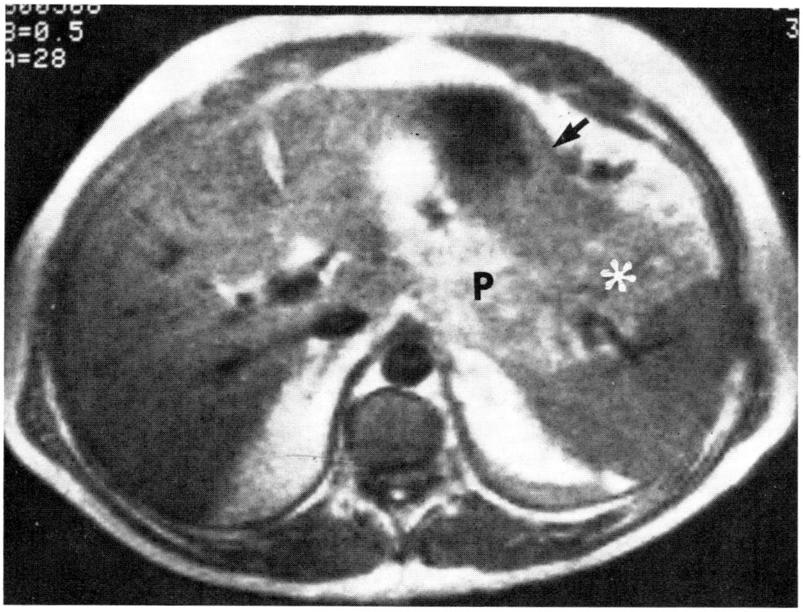

B.
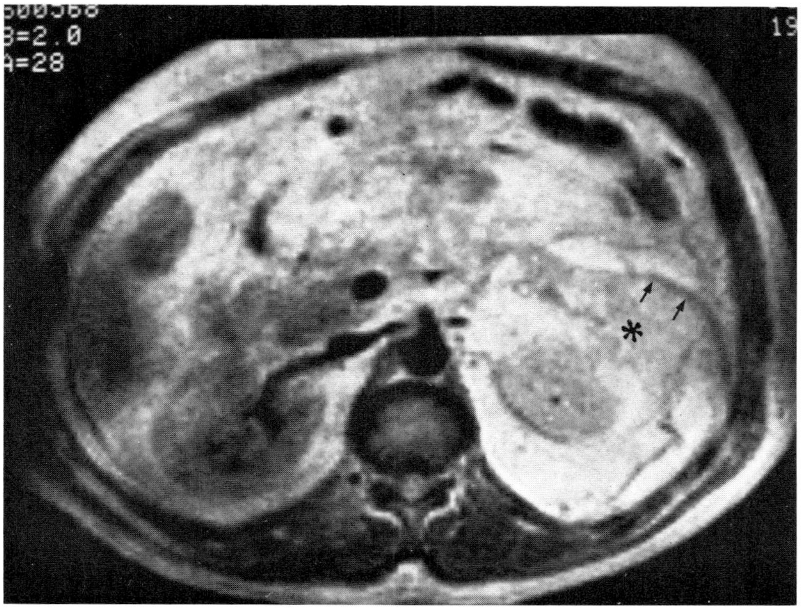

Figure 29. Acute pancreatitis.

A. Spin echo image demonstrates an enlarged pancreas (P) and loss of normal peripancreatic fat planes. The inflammatory process has spread into the lesser sac (*), decreasing the intensity of the fat and thickening the gastric wall (arrow).

B. An NMR spin echo image at lower level demonstrates spread of pancreatitis to involve the anterior pararenal space, thicken Gerota's fascia (arrows), and invade the left perirenal space.

Inflammatory Disease

In pancreatitis, pancreatic enlargement and effacement of surrounding tissue planes are seen (Fig. 29). Inflammation and edema increase the T_1 and T_2 of pancreatic tissue and are detected as increased intensity on spin echo images with long TR or TE and as decreased intensity on inversion recovery images with a TI of 277 msec. However, to date it has not been possible to correlate the degree of inflammation with T_1 or T_2 increase. The intensity of ret-

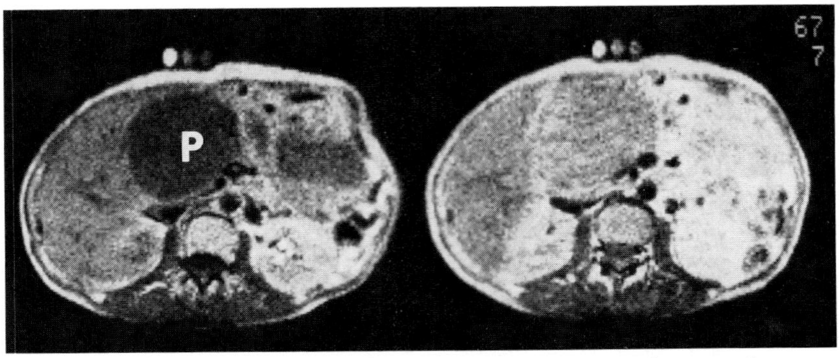

Figure 30. Pancreatic pseudocyst (P) shown as a round mass of very low intensity on spin echo image with TE of 28 msec. The pseudocyst increases in intensity on the image obtained with TE of 56 msec, reflecting the long T_1 and T_2 of the pseudocyst fluid.

roperitoneal fat bordering on inflamed pancreas often decreases, most likely because of edema, which would be expected to prolong the T_1 of fat (Fig. 29). As a result, it is difficult to delineate the pancreas in severe cases of acute pancreatitis.

Pseudocysts larger than 1.5 cm can be detected by NMR (Fig. 30); however, the margins of some cysts are not well delineated as compared with CT. Increasing TR or TE results in increased intensity of pseudocyst fluid relative to solid tissues, reflecting the long T_1 and T_2 of cyst fluid. T_1 and T_2 calculations can be used to differentiate a variety of fluids with varying composition.

Extension of pancreatitis or hemorrhage into the retroperitoneum, mesentery, or adjacent bowel can be identified morphologically or by alterations in tissue intensity. As with pseudocyst fluid, these extrapancreatic collections have markedly prolonged T_1 and T_2 and can readily be differentiated from bowel or fat by using either spin echo or inversion recovery techniques.

Pancreatic calcifications larger than 1 cm are seen as ill-defined low-intensity areas. However, smaller calcifications are not seen by NMR because of motion and volume averaging. Dilated pancreatic ducts are also difficult to visualize by NMR because they contain complex fluid and debris, which can have an intensity similar to the inflamed pancreas on spin echo images.

Ascites is readily detected by NMR when it complicates pancreatitis and can be differentiated from other biologic fluids because of its intermediate T_1 and T_2 relaxation times (Fig. 31). Pure liquids, such as found in simple renal cysts, have very long T_1 and T_2 times, while grossly impure liquids such as concentrated bile or cytosol water in solid tissues have short T_1 and T_2.

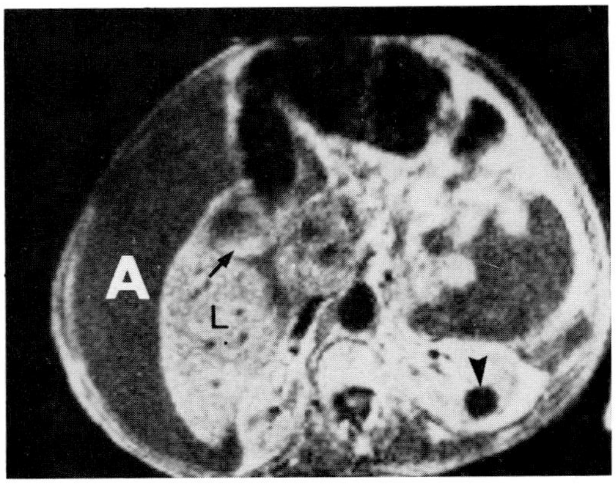

Figure 31. Pancreatitis and ascites. Ascites (A) is seen displacing the liver (L). The ascitic fluid has an intensity intermediate between that of the left renal cyst (arrowhead) and concentrated bile (arrow).

Metabolic Disease

In hemochromatosis, pancreatic iron overload can result in islet cell atrophy and diabetes. Nuclear magnetic resonance shows decreased intensity in iron-overloaded pancreatic tissue. This decrease in intensity appears to result from relaxation enhancement (T_1 and T_2 shortening) as a result of the paramagnetic influence of tissue iron on cytosol water. This mechanism has been studied in the rat liver,[10] where intensity, T_1, and T_2 can all be correlated with iron levels. Changes in the pancreas are similar and proportional to the expected pancreatic iron levels. Thus, NMR may be useful in detecting pancreatic iron overload and monitoring therapy in patients with hemochromatosis.

In patients with cystic fibrosis and exocrine pancreatic insufficiency, pancreatic atrophy and fatty replacement can be demonstrated by NMR.

REFERENCES

1. Hawkes RC, Holland GN, Moore WS, et al: Nuclear magnetic resonance (NMR) tomography of the normal abdomen. J Comput Assist Tomogr 5:613, 1981.
2. Young IR, Bailes DR, Burl M, et al: Initial chemical evaluation of a whole body nuclear magnetic resonance tomograph. J Comput Assist Tomogr 6:1, 1982.
3. Crooks L, Arakawa M, Hoenninger J, et al: Nuclear magnetic resonance whole-body imager operating at 3.5 kGauss. Radiology 143:169, 1982.
4. Margulis AR, Moss AA, Crooks LE, et al: NMR imaging of tumors of the liver. Semin Roentgenol 18:123, 1983.
5. Doyle FH, Pennock JM, Banks LM, et al: Nuclear magnetic resonance imaging of the liver: Initial experience. AJR 138:193, 1982.
6. Smith FN, Mallard JR, Reid A, et al: Nuclear magnetic imaging in liver disease. Lancet 1:963, 1981.
7. Smith FW: Two years' clinical experience with NMR imaging. Applied Radiol 12(3):29, 1983.
8. Moss AA, Goldberg HI, Stark DD, et al: Nuclear magnetic resonance imaging of hepatic tumors. Radiology (in press).
9. Alfidi RJ, Haaga JR, El Yousef SJ, et al: Preliminary experimental results in humans and animals with a superconducting whole-body nuclear magnetic resonance scanner. Radiology 143:175, 1982.
10. Stark DD, Bass NM, Moss AA, et al: Assessment of experimentally induced liver disease by NMR imaging. Radiology (in press).
11. Stark DD, Goldberg HI, Moss AA, et al: Nuclear magnetic resonance imaging of chronic liver disease. Radiology (in press).
12. Goldberg HI, Cann CE, Moss AA, et al: Noninvasive quantitation of liver iron in dogs with hemochromatosis using dual energy CT scanning. Invest Radiol 17:375, 1982.
13. Chapman RWG, Williams G, Bydder G, et al: Computed tomography for determining iron content in primary hemochromatosis. Br Med J 280:440, 1980.
14. Bacon BR, Tavill AS, Brittenhan GM, et al: Hepatic lipid peroxidation in vivo in rats with chronic iron overload. J Clin Invest (in press).
15. James TL: *Nuclear Magnetic Resonance in Biochemistry: Principles and Applications,* pp 46–48, 84–87, 298–305, 350–373. New York: Academic, 1975.
16. Hricak H, Filly RA, Margulis AR, et al: Work in progress: Nuclear magnetic resonance imaging of the gallbladder. Radiology 147:481, 1983.
17. Wheller HD: Concentrating function of the gallbladder. Am J Med 51:588, 1971.
18. Moon KL, Hricak H, Margulis AR, et al: Nuclear magnetic resonance imaging characteristics of gallstones in vitro. Radiology (in press).
19. Fukushima E, Roeder SB: *Experimental Pulse NMR: A Nuts and Bolts Approach,* pp 125–151. Reading: Addison-Wesley, 1981.
20. Smith FW, Hutchison JMS, Mallard JR, et al: Oesophageal carcinoma demonstrated by whole-body nuclear magnetic resonance. Br Med J 282:510, 1981.
21. Wesbey GE, Brasch RC, Engelstad BL, et al: Gastrointestinal contrast enhancement for NMR imaging using nontoxic oral iron solutions. Radiology (in press).
22. Hricak H, Spring DB, Moon KL, et al: Anatomy and pathology of the male pelvis as imaged by NMR. AJR (in press).
23. Goldberg HI, Gore RM, Margulis AR, et al: Computed tomography in the evaluation of Crohn's disease. AJR 140:277, 1983.
24. Mallard JR: NMR imaging at Aberdeen: A historical perspective. In *Nuclear Magnetic Resonance (NMR) Imaging,* Partain CL, James AE, Rollo FD, et al (Eds), pp 15–41. Philadelphia: W. B. Saunders, 1983.
25. Straughan K, Spencer DH, Bydder GM: NMR clinical results: Hammersmith. In *Nuclear Magnetic Resonance (NMR) Imaging,* Partain CL, James AE, Rollo FD, et al (Eds), pp 195–200. Philadelphia: W. B. Saunders, 1983.
26. Hollis DP, Saryan LA, Eggleston JC, et al: Nuclear magnetic resonance studies of cancer: VI. Relationship among spin-lattice relaxation times, growth rate, and water content of Morris hepatomas. J Natl Cancer Inst 54:1469, 1975.
27. Kiricuta I-C Jr, Simplaceanu V: Tissue water content and nuclear magnetic resonance in normal and tumor tissues. Cancer Res 13:1164, 1975.
28. Hricak H, Williams RO, Moon KL, et al: Nuclear magnetic resonance imaging of the kidney: Renal masses. Radiology 147:765, 1983.
29. Herfkens R, Davis PL, Crooks LE, et al: Nuclear magnetic resonance imaging of the abnormal live rat and correlation with tissue characteristics. Radiology 144:211, 1981.
30. Smith RW, Reid A, Hutchinson JMS, et al: Nuclear magnetic resonance imaging of the pancreas. Radiology 142:677, 1983.
31. Stark DD, Moss AA, Goldberg HI, et al: Nuclear magnetic resonance imaging of the pancreas: Normal and pathologic findings. Radiology (in press).

14

Adrenal Glands and Kidneys

HEDVIG HRICAK, M.D.
KIRK L. MOON, JR., M.D.

Nuclear magnetic resonance imaging of the upper abdomen has proved to be somewhat difficult, primarily because of degradation of image quality resulting from respiration, cardiac motion, and bowel peristalsis. The difficulty is also due to a lack of readily available NMR gastrointestinal and intravenous contrast media. Nonetheless, NMR has demonstrated an ability to provide unique and extremely informative images of the abdominal organs.[1–11] Perhaps the greatest potential of NMR in imaging of the abdomen may be in the depiction of retroperitoneal structures.[5–11]

Adrenal Glands

Normal Anatomy

The normal adrenal gland appears as an intermediate-intensity organ surrounded by high-intensity retroperitoneal fat on spin echo images. The gland is, on average, of slightly lower intensity than the liver at all TR values; however, there is significant variability in this relationship.[8] Contrast between the gland and surrounding fat is most pronounced at short TR values (0.5 sec) and diminishes somewhat as TR is increased.

The body of the right adrenal gland is positioned just dorsal to the inferior vena cava at the level where the inferior vena cava becomes an intrahepatic structure. From the body, and without a clear point of demarcation, arise a prominent medial and a less prominent lateral limb. The medial limb extends dorsally into the retroperitoneal fat between the liver on the lateral side and the diaphragmatic crura on the medial side. The NMR characteristics of the adrenal gland and diaphragmatic crura are sufficiently different for clear separation (Fig. 1). Only a small amount of retroperitoneal fat is required to distinguish the medial limb of the adrenal gland from the liver.

The right adrenal gland has a variety of configurations. Most frequently, it is linear in shape with a course parallel to diaphragmatic crus in its most cephalic portion, assuming a more horizontal position behind the inferior vena cava in its more caudal aspect (Fig. 1B). When the lateral limb of the right adrenal gland is identified, the gland assumes an inverted "V" shape. The right adrenal gland may occasionally appear as a horizontal linear structure throughout its entire length. Because there is dramatic difference between the medium-intensity adrenal gland and the low-intensity vena cava, the gland can be clearly defined and readily differentiated from the inferior vena cava (Fig. 1). The vast majority of right adrenal glands are positioned cephalad to the upper pole of the right kidney, and the lowest portion of the right adrenal gland is usually located cephalad to right renal vessels.

The shape of the left adrenal gland varies considerably. Most often it has an inverted "V" or "Y" shape, but it can also be linear or triangular in configuration. The left adrenal gland most commonly has a medial and a lateral limb that form an acute angle relative to each other (Fig. 1). The lateral and medial limbs are usually equal in length, but not infrequently the lateral limb is several millimeters longer. The medial limb of the left adrenal gland is positioned lateral to the left crus of the diaphragm. The left adrenal gland is located anterior to the ventral vertebral body margin in over 90% of the cases and is invariably positioned posterior to the anterior aortic tangent. The splenic vessels are positioned just anterior to the lateral limb of the left adrenal gland (Figs. 1A, 2). The lateral limb of the left adrenal gland can be readily distinguished from adjacent splenic vein or artery by virtue of the lack of signal from the vessels that results from blood flow.

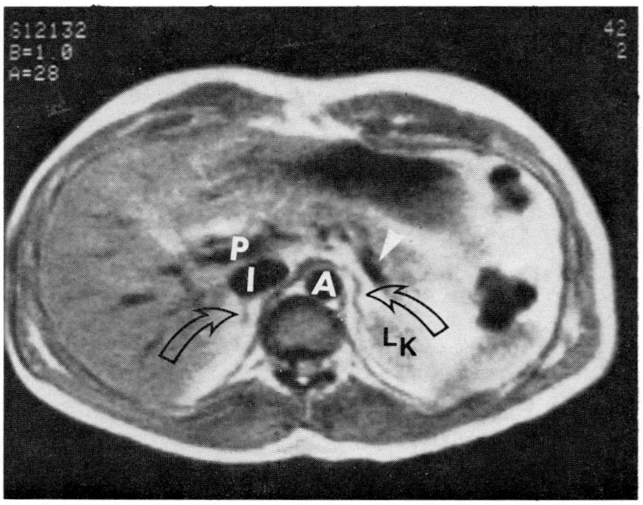

A.

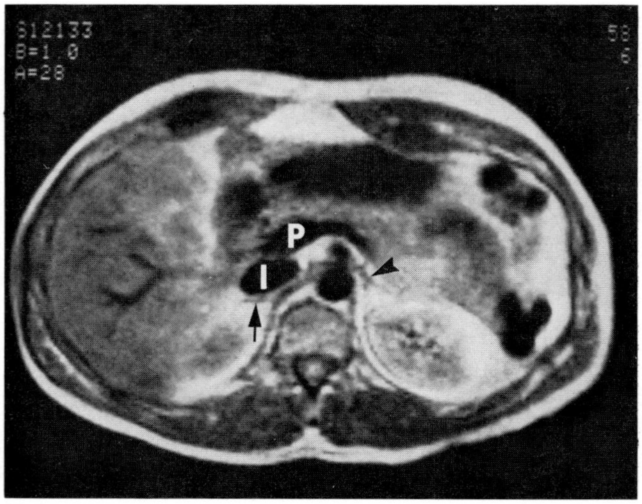

B.

Figure 1. Normal right and left adrenal glands.

A. Spin echo (TR of 1.0 sec, TE of 28 msec). The adrenal glands (arrows) are medium-intensity structures outlined by high-intensity fat. The NMR characteristics of the adrenal glands and diaphragmatic crura are sufficiently different for clear separation. Also, note high-contrast landmarks of inferior vena cava (I) and splenic vein (arrowhead). (A, aorta; P, portal vein; LK, left kidney)

B. Scan obtained 1 cm caudad shows the typical horizontal orientation of the caudal aspect of the right adrenal gland (arrow). Separation from the inferior vena cava (I) is clear. The medial and lateral limbs of the left adrenal gland form an acute angle relative to each other, assuming an inverted "V" configuration (arrowhead). The confluence of the splenic and portal vein (P) is well demonstrated.

Nuclear magnetic resonance appears to depict the normal adrenal glands with a frequency that is comparable to CT and greater than ultrasound in the adult patient. However, in a newborn, ultrasound should remain as a primary scanning modality.[12] Since the adrenal glands and the tissues surrounding them have different relaxation characteristics, it might be expected that NMR would be more successful than CT in discriminating the normal gland from adjacent tissues in patients with a paucity of retroperitoneal fat.

Pathology

In the assessment of adrenal hyperplasia, many measuring methods have been proposed. They range from quantifying adrenal gland length as determined from continuous imaging sections to measuring thickness of the adrenal gland perpendicular to the long axis of the gland (thickness is defined as the greatest thickness measured in any site). These measurements are only an approximation, and great variations exist from patient to patient. We prefer the estimation of the thickness of the adrenal gland as compared to the thickness of the diaphragmatic crus. In the normal adrenal gland, thickness of the gland will be less than or equal to that of the ipsilateral crus of the diaphragm. A diffuse bilateral adrenal gland enlargement with preservation of shape and configuration of gland has been the second most frequent feature of ACTH-dependent Cushing's syndrome as visualized by CT. However, normal adrenal glands have been reported as the most frequent CT finding in those patients. It will be of interest to see if NMR might be able to depict cortical hyperplasia prior to gross enlargement of the gland. In several cases examined to date, NMR appears to have differentiated the hypertrophic adrenal cortex from the lower-intensity adrenal medulla (Figs. 2, 3). The NMR display of adrenal tumors (metastatic and primary lesions, cortical adenoma, nonfunctioning ad-

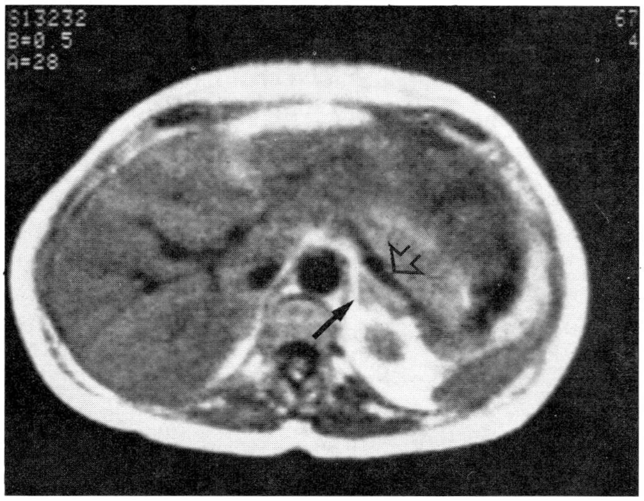

A.

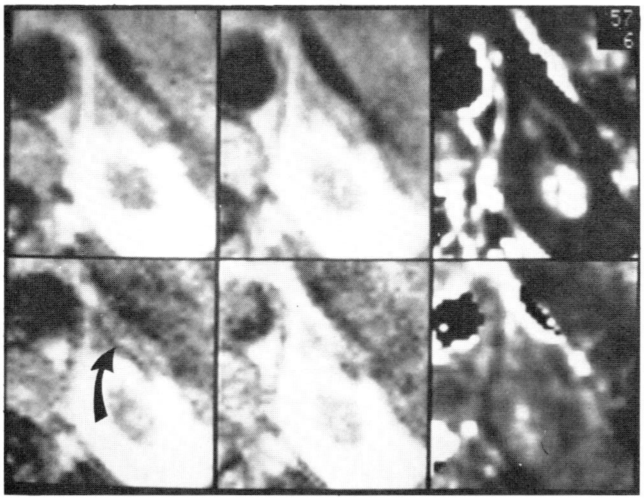

B.

Figure 2. Hyperplasia of the left adrenal gland.

A. Spin echo image through the upper abdomen obtained with TR of 0.5 sec and TE of 28 msec. There is an abnormal, enlarged, inverted "V"-shaped left adrenal gland (arrow). The splenic vessels (open arrowhead) are positioned just anterior to the lateral limb of the left adrenal gland.

B. Top row, left to right: TR of 0.5 sec, TE of 28 msec; TR of 1.0 sec, TE of 28 msec; calculated T_1 image. Bottom row, left to right: TR of 0.5 sec, TE of 56 msec; TR of 1.0 sec, TE of 56 msec; calculated T_2 image. Overall size of the left adrenal gland is best demonstrated in the upper left (0.5-sec/28-msec) image. In the bottom left image (0.5-sec/56-msec), the central portion of the adrenal gland is of lower intensity (arrow) than the outer rim. As seen on calculated T_1 images, the central portion (medulla) has a prolonged T_1 relaxation value as compared with the surrounding cortex.

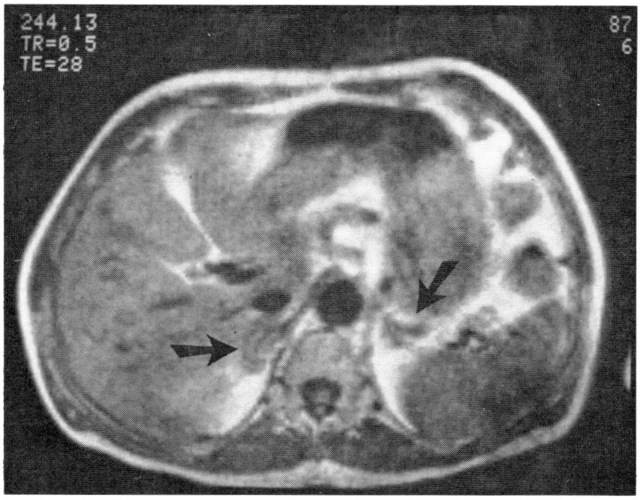

A.

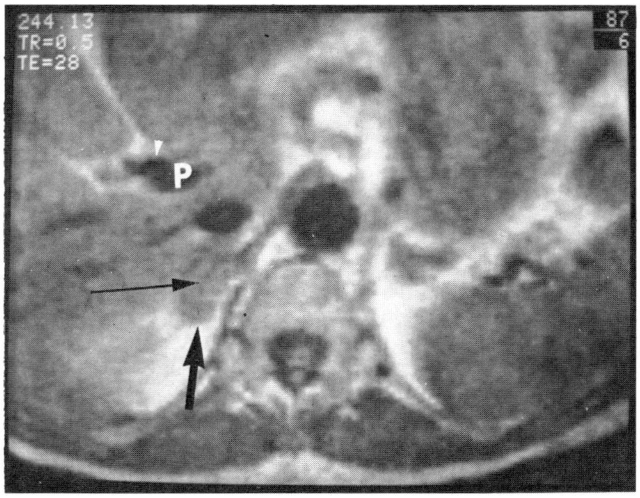

B.

Figure 3. Bilateral adrenal hyperplasia.

A. Spin echo image (TR of 0.5 sec, TE of 28 msec) through the upper abdomen. Both right and left adrenal glands (arrows) are enlarged.

B. Enlargement of the area of interest emphasizes differentiation between higher-intensity adrenal cortex (slim arrow) and lower-intensity adrenal medulla (wider arrow). Differentiation between adrenal cortex and medulla is seen in both right and left adrenal glands. Note the extension of the enlarged right adrenal cortex to the posterior aspect of the inferior vena cava. The right portal vein (P) and common hepatic duct (arrowhead), surrounded by high-intensity adipose tissue present in the fissure of the ligamentum venosum, are demonstrated.

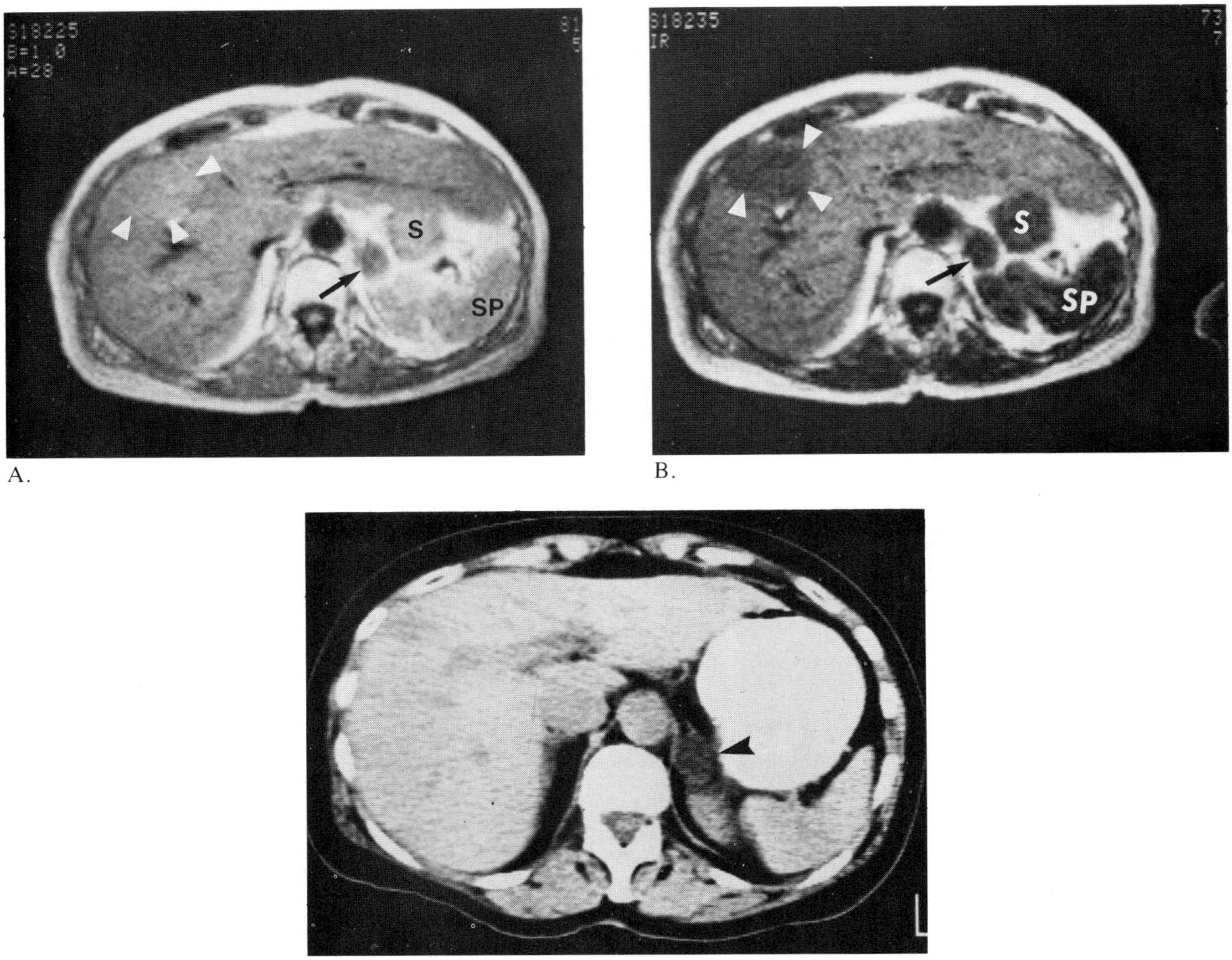

A. B. C.

Figure 4. Carcinoid tumor metastatis to left adrenal gland and the liver.
A. On the spin echo image (TR of 1.0 sec, TE of 28 msec) the lesion in the left adrenal gland (arrow) is of low intensity. The hepatic metastases (arrowheads) are of higher intensity than the surrounding liver. (S, stomach; SP, spleen; LK, left kidney)
B. The inversion recovery image (TR of 1.0 sec, TI of 420 msec) exhibits greater tissue contrast, and both adrenal lesion (arrow) as well as metastatic liver lesion (arrowheads), having longer T_1 values, appear with lower intensity. (S, stomach; SP, spleen)
C. Comparison with CT scan through the same anatomic section showing a low attenuation of left adrenal lesion (arrowhead).

renal tumors, pheochromocytoma, and myelolipoma) is similar to that shown by CT (Figs. 4, 5).[8] Nuclear magnetic resonance is effective in assessing the adjacent organs and can depict vascular compressions and secondary organ congestion, as illustrated in Figure 5.

While CT is currently the imaging procedure of choice in the evaluation of the adrenal glands, NMR will assume a prominent role in adrenal imaging because it demonstrates normal and pathologic adrenal anatomy in a fashion comparable to CT, but it is free of radiation and possesses superior soft-tissue contrast.

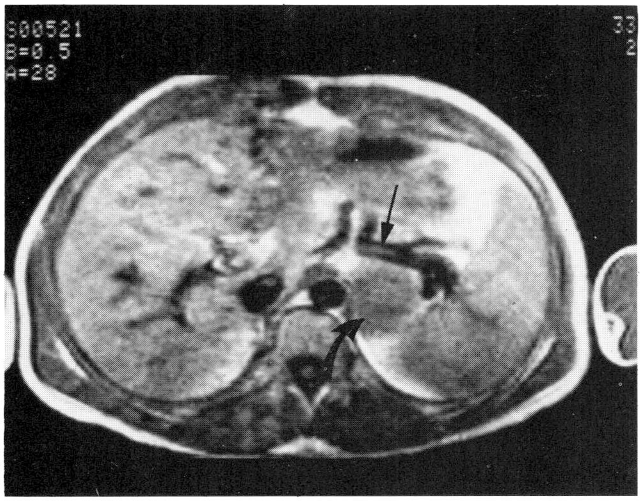

A.

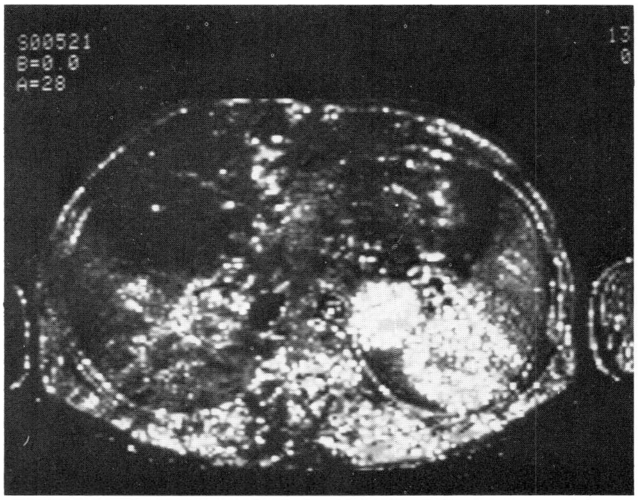

B.

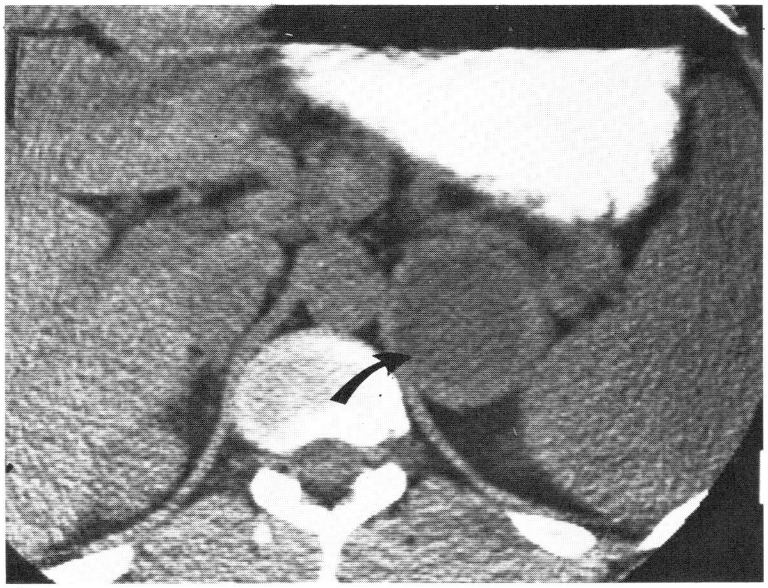

C.

Figure 5. Pheochromocytoma of the left adrenal gland.

A. On spin echo image (TR of 0.5 sec, TE of 28 msec), tumor in the left adrenal bed (curved arrow) is of inhomogeneous low intensity. Also, note the difference in intensity of the spleen, with the posterior part of the spleen being of considerably lower intensity than its anterior portion. There is a high-intensity signal within the splenic vein (straight arrow). This may be due to slow blood flow in this region.

B. Calculated T_1 image demonstrating a long T_1 value of the pheochromocytoma and also a long T_1 value of the posterior half of the spleen. At surgery, there was congestion of the posterior aspect of the spleen.

C. Comparison CT scan showing cystic nature of the tumor (curved arrow). The spleen has a homogeneous CT appearance.

Kidneys

Normal Anatomy

The anatomic features of the normal kidney are clearly defined by NMR.[4,6,10,11] On spin echo images, the renal hilum is imaged as a high-intensity area because of the presence of hilar adipose tissue (with its short T_1 and long T_2 values). The renal veins and arteries are seen as low-intensity tubular structures traversing the hilar region (Fig. 6). The low intensity of the vessel lumen is due to rapid blood flow. The pelvocaliceal system and the ureters are of low intensity because of presence of urine, which has long T_1 and T_2 relaxation values. On images obtained during increased diuresis, mild dilatation of pelvocaliceal system can be appreciated. On spin echo images, the renal cortex and medulla can be easily discriminated when imaging is performed with a short pulse sequence interval (TR of 0.5 sec). T_1 differences are accentuated on images obtained at short TR values. Since the medulla has a longer T_1

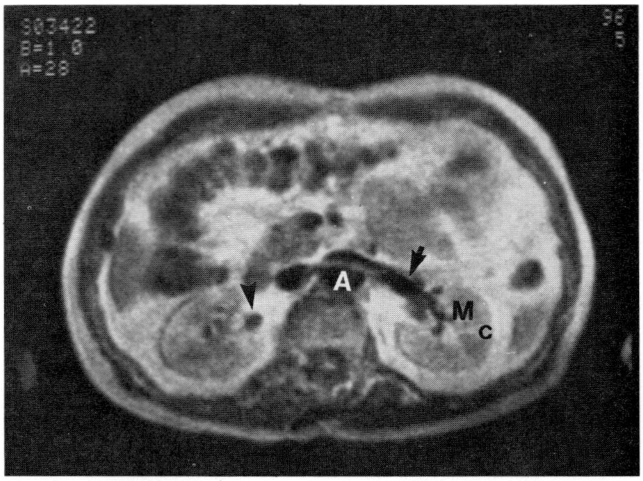

Figure 6. Normal kidneys. Spin echo image (TR of 1.0 sec, TE of 28 msec). There is good differentiation between higher-intensity cortex (c) and lower-intensity medulla (M). The left renal vein (arrow) is seen anterior to the aorta (A). A lower-intensity circle medial to the right kidney represents a normal ureter (arrowhead).

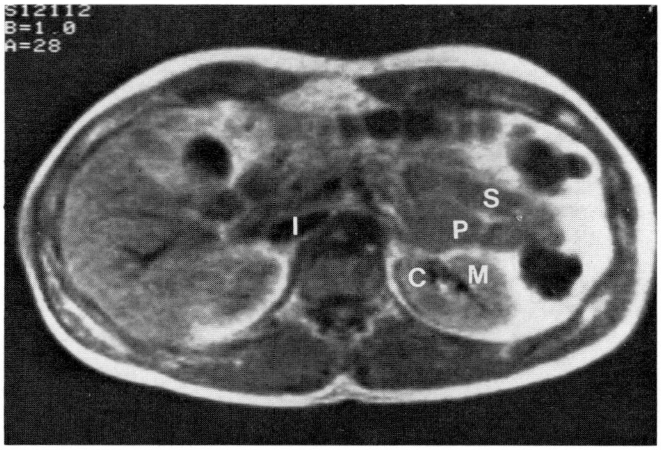

A.

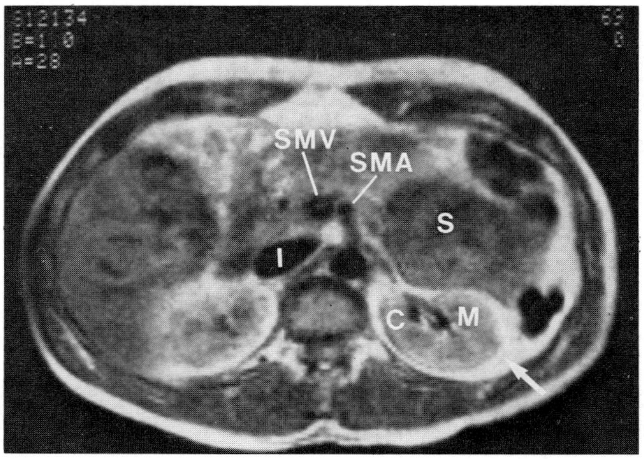

B.

Figure 7. Hydration challenge varying the appearance of normal kidneys. Images are obtained with spin echo sequence in which TR is 1.0 sec and TE is 28 msec.

A. Image obtained following 20 hr of dehydration. There is a faint differentation between cortex (C) and medulla (M). Note the collapsed stomach (S); the retrogastric fat outlines the body and tail of the pancreas (P). (I, inferior vena cava)

B. Image through the same anatomic level obtained following hydration. Renal cortex (C) and medulla (M) are clearly differentiated. There is increase in the renal size and distention of the inferior vena cava (I). Gerota's fascia (arrow) is seen as a low-intensity line dividing perirenal from pararenal space. Note superior mesenteric artery (SMA) and superior mesenteric vein (SMV). (S, fluid-distended stomach) than the cortex does, the intensity of the medulla is considerably lower than that of the cortex on short TR images. The cortex emits a higher-intensity signal than the medulla, but cortical intensity is lower than either the hilar or perirenal adipose tissue. With the prolongation of the pulse sequence interval, corticomedullary differentiation becomes less prominent, and at the TR of 2.0 sec it is almost non-existent. When contrast between cortex and medulla is expressed in percentages, a 48% contrast difference is seen when utilizing a pulse sequence of 0.5 sec, and a 3% contrast is present at a pulse sequence of 1.5 sec. The cortex-medulla contrast difference becomes negligible at a pulse sequence of 2.0 sec. The prominence of the medulla is also dependent on the state of hydration.[6] With good hydration and increased diuresis, the medulla is somewhat swollen, and therefore more prominent (Fig. 7). Perirenal fat is imaged as a high-intensity area because of the relatively short T_1 and relatively long T_2 relaxation values of the adipose tissue. Gerota's fascia is visible as a lower-intensity line dividing the perirenal from pararenal space.

In summary, the optimal NMR spin echo scanning sequence for display of renal anatomy is the sequence in which resolving power of the image is at its optimum (Figs. 6, 7). The resolving power of NMR images represents the ability to provide information that will allow the confident differentiation of an anatomic point from the surrounding background. The resolving power depends on signal-to-noise level and tissue contrast, with tissue contrast representing the intensity of the organ over intensity of the background.[13] With prolongation of the pulse sequence interval, the signal-to-noise ratio improves, but the cortex-medulla tissue contrast decreases; therefore, the optimal resolving power of the image appears to be at the pulse sequence interval of 1.0 sec, where the balance between signal-to-noise level and soft-tissue contrast is optimal.

On an inversion recovery image obtained with a TR of 1.0 sec and TI of 420 msec, the renal cortex appears significantly brighter than the medulla, and the corticomedullary differentiation is usually more striking than on any spin echo images (Fig. 8). However, signal levels obtained with this inversion recovery technique are relatively low, and the resolving power is degraded as a result. On inversion recovery images, perirenal and hilar fat remain bright because of their short T_1 value. The urine in the renal pelvis and ureter remains of low intensity because of its long T_1 value.

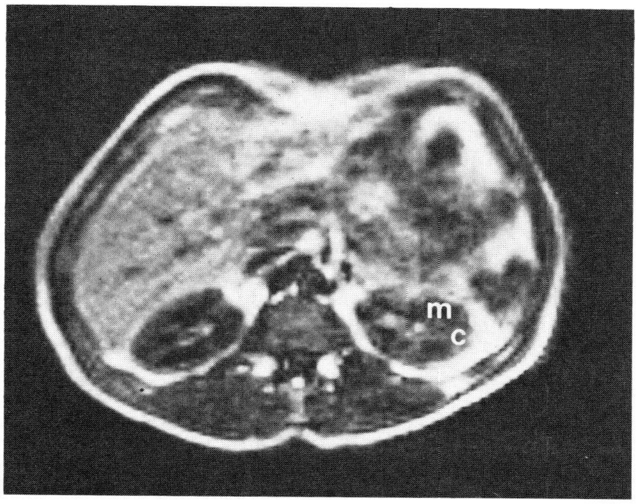

Figure 8. Normal kidney on inversion recovery image with TR of 1.0 sec and TI of 420 msec. The differentiation of the renal cortex (c) and medulla (m) is accentuated, but the signal level obtained with this inversion recovery technique is relatively low; as a result, resolving power of the image is degraded.

Pathology

Pelvocaliectasis is readily detectable by NMR, and the dilated ureter can be easily followed.[6] When pelvocaliectasis was caused by obstruction, no differentiation between cortex and medulla could be obtained regardless of sequence or technique utilized (spin echo or inversion recovery) (Fig. 9). The dilated ureter is of low intensity because of the presence of urine. The distended ureter can be differentiated from bowel, lymph nodes, and blood vessels by analyzing the multiple spin echo intensity images. With x-ray CT, if the kidneys are non-excreting, such differentiation is sometimes difficult. When obstruction is present, NMR, like CT, will determine the cause of obstruction.

When urinary outlet obstruction is caused by retroperitoneal fibrosis, NMR images will reveal the extent of the fibrous mass and plaque and show the relationship of the fibrosis to the major retroperitoneal and pelvic vessels (Figs. 10–13). Furthermore, variation in intraluminal signal on NMR images obtained with multiple spin echo parameters may suggest the presence of abnormal flow in vessels constricted by retroperitoneal fibrosis (Fig. 11). To obtain such information with CT, bolus injection of contrast medium is necessary (Fig. 12). The high contrast between flowing blood (giving little or no NMR signal) in the vessel and the adjacent soft tissue suggests the advantage of NMR imaging in defining the effect of retroperitoneal process upon the inferior vena cava

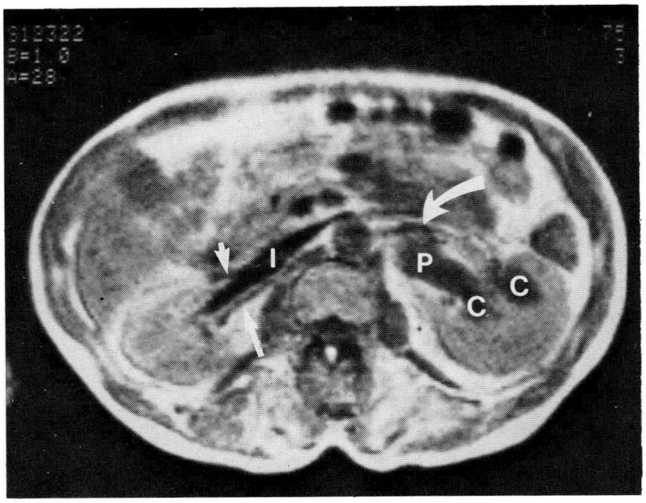

A.

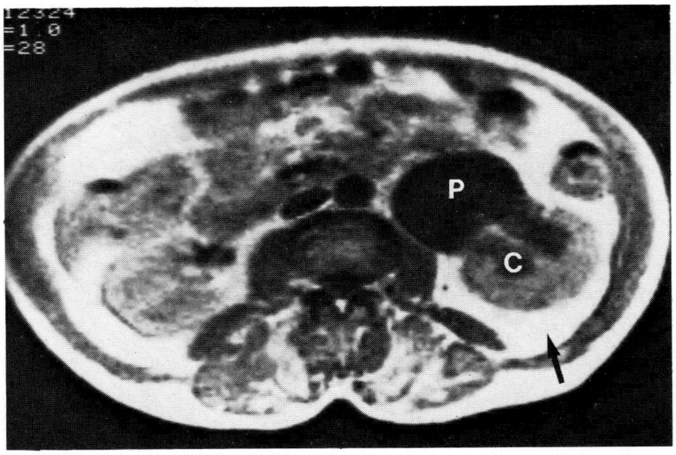

B.

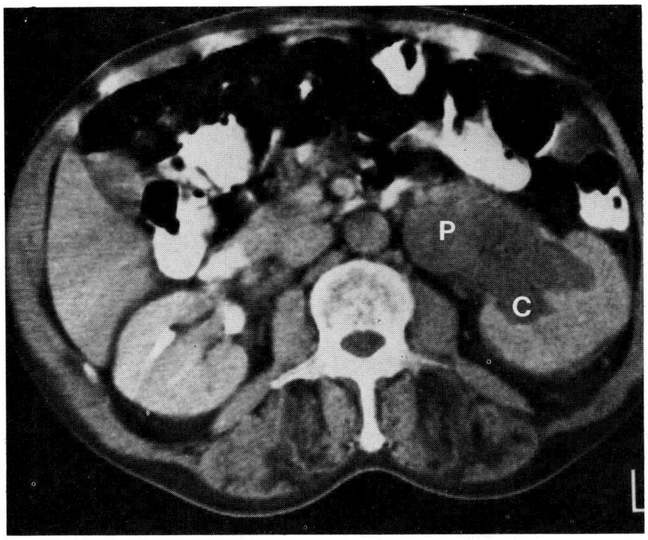

C.

Figure 9. Left ureteropelvic junction obstruction.

A. Spin echo NMR image (TR of 1.0 sec, TE of 28 msec). Right kidney is normal, with clear cortical-medullary differentiation. Right renal artery (long arrow) and renal vein (short arrow) entering inferior vena cava (I) are imaged. The calices (C) and pelvis (P) in the left kidney are dilated. (curved arrow, left renal artery)

B. Image obtained 2 cm caudad showing markedly dilated extrarenal pelvis (P) on the left. Dilated calices (C) are seen as well. No cortical-medullary differentiation on the left is seen, while there is a clear cortical-medullary differentiation of the right kidney. (arrow, Gerota's fascia)

C. Comparison CT scan showing a markedly dilated extrarenal pelvis (P) and dilated calices (C) of the left kidney.

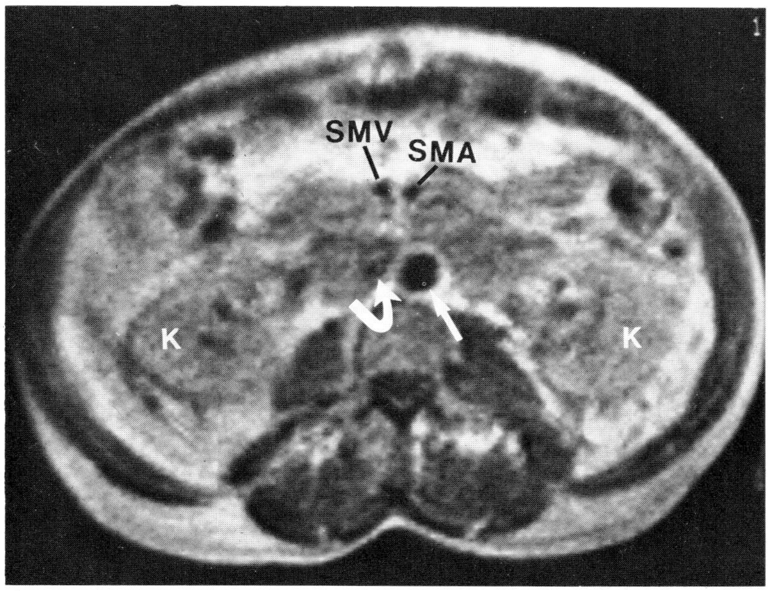

A.

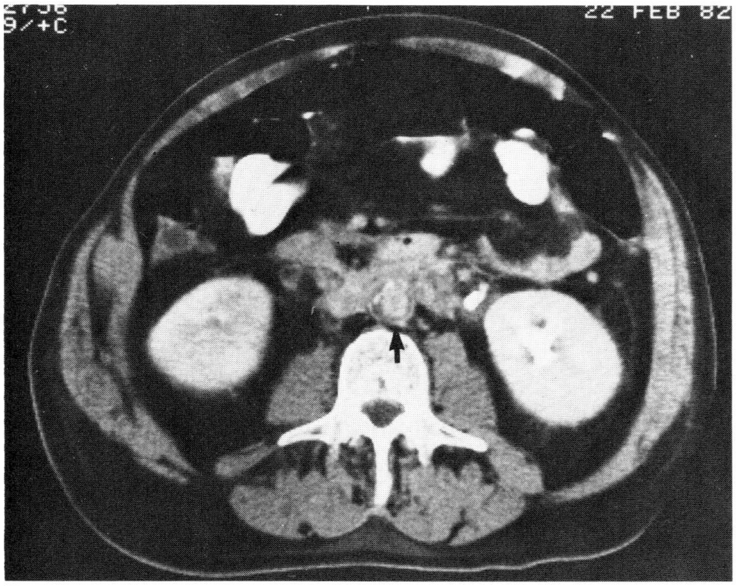

B.

Figure 10. Retroperitoneal fibrosis.

A. An NMR image through the upper abdomen. Spin echo technique (TR of 1.0 sec, TE of 28 msec). Fibrotic plaques are localized to periaortic area. Superior mesenteric artery (SMA) and superior mesenteric vein (SMV) are displaced anteriorly. Aorta (straight arrow) is not involved, and blood flow through the aorta is unaltered. Inferior vena cava (curved arrow) is compressed. (K, kidneys)

B. Corresponding CT image. Aortic wall is calcified (arrow). There is fibrotic plaque around the aorta. No information about the flow through the aorta or inferior cava can be obtained.

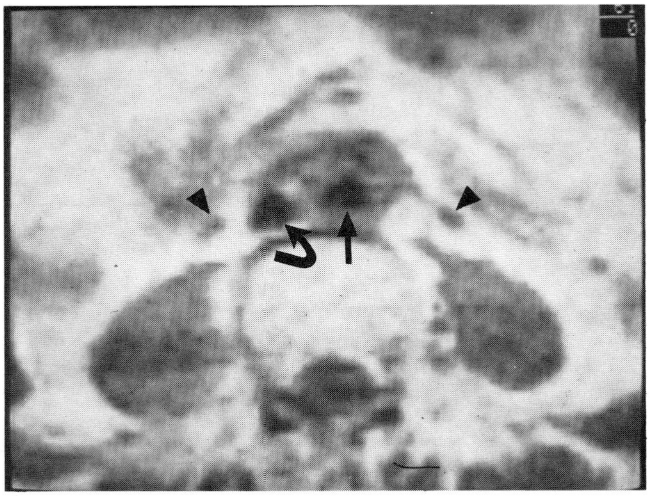

A.

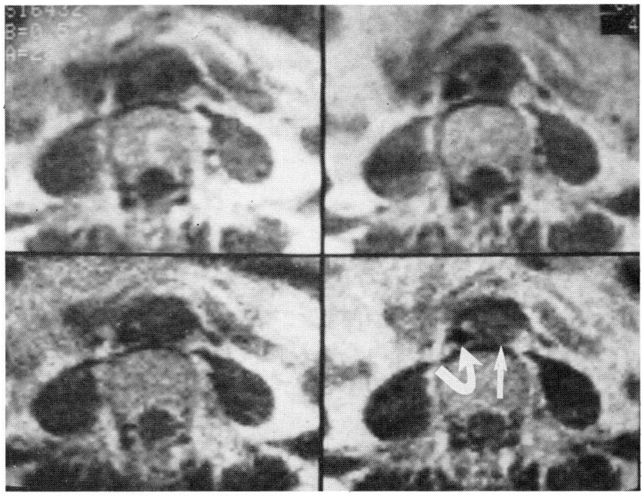

B.

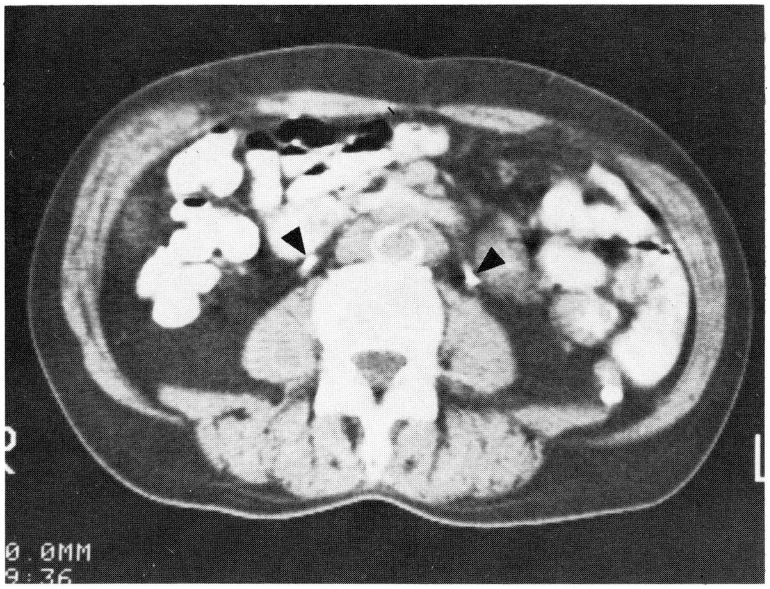

C.

Figure 11. Retroperitoneal fibrosis.

A. Axial scan above aortic bifurcation. There is retroperitoneal plaque surrounding great vessels. Both aorta (straight arrow) and inferior vena cava (curved arrow) are well seen. (arrowheads, bilateral ureteral stents)

B. Images at the same cross-sectional level obtained with different spin echo imaging parameters. From left to right first in the top row and then the bottom: 0.5 sec/28 msec; 1.0 sec/56 msec; 0.5 sec/56 msec; 1.0 sec/56 msec. On images obtained with longer echo delay (TE of 56 msec) there is signal in the lumen of the aorta (arrow). This suggests presence of slow flow in the area of aortic constriction. Fibrotic plaque produces no flow alteration in the inferior vena cava (curved arrow).

C. Corresponding CT scan. Aortic wall is calcified. Plaque around the aorta is well seen, but no information about the inferior vena cava or flow through the aorta could be obtained. Bilateral ureteral stents are seen (arrowheads).

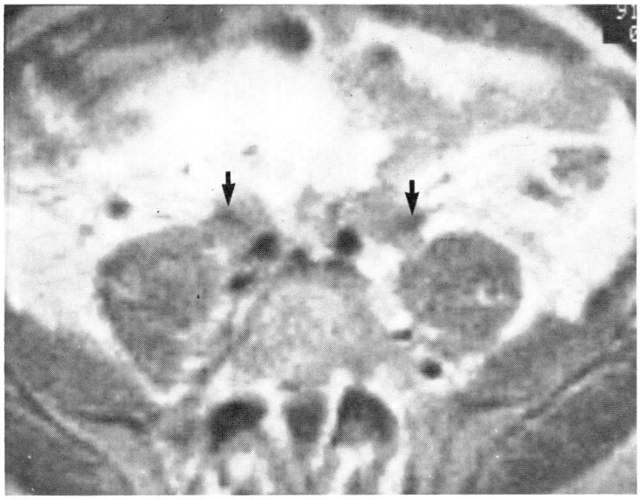

A.

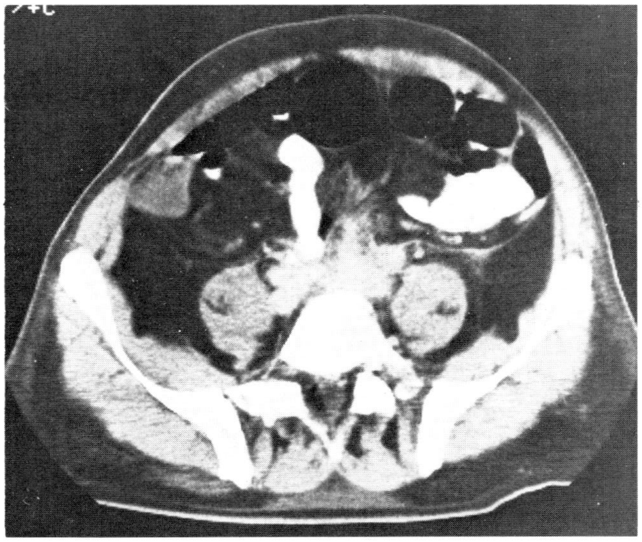

B.

Figure 12. Retroperitoneal fibrosis.
A. Axial scan below the level of aortic bifurcation. Retroperitoneal plaque is seen surrounding iliac vessels. Both common iliac arteries and veins are well seen. Ureteral stents are also seen (arrows).
B. Corresponding CT image following bolus injection of a contrast medium shows enhancement of the fibrous tissue, but no information about the underlying vessels could be obtained.

and aorta. Extrinsic compression of the aorta and inferior vena cava is well demonstrated by NMR. Furthermore, the T_1 and T_2 relaxation values of retroperitoneal fibrosis are distinctly different from adjacent psoas muscle or adipose tissue.[9]

Another clinical problem related to pathology in the renal hilum is the difficulty in differentiating pelvic sinus lipomatosis and peripelvic cysts. Because fluid (long T_1) and fat (short T_1) have opposite NMR relaxation parameters, NMR readily distinguishes between these two entities (Figs. 13, 14). Fluid and fat also have opposite characteristics by sonography.[14] However, while peripelvic cysts can be clearly differentiated from sinus lipomatosis, often sonographically, peripelvic cysts are indistinguishable from pelvocaliectasis. The differentiation between peripelvic cysts and sinus lipomatosis is superior on NMR as compared with x-ray CT, where partial volume averaging may result in ambiguous findings.[6,14]

In renal medical diseases, NMR promises to make a significant contribution.[6,11] Renal volume can be calculated, and the NMR characteristics of the renal parenchyma can be assessed. Since the incidence of acquired cystic disease and renal neoplasm is increased in chronic renal failure patients,[15] NMR's ability to make a radiologic evaluation without iodine contrast media should make it most useful (Figs. 15, 16).

The ability of NMR to distinguish solid from cystic renal lesions appears similar to that of contrast-enhanced x-ray CT.[1,6,10] Nuclear magnetic resonance, however, is superior to CT in its ability to characterize the nature of the cystic fluid.[6] Hemorrhagic and infected cysts have been shown to have strikingly different spin echo intensities as compared with simple renal cysts. The simple renal cyst (i.e., a cyst that is neither infected nor had bleeding within it) has long T_1 and T_2 relaxation values (Figs. 17–19). Therefore, on spin echo images with short pulse sequence intervals (TR of 0.5 sec) it appears of low intensity. As the pulse sequence interval increases, the intensity signal from the fluid within the cyst changes. It becomes brighter with a longer pulse sequence interval, but it also becomes brighter with a longer TE value because of long T_2 relaxation values (Fig. 17) of the simple cyst. The cyst is homogeneous and has a sharp outline toward the renal parenchyma (Fig. 18). The wall of the cyst cannot be measured (Figs. 17, 18). All hemorrhagic cysts studied to date are of high intensity on all pulse

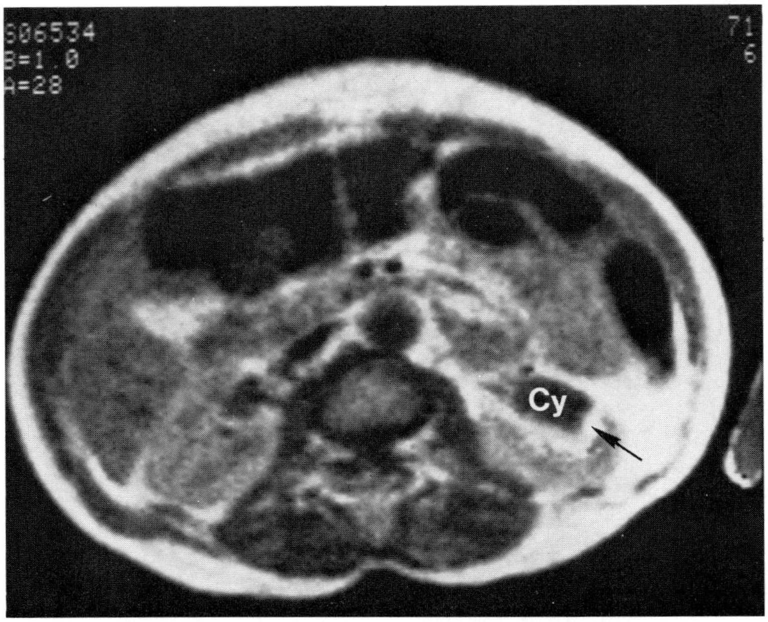

Figure 13. Left parapelvic cyst. Spin echo image obtained with TR of 1.0 sec, TE of 28 msec. Fluid collection (Cy) is seen in the hilar region of the left kidney. High-intensity hilar adipose tissue (arrow) is seen interposed between the fluid collection and the renal parenchyma.

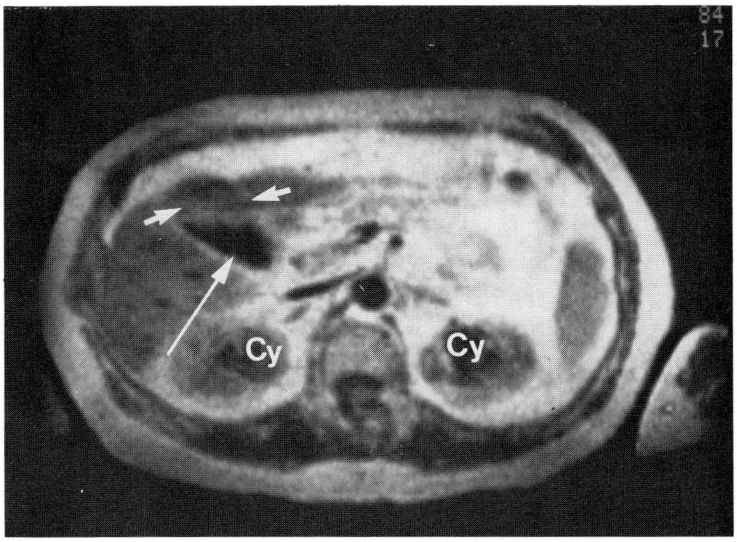

Figure 14. Bilateral peripelvic cyst (Cy). Incidental finding is gallstones in the dependent portion of the gallbladder (long arrow). The stones are of low intensity and are seen in the dependent portion of the gallbladder. Gallbladder bile (small arrows) is seen above the gallstones.

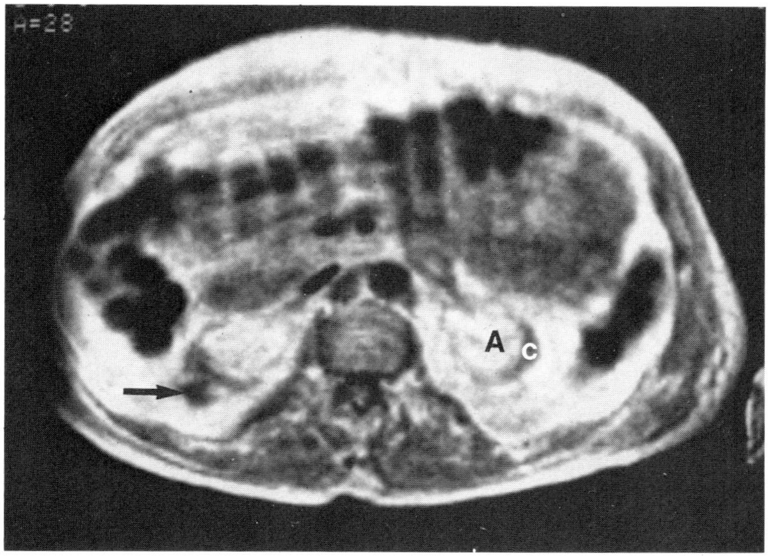

Figure 15. Chronic glomerulonephritis. Both kidneys are small. There is marked thinning of the renal cortex (c) and abundant hilar adipose tissue (A). A small cyst (arrow) is seen arising from the posterior lateral aspect of the right kidney.

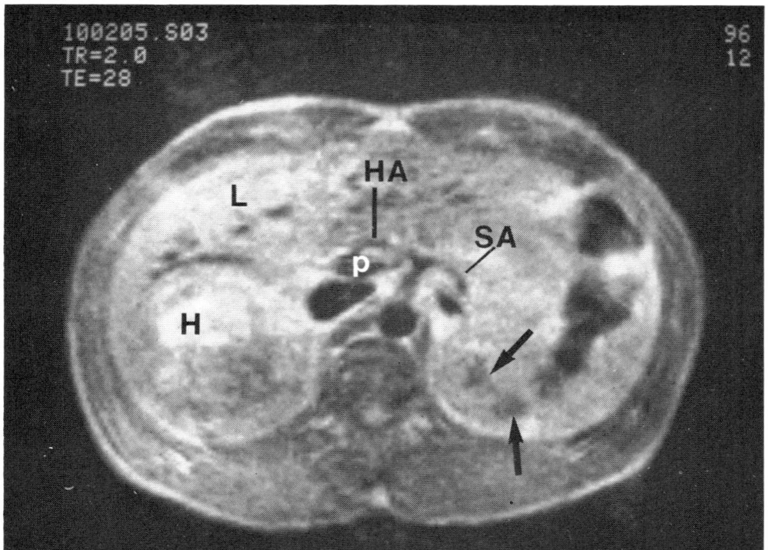

Figure 16. Chronic renal failure patient with history of 5 yr on hemodialysis. Spin echo image (TR of 2.0 sec, TE of 28 msec). The left kidney is small and has numerous simple cysts (arrows) within it. There is a large inhomogeneous mass in the right renal bed representing renal cell carcinoma. High-intensity area (H) corresponded to area of acute hemorrhage. Good demonstration of celiac axis dividing into splenic (SA) and hepatic (HA) artery. Note the good-quality diagnostic NMR image despite the lack of adipose tissue in this asthenic patient. (L, liver; p, portal vein).

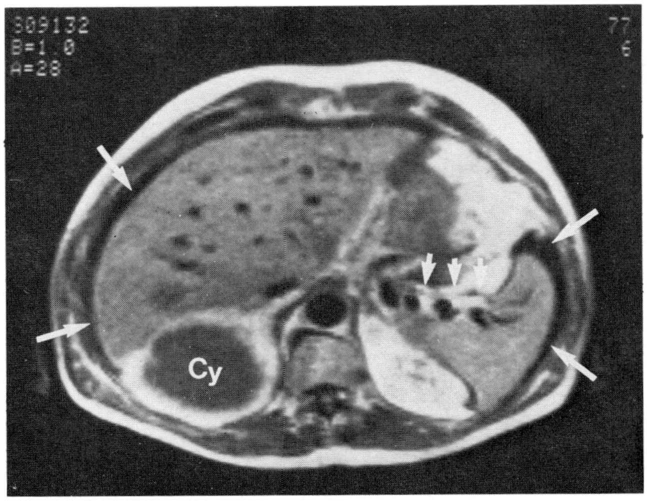

A.

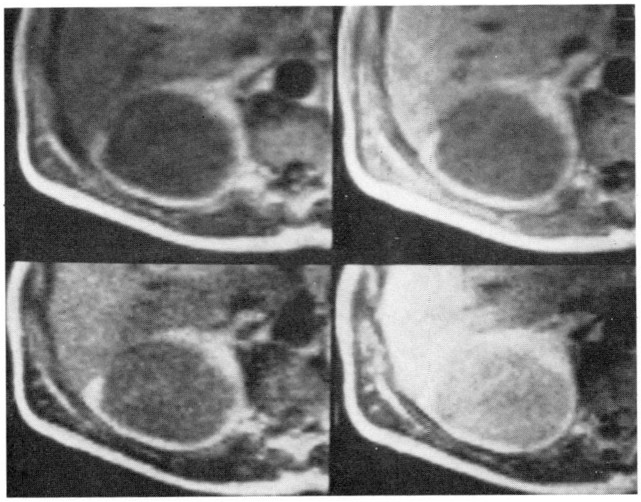

B.

Figure 17. A simple renal cyst in right kidney.

A. Spin echo image through the upper abdomen (TR of 1.0 sec, TE of 28 msec). Right renal cyst (Cy) is of homogeneous low intensity. Walls of the cyst are not measurable. Ascites (long arrows) is seen as a low-intensity band surrounding the liver and spleen. Note the tortuous splenic artery (small arrows).

B. Different-intensity spin echo images through the right renal cyst from left to right, first in the top row and then in the bottom: 0.5 sec/28 msec; 1.0 sec/28 msec; 0.5 sec/56 msec; 1.0 sec/56 msec. Contents of the cyst show a relative increase in intensity from the 0.5-sec to the 1.0-sec image because of the long T_1 value of the fluid (T_1 of 2.0 sec). There is also increase in intensity from images with TE of 28 msec to images with 56 msec because of long T_2 value of the fluid (T_2 of 200 msec).

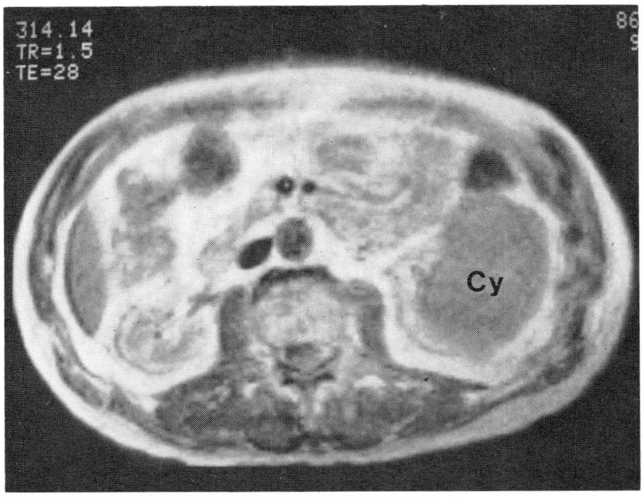

A.

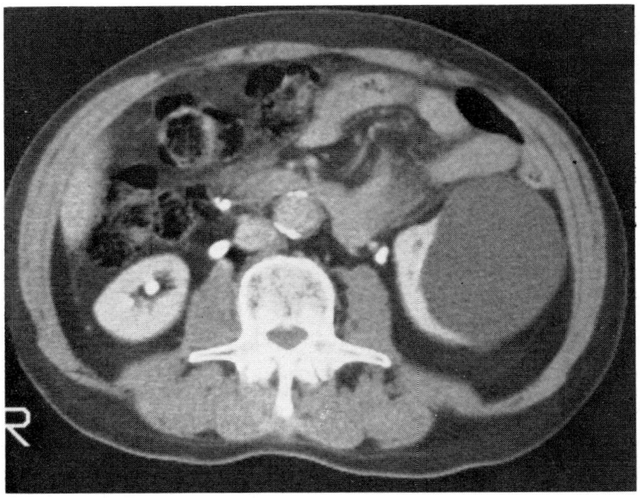

B.

Figure 18. Simple renal cyst.

A. Simple renal cyst, left kidney. Spin echo image (TR of 1.5 sec, TE of 28 msec). The cyst has a homogeneous low-intensity signal. The wall of the cyst cannot be measured.

B. Corresponding CT scan showing simple cyst of the left kidney.

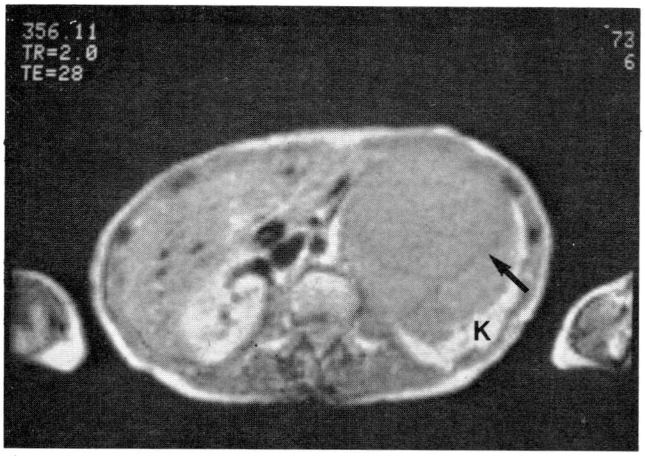

A.

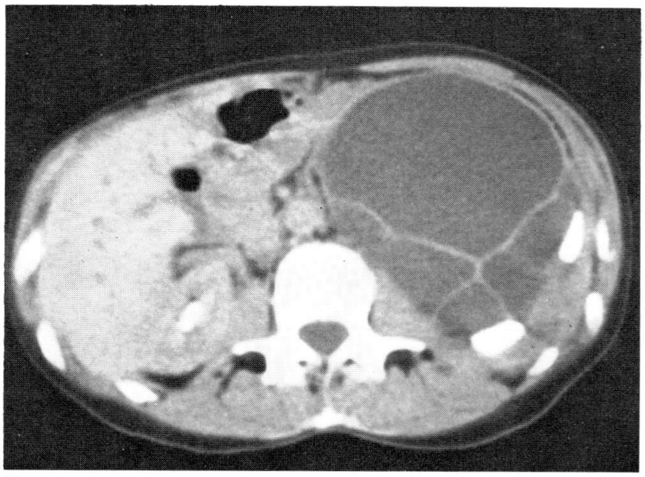

B.

Figure 19. Multilocular cystic nephroma.
A. Spin echo image (TR of 2.0 sec, TE of 28 msec). There is large multiseptate cyst arising from the left kidney. Relaxation values of the cystic fluid indicate simple cyst. Septa are seen as a low-intensity band (arrow) traversing the cyst. (K, displaced left kidney)
B. Corresponding CT scan having characteristic appearance for multilocular cystic nephroma of the left kidney.

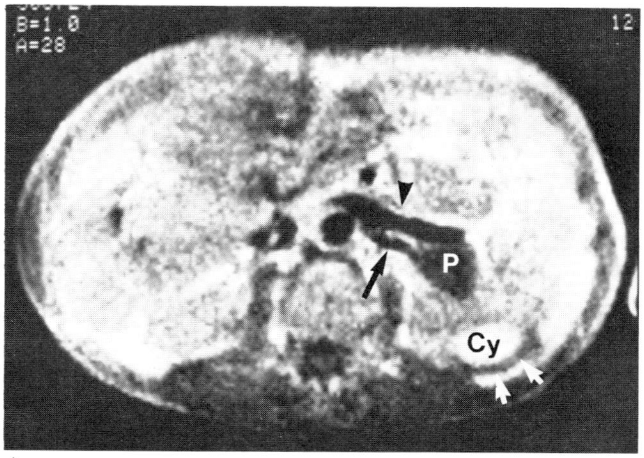

A.

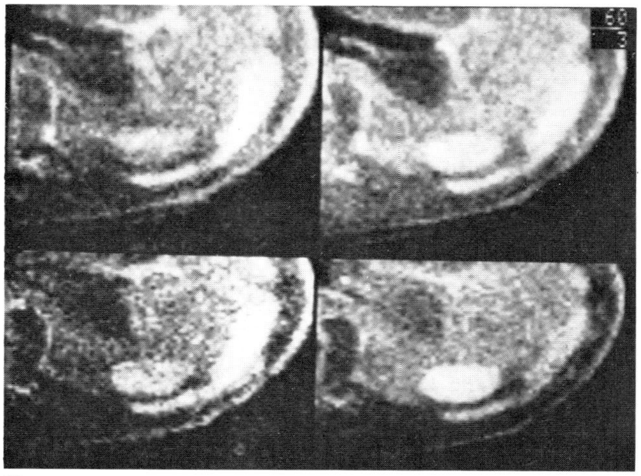

B.

Figure 20. Hemorrhagic cyst, left kidney.
A. Spin echo image (TR of 1.0 sec, TE of 28 msec). Left renal pelvis (P) is dilated. Both left renal artery (black arrow) and vein (arrowhead) are seen. There is a high-intensity cyst (Cy) in the posterior aspect of the left kidney. The low-intensity outer margin of the cyst represents peripheral wall calcifications (white arrows).
B. Spin echo images. From left to right, first in the top row and then the bottom: 0.5 sec/28 msec; 1.0 sec/28 msec; 0.5 sec/56 msec; 1.0 sec/56 msec. The intensity change of the hemorrhagic cyst is markedly different from that of simple cyst (see Fig. 17). The cyst is consistently of high intensity on all pulse sequence intervals, showing a slight further increase in intensity with a longer TR value.

sequence intervals and have shown a slight increase in intensity on longer pulse sequence intervals (Fig. 20). The ability of NMR to reveal information about the nature of the cyst content will be of value in the clinical investigation of acute flank pain in patients with suspected hemorrhagic or infected renal cysts. This is of great utility in patients with polycystic disease (Figs. 21, 22). These patients often present with acute flank pain, and knowledge of the site of intracystic hemorrhage is of great assistance.

The NMR appearance of renal cell carcinoma as studied to date shows a spectrum of findings on spin echo images.[1,5,7,10] The intensity of tumors ranged from low to very high. A feature common to all renal neoplasms has been a prolongation of T_1 as compared with the adjacent normal renal parenchyma. In the absence of necrosis or hemorrhage, clear-cell carcinoma exhibited slightly higher-intensity signals than normal cortex, while the granular cell carcinoma was of low intensity (Figs. 23, 24). On inversion recovery images, renal cell carcinomas exhibited lower-intensity signal than the surrounding renal

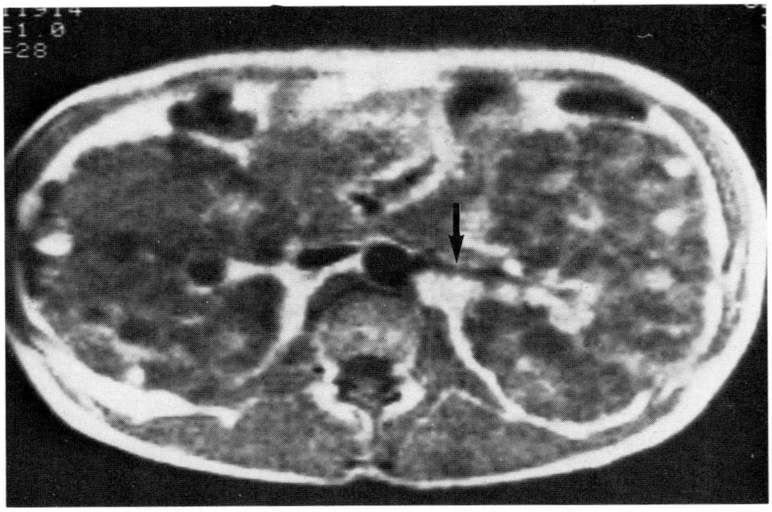

Figure 21. Advanced polycystic kidney disease. Spin echo image (TR of 1.0 sec, TE of 28 msec). Both kidneys are markedly enlarged, and the renal parenchyma is replaced by various-sized cysts. Note the different signal intensity in the cysts, ranging from low values for simple cysts to high-intensity signals suggestive of hemorrhagic cysts. Left renal artery (arrow) is seen entering the aorta.

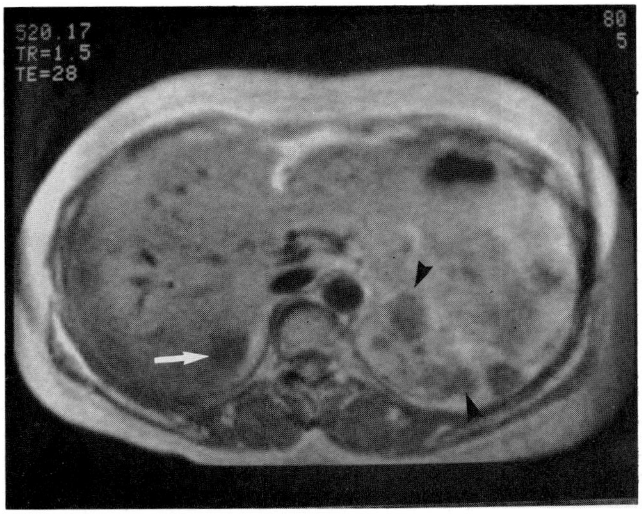

A.

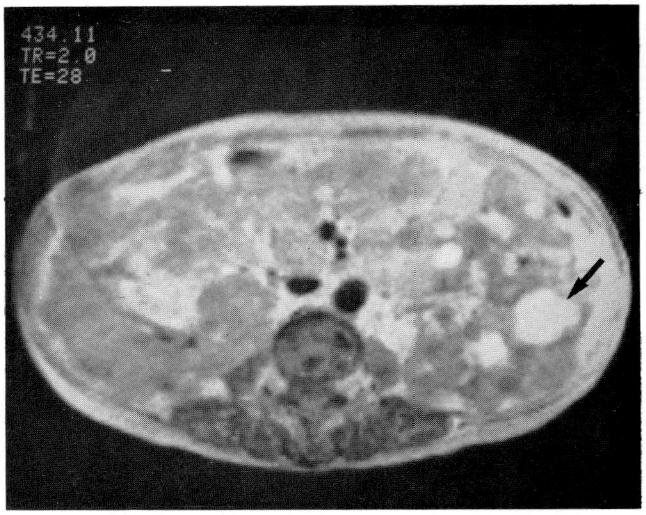

B.

Figure 22. Polycystic kidney and liver disease.

A. Spin echo image (TR of 1.5 sec, TE of 28 msec). There is a simple liver cyst (arrow). There are numerous simple cysts within the left kidney (arrowheads).

B. Scan obtained 6 cm caudad shows both kidneys to be markedly enlarged. The renal parenchyma is replaced by various-sized cysts. The intensity signal of the cysts ranges from low- to high-intensity hemorrhagic cysts. The largest hemorrhagic cyst (arrow) is seen along the lateral aspect of the left kidney. The patient experienced acute left flank pain 2 days prior to NMR study.

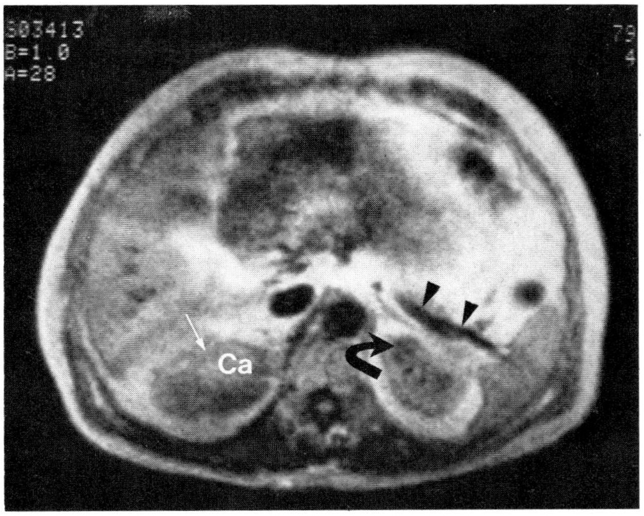

A.

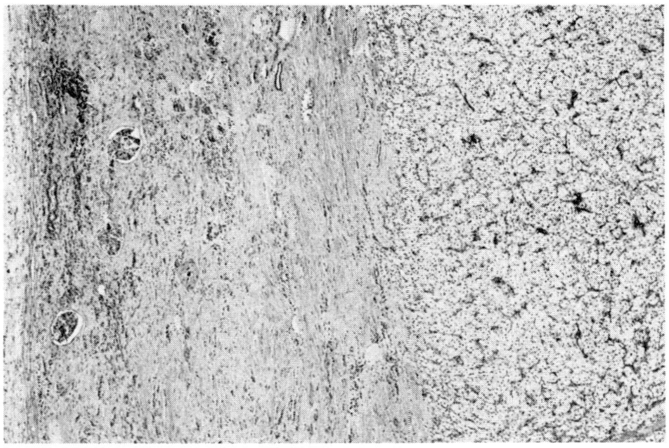

B.

Figure 23. Clear-cell renal cell carcinoma, right kidney; simple renal cyst, left kidney.

A. Spin echo image (TR of 1.0 sec, TE of 28 msec). Renal cell carcinoma (Ca) is imaged with high intensity. There is a lower-intensity band (straight arrow) interposed between the tumor and the normal renal parenchyma. This represents tumor pseudocapsule. Incidentally noted is simple cyst (curved arrow) in the left kidney that is imaged as a round low-intensity structure. (arrowheads, splenic vein)

B. Photomicrograph (hematoxylin and eosin × 30) showing a clear-cell renal carcinoma and adjacent tumor pseudocapsule. The tumor pseudocapsule consists of normal renal parenchyma compressed by expanding tumor and deposition of collagen tissue. At the periphery of the pseudocapsule there are few residual cortical tubules and glomeruli.

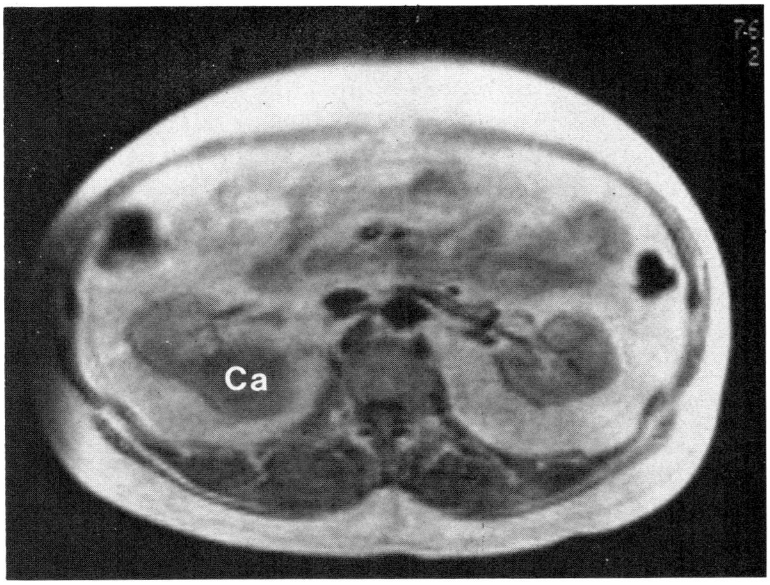

Figure 24. Granular cell carcinoma, right kidney.

A. Spin echo image (TR of 1.0 sec, TE of 28 msec). Renal cell carcinoma (Ca) is imaged as an exophytic inhomogeneous lower-intensity lesion arising from the posterior medial aspect of the right kidney.

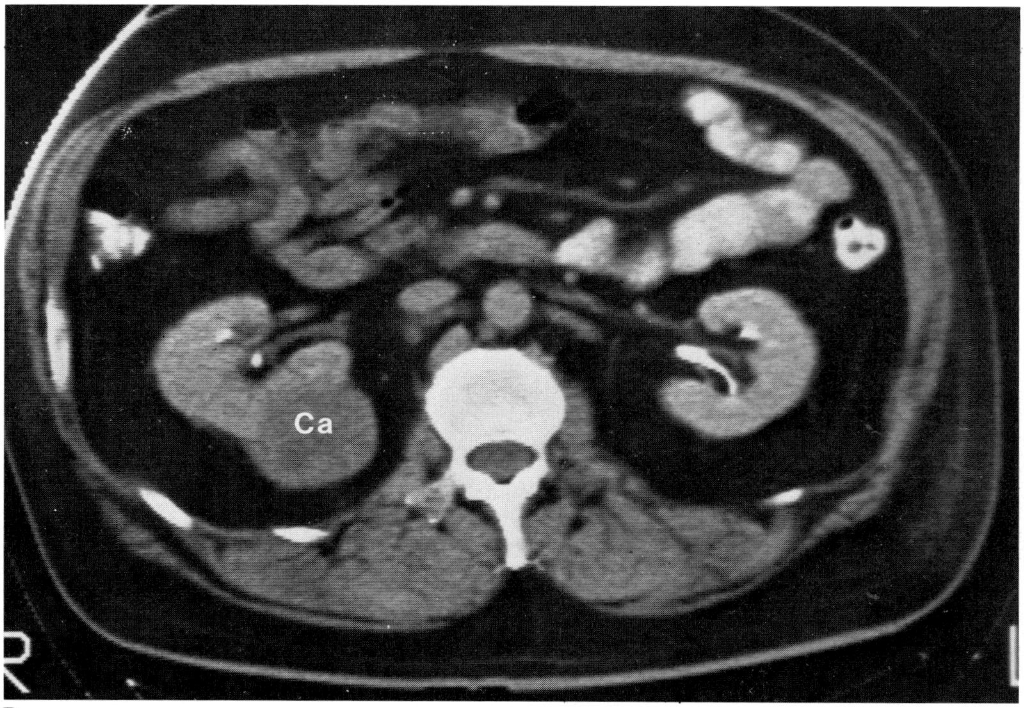

B.

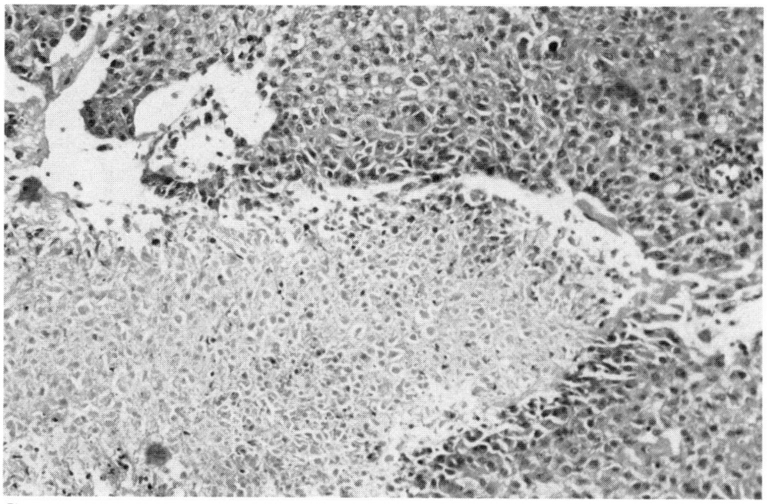

C.

B. Corresponding contrast-enhanced CT scan showing a solid low-attenuation lesion (Ca) in the right kidney.
C. Photomicrograph (hematoxylin and eosin × 375) showing viable tumor cells (granular cell type) with broad areas of necrosis.

parenchyma presumably because of the prolonged T_1 values of the neoplastic tissue. A tumor pseudocapsule could be detected with spin echo technique as a low-intensity line interposed between the tumor and the normal renal parenchyma.[7] In our preliminary experience in the staging of renal neoplasms, NMR results have approached those of CT and have provided additional information about intrarenal vein involvement (Fig. 25). Nuclear magnetic resonance may be superior to CT in evaluating the extension of tumor thrombus when the thrombus extends into the intrahepatic portion of the inferior vena cava (Fig. 25). It can distinguish between blood clot or blood stasis and tumor thrombus (Fig. 25). A detailed evaluation of tumor extension with special emphasis on involvement of the intrahepatic portion of the

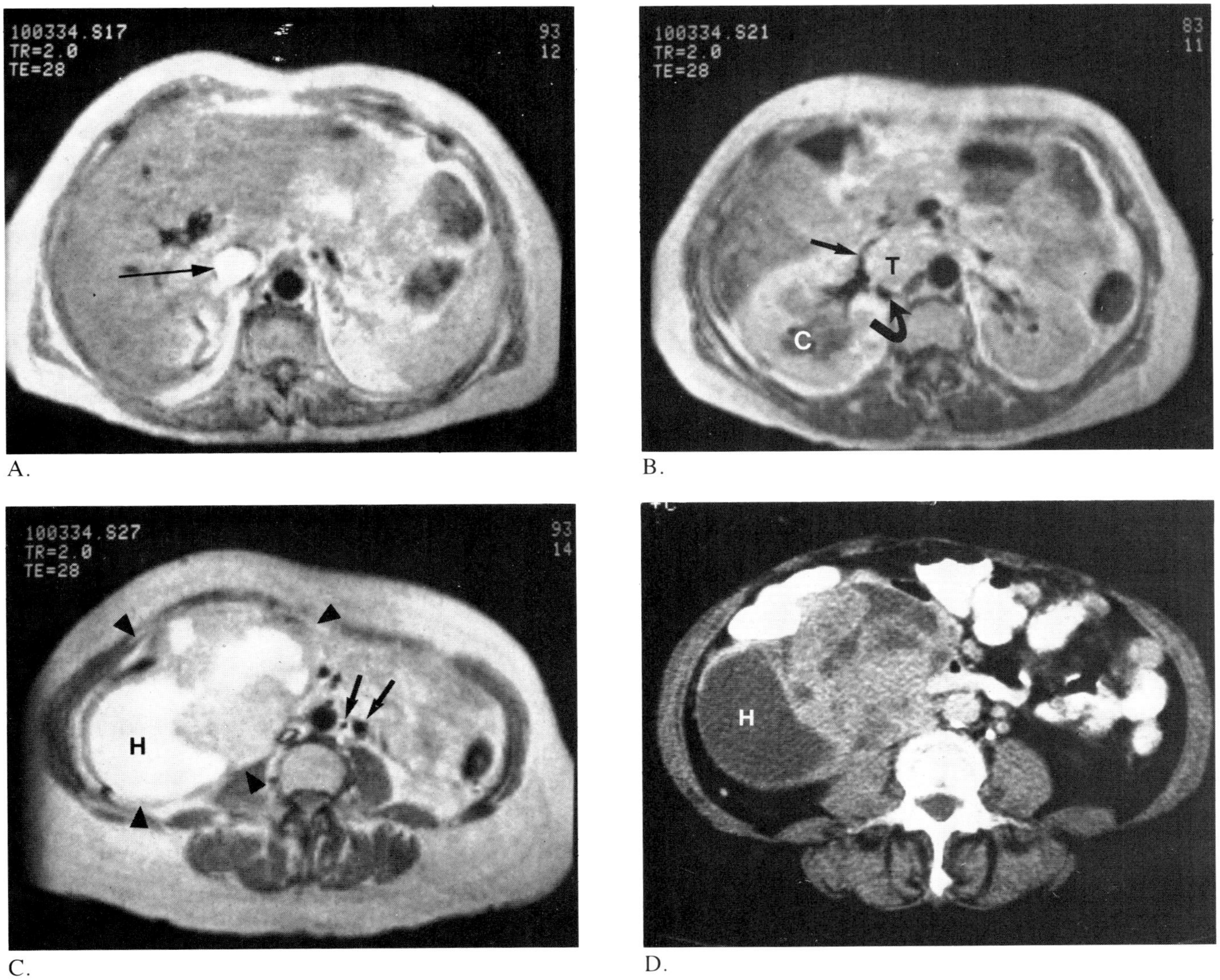

Figure 25. Leiomyosarcoma of the inferior vena cava.

A. Spin echo image (TR of 2.0 sec, TE of 28 msec) through the upper abdomen. There is high-intensity signal (arrow) within inferior vena cava, representing a blood clot.

B. Spin echo image (TR of 2.0 sec, TE of 28 msec) through mid-abdomen showing both right and left kidney. There is widening of the calices (C) of the right kidney. Right renal vein (straight arrow) is imaged with low intensity, indicating persistent blood flow. Also seen is right renal artery (curved arrow). Tumor (T) within inferior vena cava is imaged with medium intensity.

C. Large retroperitoneal mass (arrowheads) with inhomogeneous signal intensity. A high-intensity area (H) represents area of hemorrhage. Note rich collateral vessels (arrows) in the left periaortic region.

D. Corresponding CT scan. The CT attenuation value of area of hemorrhage was that of simple fluid (5 H).

inferior vena cava in which liver resection is necessary is of extreme clinical importance. In stage IV renal neoplasms, when there is direct extension of the tumor to adjacent organs, NMR effectively demonstrates extension into diaphragmatic crus or psoas muscle. By using a short pulse sequence interval on spin echo technique or an inversion recovery technique, one can produce NMR images that are extremely helpful in exact delineation of the tumor extension.

A drawback of NMR in the assessment of renal masses (both solid and cystic) and nephrolithiasis is the poor definition of calcifications on all types of NMR imaging techniques. Calcifications have no NMR signal because of the low density of mobile protons in calcific deposits. Even when calcifications are seen, they cannot be readily differentiated from dense fibrous tissues. While this represents a problem in evaluation of nephrolithiasis, this may not represent a significant disadvantage in evaluating renal masses with NMR because the superior soft-tissue contrast and ability to analyze the chemical composition of fluids offer information never before obtained by any other radiologic modalities.

REFERENCES

1. Kaufman L, Crooks LE, Margulis AR (Eds): *Nuclear Magnetic Resonance Imaging in Medicine*. New York: Igaku-Shoin, 1981.
2. Gamsu G, Webb WR, Sheldon P, et al: Nuclear magnetic resonance imaging of the thorax. Radiology 147:473, 1983.
3. Herfkens RJ, Higgins CB, Hricak H, et al: Nuclear magnetic resonance imaging of the cardiovascular system: Normal and pathologic findings. Radiology 147:749, 1983.
4. Young IR, Bailes DR, Burl M, et al: Initial clinical evaluation of a whole body nuclear magnetic resonance (NMR) tomograph. J Comput Assist Tomogr 6:1, 1982.
5. Alfidi RJ, Haaga JR, El Yousef SJ, et al: Preliminary experimental results in humans and animals with a superconducting, whole-body nuclear magnetic resonance scanner. Radiology 143:175, 1982.
6. Hricak H, Crooks L, Sheldon P, et al: Nuclear magnetic resonance imaging of the kidney: Part I. Radiology 146:425, 1983.
7. Hricak H, Williams RD, Moon KL: Nuclear magnetic resonance imaging of the kidney: Part II. Renal masses. Radiology 147:765, 1983.
8. Moon KL, Hricak H, Crooks LE, et al: Nuclear magnetic resonance imaging of the adrenal gland: A preliminary report. Radiology 147:155, 1983.
9. Hricak H, Higgins CB, Williams R: NMR imaging in retroperitoneal fibrosis. AJR 141:35, 1983.
10. Smith FW, Hutchison JMS, Mallard JR, et al: Renal cyst or tumor— Differentiation by whole body NMR imaging. Diagn Imaging 50:61, 1981.
11. Smith FW, Reid A, Mallard JR, et al: NMR tomographic imaging in renal disease. Diagn Imaging 51:209, 1982.
12. Oppenheimer DA, Caroll BA, Yousem S: Sonography of the normal neonatal adrenal gland. Radiology 146:157, 1983.
13. Crooks LE, Arakawa M, Hoenninger J, et al: High resolution NMR imaging. Radiology (in press).
14. Davidson AJ, Hricak H: Comment to letter to editor: The myth of anechoic renal sinus fat. Radiology 147:598, 1983.
15. Ishikawa I, Saito Y, Onouchi Z, et al: Development of acquired cystic disease and adenocarcinoma of the kidney in glomerulonephritis chronic hemodialysis patients. Clin Nephrol 14:1, 1980.

15

Pelvis

HEDVIG HRICAK, M.D.

NORMAL ANATOMY
MALE PELVIS
- Prostate Gland
- Seminal Vesicles
- Penis
- Inguinal Canal
- Testis

FEMALE PELVIS
- Uterus
- Cervix
- Vagina
- Ovary

PATHOLOGY OF THE LOWER GENITOURINARY TRACT
- Urinary Bladder

PATHOLOGY OF THE PROSTATE GLAND
PATHOLOGY OF FEMALE REPRODUCTIVE ORGANS

Before the advent of the newer imaging modalities capable of topographic display of pelvic anatomy, the diagnostic evaluation of the lower genitourinary tract was limited to contrast studies of the lower ureters, urinary bladder, and urethra. The limitations of the conventional radiologic methods, particularly their inability to directly display soft-tissue planes, resulted in a large number of exploratory laparotomies. Ultrasound has increased the diagnostic yield in the pelvic organs. The depiction of pelvic pathology and the distinction between cystic and solid masses by sonography are accurate.[1] There are limitations of the technique, primarily the inability of diagnostic ultrasound to visualize bowel, mesentery, fat planes, and the pelvic fossa. The introduction of CT has further improved the diagnostic accuracy in the evaluation of the pelvis, because soft-tissue planes and viscera are directly displayed.[2,3] The experience with CT, however, has demonstrated several important limitations— in particular, based on the CT density differences, normal prostatic tissue cannot be differentiated from prostatic adenoma, prostatitis, or carcinoma.[2,3] The CT evaluation of pathology at the bladder base is limited;[3,4] the accuracy of CT in detecting lymph node metastases varies from series to series. In an article evaluating lymph node involvement by radiologic modalities including lymphangiography and CT scans, a positive correlation of only 79% was obtained.[5] Therefore, open biopsy of pelvic lymph nodes is still a leading mode in the staging of pelvic malignancies.[5,6] One further limitation of CT in imaging the pelvis is the occasional degradation of the image by beam-hardening artifacts.

Clinical experience with NMR imaging is accumulating at a rapid pace.[7–11] A wealth of information obtained by NMR is opening a new era that brings hope for the solution of many clinical problems, such as female infertility, early detection of prostatic malignancy, and a more reliable method for the staging of pelvic malignancies.[7] The noninvasive nature of NMR, its ability to obtain direct multiplanar images, the superior soft-tissue contrast resolution (soft-tissue contrast by NMR exceeds 500%, as compared with only about 7% by x-ray CT), its ability to provide information about chemical composition, and the absence of beam-hardening artifacts all make NMR a highly suitable modality for pelvic imaging. These advantages are further enhanced because the quality of pelvic images is not degraded by respiratory motion.

Normal Anatomy

Diagnostic evaluation of three-dimensional NMR imaging requires an understanding of pelvic anatomy and the anatomic relationships of the pelvic viscera. The symmetric muscular walls of the pelvis provide helpful landmarks in the evaluation of various anatomic structures and their variants (Fig. 1). On the gray scale of NMR intensity, skeletal muscles, being striated muscles, are imaged with low intensity.[9] Striated muscles have moderately long T_1 and short T_2 relaxation values. With short repetition times (TR), the low-intensity signal of the striated muscle is the

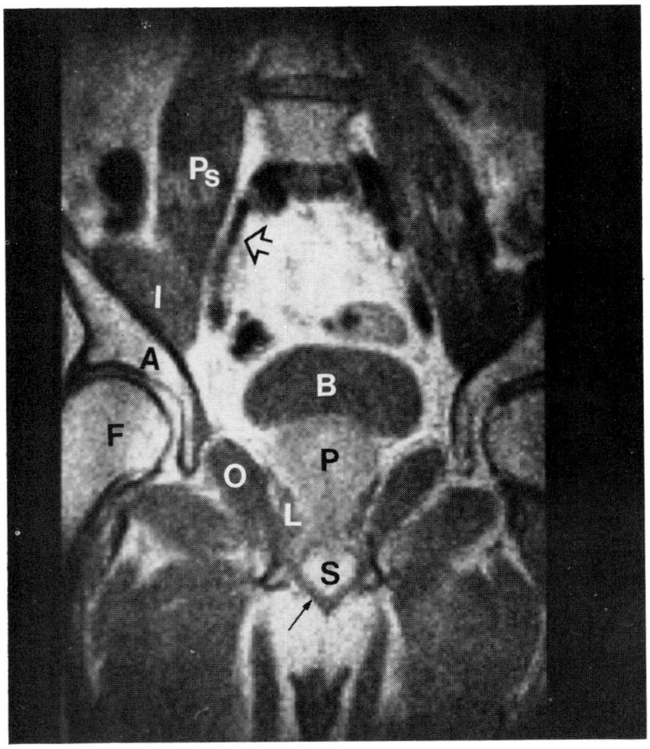

Figure 1. Normal anatomy, NMR coronal image. Spin echo (TR of 1.5 sec, TE of 28 msec). There is clear depiction of the musculoskeletal system. Small wedge of adipose tissue imaged with high intensity is seen separating obturator internus (O) and levator ani (L) muscle. There is clear separation between levator ani muscle (L) and base of the prostate gland (P). Prostate gland is pressing upon the bladder floor (B). Bulbus of the penis (corpus spongiosum, S) is imaged with high intensity. In the femoral head (F), cortical bone is imaged as a low-intensity line surrounding high-intensity bone marrow because of the lack of mobile hydrogen. (A, acetabulum; Ps, psoas major; I, iliacus; small arrow, bulbospongiosus; open arrow, common iliac vein)

result of long T_1 relaxation time, which does not permit full recovery of magnetization between successive excitations. There is a relative increase in signal intensity with longer TR values, but the striated muscle remains less intense than surrounding tissue. With the increasing echo delay time (TE of 56 msec) the relative intensity of the striated muscle (due to short T_2 relaxation value) will decrease as compared with the adjacent soft tissue. In order of increasing signal intensity, the NMR gray scale shows air and cortical bone with the lowest-intensity signal, followed by ligaments and tendons, normal body fluids, striated muscle, smooth muscle, viscera, brain, spinal cord, bone marrow, and fat as the brightest tissue.[9]

The pelvic blood vessels form important anatomic landmarks. The advantage of NMR lies in its ability to clearly demonstrate blood vessels in the three obtainable planes of imaging: the coronal, sagittal, and transaxial. The signal intensity from flowing blood is a function of the volume and velocity of hydrogen nuclei present. It is also affected by instrument parameters and by the position of the imaging plane in the multisection sequence. For the purpose of the following discussion of the anatomic display of pelvic vessels, the subject will be simplified because only patent blood vessels with normal flow velocity, which are imaged as a channel of low intensity, are easily distinguished from the surrounding adipose tissue, lymph nodes, nerves, and pelvic muscles.[10,11]

The recognition of normal or mildly enlarged pelvic lymph nodes is more difficult in the pelvis than in the abdomen or chest. Although the nodes are generally located along the course of the iliac vessels, there is a great variability in their location and range of normal diameter.[10] Furthermore, the transaxial plane of orientation of the distribution of pelvic arteries, veins, and lymph nodes imposes further difficulties if only the transaxial plane of scanning is used. The advantage of NMR is its ability to directly image in coronal and sagittal section, rendering better detection of lymph nodes. Even the presacral nodes, which are suboptimally evaluated by any other radiologic modality (lymphangiography or x-ray CT), are easily displayed on three-dimensional NMR images. The NMR intensity of lymph nodes varies with the scanning technique utilized. The lymph nodes have a long T_1 (as compared with viscera) and a relatively long (as compared to striated muscle) T_2 relaxation time. With short TR values the lymph nodes are imaged with higher intensity signal than the striated muscle is. Although the nodes have long T_1 relaxation times, and therefore the initial signal intensity is low because of slow decay (long T_2 relaxation time), the signal intensity will not decrease during the 28-msec echo time (TE). The longer the TE value, the greater the relative increase of signal. With the longer TR, the additional component of increase in signal intensity over that seen with short TR is the reflection of long T_1 value of the lymph nodes. With the pulse sequence interval in which TR is 2.0 sec and TE is 56 msec, the intensity of pelvic lymph nodes becomes very similar to that of surrounding adipose tissue (Fig. 2). Unlike CT, there is always a clear differentiation between the pelvic nodes and the low-intensity patent blood vessels (Fig. 3). As with CT, when the lymph node is larger than 2 cm, malignant involvement is highly suspected. However, in this early experience it is not clear if in the evaluation of the lymph nodes of normal size, NMR will be able to distinguish between normal inflammation or nodes affected by malignancy.

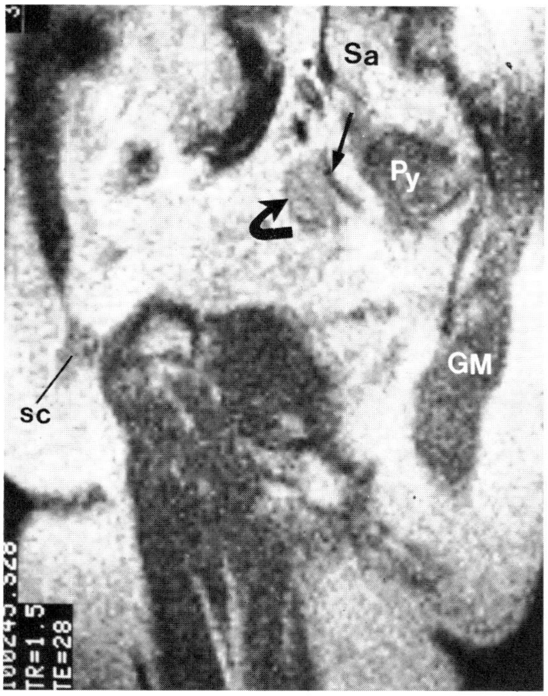

A.

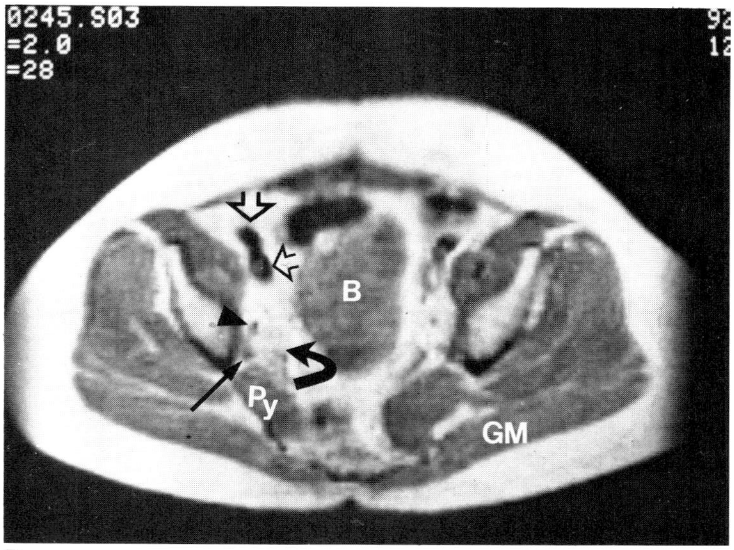

B.

Figure 2. Lymph node metastases from prostatic carcinoma.

A. Sagittal spin echo image (TR of 1.5 sec, TE of 28 msec). The enlarged lymph node (curved arrow) is imaged with medium intensity. It is seen anterior to internal iliac vein (straight arrow). (B, urinary bladder; GM, gluteus maximus; Py, pyriform muscle; Sa, sacrum; sc, spermatic cord)

B. Axial spin echo image (TR of 2.0 sec, TE of 28 msec). The signal intensity of the enlarged lymph node (curved arrow) is increased with the longer TR value and now blends with the surrounding adipose tissue. The node is located medial and anterior to the internal iliac vein (straight solid arrow) and medial to internal iliac artery (arrowhead). Note a good display of external iliac artery (large open arrow) and vein (small open arrow). On both scans there is good display of skeletal muscles. Pyriform muscle (Py) is seen anterior to the sacrum (Sa).

The urinary bladder is a helpful landmark in analyzing the male or female pelvis. For optimal pelvic scanning, unless medically contraindicated, the urinary bladder should be distended. While filled with urine, the bladder is easily demonstrated on axial, sagittal, and coronal images. Urine is a relatively pure liquid whose primary component is water. The T_1 and T_2 relaxation times of urine are, therefore, long. The NMR intensity of the urine markedly changes with the data acquisition technique and variations in the TR and TE instrument parameters. With spin echo technique, as the repetition rate (TR) increases, the recovery of magnetization increases, and, consequently, the NMR signal intensity from urine (long T_1) becomes bright and changes from a low- to medium-intensity signal (black to gray). When the echo delay time (TE) is increased because of the long T_2 value of urine, the signal intensity will further increase. These variations result in the change of the intensity tissue characteristics and permit the distinction between urine and bladder wall when long TR and TE values are utilized (Fig. 4).

The perivesical fat planes are readily distinguished. The adipose tissue has a short T_1 and a relatively long T_2 relaxation parameter. Even on images obtained with short TR values, there is relatively high signal intensity seen from the tissues with short T_1 relaxation times. As the TR value increases, the magnetization increases and, consequently, the NMR signal becomes more intense. On images obtained with longer TE values (TE of 56 msec), adipose tissue maintains a high signal intensity. However, its absolute signal has actually decreased.

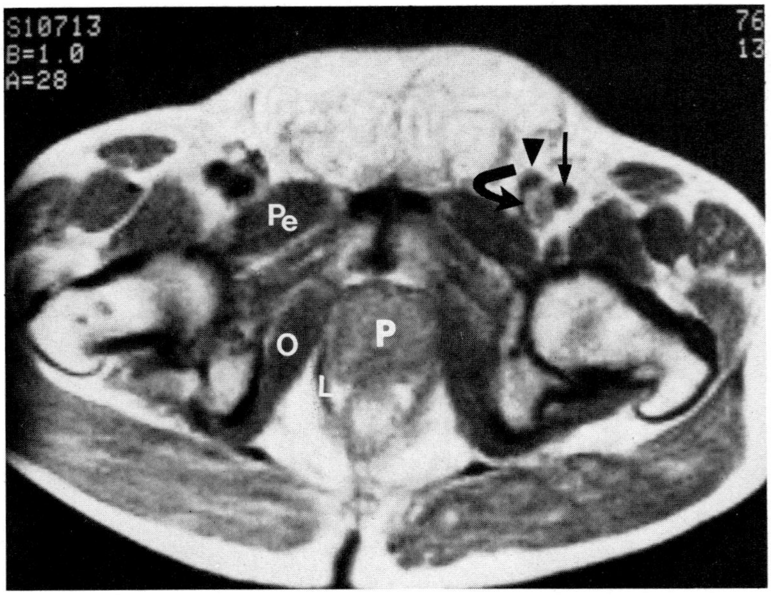

Figure 3. Large inflammatory lymph node (large arrow). Spin echo (TR of 1.0 sec, TE of 28 msec). A large biopsy-proven inflammatory lymph node is seen in the left inguinal region. With this imaging sequence the node is displayed with medium intensity. The gland is enlarged, but it is imaged with homogeneous medium-intensity signal, indicating benign nodular hyperplasia. Levator ani (L) can be separated from the prostatic parenchyma. (arrowhead, femoral vein; straight arrow, femoral artery, Pe, pectinus; O, obturator internus; P, prostate gland)

The relationship of the bladder to the surrounding bony structures and viscera depends on its distention. Between the bladder and pubic bones lie the retropubic space (space of Retzius), containing adipose tissue of high intensity and, more caudad, the low-intensity puboprostatic and pubovesical ligaments. Ligaments, because of the presence of connective tissue, which has low water and therefore hydrogen content, are always imaged with very low intensity. As the bladder fills, the organ comes to occupy more of the retropubic space; further filling results in the superior surface of the bladder extending above the pubis. The superior surface of the bladder and its median portion are covered by the peritoneum and are related to the mesentery and intestines. With bladder distention the peritoneum is displaced from the anterior abdominal wall. The anterior aspect of the distended bladder is in direct contact with the abdominal wall without intervention of the peritoneum. It is therefore possible to characterize the fully distended bladder suprapubically without entering the retroperitoneum.

If the bladder is fully distended, the bowel loops will be displaced out of the pelvis, but the mesenteric fat and/or the small bowel loops may still produce a volume averaging artifact at the dome of the bladder (Fig. 5). By analyzing different pulse sequence intervals and multiple adjacent sections, the bowel loops and mesentery at the bladder dome can be identified and distinguished from a lesion. During respiration, no significant motion of the pelvic organs is present. The respiratory and peristaltic motion of the bowel and mesentery outside the pelvis, which rapidly change position, will frequently produce blurred, indistinct margins (Fig. 5). The ability to obtain additional sagittal and coronal images further permits the identification of volume averaging artifacts at the bladder dome.

Posterior to the bladder in the female are the cervix of the uterus and the vagina (Figs. 6, 7). The extravaginal portion of the cervix lies against the superior part of the base of the bladder, while the anterior vaginal wall is in contact with most of the remaining area of the bladder base. The base or posterior surface of the bladder in males is related to the seminal vesicles, the ampulla of the rectum, and the rectovesical pouch of the peritoneum (Fig. 8).[12]

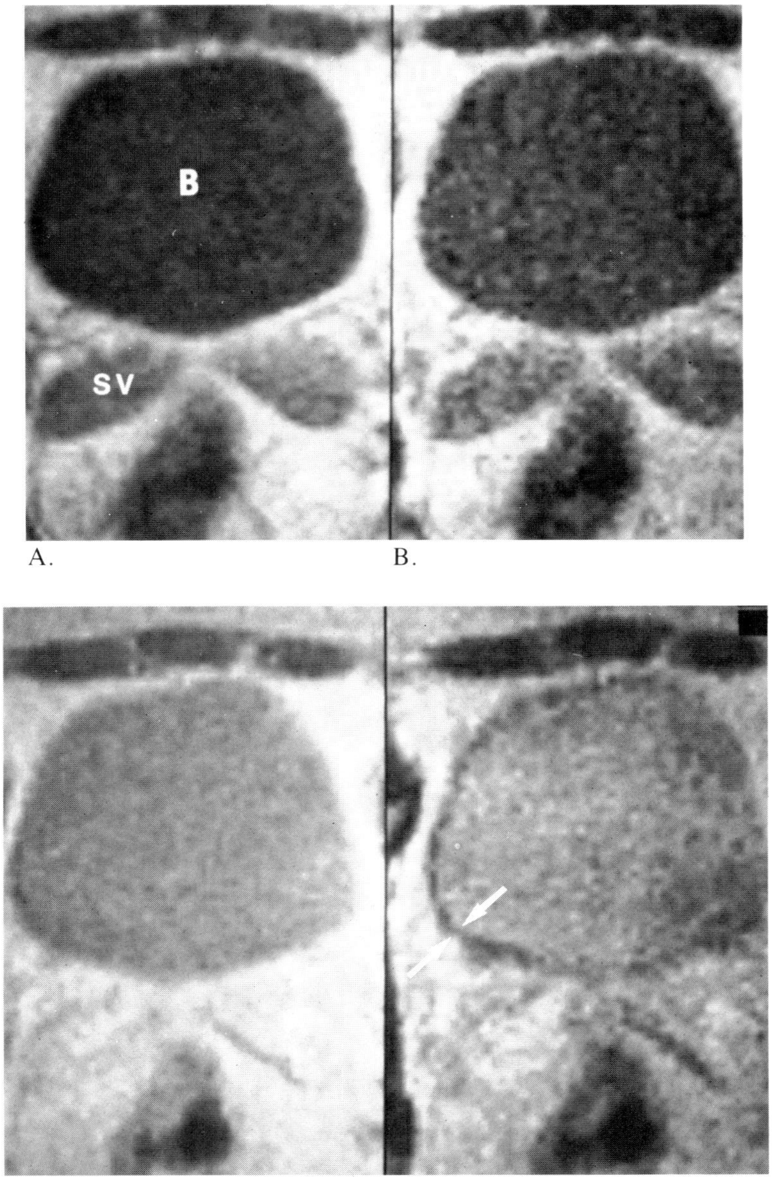

Figure 4. Fluid relaxation pattern on spin echo images: A, 0.5 sec/28 msec; B, 0.5 sec/56 msec; C, 2.0 sec/28 msec; D, 2.0 sec/56 msec. Urine-filled bladder (B). With increase in TR value, urine, the primary component of which is water, has long T_1 and T_2 relaxation times. On spin echo technique, as the TR increases the recovery of magnetization increases and, consequently, signal intensity from urine changes from low-intensity (A) to medium high-intensity signal (C) (TR of 2.0 sec). There is also increase in signal intensity from the urine with longer echo time. Such variation in signal intensity permits the distinction between urine and bladder (arrows in D) when long TR and TE values are utilized. Similarly noted is the change of intensity of the seminal vesicle (Sv) with the change in TR and TE values.

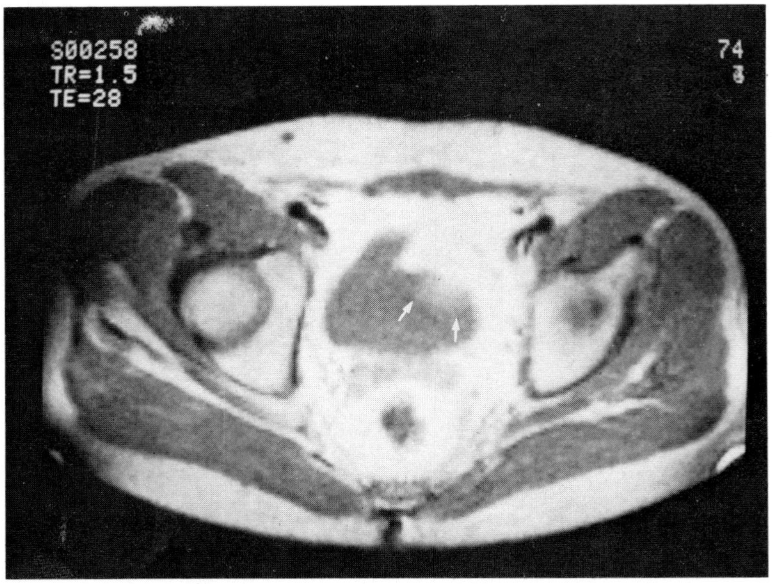

Figure 5. Volume averaging artifact due to mesentery and bowel at the dome of the urinary bladder. Spin echo (TR of 1.5 sec, TE of 28 msec). The high-intensity area at the anterior portion of the bladder represent mesentery. The artifact representing mesentery and bowel artifacts has blurred edges along its posterior aspect (small arrows). The bladder is moderately distended.

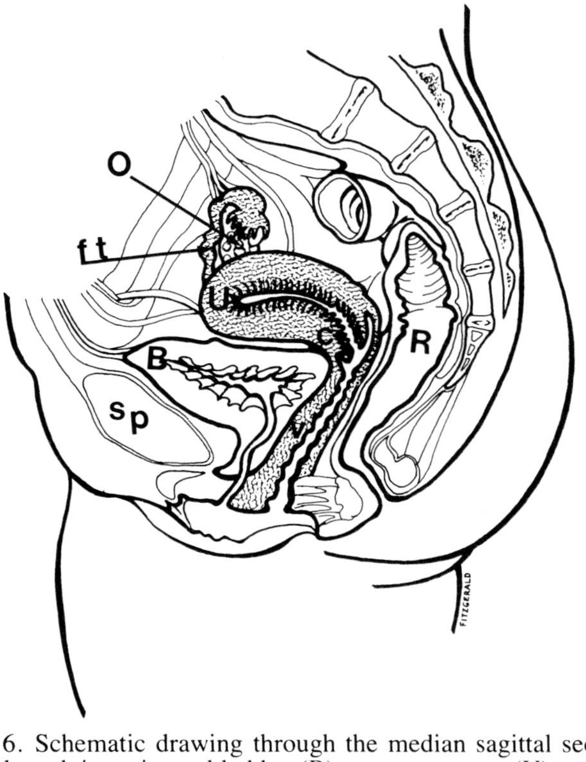

Figure 6. Schematic drawing through the median sagittal section of a female pelvis: urinary bladder (B); corpus uterus (U); cervix (C); vagina (V); fallopian tube (ft); ovary (O); rectum (R); and symphysis pubis (sp).

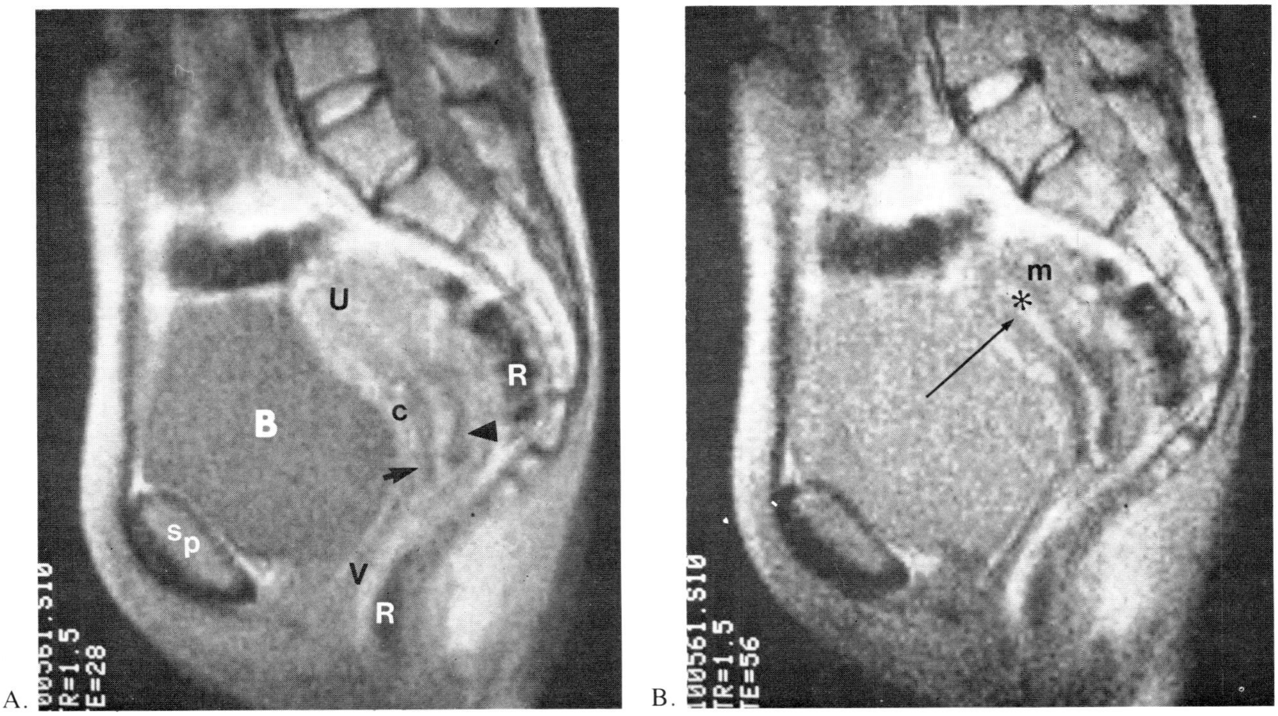

Figure 7. Spin echo images of midline sagittal section of the female pelvis corresponding to schematic drawing of Figure 6. In A, TR is 1.5 sec, TE is 28 msec. In B, TR is 1.5 sec, TE is 56 msec. The urinary bladder (B) is distended. Posterior to the bladder is the corpus uterus (U), cervix (C), and vagina (V). Air within the rectum (R) is seen posterior to the cervix and vagina. In B, obtained with longer TE (56 msec), there is clear differentiation between myometrium (m) and the endometrium. The endometrium can be further divided into lower-intensity zone basalis (long arrow) and cyclic endometrium (*). Note that the NMR characteristics of the cervix are different from the corpus uterus. Also, the anterior cervical lip (short arrow) projects lower than the posterior cervical lip (arrowhead). (Sp, symphysis pubis)

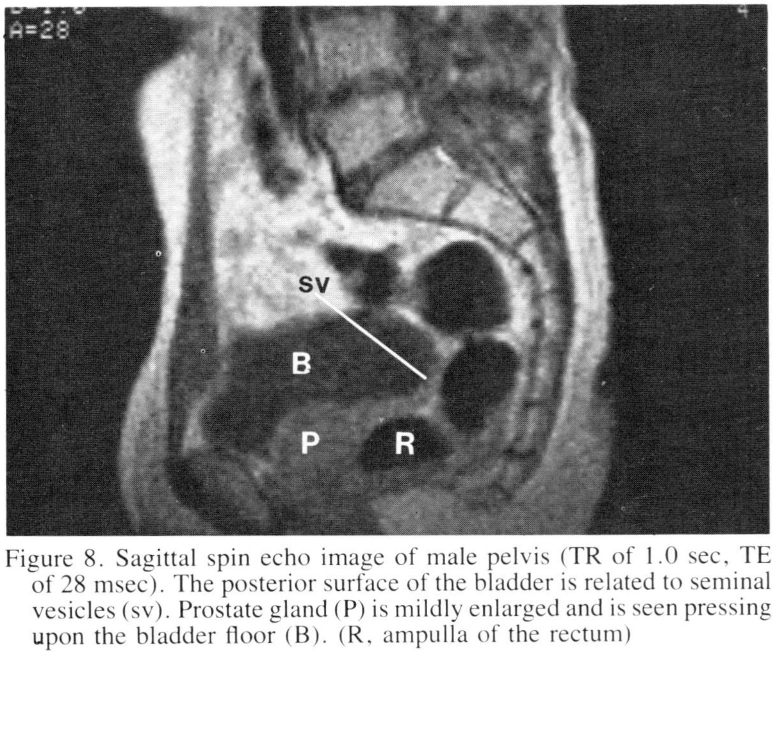

Figure 8. Sagittal spin echo image of male pelvis (TR of 1.0 sec, TE of 28 msec). The posterior surface of the bladder is related to seminal vesicles (sv). Prostate gland (P) is mildly enlarged and is seen pressing upon the bladder floor (B). (R, ampulla of the rectum)

Male Pelvis

Prostate Gland

The description of prostatic anatomy was based for many years on a study done in 1912 by Lowsley.[13] He postulated the existence of glandular subdivision lobes in the adult prostate derived from the anatomy of the fetal and newborn prostate gland. The morphological appearance of the adult gland itself was not investigated. A recent study by McNeal revealed that the adult prostate gland does not have lobes.[14] There are two main regions of the glandular prostate. These are the *peripheral zone,* constituting approximately 75% of the gland, and the *central zone,* occupying about 20% of the volume of the glandular tissue. The anatomic distinction between the central and peripheral zone is clinically important because the peripheral zone is almost exclusively the site of origin of prostatic carcinoma. Approximately 5% of the prostate gland is present in the *transitional zone*. The glandular tissue of the transitional zone is histologically identical to that of the peripheral zone, and the anatomic separation of these two zones is accentuated by the intimate relationship of the transitional zone to the periprostatic sphincter. The transitional zone is the site of origin of benign prostatic hyperplasia, while the adjacent and closely related peripheral zone is prone to carcinoma.[14]

Anteriorly, the fibromuscular band covers the anterolateral surface of the glandular prostate. Given the understanding of the anatomic distribution of prostatic disease, the imaging modality capable of displaying these anatomic zones and imaging the gland in three dimensions is invaluable. On NMR, the prostate gland is well displayed on axial, sagittal, and coronal planes (Figs. 1, 3, 8).[7,15–18] Further value of NMR imaging is due to its superior soft-tissue contrast resolution and ability to differentiate the anatomic zones of the prostate.

The prostate gland is displayed on spin echo images with medium-intensity signal that increases in intensity with longer TR and TE values. The tissue differentiation between prostatic zones is seen on longer TR and TE values. This is probably a function of the difference in T_2 relaxation parameters of the prostatic zones since the T_1 relaxation times do not show significant discrepancy. On the coronal images (spin echo sequence with TR of 1.5 or 2.0 sec) the central zone of the prostate gland can be distinguished from the peripheral zone. The NMR intensity of the central zone is less than that of the surrounding peripheral zone. This is probably secondary to the presence of striated muscle (which causes shortening of T_2) within

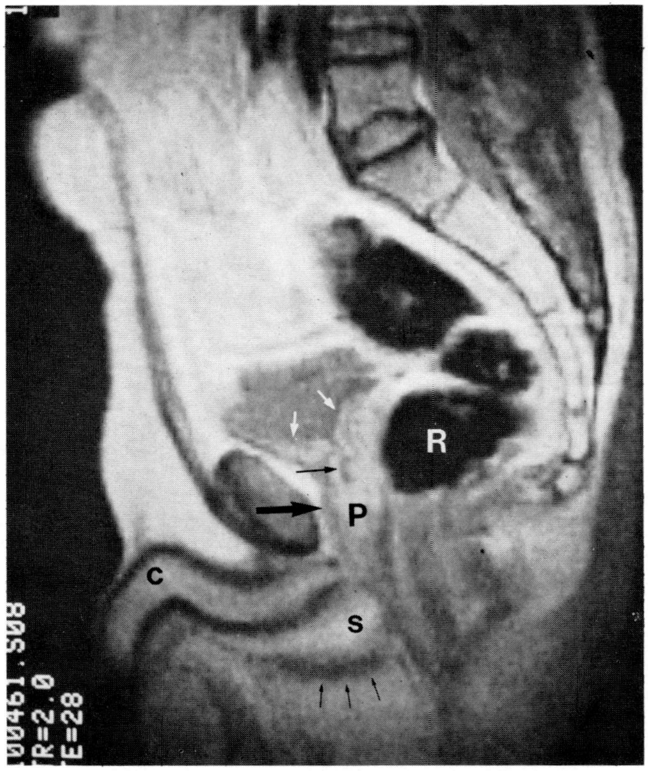

Figure 9. Status post-TURP. Spin echo (TR of 2.0 sec, TE of 28 msec). The anterior fibromuscular band (large black arrow) is imaged with low intensity and is clearly separated from the higher-intensity prostate gland (P). Note that the proximal prostatic urethra (single small black arrow) is dilated as a consequence of transurethral prostatic resection. Note minor muscular hypertrophy of the bladder trigonum (white arrows). Good display of bulbous penis (corpus spongiosum: S) and bulbospongious muscle (small black arrows). (C, corpus cavernosum; R, distended rectal ampulla)

the central zone. On the sagittal scan, there is clear depiction of the low-intensity anterior fibromuscular band. The low intensity of the fibromuscular ring is due to the large component of connective tissue (Fig. 9). The prostatic parenchyma is clearly differentiated from the surrounding levator ani muscle when the images are obtained with TR of 1.0 sec and greater. The parenchymal tissue is of higher intensity than the striated muscle. The difference is even more pronounced with longer TE values. The differentiation between the base of the prostate gland and levator ani muscle is seen in all imaging planes, but it is best appreciated on coronal sections (Fig. 1). The interface between the bladder base and the prostate gland is also enhanced on coronal and sagittal scans. Because of different NMR tissue properties, the bladder floor can be clearly separated from the dome of the prostate gland, although no fat plane is present (Figs. 1, 8). The relationship of the prostate gland to the symphysis pubis is clear (Figs. 3, 8, 9). The preprostatic fat in the space of Retzius is imaged with a high-intensity signal. The puboprostatic ligament attached to the apex of the prostate gland is imaged with low intensity. The distinction between the gland and the rectum is obtained by the ability of NMR to depict the Denonvilliers' fascia.[15] The fascia is of lower intensity than either the rectal wall or the prostate gland. The low intensity of the fascia is due to the presence of collagen tissue, which exhibits a lack of water and, hence, hydrogen. The low intensity of the fascia appears more pronounced on the second spin echo. A small amount of retroperitoneal fat is present between the dome of the prostate gland and the rectum, making a clear delineation in that location.

Seminal Vesicles

The seminal vesicles are convoluted pouches embedded within the retroperitoneal fat. The seminal vesicles and the ampulla ductus deferens are closely applied to the posterior surface of the bladder base, diverge from each other like the limbs of the letter "V," and are oriented to each other at approximately a 30° angle. These structures are covered by the peritoneum, which continues for a short distance from the superior surface of the bladder under its base. The peritoneum caps the ampulla ductus deferens and seminal vesicles before being reflected onto the anterior surface of the rectum (pouch of Douglas) (Fig. 10). The ability of NMR to differentiate seminal vesicle wall and fluid depends on the size of the seminal vesicle and the TR and TE instrumentation parameters (Figs. 10, 11). With the spin echo technique and a short TR (TR of 0.5 or 1.0 sec), the seminal vesicles are displayed with a homogeneous medium-intensity signal. The differentiation between fluid content and seminal vesicle wall is not possible. With the same pulse sequence interval, but a longer TE value, there is distinction between the higher-intensity seminal vesicle fluid and the lower-intensity seminal vesicle wall. The seminal vesicle wall-fluid differentiation is possible on the first echo image (TE of 28 msec) and longer TR values (TR of 1.5 and 2.0 sec). On the image with TR of 2.0 sec and TE of 56 msec there is considerable increase in the intensity of the seminal vesicle, and the intensity signal from the seminal vesicle is similar to that of the surrounding adipose tissue; the seminal vesicle wall and the closely applied peritoneal coverage, however, are imaged as a low-intensity stripe (Fig. 11).[7] The assessment of the seminal vesicles is enhanced by multiplanar imaging because the symmetric position of the seminal vesicles can be altered by a distended rectum or when the scanning is performed in the prone position (Figs. 8, 10). Therefore, axial scans may create a false impression of unilateral seminal vesicle enlargement or obliteration of the normal angle between the seminal vesicle and the posterior wall of the bladder.

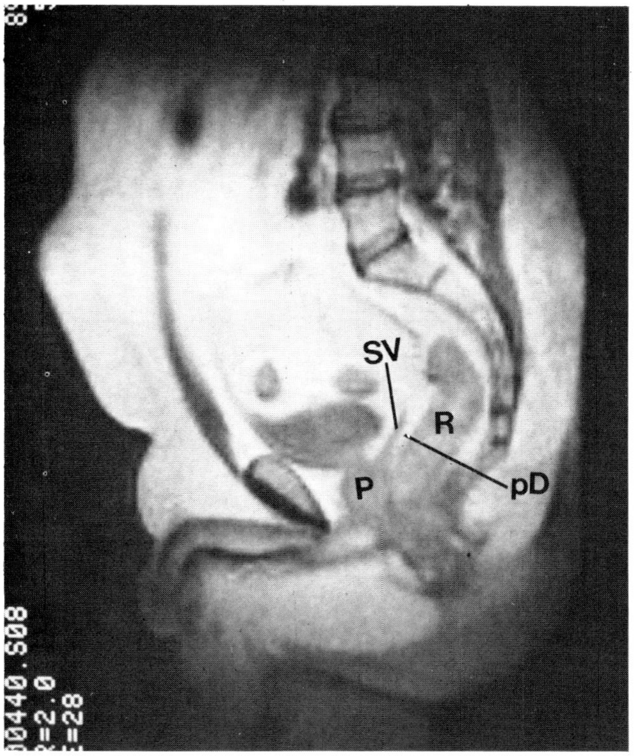

Figure 10. Sagittal scan of male pelvis. Spin echo (TR of 2.0 sec, TE of 28 msec). The seminal vesicles (SV) are seen embedded within the retroperitoneal fat. They are closely applied to the posterior surface of the bladder and oriented at approximately a 30° angle. Seminal vesicles in this patient on estrogen therapy are thin, and no fluid could be detected within them even with long TR and TE values. Because the rectum (R) is empty, there is a clear display of the adipose tissue within pouch of Douglas (pD).

Penis

The penis is composed of two corpora cavernosa and the corpus spongiosum traversed by the urethra. The corpora cavernosa are well displayed by NMR from their attachment to the ischium just below the urethrogenital diaphragm to their extension within the penis (Figs. 9, 12). The corpora cavernosa are of medium intensity with short TR and increase in intensity with longer TR values. Two crura of the corpora cavernosa are seen separated by a low-in-

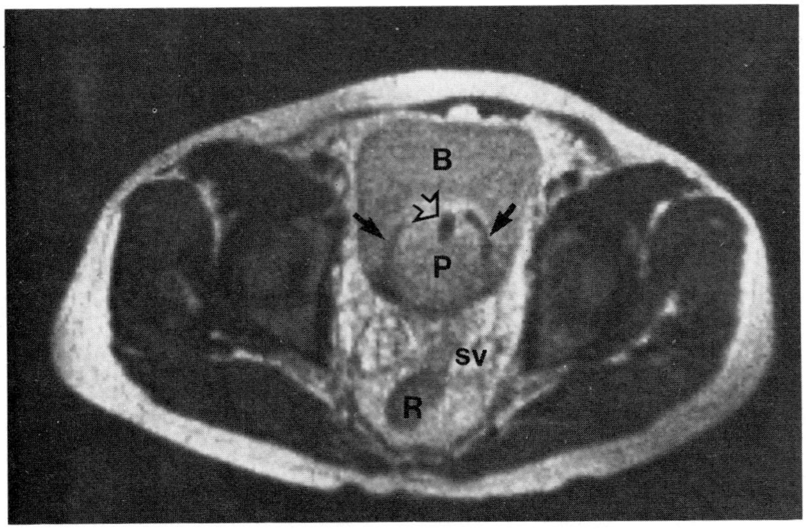

Figure 11. Benign nodular hyperplasia. Spin echo axial scan (TR of 2.0 sec, TE of 56 msec). The intravesical portion of the prostate gland (P) is enlarged, protruding into the urinary bladder (B). Note lower-intensity surgical pseudocapsule (black arrows) seen surrounding the adenomatous gland. Foley catheter (open arrow) is seen in the prostatic urethra. Seminal vesicles (sv) are generous. There is clear differentiation between high-intensity seminal vesicle fluid and low-intensity seminal vesicle wall. (R, rectum)

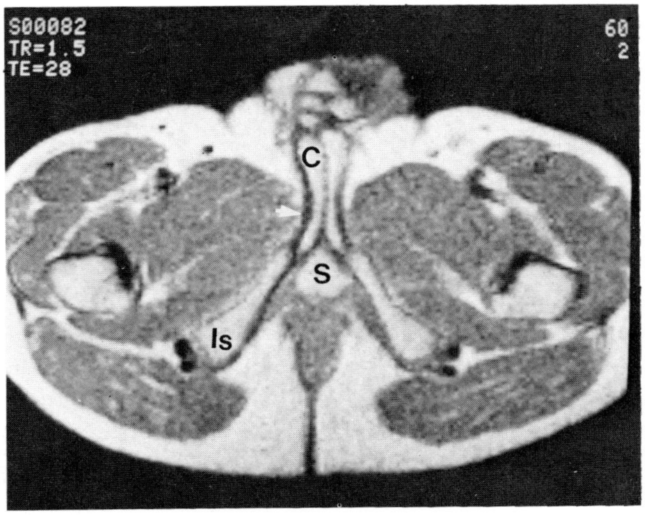

Figure 12. Transaxial scan through the lower male pelvis. Spin echo (TR of 1.5 sec, TE of 28 msec). The cylindrical masses of corpus cavernosum are imaged with high intensity. Their attachment to the ischium (Is) is clearly seen. Two corpora cavernosa (C) are separated by septum. Corpora cavernosa muscle (white arrow) is imaged with low intensity (S, corpus spongiosum)

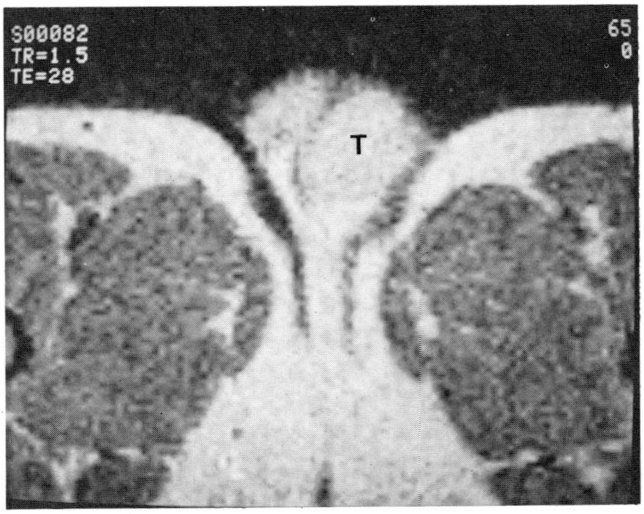

Figure 13. Transaxial scan. Spin echo (TR of 1.5 sec, TE of 28 msec). Testis (T) is imaged with medium high intensity.

tensity septum. The corpora cavernosa muscle, being a striated muscle, is imaged with low intensity and is delineated from the cavernous parenchyma. The corpus spongiosum has an intensity signal similar to that of the corpora cavernosa and can be distinguished from the corpora cavernosa by its anatomic location and its surrounding of the tunica albuginea, which, because of the presence of collagen tissue, is imaged as a low-intensity line. The surrounding bulbospongiosus muscle is imaged with low intensity (Figs. 1, 9). The distinction between the muscle and the corporal parenchyma is not possible by any other imaging modality.[15]

Inguinal Canal

The ductus deferens and the internal spermatic vessels within the inguinal canal can be depicted by nuclear magnetic resonance. Patent blood vessels are imaged with low intensity because of blood flow. The ductus deferens emits a signal of medium intensity, which increases with longer pulse intervals.

Testis

The testis images with a medium-intensity signal, which increases with longer TR values (Fig. 13). A definitive distinction between the testis and epididymis has not been observed.

Female Pelvis

Uterus

The uterus is divided into corpus, isthmus, and cervix. The gross configuration of the uterus changes throughout adult life. The premenarchal uterus is markedly different from the uterus during the reproductive or the postmenopausal years.[19] The change in the uterine configuration involves the change in the relationship of the uterine segments. The cervical-fundal ratio changes from 35:1 in the neonatal period to a point where the cervix accounts for more than half of the uterine length in the premenarchal uterus, until approximately the thirteenth year.[19] During adult life, the corpus uterus accounts for the greatest proportion of the uterine length.[19,20]

By three-dimensional NMR images the position of the uterus, posterior to the bladder and anterior to the rectum, is easily outlined (Figs. 6, 7).[21] Its configuration and position are affected by the degree of bladder or rectal distention. The relationships of the bladder, uterus, and rectum may vary because the size and position of the uterus are subject to alterations (Fig. 14). The nongravid uterus is normally inclined forward and upward and lies adjacent to the posterior part of the superior surface of the bladder. However, the body of the uterus may be retroverted and directed toward the rectum and sacrum, thereby leaving the anterosuperior surface of the bladder in contact with the intestine. The bladder indentation by the uterus should be evaluated with caution because it varies with the degree of bladder distention (Fig. 14).[21] If only sagittal scans are utilized, scanning with a distended urinary bladder is preferred in order to avoid a false-positive diagnosis of uterine enlargement. However, if three-dimensional images are available, uterine volumes can be obtained, making the estimation of uterine enlargement more accurate. In estimating the uterine volume, the patient's age, precise clinical history, and number of pregnancies play an important role. The uterine weight increases with increased parity and with age up to 49 yr. At puberty, the nulliparous uterus has an average length of 5.5–8 cm, a width of 4.5–6 cm, and an anteroposterior diameter of 2.5–3.5 cm. Since the specific gravity of uterine tissue is considered to be equal (1:1), the uterine weight can be calculated from the volume. At puberty, the uterus weighs between 40 and 50 gm. The parous uterus is considerably larger than the nulliparous one and weighs between 50 and 80 gm.[19,20,22] Ten grams can be added to the weight of the uterus for each of the first three births, and 50 gm for four births or more. The average measurements of a parous uterus are as follows: length, 7–9 cm; width, 4.5–6 cm; and anteroposterior diameter, 2.5–3.5 cm. Beyond 130 gm the uterus is probably abnormally enlarged regardless of age or parity.[20,22]

By varying imaging parameters the two distinct parts of the uterus (myometrium and endometrium) can be separated by NMR (Figs. 7, 14, 15). The myometrium, composed of smooth muscle, is imaged on spin echo and short pulse sequence intervals (TR) with a medium-intensity signal. An increase in the signal intensity is obtained with longer TR values, and there is also an increase in signal intensity with a longer echo delay time (TE of 56 msec). In our limited experience, the NMR characteristics of the myometrium are similar throughout life, and no definite difference was seen in the NMR signal intensity of the myometrium between the premenarchal, reproductive, and postmenopausal uterus. The myometrium during pregnancy was not studied.

By NMR, the endometrium can be clearly separated from the myometrium. The NMR differentiation between endometrium and myometrium depends on the scanning sequence utilized (Fig. 15). It is more pronounced with longer TR values and longer echo delay times (TE) than with shorter values.

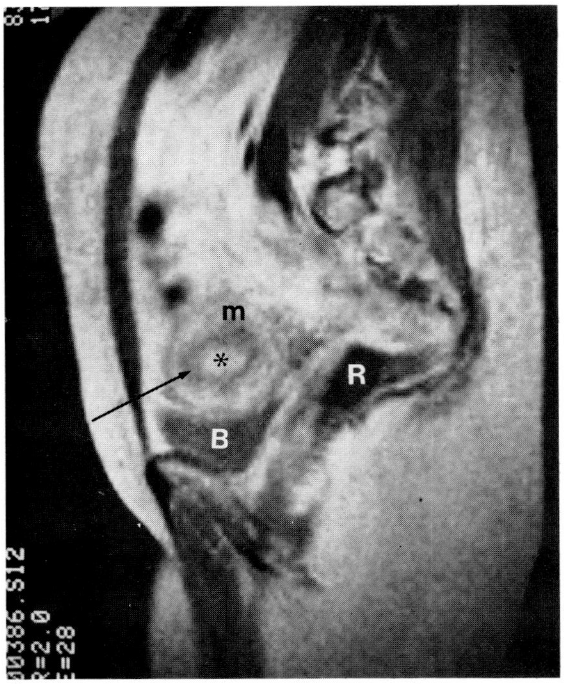

A.

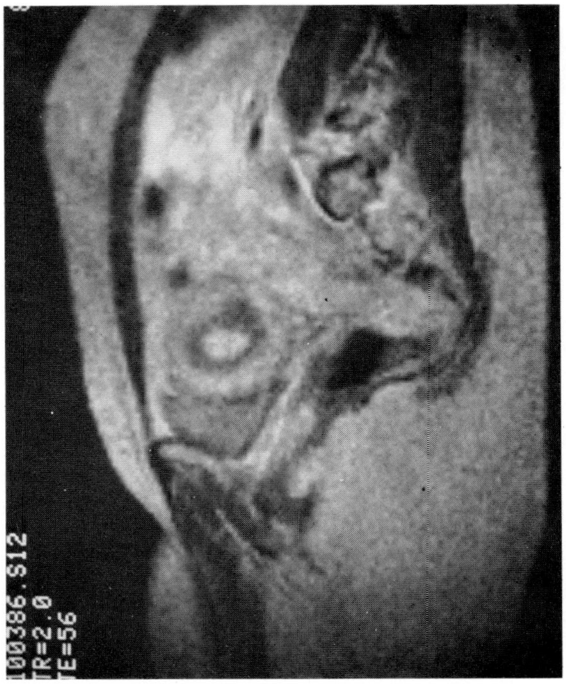

B.

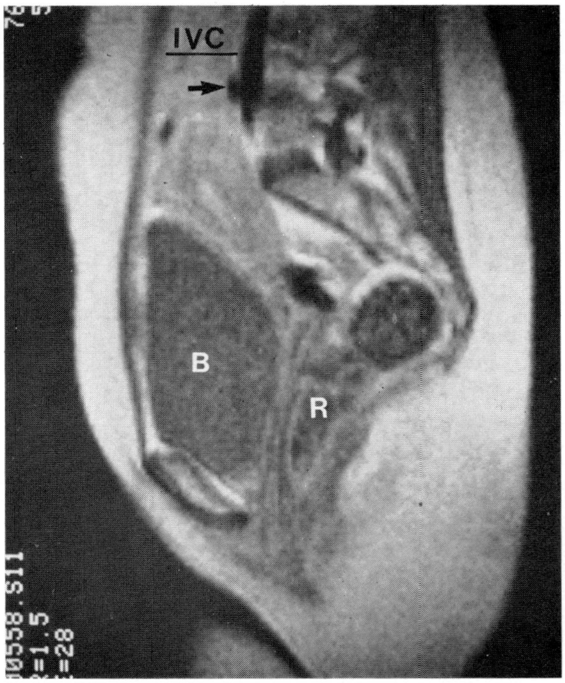

C.

Figure 14. Variation in the configuration of corpus uterus in relation to bladder and rectum distention.

A. Spin echo (TR of 2.0 sec, TE of 28 msec). Only a small amount of urine is present within the bladder (B). Rectum (R) has a small amount of air, but the distal part of the rectum is collapsed. The corpus uterus is globular in configuration, indenting the dome of the bladder. Myometrium (m) and endometrium can be clearly separated. Furthermore, there is a clear separation between the two zones of the endometrium: the lower intensity zone basalis (arrow) and the higher-intensity (*) zone functionalis.

B. Sagittal scan obtained with longer echo delay time (TE of 56 msec). The distinction between myometrium and endometrium is even more pronounced.

C. Sagittal scan obtained in the same patient showing bladder (B) and rectum (R). Spin echo (TR of 1.5 sec, TE of 28 msec). Fully distended bladder is elevating corpus uterus. Corpus uterus is now pear-shaped. While the differentiation between the myometrium and endometrium is possible, the differentiation is less apparent than on the image with longer TR values. Note the right common iliac artery (arrow) seen anterior to distal portion of the inferior vena cava (IVC).

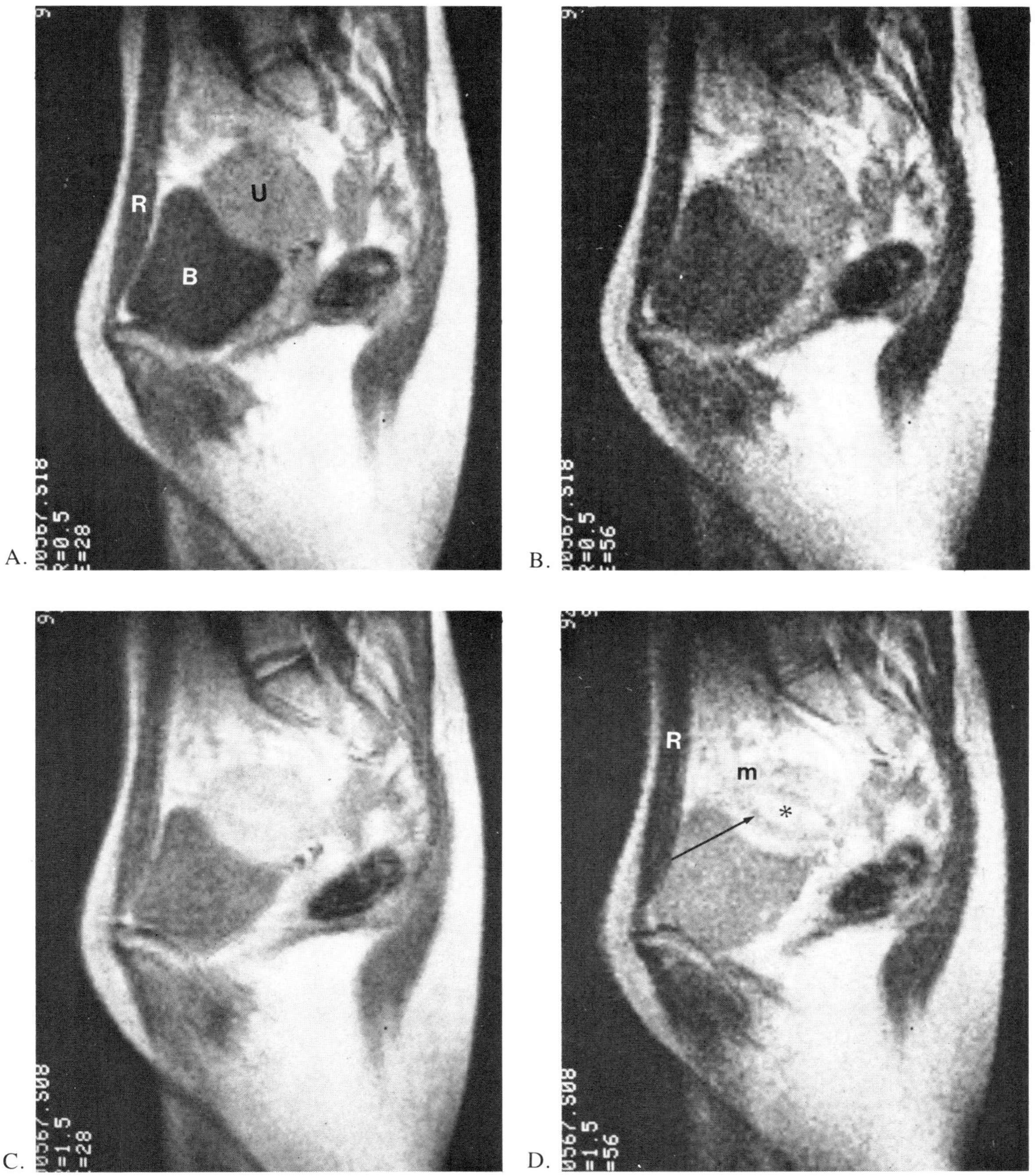

Figure 15. Spin echo sagittal image through the female pelvis. Variations in the corpus uterus displayed as a function of TR and TE instrument parameters. In A, TR of 0.5 msec, TE of 28 msec. In B, TR of 0.5 sec, TE of 56 msec. In C, TR of 1.5 sec, TE of 28 msec. In D, TR of 1.5 sec, TE of 56 msec. In A, corpus uterus (U) is imaged with a medium intensity. The signal intensity is higher as compared with striated muscle (rectus muscle: R) or urine-filled urinary bladder (B). Myometrium and endometrium cannot be separated.

In B, TR parameter is also 0.5 sec, but TE is 56 msec. There is increase in intensity of urine within the bladder, and there also is increase in intensity of the uterine corpus.

In C, (TR of 1.5 sec, TE of 28 msec), there is considerable increase in intensity from the uterine corpus. Myometrium can be separated from the endometrium, but the separation is less distinct than in D, which is obtained with the same TR value but longer echo delay time (TE of 56 msec). The myometrium (m) is now of high intensity. It is of considerably higher intensity than the rectus muscle (R). The endometrium can be divided into a lower-intensity zone basalis (long arrow) and the high-intensity signal zone functionalis (*).

There is marked difference in the histologic appearance of the endometrium in different stages of life.[20, 23] The NMR characteristics of the endometrium reflect the histologic changes and differ from stage to stage. The endometrium of the premenarchal uterus is thin, populated by inactive glands, and set with inactive stroma. By NMR, the premenarchal uterine endometrium is imaged as a low-intensity band outlined against the uterine cavity by a pencil-thin high-intensity line. This distinction is seen only when long TR and TE values are used. The low-intensity band of the endometrium was distinguishable from the higher-intensity myometrium on both TE images if the TR value was 1.0 sec or greater.

The fundal endometrium of the reproductive years consists of two distinct zones: the basalis, which is not hormonally sensitive (consisting of weak proliferative glands and surrounding spindle stroma immediately adjacent to the myometrium), and the superficial portion, known as the functionalis or cyclic endometrium because it responds to circulating sex hormones. The basalis cells and the proliferative cells of the functionalis are morphologically identical.[20, 23] By NMR, there is a difference between the basalis zone, which is imaged with low intensity, and the cyclic endometrium or functionalis zone, which is imaged with high intensity. The ability to differentiate two zones of the endometrium as well as to differentiate endometrium from myometrium depends on the scanning sequence utilized (Figs. 7, 14, 15). On spin echo technique with short TR values (TR of 0.5 sec) no differentiation between the myometrium and endometrium or the two zones of the endometrium is possible. This is true whether one uses TE of 28 or 56 msec. The differentiation between zones of uterus is indistinct with TR of 1.0 sec, but is clear when the scanning sequence is performed with TR of 1.5 sec, and even more so with TR of 2.0 sec. Also, with longer TR values the differentiation is more apparent on the second echo delay time (TE of 56 msec).

In our preliminary experience, there are changes in the appearance of the endometrium during the menstrual period. It is known that the width of the myometrium is constant throughout the menstrual period, averaging 1.5 to 2.5 cm. However, the thick-

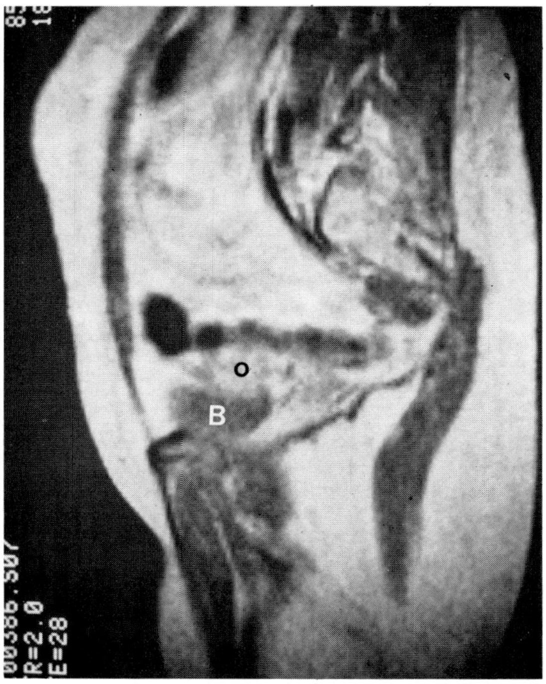

A.

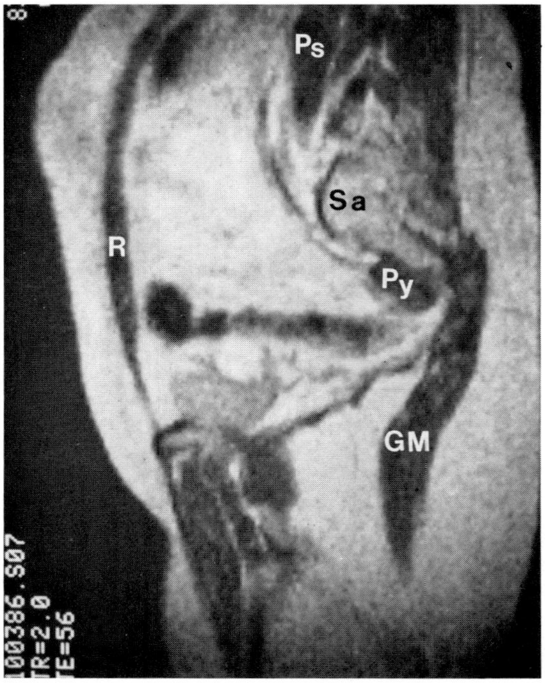

B.

Figure 16. Female pelvis. Sagittal scans with the same TR (2.0 sec) but different TE values: 28 msec in A, 56 msec in B. In A, the ovary (o) is seen superior to the dome of the bladder (B). The ovary has medium high-intensity signal. In B, the second echo image (TE of 56 msec), there is considerable increase in the signal intensity from the ovary. The signal intensity of the ovary is now similar to that of adipose tissue. Note the superior display of the skeletal muscle by NMR. Pyriform muscle (Py) is in front of the sacrum (Sa). (R, rectus abdominis; Ps, psoas major; GM, gluteus maximus) ness of the endometrium varies from 1 to 8 mm.[20] The cyclic endometrium in the reproductive years presents in two phases: the proliferative endometrial phase (ovarian follicular phase) and the endometrial secretory phase (ovarian luteal phase). During the proliferative phase, the endometrium grows rapidly in response to circulating estrogens. It triples in thickness from approximately 1 to 3 mm. Gross examination of the secretory endometrium reveals it to be 3 to 5 mm thick and to have a creamy yellow appearance. In the absence of estrogen stimulation, the endometrium becomes thin and inactive. The inactive endometrium measures less than 1 mm in thickness. The change of width of the functionalis zone during the menstrual period can be detected by NMR. The cyclic endometrium remains of high intensity throughout the cycle, but it appears to become wider during the ovarian luteal phase.

The postmenopausal uterus involutes and atrophies. This is reflected not only in a decrease in weight and dimension, but there is also a change in the appearance of the endometrium. On NMR, the postmenopausal endometrium is seen as a low-intensity band. The high-intensity zone of the cyclic endometrium is not seen.

Cervix

The cervix is the most inferior portion of the uterus protruding into the upper vagina. The vagina is fused circumferentially and obliquely around the distal part of the cervix. The axis cervix can be divided into anterior and posterior lips. The anterior lip is shorter and projects lower than its posterior counterpart (Fig. 17). The cervix in the adult nulliparous nongravida measures between 2.5–3 cm in length, and in its normal position is slightly angulated downward and backward. In the postmenopausal uterus the cervix is atrophic and elongated.

By NMR, there is a clear distinction between the cervix and corpus uterus. The ability to differentiate these two parts of the uterus, however, depends on the sequence utilized. On the short pulse sequence interval (TR of 0.5 sec) the intensity of the cervical portion blends with that of the corpus uterus. With a prolonged pulse sequence interval, specifically, when the pulse sequence interval is 1.5 or 2.0 sec,

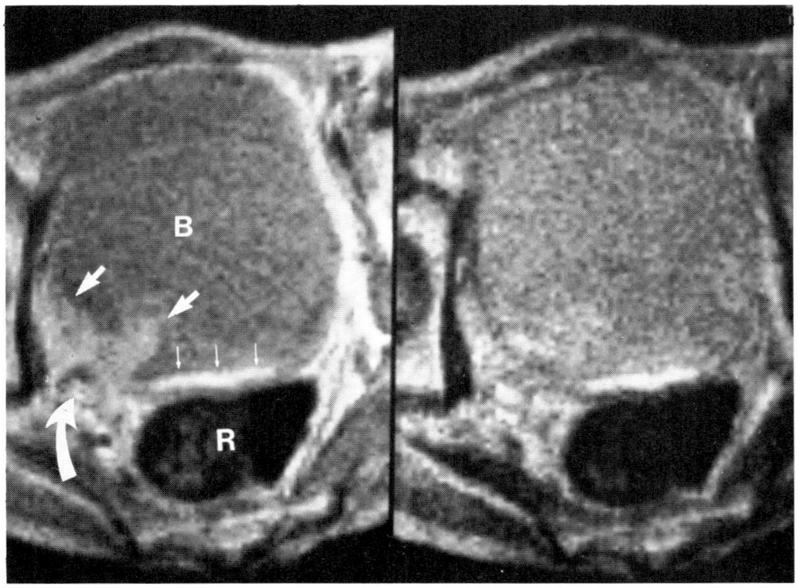

Figure 17. Transitional cell carcinoma of the urinary bladder. In A, transaxial spin echo image (TR of 1.5 sec, TE of 28 msec). In B, TR of 1.5 sec, TE of 56 msec. Broad-base transitional cell carcinoma (large arrows) is imaged with a higher intensity signal than bladder wall. In A, there is marked contrast between high-intensity neoplasm and medium-intensity urine within the urinary bladder (B). At the posterior bladder wall (small arrows) there is a high-intensity area that at surgery corresponded to neoplastic infiltration with a pronounced mucosal edema. The tumor was extending into the right ureter. The right ureter (curved arrow) is dilated. In B, the density of the neoplasm blends with urine of increased intensity. Delineation of the neoplasm is better appreciated with a shorter TE value. (R, rectum)

there is a clear distinction between the two segments. The anatomic relation between cervix, bladder, and rectum, as well as delineation of the anterior and posterior cervical lips, is excellently displayed on the sagittal image (Figs. 6, 7).

The cervix has three distinct zones as imaged by NMR. The outer zone is of medium to high intensity, similar to that of the myometrium of the corpus uterus. The zone adjacent to it is of low intensity, similar to the basalis of the corpus, but is wider and more pronounced (Fig. 15). The high-intensity central zone is very narrow. In our initial experience there is no difference in the appearance of the cervix throughout the menstrual cycle.

Vagina

The vagina extends from the vestibule to the uterus, lying dorsal to the urinary bladder and ventral to the rectum (Fig. 7). Its axis forms an angle of more than 90° with that of the uterus. It measures 6 to 7 cm along its ventral wall and 9 cm along its dorsal wall. The vagina surrounds the axis cervix but touches the uterine wall.

The clear identification of the vagina on axial NMR scans, as in CT, might be difficult. There is an excellent anatomic display of the relationship between the bladder, vagina, and rectum on NMR coronal scans, and an even better one on sagittal scans (Fig. 18). The intensity of the wall of the vagina on the short pulse sequence interval is similar to that of the uterus, including axis cervix. However, with a long pulse sequence interval, it is possible to differentiate between the vagina imaged with medium intensity and the vaginal lumen imaged as a low-intensity line. The distinction between the distended bladder and the vagina is improved with long pulse sequence intervals.

Such a superior display of the vagina on the NMR sagittal image should have the potential for depicting any local extension of tumors of the vagina. Currently, the exact delineation of spread of the malignancy by either ultrasound or CT has limitations.

Ovary

The ovary can be seen on either axial, sagittal, or coronal images. Like other tissues, the intensity of the ovaries changes, depending on the sequence utilized (Fig. 16). They are of low to medium intensity on short pulse sequence intervals and remain of higher intensity with longer pulse sequence intervals. With a long pulse sequence interval and a long echo delay time (TE of 56 msec) the intensity of the ovaries increases and sometimes becomes brighter than that of the surrounding fat. Therefore, for clear separation of the ovary from surrounding adipose tissue, short pulse sequence intervals are preferred.

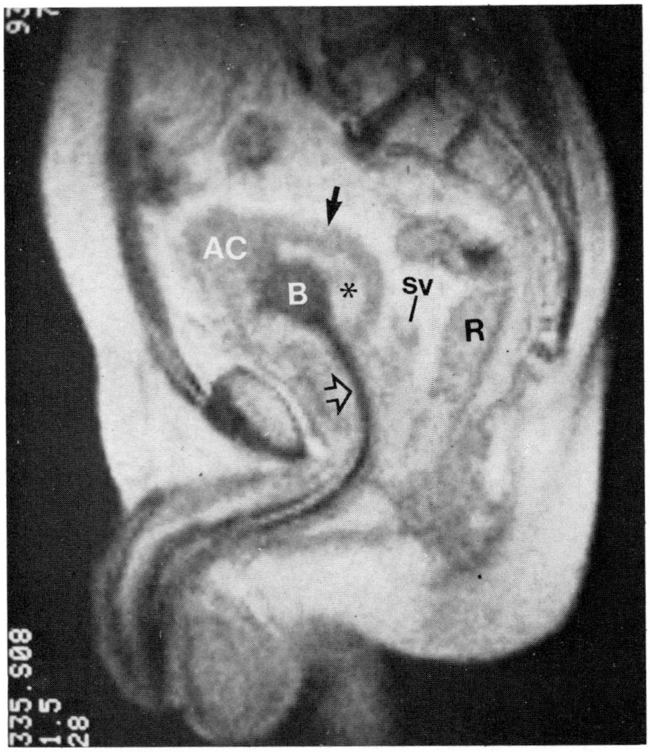

Figure 18. Transitional cell carcinoma in the urinary bladder (*) and adenocarcinoma (AC) in the urachal cyst. Sagittal spin echo image (TR of 1.5 sec, TE of 28 msec). The bladder wall is markedly thickened and has two differential zones of signal intensity. The inner higher-intensity zone corresponded to area of sessile transitional cell carcinoma (*). The carcinoma infiltrated the bladder wall but the deep detrusor muscles (solid black arrow) were not involved. Medium-intensity mass at the dome of the bladder represents adenocarcinoma within urachal cyst. There is a Foley catheter (open arrow) within the urethra. Prostate gland is enlarged. Note seminal vesicles (sv) and rectum (R). (B, bladder lumen)

Pathology of the Lower Genitourinary Tract

Urinary Bladder

With the urinary bladder distended, normal bladder thickness or any pathologic thickening of the bladder is easily recognized. As previously described, this is best appreciated with longer pulse sequence intervals. Bladder-wall thickening secondary to long-standing obstruction has a uniform medium-intensity signal. In our preliminary experience, bladder neoplasms, whether sessile or pedunculated, were depicted and delineated by NMR (Figs. 17–19). The superior ability of NMR in soft-tissue characterization

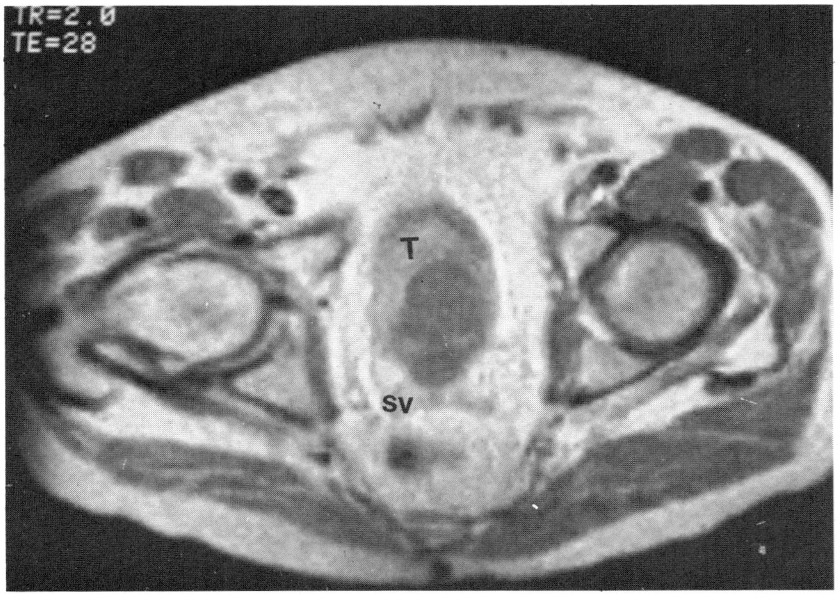

A.

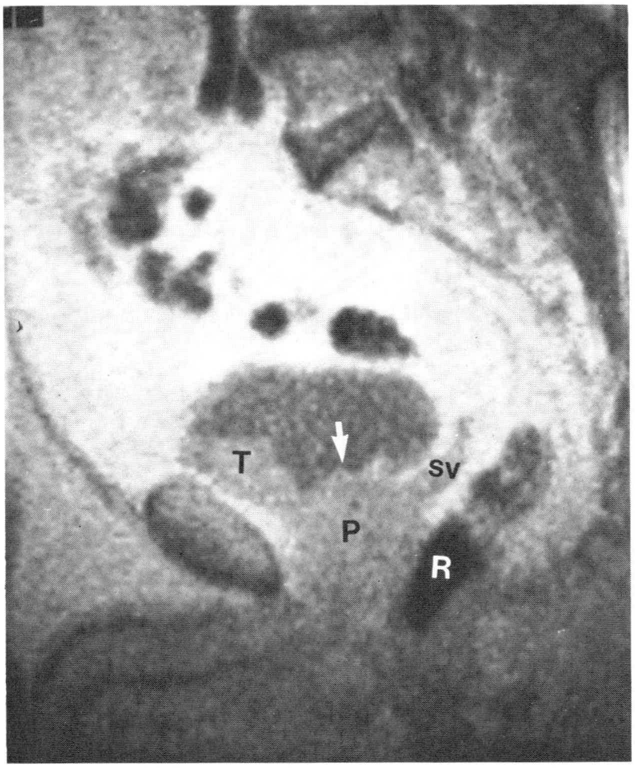

B.

Figure 19.

A. Transitional cell carcinoma of the urinary bladder. Axial scan spin echo (TR of 2.0 sec, TE of 28 msec). Tumor (T) is imaged with high intensity and is seen occupying the anterior and lateral portion of the bladder. Tumor signal intensity is considerably higher than that of urine within the bladder. Tumor extension to the bladder base on the axial scan was not appreciated. (sv, seminal vesicles)

B. Sagittal scan obtained with spin echo (TR of 2.0 sec, TE of 28 msec). The tumor (T) in the anterior portion of the bladder is well seen, but also appreciated is tumor extension along the bladder base (white arrow). This was correlated with findings at cystoscopy. Prostate gland (P) is enlarged. (sv, seminal vesicles; R, rectum) allows the differentiation of the sessile-type transitional cell carcinoma from the surrounding bladder wall (Fig. 18). The ability of NMR to obtain direct sagittal scans enhances its advantage in the evaluation of the tumor located at the bladder base, an area not optimally seen on CT images (Fig. 19). By combining images from two different planes, NMR can better delineate the extent of the neoplasm (Fig. 19). Further experience is needed in the staging of bladder neoplasms, with special emphasis on the capability of NMR to recognize early perivesical involvement. The sensitivity of NMR in the preoperative staging and defining of the depth of tumor extension into the submucosa and deep muscular layers needs to be explored also.

Pathology of the Prostate Gland

Benign prostatic hyperplasia is seen as an enlarged gland with a medium-intensity signal (Figs. 8, 11, 20). The signal intensity varies depending on the sequence utilized: it ranges from a low-intensity signal with a short pulse sequence interval to medium high-intensity with a long pulse sequence interval. In the evaluation of nodular hyperplasia, multiplanar NMR images permit not only the accurate volumetric measurement, but also the display of the intravesical portion of the gland (Fig. 20). Awareness of the extension of the intravesical portion of the gland is of extreme importance prior to surgical planning. The thick surgical pseudocapsule surrounding the adenomatous tissue is clearly depicted on any imaging plane (Figs. 11, 20). The uneven extension of the gland with the asymmetric location of the urethra can also be seen. The anterior fibromuscular band of the prostate gland retains its slight concave configuration, and on sagittal images the lobularity of the posterior and intravesical part of the gland is appreciated (Fig. 20). Even minor thickening of the bladder trigonum or mucosal edema secondary to urinary distention can be seen on sagittal and coronal images (Fig. 9).

Prostatic carcinoma is the second most common cause of cancer death among American men. Yet the tumor is slowly progressive, and if detected while truly confined to the prostate gland, it is potentially curable. Currently, no simple screening test for prostatic carcinoma is available. Screening for asymptomatic patients with potentially curable prostatic carcinoma relies almost exclusively upon digital rectal examination (Table I). However, 80% of the patients have advanced tumors when first diagnosed.[5,6] Once the diagnosis is made (needle biopsy following rectal examination is the standard method), accurate tumor

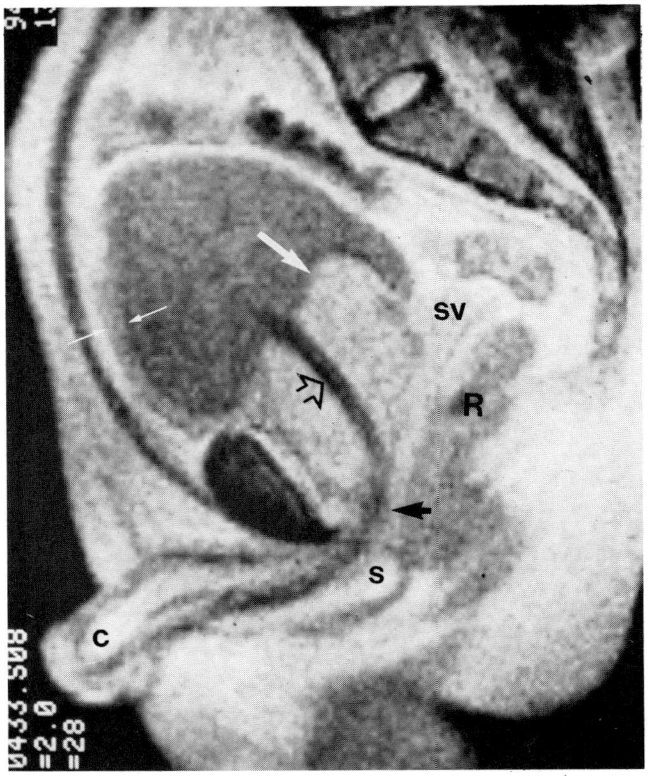

Figure 20. Benign nodular hyperplasia. Prostate gland is enlarged. Specifically, there is enlargement of the posterior and intravesical (large white arrow) portion of the prostate gland protruding within the bladder lumen. The urinary bladder is well distended, and thickening of the bladder wall is appreciated (small white arrows). A Foley catheter is seen in the prostatic urethra (open arrow), membranous urethra (solid black arrow) and penile urethra (crossing through the corpus spongiosum: s). (c, corpus cavernosum; sv, seminal vesicle; R, rectum)

staging to determine the extent of the disease is essential in planning the proper therapeutic approach. In addition to the definition of the local extension of the disease, the main task is to assess pelvic lymph node involvement. Reliance on digital examination alone may result in serious errors in staging (Table II).[5]

In the evaluation of prostatic carcinoma, in our preliminary experience, NMR appears to be able to depict the neoplastic tissue even when the pathologic condition is confined within the gland (stage A2, B1, and B2 carcinoma) (Figs. 21, 22). The prostatic carcinoma appears to have a higher-intensity signal than that of the surrounding gland. In the prediction of clinical outcome, current literature emphasizes the importance of tumor volume in addition to the tumor being confined within the prostate gland. The precise volume measurements of the primary lesion may prove to be the most significant in selecting

TABLE I

Clinical Staging of Prostatic Cancer*

*Clinical Stage***	*Digital Rectal Examination*	*Acid Phosphatase*	*Bone Scan/ Survey*
A	Tumor not detected	N†	N
B	Tumor confined to the prostate		
B1	Tumor less than 1.5 cm in diameter	N	N
B2	Tumor greater than 1.5 cm in diameter	N	N
C	Disease locally extensive (periprostatic tissue, seminal vesicles, urinary bladder,urethra)	N or +‡	N
D	Distant metastases (any size of tumor)	N or +	N or +

(Modified, with permission, from Klein LA: Prostatic carcinoma. N Engl J Med 300:824, 1979.)
*Staging classification of Whitmore[24] as modified by Klein.[5]
**Assignment of clinical stage is based on digital rectal examination, acid phosphates, and bone survey or scan.
†N = normal.
‡+ = positive.

TABLE II

Clinical Stage	*Frequency of Lymph Node Metastases (%)**
A1	5
A2	25
B1	8–21
B2	4–45
C	40–80
D	100

*Frequency of lymph node metastases makes it clear that assignment of clinical stage without regard to regional lymphatics is erroneous.

potentially curable patients. The capability of direct three-dimensional imaging increases the diagnostic yield in tumor volume estimation. In the case of advanced tumor spread, a combination of multiplanar imaging and superior soft-tissue contrast resolution by NMR promises improved detection of the periprostatic tissue, seminal vesicles, and/or bladder base. Axial scans alone can be misleading in the evaluation of seminal vesicle involvement (sometimes there is asymmetric position of the vesicles). Analysis of the seminal vesicles in at least two projections is more accurate. Furthermore, change in intensity in the seminal vesicles is of even more importance in evaluating involvement by the neoplasm.[7,15] Posterior extension of prostatic carcinoma is a rare occurrence, and can be best appreciated on sagittal NMR scans because of the clear display of Denonvilliers' fascia.

In conclusion, the potential of NMR imaging of the prostate gland appears to be in its ability to detect pathologic conditions while they are confined to the gland. While the disruption of the normal gland architecture has been depicted by NMR in this early stage, it is not known if a neoplastic nodule can be distinguished from chronic prostatitis. More experience is needed on a larger number of patients of all ages with normal glands, chronic prostatitis, nodular hyperplasia, and neoplasm.

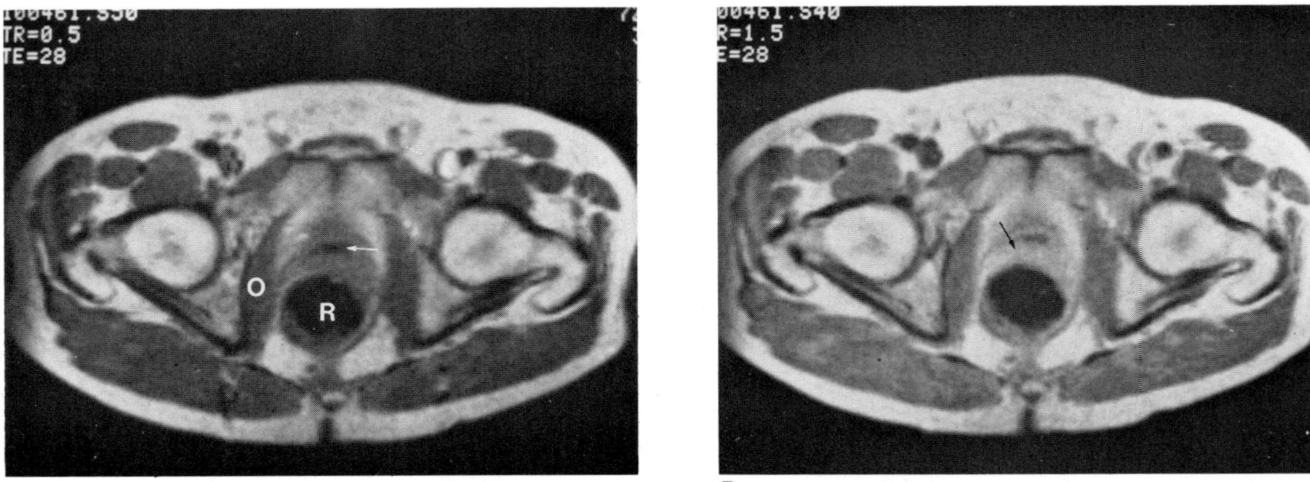

A. B.

Figure 21. Status post-TURP; A2 prostatic carcinoma. Transaxial scan: in A, TR of 0.5 sec, TE of 28 msec; in B, TR of 1.5 sec, TE of 28 msec. In A, with a short TR value, the residual prostate gland is imaged with medium high-intensity signal. It is higher in intensity than the striated skeletal muscle, that is, obturator muscle (O). The prostatic urethra (white arrow) is dilated secondary to TURP. (R, rectum). In B, imaged with longer TR (1.5 sec), there is increase in the intensity of the prostatic tissue. However, the increase in intensity is uneven, with a higher area seen in the right and posterior portion. As confirmed by surgery, it was a residual prostatic carcinoma (black arrow). Also, note a high-intensity halo around the prostatic urethra. This is probably due to edema secondary to recent surgery and remains of the homogeneous high intensity with short and long TR value.

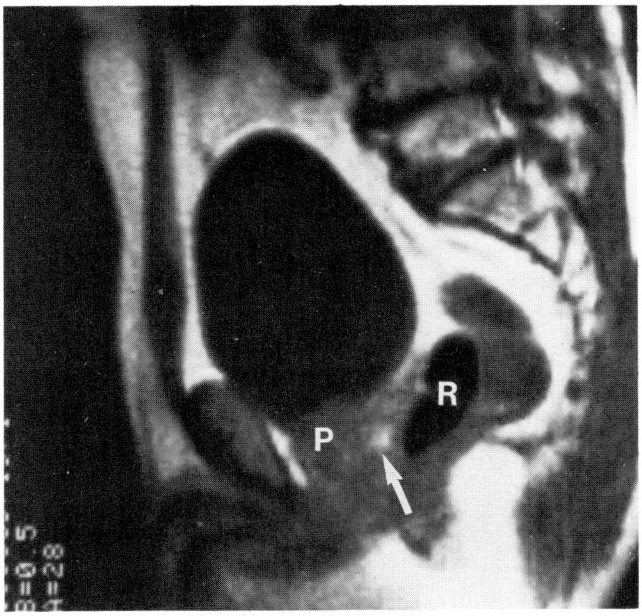

Figure 22. Prostatic carcinoma, clinical stage B1. Sagittal spin echo image (TR of 0.5·sec, TE of 28 msec). Prostate gland (P) is enlarged. Along its posterior aspect there is localized high-intensity area (white arrow), which at surgery was proved to be small localized adenocarcinoma of the prostate gland. (R, rectum)

Pathology of Female Reproductive Organs

Our experience with pathology of female reproductive organs is very limited. We encountered a case of adenomyosis (Fig. 23). Adenomyosis refers to the presence of endometrial glands and stroma deep within the myometrium. By NMR, it was imaged as a localized low-intensity area within the wall of the myometrium. The signal intensity of the adenomyosis was similar to that of the basalis zone of the endometrium. Leiomyomas, however, were imaged with a signal intensity similar to that of the myometrium, and the diagnosis was suggested by the uneven thickness of the myometrium and enlargement of the uterus. As correlated with ultrasound, no hemorrhage or necrosis was present within the leiomyomas. Patients with folicular ovarian cysts and dermoid cysts were studied. The dermoid cyst was imaged with low signal intensity; the cystic fluid relaxation was similar to that of urine. Because endometriosis was a clinical concern in that patient, the fact that the fluid's relaxation parameters were similar to water excluded the possibility of endometrioma. At surgery, a thin wall cyst with clear

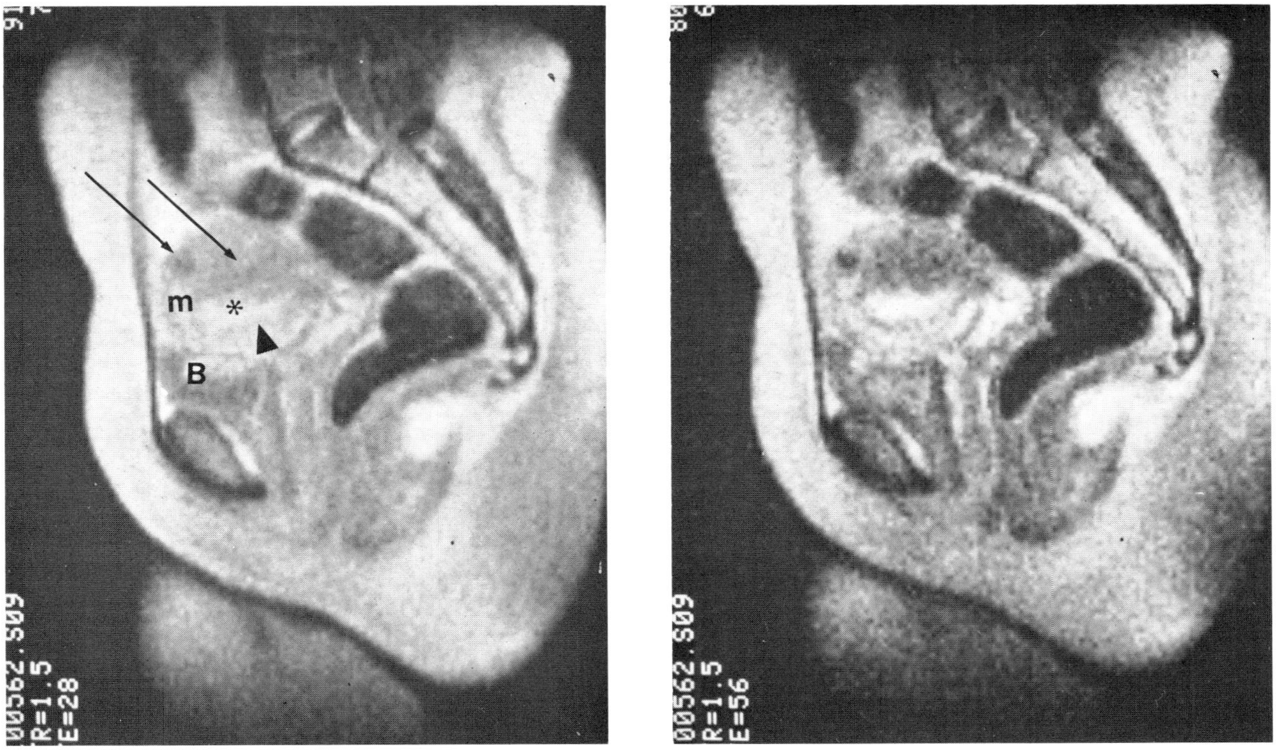

A. B.

Figure 23: Adenomyosis uterus. Sagittal spin echo scan. In A, TR of 1.5 sec, TE of 28 msec. In B, TR of 1.5 sec, TE of 56 msec. Only a small amount of urine is present within the bladder (B). Corpus uterus is seen superior to the bladder. Myometrium (m) is of higher intensity than the basalis zone of the endometrium (arrowhead). The differentiation is better appreciated with longer TE values. Zone functionalis (*) is of higher intensity than either myometrium or basalis zone. Within the myometrium there are two zones of low intensity (arrows). The intensity of that portion is similar to that of normal zone basalis (arrowhead) of the endometrium.

fluid (no protein or lipid content found) was encountered. A clump of hair was adherent to the wall.

These are only preliminary observations; further studies on a larger number of women are needed. Nuclear magnetic resonance promises to play an important role in the evaluation of the female pelvis from infancy onward. Pubertas praecox, female infertility, or the problems of repeated spontaneous abortions during the first trimester will undoubtedly also be studied by this modality. In addition, NMR will assume a prominent role in the clinical management of pelvic malignancies.

Acknowledgment

I wish to acknowledge my gratitide to Gloria Riggs and Philip E. Sheldon for their assistance and timeless dedication in the preparation of this chapter.

REFERENCES

1. Walsh JW, Taylor KJW, Wasson JF, et al: Gray-scale ultrasound in 204 proven gynecologic masses: Accuracy and specific diagnostic criteria. Radiology 130:391, 1979.
2. Golimbu M, Morales P, Al-Askari S, et al: CAT scanning and staging of prostatic cancer. Urology 23:305, 1981.
3. Denkhaus H, Dierkopf W, Grabbe E, et al: Comparative study of suprapubic sonography and computed tomography for staging of prostatic carcinoma. Urol Radiol 5:1, 1983.
4. Watanabe H, Igari D, Tanahashi Y, et al: Transrectal ultrasonography of the prostate. J Urol 114:734, 1975.
5. Klein LA: Prostatic carcinoma. N Engl J Med 300:824, 1979.
6. Fair WR, Kadmon D: Carcinoma of the prostate: Diagnosis and staging. Current status and future prospects. World J Urol 1:3, 1983.

7. Hricak H, Spring DB, Moon KL Jr, et al: Anatomy and pathology of the male pelvis as imaged by NMR. AJR (in press).
8. Kaufman L, Crooks LE, Margulis AR (Eds): *Nuclear Magnetic Resonance Imaging in Medicine,* pp 11–52. New York: Igaku-Shoin, 1981.
9. Herfkens R, David PL, Crooks LE, et al: NMR imaging of the abnormal live rat and correlation with tissue characteristics. Radiology 141:211, 1981.
10. Kaufman L, Crooks LE, Sheldon PE, et al: Evaluation of NMR imaging for detection and quantification of obstructions in vessels. Invest Radiol 17:554, 1982.
11. Kaufman L, Crooks L, Sheldon P, et al: The potential impact of nuclear magnetic resonance imaging on cardiovascular diagnosis. Circulation 67:251, 1983.
12. Gosling JP, Dixon JS, Humpherson JR: *Functional Anatomy of the Urinary Tract: An Integrated Text and Colour Atlas*. Baltimore: University Park Press, 1982.
13. Lowsley OS: The development of the human prostate gland with reference to the development of other structures at the neck of the urinary bladder. Am J Anat 13:299, 1912.
14. McNeal JE: The prostate gland: Morphology and pathobiology. Monogr Urol 4:3, 1983.
15. Hricak H, Tanagho EA, Williams RD, et al: The value of multiplanar NMR imaging of the male pelvis. Presented at the 2nd Annual Meeting of the Society of Magnetic Resonance Imaging in Medicine, 1983.
16. Smith WF: Two years' clinical experience with NMR imaging. Applied Radiol 12(3):29-42, 1983.
17. Bryan PJ, Butler HE, Haaga JR, et al: NMR scanning of the pelvis. Presented at the 83rd Annual Meeting of the American Roentgen Ray Society, 1983.
18. Hricak H, McNeal JE: Prostate gland: Anatomic-MRI correlation. (In preparation.)
19. Hendrickson MR, Kempson RL: Embryology and anatomy. In *Surgical Pathology of the Uterine Corpus,* pp 1–16. Philadelphia: W. B. Saunders, 1980.
20. Demopoulos RI: Normal endometrium. In *Pathology of the Female Genital Tract,* Blaustein A (Ed), pp. 235–278. New York: Springer-Verlag, 1982.
21. Hricak H: The role of NMR imaging in the female pelvis: Preliminary results. (In preparation.)
22. Landlois PL: The size of the normal uterus. J Reprod Med 4:31, 1970.
23. Hendrickson MR, Kempson RL: The normal endometrium: The baseline. In *Surgical Pathology of the Uterine Corpus,* pp 36–98. Philadelphia: W. B. Saunders, 1980.
24. Whitmore WF Jr: Hormone therapy in prostatic cancer. Am J Med 21:697, 1956.

16

Musculoskeletal System

HARRY K. GENANT, M.D.
KIRK L. MOON, JR., M.D.
NEIL I. CHAFETZ, M.D.
CLYDE A. HELMS, M.D.

Recent reports have documented the ability of NMR to demonstrate normal anatomy and pathologic conditions in a variety of cerebral, cardiovascular, and abdominal applications.[1–6] To date, however, the role of NMR in the diagnosis of musculoskeletal disorders has not been extensively explored.[7,8] In order to determine potentially useful applications of the modality in musculoskeletal diagnosis, we have reviewed and reported our early experience with NMR imaging of healthy subjects and patients with musculoskeletal disorders.[9,10]

Imaging Techniques Utilized

All subjects included in this review were examined in the NMR imager at the University of California, San Francisco Radiologic Imaging Laboratory. An elliptical body coil with a 50- × 58-cm aperture was used in all cases. The spatial resolution of the device was 2.1 × 2.1 mm in the configuration used to image most patients in this study. Each individual imaging plane has a section thickness of 7 mm. Adjacent sections were centered 12 mm apart, resulting in a 5-mm non-imaged gap between sections. The display matrix was composed of 256 horizontal × 128 vertical elements.

All images were produced by using a spin echo pulse sequence and a two-dimensional Fourier transform image construction technique.[1] A variety of spin echo imaging sequences were used, the details of which are summarized in Table I. In any given case one to three of the scanning sequences were used to examine the region of interest. Signal intensity images of the first (28-msec) and second (56-msec)

TABLE I

Spin Echo Imaging Sequences

Pulse Sequence Interval (msec)	*Echo delay Times (msec)*	*No. of Sections*	*Imaging Time (min)*
500	28, 56	5	4.5
1000	28, 56	10	9
1500	28, 56	15	13.5
2000	28, 56	20	18

spin echoes were generated for each of the pulse sequence intervals chosen. For simplicity of reference, these images will be referred to in the following format: spin echo pulse sequence interval/echo delay time. Thus an intensity image of the first (28-msec) spin echo obtained at a pulse sequence interval of 1.0 sec (1000 msec) will be referred to as a spin echo 1000/28 image. T_1 maps of individual anatomic sections were computer generated by comparison of intensity images of a given section obtained at two different pulse sequence intervals (e.g., by comparing the spin echo 500/28 image with the spin echo 1000/28 image). T_2 maps of individual anatomic sections were generated by comparison of the first and second spin echo intensity images at a single pulse sequence interval (e.g., by comparing the spin echo 1000/28 image with the spin echo 1000/56 image). T_1 and T_2 values could be determined for any desired region of interest on each imaging section.

Patient Populations Examined

Besides patients with no known musculoskeletal disorders who were examined in a variety of anatomic sites (lumbosacral spine, hips, knees, feet, etc.), patients with a broad spectrum of musculoskeletal disorders were also examined. The latter suffered from low back pain syndrome, suspected avascular necrosis of the femoral head, and skeletal neoplasms. The latter category included metastases, chondrosarcoma, ameloblastomata, osteochondromata, soft-tissue tumors, sacral bone cyst, and thalassemia with excessive marrow iron distribution. Computed tomographic scans were available for review and comparison on almost all patients. Conventional radiographs were also available in most cases. T_1 and T_2 measurements were obtained on normal tissues from a variety of sites and on all pathologic tissues that became available.

Clinical Findings

Although the absolute intensities of normal tissues with spin echo techniques change significantly as the timing parameters of the imaging sequences are changed, the relative spin echo gray scale rankings of the various tissues remain quite constant. They are presented in Table II. The T_1 and T_2 values obtained on normal tissues are presented in Table III. The NMR appearances of normal and pathologic tissues will be presented on a region-by-region basis.

TABLE II

Nuclear Magnetic Resonance Spin Echo Gray Scale in Descending Order of Brightness*

Fat
Marrow and cancellous bone
Brain and spinal cord
Viscera
Muscle
Fluid-filled cavities
Ligaments and tendons
Blood vessels with rapid flow
Compact bone
Air

*The brighter the image, the shorter the T_1 and/or the longer the T_2. The darker the image, the longer the T_1 and/or the shorter the T_2 and/or the proton density is very low.

TABLE III

T_1 and T_2 Values Obtained from Normal Tissues

Tissue	*T_1 msec(±S.D.)*		*T_2 msec(±S.D.)*		*No. of Samples*
Muscle	541	(141)	35	(6)	13
Subcutaneous fat	218	(68)	61	(18)	12
Vertebral marrow	420	(112)	50	(12)	16
Intervertebral disc	2,623	(2,030)	48	(7)	4
Ligament/tendon	864	(206)	107	(16)	3

Lumbosacral Spine

Normal Anatomy. In the transaxial images of the lumbosacral spine (Fig. 1), all of the following anatomy is well seen: disc material, vertebral body cortex and medullary cavity, epidural fat and vessels, thecal sac, spinal cord (in the upper lumbar region), and nerve roots. Axial images through intervertebral disc material commonly demonstrate a sharp distinction between annulus fibrosus (appearing dark) and nucleus pulposus (appearing bright). In the direct sagittal (Fig. 2) and parasagittal images, the vertebral bodies, intervertebral disc material, spinal cord, and nerve roots (traversing the neural foramina) are clearly defined.

Anatomic resolution of the CT images is somewhat superior to that of the transaxial NMR images in all cases. Anatomic resolution of the direct sagittal NMR images is slightly superior to that of reformatted sagittal CT images, and the sagittal NMR images include more anatomic territory in the craniocaudal dimension than the reformatted CT images do.

Lumbar Disc Protrusion. Nuclear magnetic resonance demonstrates abnormalities in the majority of patients with CT evidence of lumbar disc protrusion (Figs. 3–6). The few cases with no NMR abnormality all had mild, broad-based disc protrusion on CT. In addition to demonstrating disc protrusion, extruded disc fragments, and encroachment upon the neural foramina by protruding disc material, NMR also demonstrates loss of the normally bright signal from the nucleus pulposus in the majority of patients.

Hip Joints

Normal Anatomy. In the patients with normal hip joints (Fig. 7), cortical bone and fibrous capsular structures are invariably dark and quite well defined. The medullary cavities of the femoral head and pelvic bones have a bright NMR signal, presumably because of the high fat content of the marrow. To date, the central portions of the femoral heads are invariably of slightly lower intensity than the outer portions.[3] Since bone is known to produce a lower-intensity NMR signal than marrow, these regions of decreased NMR signal may represent areas where cancellous bone is most prominent. Muscles, ligaments, fat, and blood vessels surrounding the hip joint are all well defined.

Avascular Necrosis of the Femoral Head. Patients with suspected avascular necrosis of the femoral head were examined (Figs. 8–10). Nuclear magnetic resonance demonstrated abnormalities in all patients with definite avascular necrosis and demonstrated normal findings in a patient subsequently shown not to have the disease. Characteristic abnormalities on the NMR image were overall decrease in the intensity of the involved femoral head, accentuated inhomogeneity of the medullary cavity, and loss of the normally sharp cortical definition. One patient had an intermediate-intensity halo surrounding the involved joint that appeared to represent synovial hypertrophy or effusion (Fig. 8C). An arthrocentesis was not performed.

Knee Joints

Normal Anatomy. Bony cortex, medullary cavity, muscles, ligaments, tendons, and blood vessels were all well seen in two patients with three normal knee joints (Figs. 11, 12). The cruciate ligaments were particularly well seen, appearing as dark structures outlined by high-intensity fat in the intercondylar space. Articular cartilage could not be differentiated from subchondral cortex. Meniscal cartilage could be faintly seen as an intermediate-intensity structure on several transaxial images and was well demonstrated on coronal sections.

Feet

Normal Anatomy. Bony cortices, medullary cavities, tendons, ligaments, and blood vessels were accurately defined in the feet that were examined (Fig. 13).

Bone Tumors

In all patients with bony metastases, the intensity of the medullary cavity was abnormally low in regions that contained metastatic foci. In one patient with a bulky metastasis involving the T11 vertebral body, direct sagittal NMR images demonstrated extension of the tumor into the spinal canal with involvement of the spinal cord (Fig. 14). The T_1 and T_2 values of the metastases were significantly longer than those of the adjacent vertebrae.

One patient with a chondrosarcoma of the L2 vertebra was examined (Fig. 15). The tumor appeared to be of higher intensity than uninvolved bone.

Two patients with benign osteochondromas were examined (Fig. 16). The cartilaginous cap could be easily discriminated from adjacent muscle in both cases.

(text resumes on page 270)

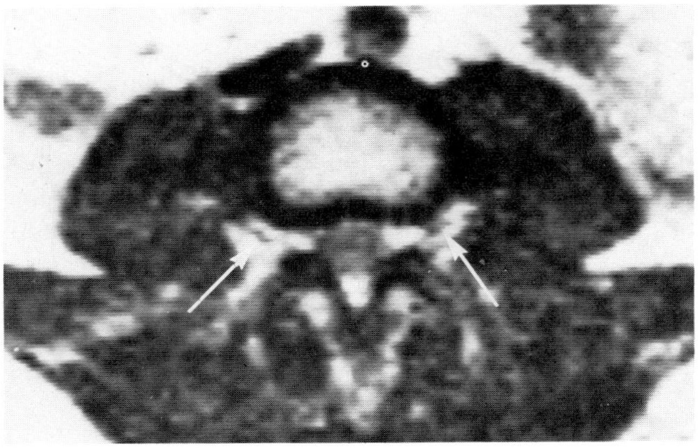

A.

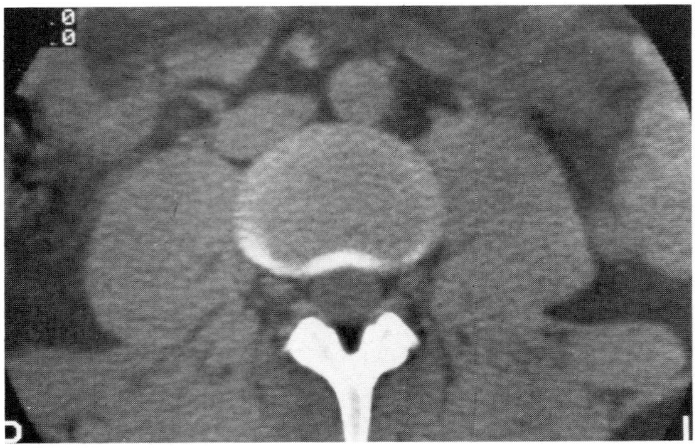

B.

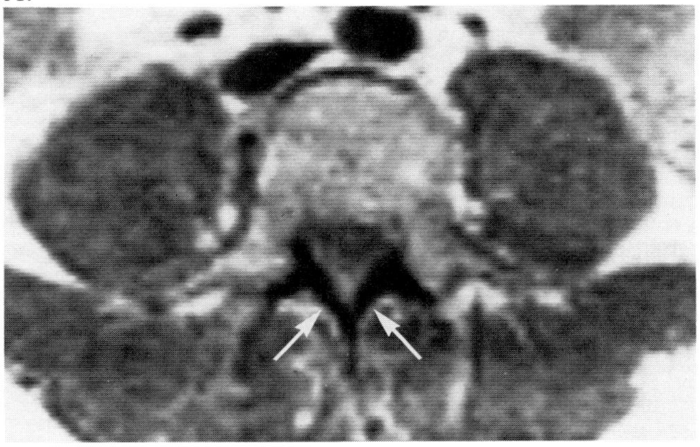

C.

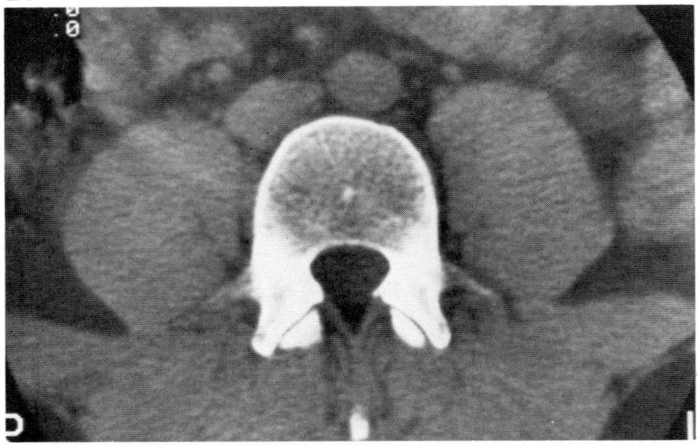

D.

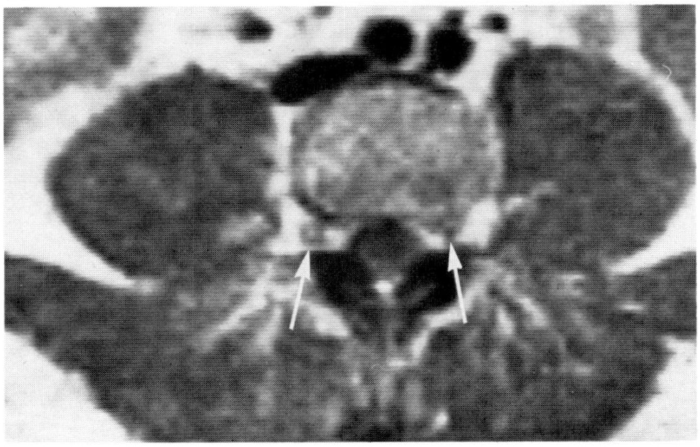

E.

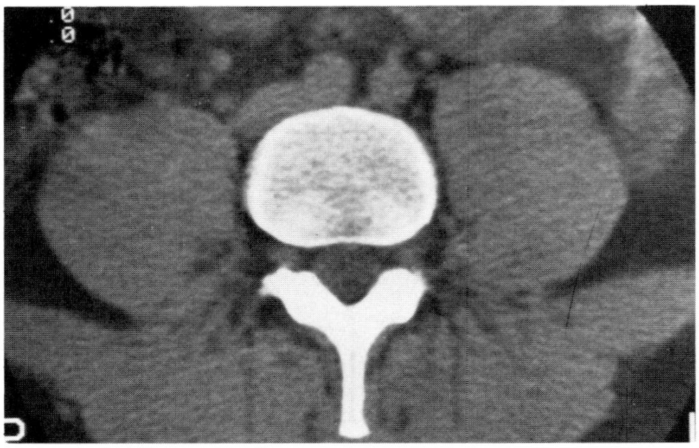

F.

Figure 1. Axial views of normal lumbosacral spine at the L3–4 level, with NMR (left column) and CT (right column) correlation. All NMR images are spin echo 1500/28.

A and B. Sections through the L3–4 intervertebral disc. Disc material has a high-intensity center and a low-intensity outer ring. Cortical bone in the posterior elements is black. The thecal sac and L3 nerve roots (arrows) are outlined by high-intensity epidural fat. The aorta and inferior vena cava are dark because of rapid blood flow.

C and D. Sections through the L4 pedicles. Very little epidural fat is present at this level. The ligamentum flavum (arrows) appears very dark.

E and F. Sections through the L4 body below the pedicles. The L4 nerve roots (arrows) and thecal sac are outlined by epidural fat.

(Reproduced, with permission, from Moon KL Jr, Genant HK, Helms CA, et al: Musculoskeletal applications of nuclear magnetic resonance. Radiology 147:161, 1983.)

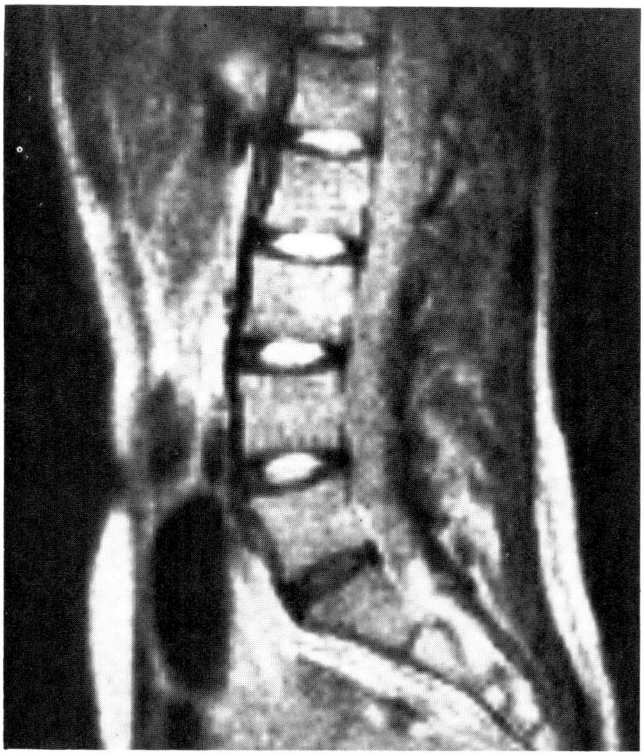

A.

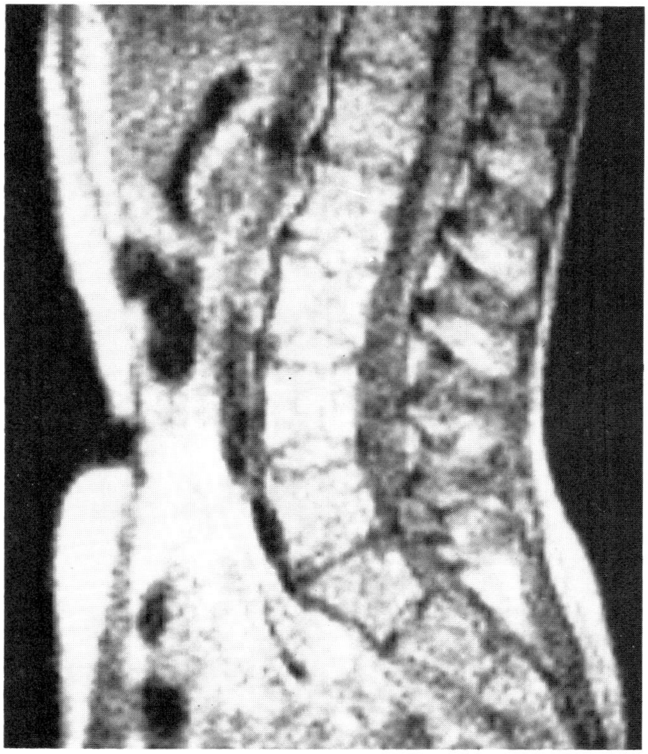

B.

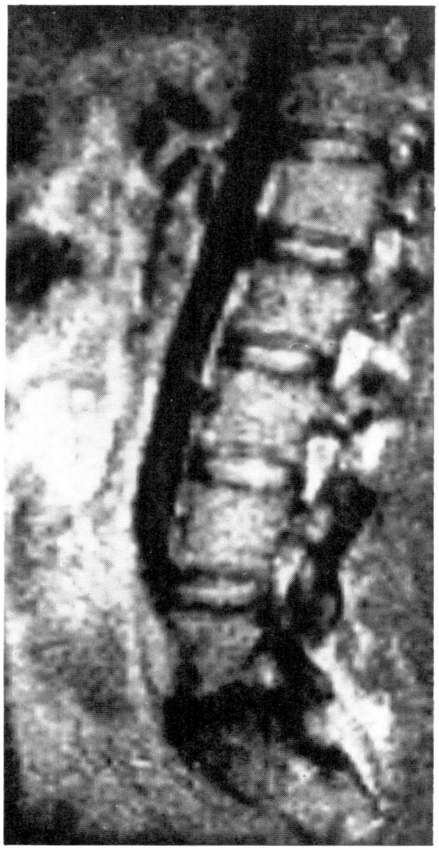

C.

Figure 2.

A. Near-midline parasagittal spin echo 2000/28 image of normal lumbosacral spine. The anterior longitudinal ligament combines with the anterior bony cortices of the vertebral bodies to produce the dark line coursing down the anterior aspect of the spine. The intervertebral disc material has a bright center (presumably, nucleus pulposus) and a darker outer rim (presumably, annulus fibrosus).

B. Midline sagittal spin echo 1500/28 image. The spinal cord is visible as an intermediate-intensity structure outlined by darker cerebrospinal fluid.

C. Parasagittal spin echo 1500/28 image through the left neural foramina. The neural foramina are highlighted by bright epidural fat. The nerve roots are visible as lower-intensity structures within the neural foramina inferior to the pedicles. The long dark structure anterior to the spine is the aorta.

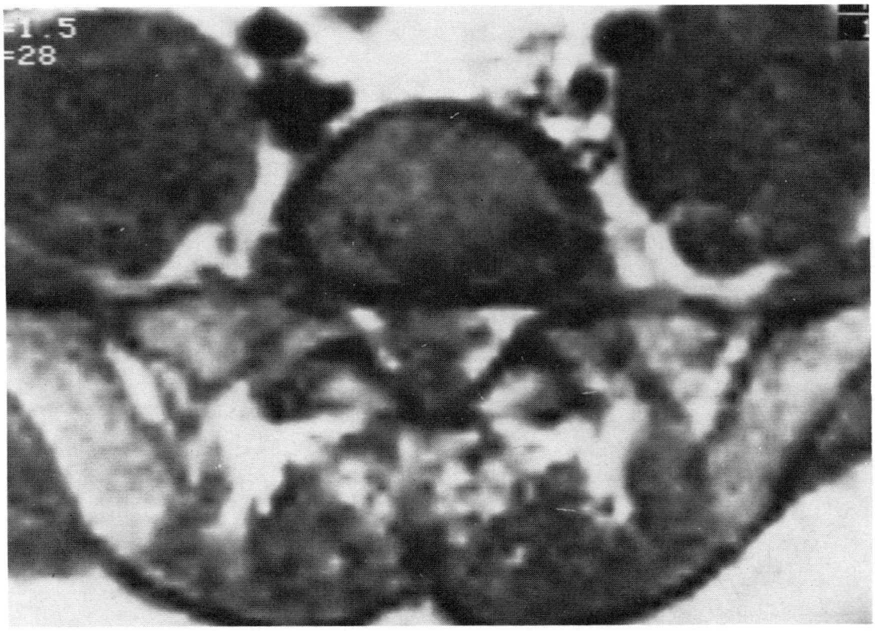

A.

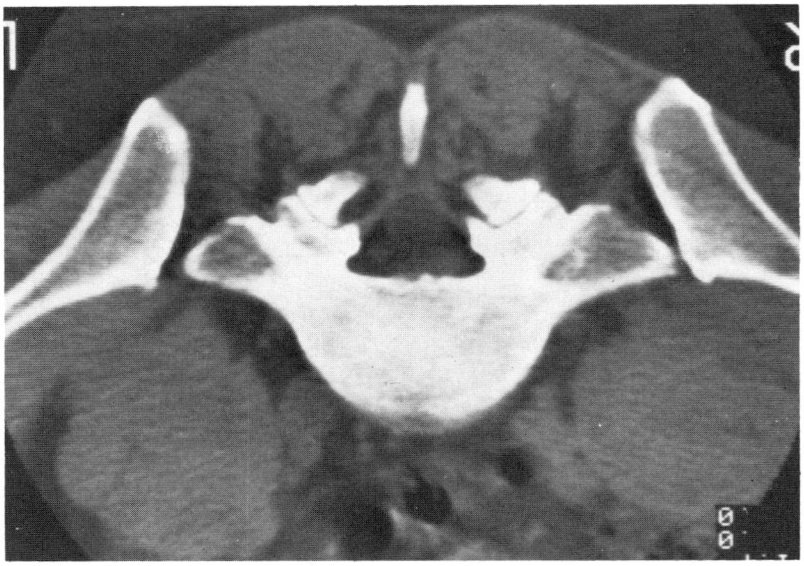

B.

Figure 3.
A and B. Axial spin echo 1500/28 (top) and CT images of lumbosacral spine at S1 level showing a soft-tissue density in the anterior portion of the spinal canal that impinges on the thecal sac and the right S1 nerve root. The epidural fat is asymmetric, and the right nerve root and thecal sac are displaced from their normal position.

(Reproduced, with permission, from Moon KL Jr, Genant HK, Helms CA, et al: Musculoskeletal applications of nuclear magnetic resonance. Radiology 147:161, 1983.)

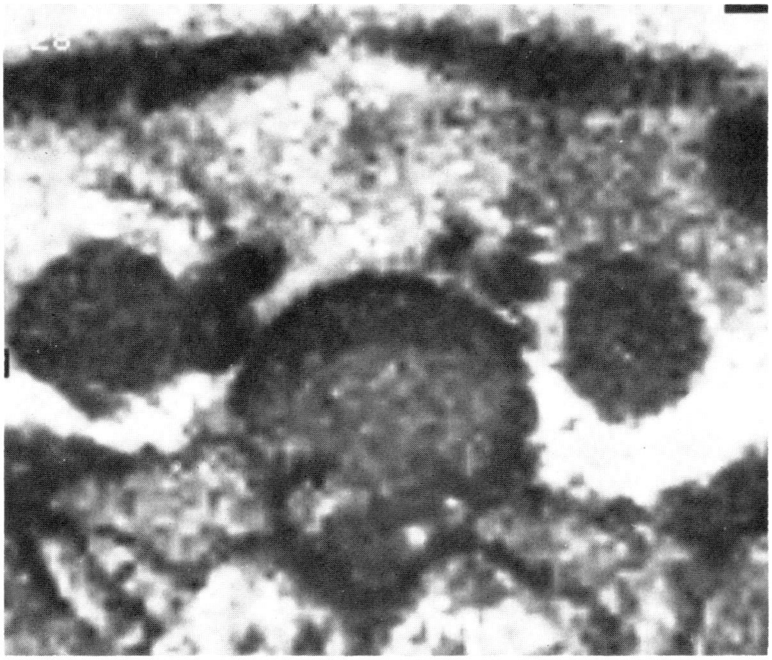

A.

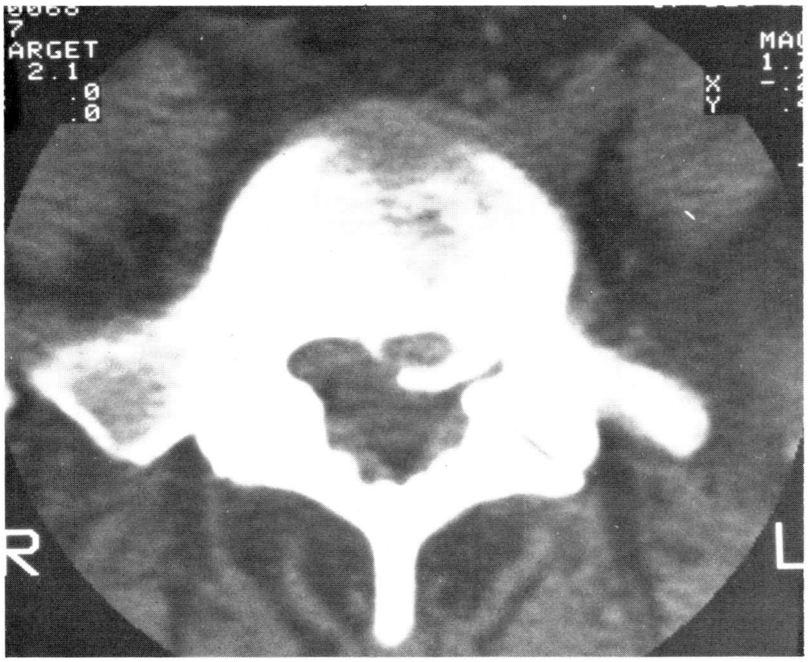

B.

Figure 4.
A and B. Axial spin echo 1500/28 (top) and CT images of lumbosacral spine at S1 level showing a large, partially calcified disc herniation that impinges on the thecal sac and the left S1 nerve root. The epidural fat is asymmetric, and the left nerve root and the thecal sac are displaced from their normal position.

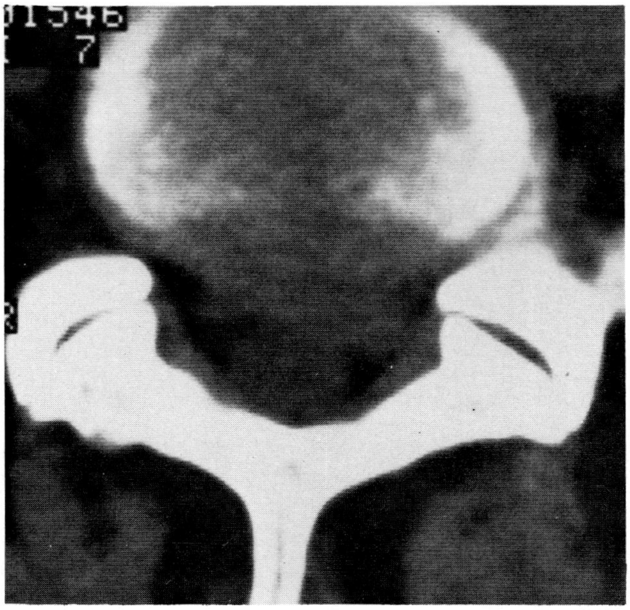

A.

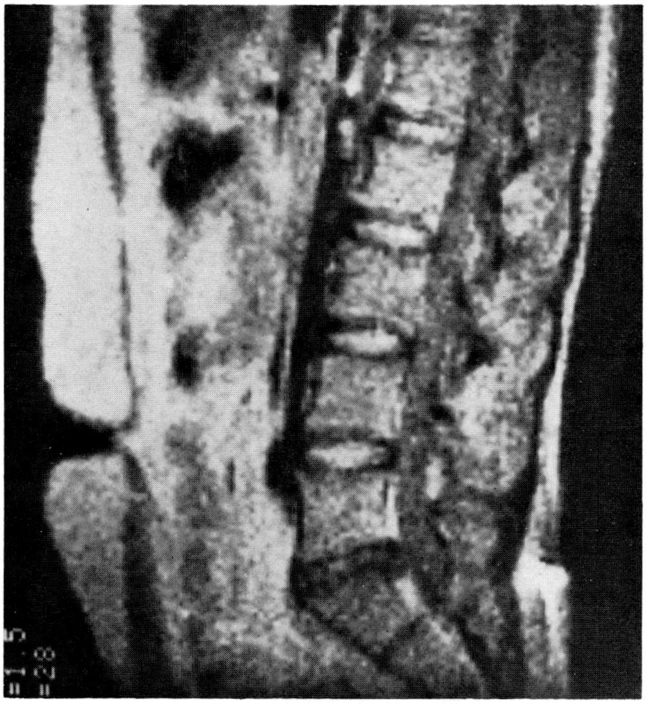

B.

Figure 5. Left posterolateral disc protrusion at L5–S1 level in a patient 3 days after chymopapain injection.
A. Axial CT image through the L5–S1 disc showing prominent disc bulging with compression of the thecal sac and left L5 nerve root.
B. Parasagittal spin echo 1500/28 NMR image showing abnormally dark disc at L5–S1 level and prominent posterior bulge of disc material. Epidural fat is present above and below the lesion, but is displaced at the level of the protrusion.

(Reproduced, with permission, from Moon KL Jr, Genant HK, Helms CA, et al: Musculoskeletal applications of nuclear magnetic resonance. Radiology 147:161, 1983.)

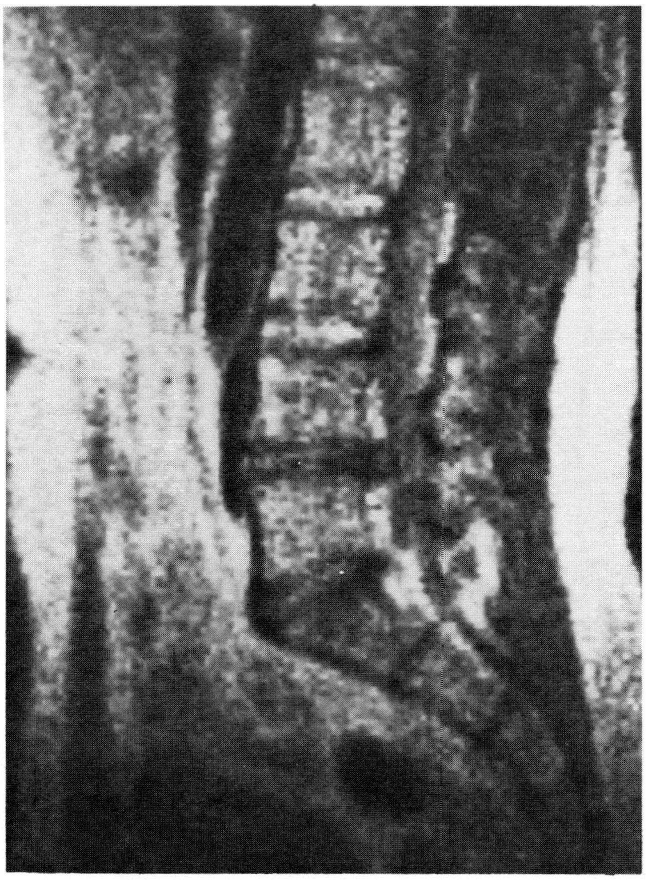

Figure 6. Sagittal 1500/28 image shows a large, low-intensity, posteriorly herniated L4–5 disc effacing epidural fat and displacing the thecal sac.

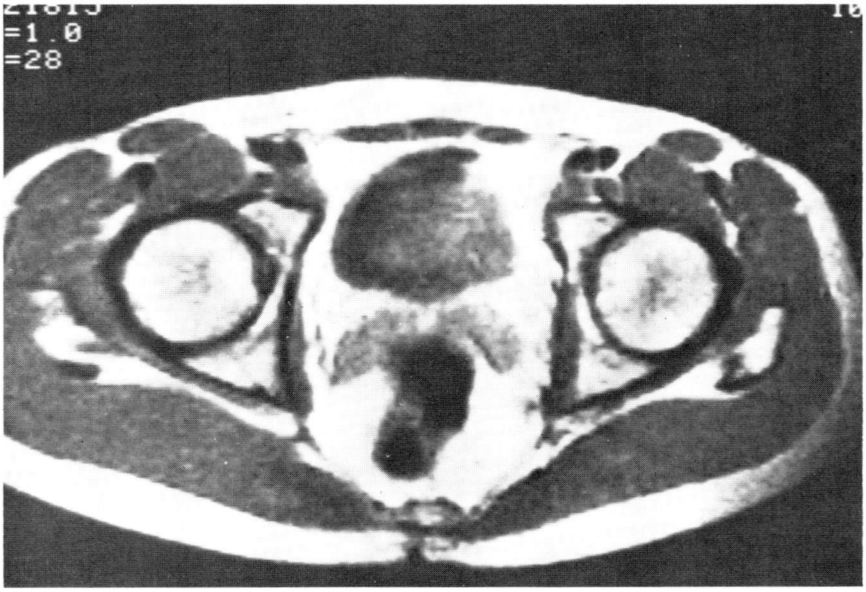

A.

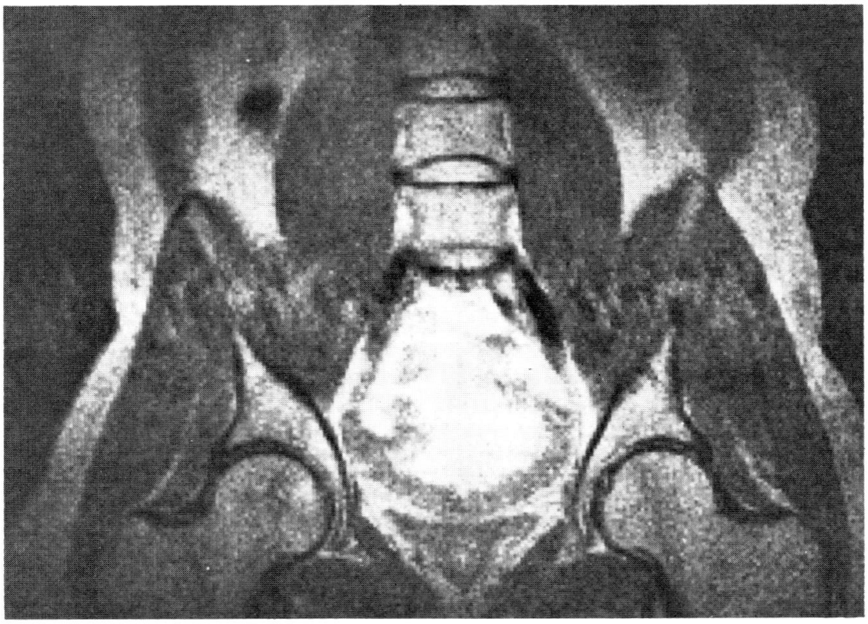

B.

Figure 7.

A. Transaxial spin echo 1000/28 image of the hip joints of a normal volunteer. A low-intensity band is present in each femoral head and may be due to the presence of trabecular bone.

B. Coronal spin echo 2000/28 image of the hip joints and lower lumbar spine of a normal volunteer.

(continued on next page)

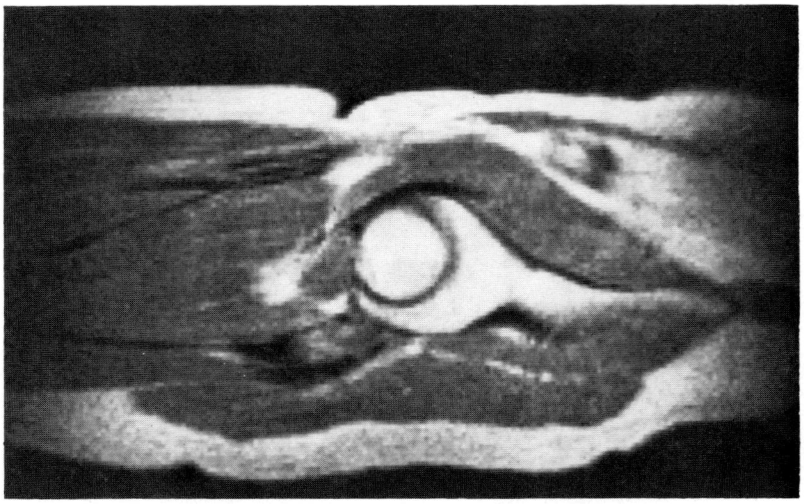

C.

C. Sagittal spin echo 2000/28 image of a normal hip joint.
(Fig. 7A reproduced, with permission, from Moon KL Jr, Genant HK, Helms CA, et al: Musculoskeletal applications of nuclear magnetic resonance. Radiology 147:161, 1983.)

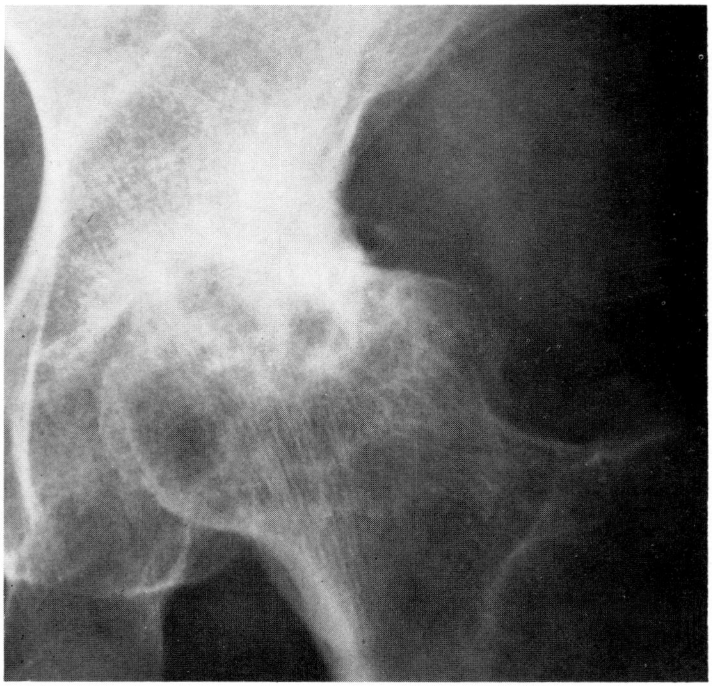

A.

Figure 8. Advanced avascular necrosis of the left femoral head.
A. Conventional radiographic findings include marked joint-space narrowing, deformity of the femoral head, bony sclerosis on both sides of the joint, and subchondral cyst formation in the femoral head.

(continued on next page)

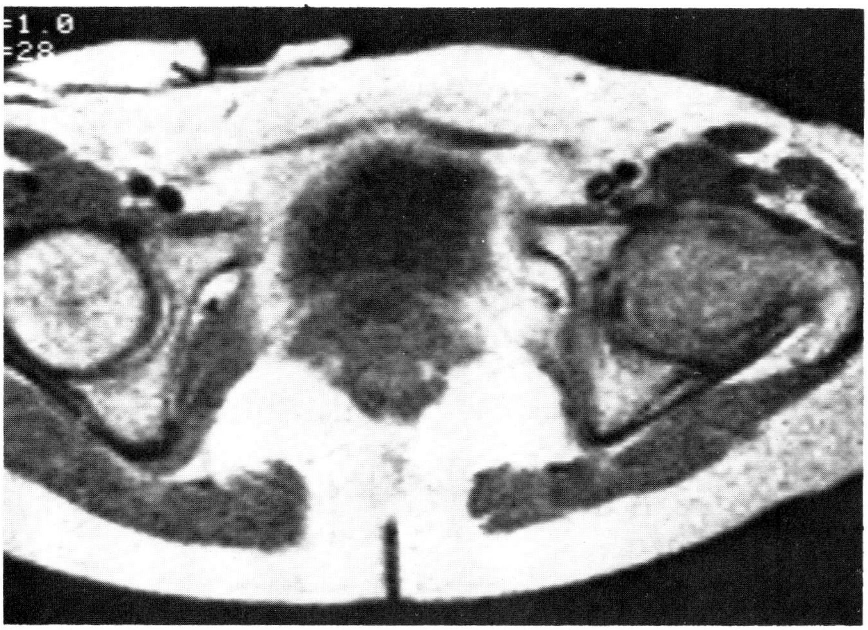

B.

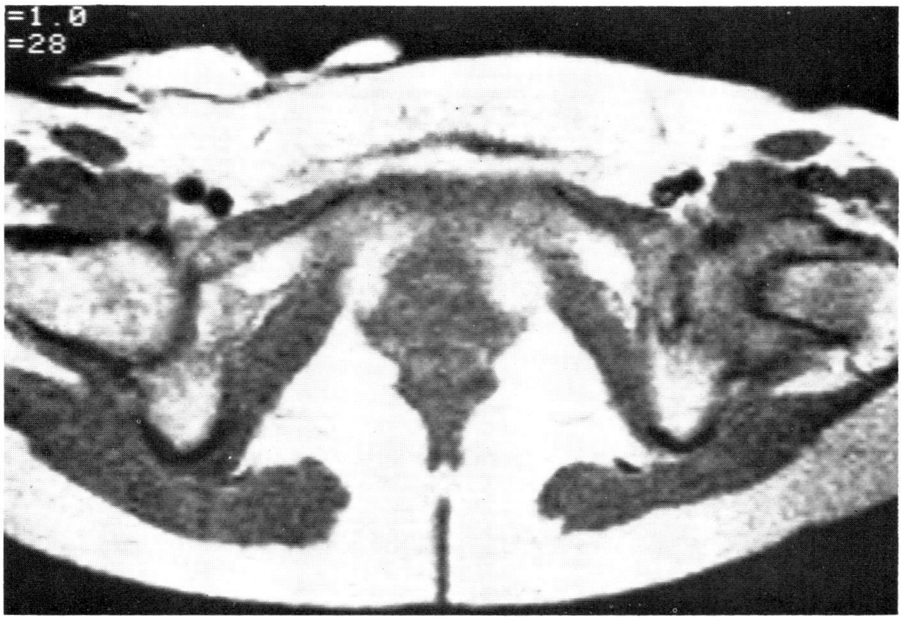

C.

B. Axial spin echo 1000/28 image through the midportion of the femoral heads. The left femoral head shows loss of the normally sharp cortical definition and a lower signal intensity than normal. A halo of low-intensity material surrounds the femoral head. This might represent thickened joint capsule, synovial hypertrophy, or synovial effusion.

C. Axial spin echo 1000/28 image 12 mm caudal to B, showing the low-intensity halo surrounding the left femoral neck in a distribution consistent with the synovial space.

(Reproduced, with permission, from Moon KL Jr, Genant HK, Helms CA, et al: Musculoskeletal applications of nuclear magnetic resonance. Radiology 147:161, 1983.)

A.

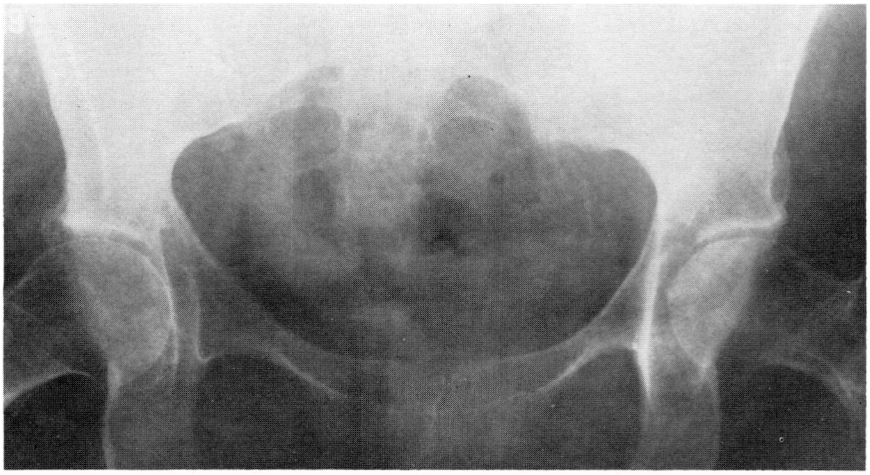

B.

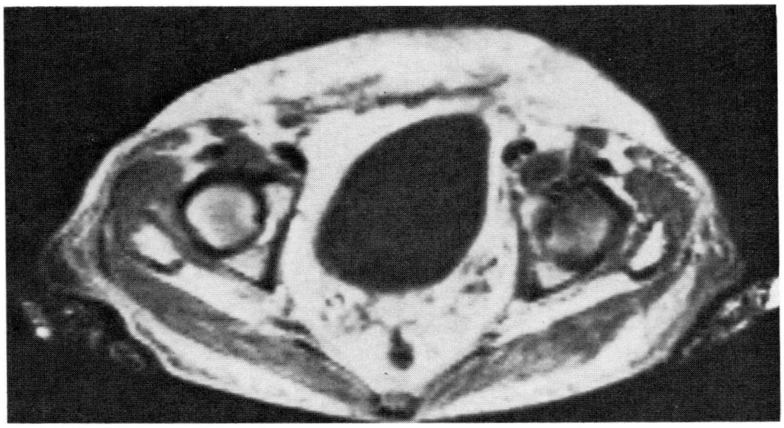

C.

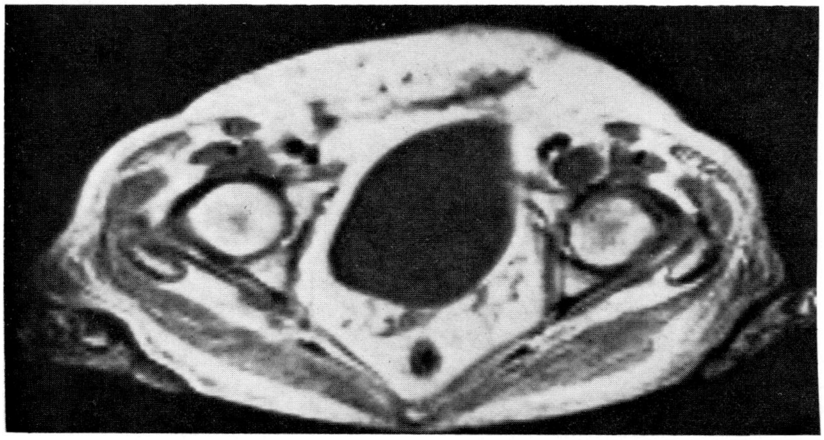

Figure 9. Subtle avascular necrosis of the left femoral head.

A. Slight joint-space narrowing, cortical irregularity, and subchondral sclerosis are present in the superior aspect of the left femoral head.

B. Axial spin echo 1000/28 image shows loss of sharp cortical definition, decreased signal intensity, and accentuated inhomogeneity in the superior aspect of the left femoral head. The right hip is normal.

C. Axial spin echo 1000/28 image centered 12 mm caudal to B. The central portion of the left femoral head appears less involved than the superior aspect seen in B.

(Reproduced, with permission, from Moon KL Jr, Genant HK, Helms CA, et al: Musculoskeletal applications of nuclear magnetic resonance. Radiology 147:161, 1983.)

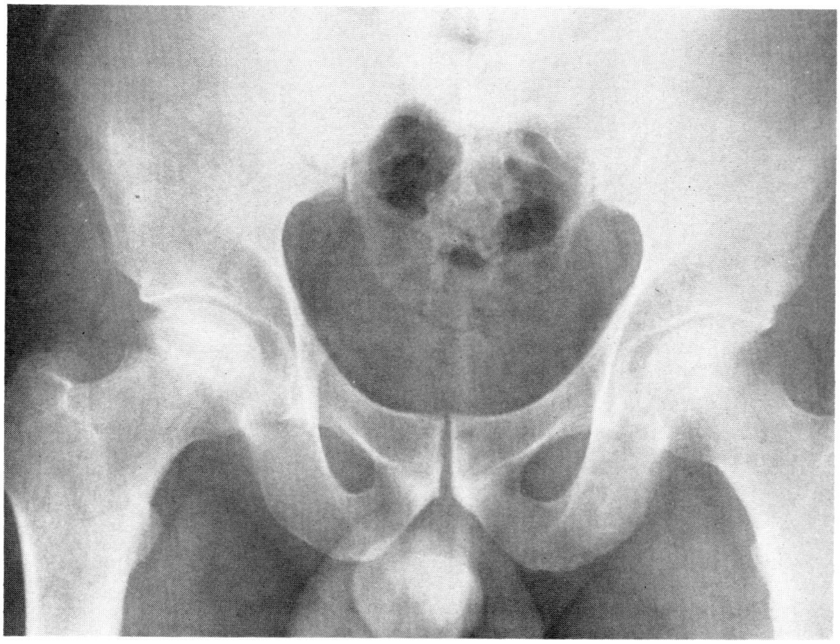

A.

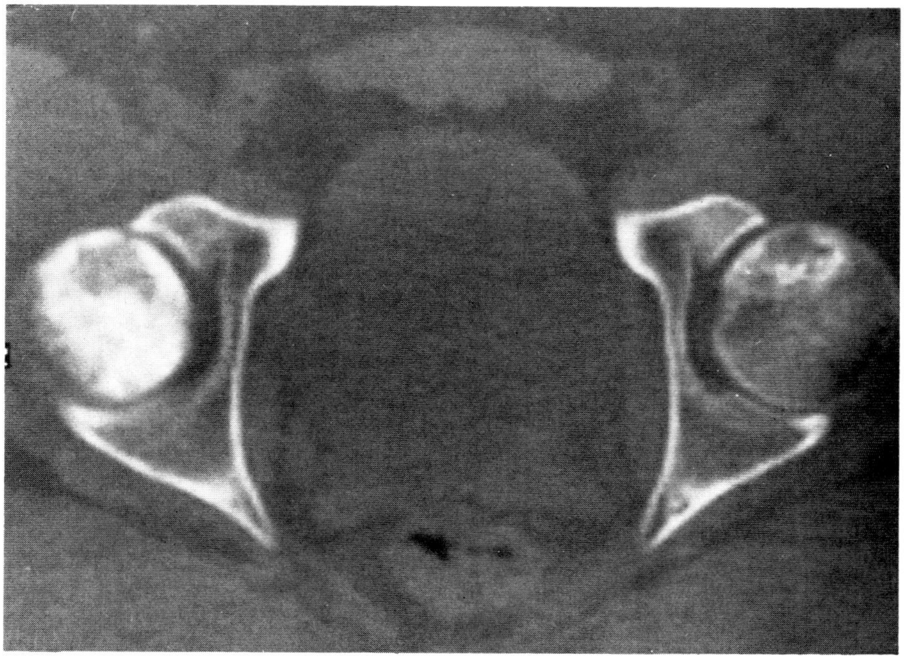

B.

Figure 10. Avascular necrosis of the femoral heads.
A. Conventional radiograph shows cortical irregularity and subchondral sclerosis in the superior aspect of the right femoral head.
B. Axial CT image shows evidence of bilateral involvement.

(continued on next page)

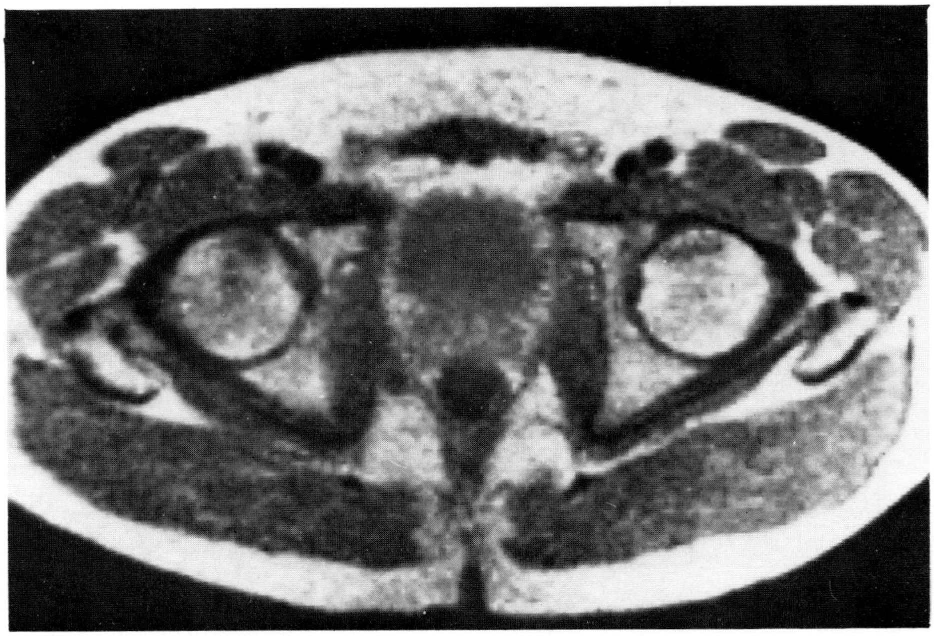

C.

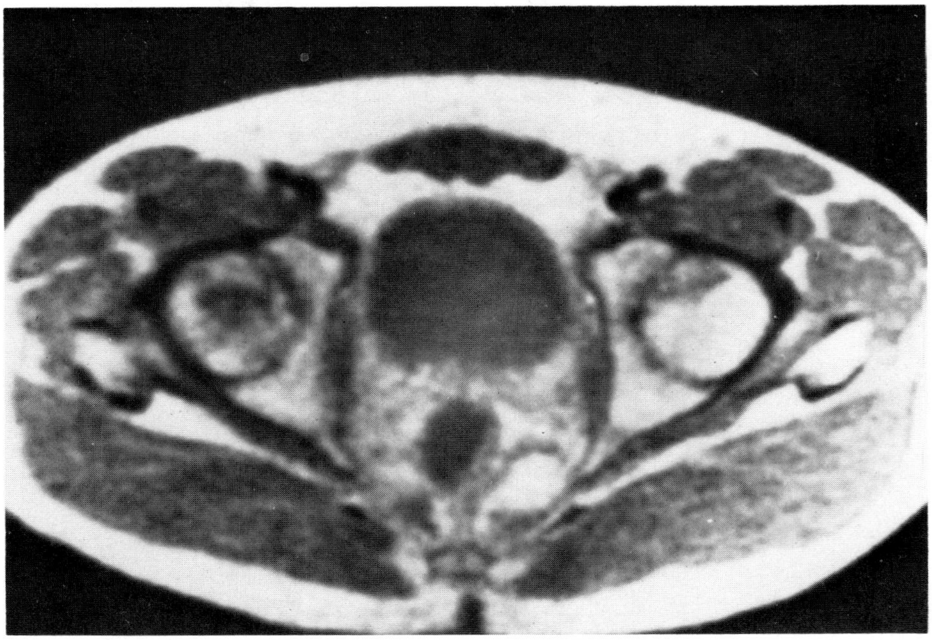

D.

C and D. Corresponding axial spin echo 1000/28 image shows decreased signal intensity and accentuated inhomogeneity in the superior aspect of both femoral heads.

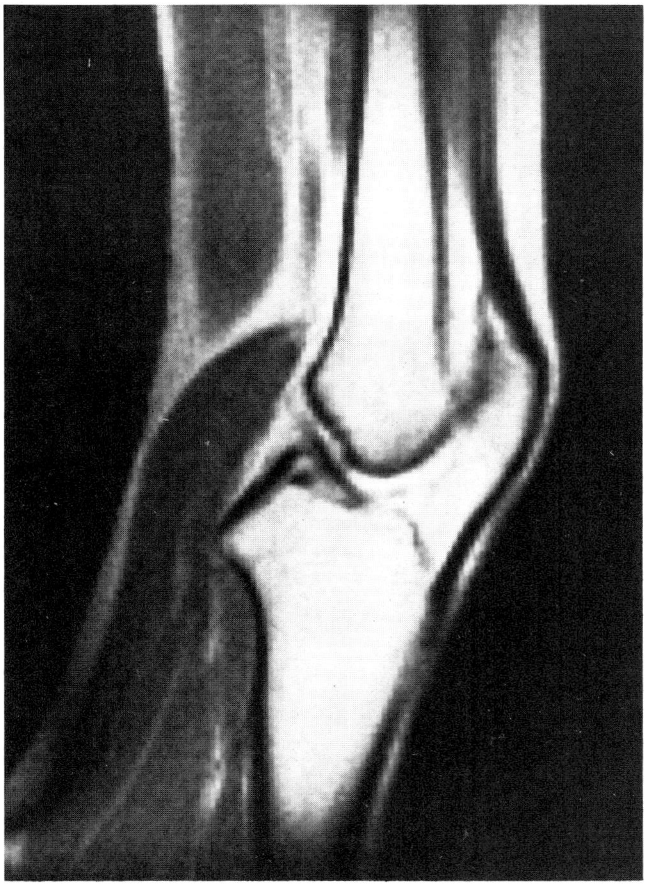

A.

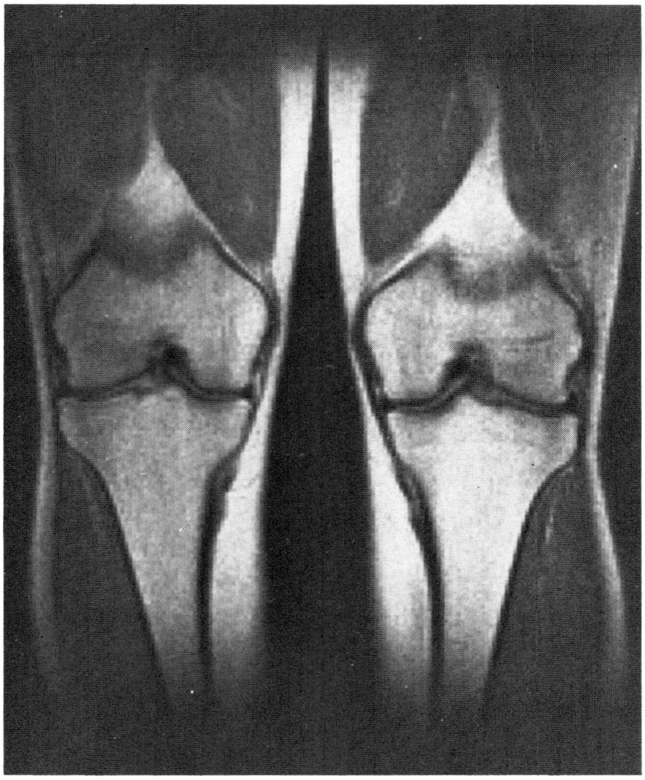

B.

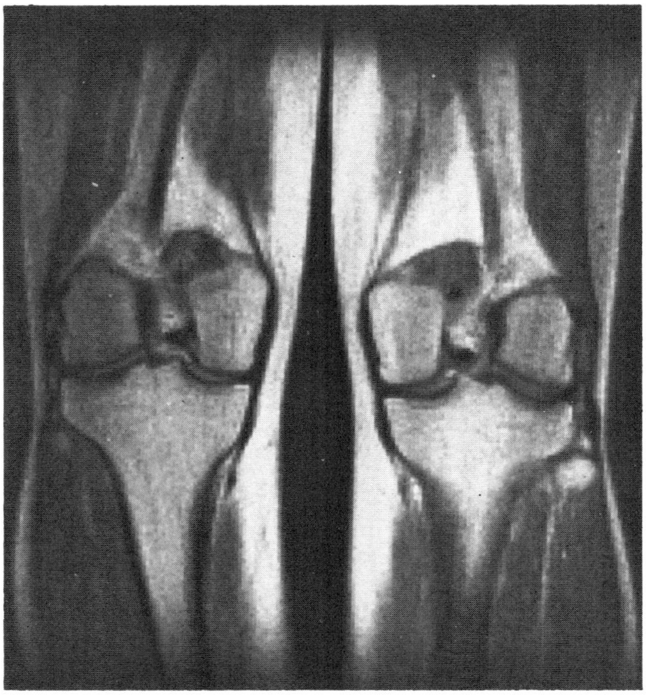

C.

Figure 11.

A. Sagittal spin echo 2000/28 image of the knee of a normal volunteer. The quadriceps tendon and patellar, anterior, and posterior cruciate ligaments are well defined.

B and C. Coronal sections through the femoral condyles. The meniscal cartilages are imaged as intermediate-intensity structures within the joint space. The popliteal arteries are visible as dark tubular structures above the condyles.

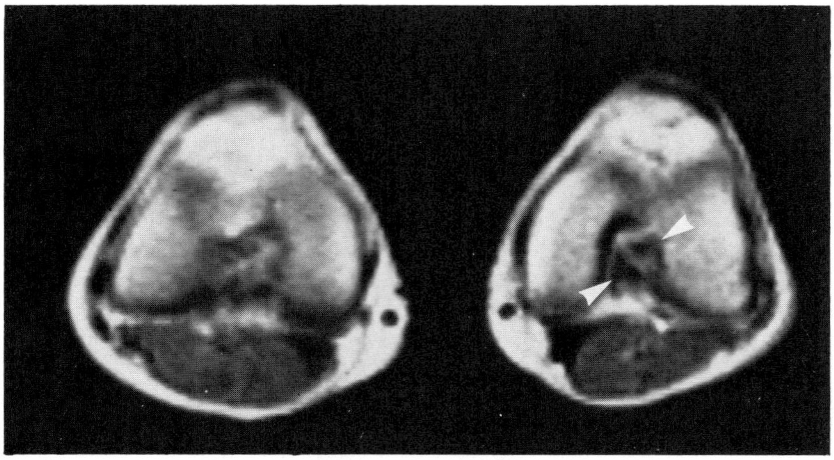

A.

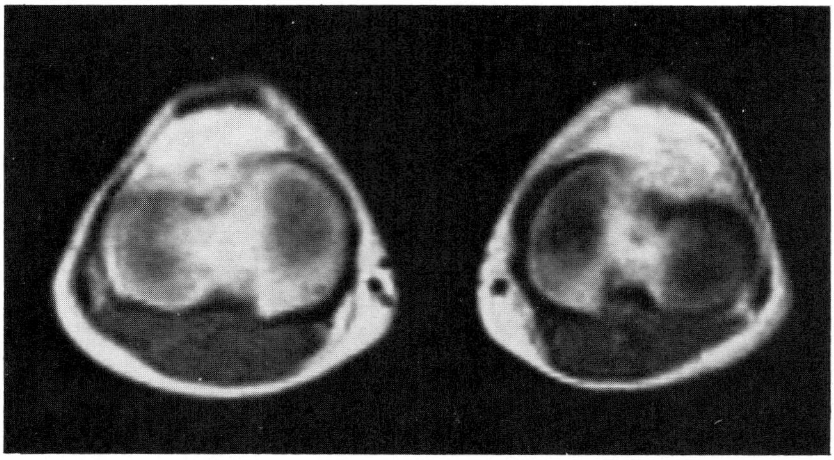

B.

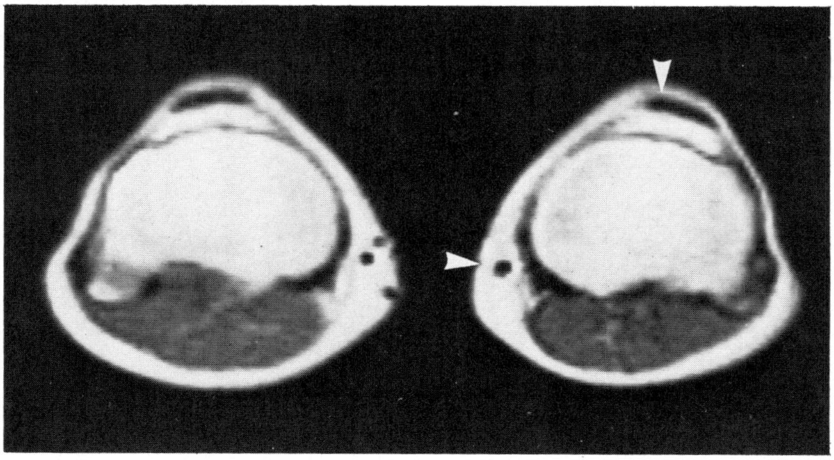

C.

Figure 12.

A. Axial spin echo 1000/28 image of the same knee as in Figure 11 at the level of the femoral condyles. The cruciate ligaments are visible in the intercondylar notch of the left knee (arrowheads).

B. Axial section at the level of the meniscal cartilages showing indistinct visualization of these cartilages.

C. Axial section through tibial plateau. The patellar ligaments (vertical arrowhead) and saphenous veins (horizontal arrowhead) are well seen.

(Reproduced, with permission, from Moon KL Jr, Genant HK, Helms CA, et al: Musculoskeletal applications of nuclear magnetic resonance. Radiology 147:161, 1983.)

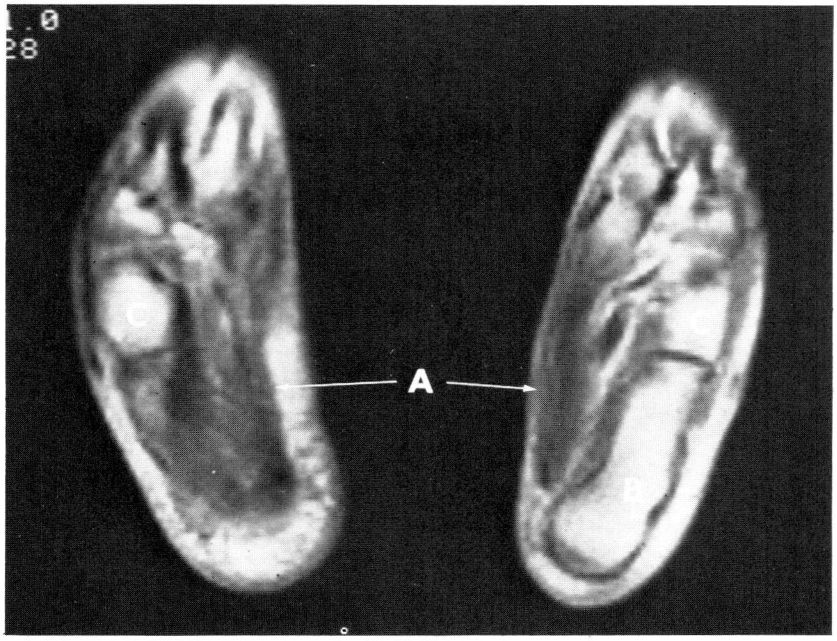

A.

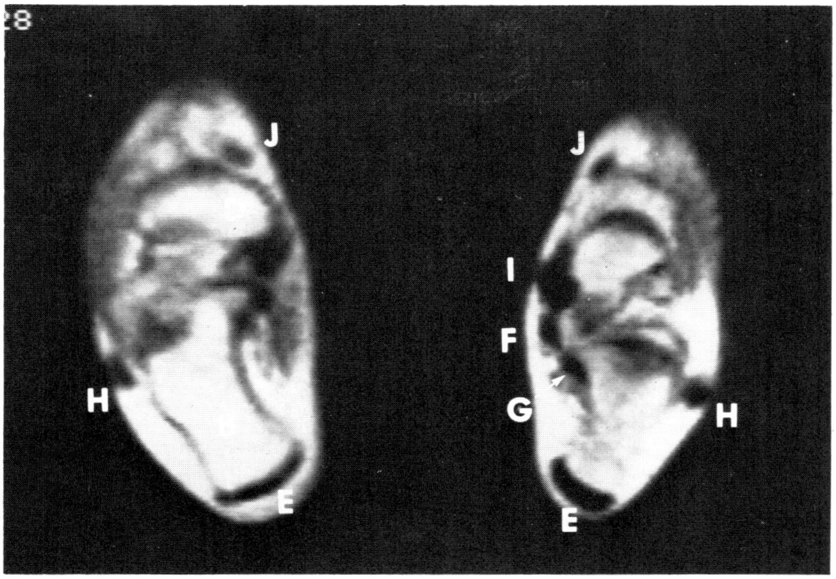

B.

Figure 13.

A–C. Serial axial spin echo 1000/28 images (Figs. A and C) and 500/28 image (Fig. B) of a normal foot showing plantar fascia and muscles (A), calcaneus (B), tarsal cuboid (C), tarsal navicular (D), Achilles tendon (E), peroneus tendons (F), tendon of the flexor hallucis longus (G), tendons of the posterior tibial muscle and long flexors of the toes (H), deltoid ligament (I), and tendon of the anterior tibial muscle (J).

(Reproduced, with permission, from Moon KL Jr, Genant HK, Helms CA, et al: Musculoskeletal applications of nuclear magnetic resonance. Radiology 147:161, 1983.)

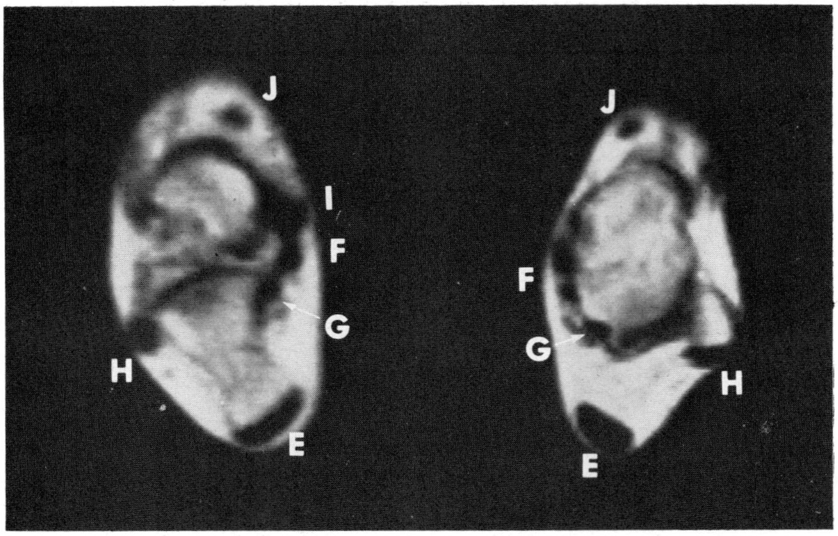

C.

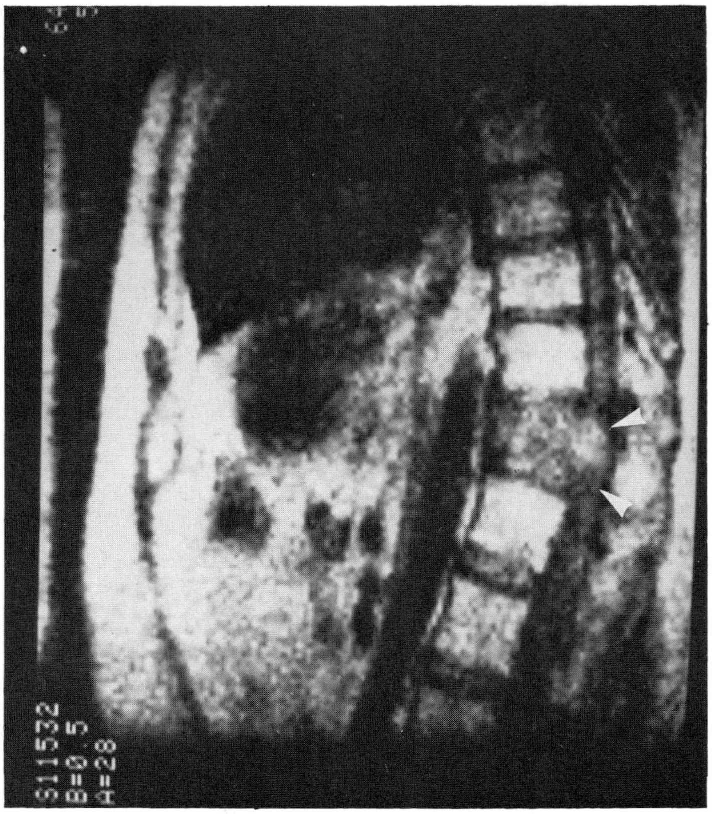

Figure 14. A spin echo 500/28 image of a metastasis to the T11 vertebral body. The cortices of the body have been destroyed. The tumor mass extends posteriorly into the spinal canal and compresses the spinal cord (arrows). The medullary cavity of the involved vertebra has a lower signal intensity than that of adjacent normal vertebrae.

(Courtesy of Dr. Michael Brant-Zawadski. Reproduced, with permission, from Moon KL Jr, Genant HK, Helms CA, et al: Musculoskeletal applications of nuclear magnetic resonance. Radiology 147:161, 1983.)

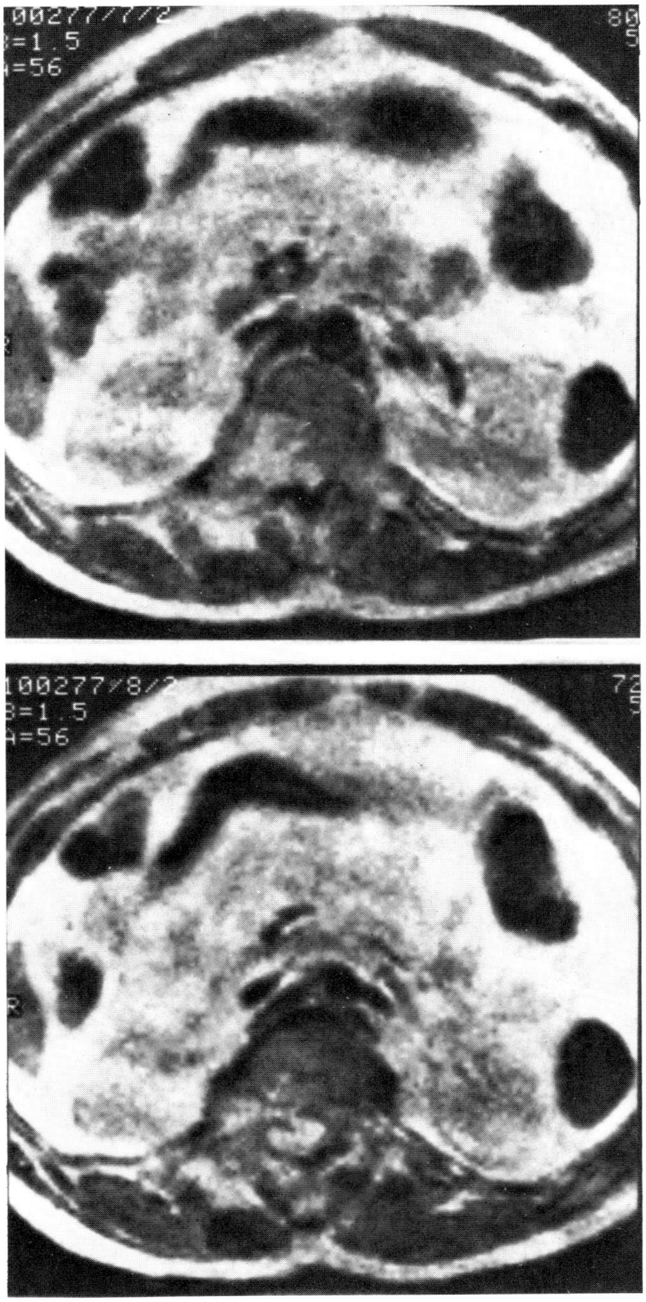

Figure 15. Two adjacent spin echo 1500/56 images of a chondrosarcoma involving the L2 vertebral body. The intensity of the tumor is higher than that of adjacent vertebral marrow.

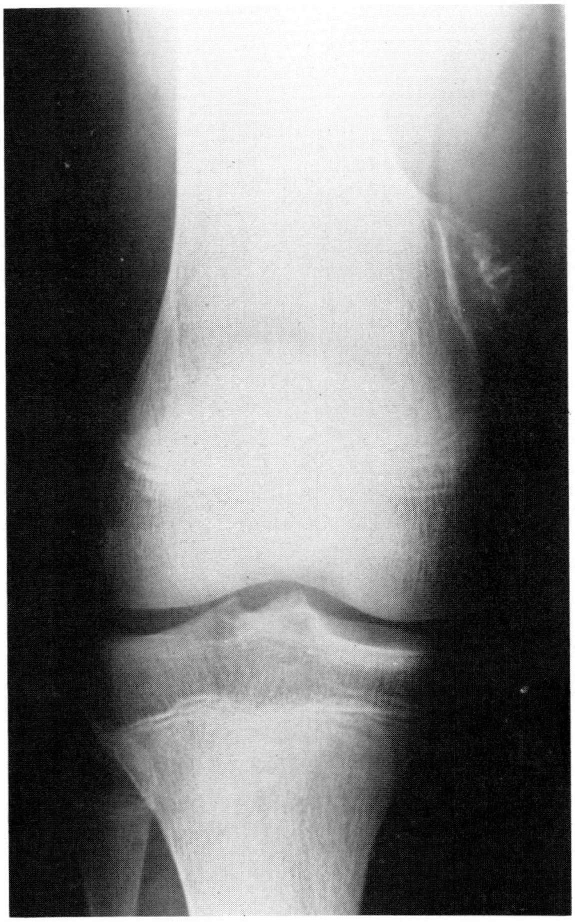

A.

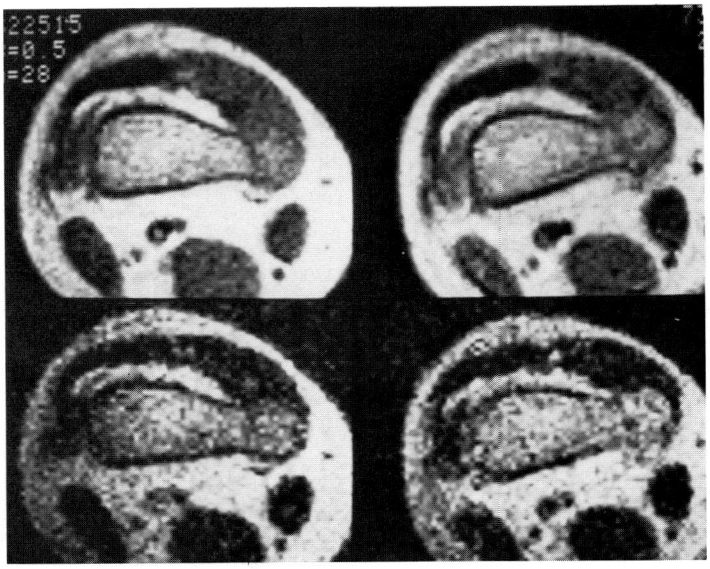

B.

Figure 16. Osteochondroma of the distal femur.
A. Conventional radiograph.
B. Four spin echo intensity images. Top, left to right: 500/28, 1000/28. Bottom, left to right: 500/56, 1000/56. The head of the osteochondroma is difficult to distinguish from adjacent quadriceps muscle on the 500/28 image. This distinction is straightforward on the other three intensity images, reflecting the different relaxation rates of the osteochondroma and adjacent muscle. The quadriceps tendon can be clearly distinguished from quadriceps muscle because of lower signal intensity relative to muscle.
(Reproduced, with permission, from Moon KL Jr, Genant HK, Helms CA, et al: Musculoskeletal applications of nuclear magnetic resonance. Radiology 147:161, 1983.)

Soft-Tissue Tumors

One patient with a large cavernous hemangioma of the abdominal wall (Fig. 17) and one patient with a recurrent liposarcoma of the distal thigh (Fig. 18) were examined. Anatomic depiction of the lesions was similar on both CT and NMR.

Other Conditions

A variety of other normal anatomic sites (neck, shoulder, and upper extremities) were examined and gave results similar to those described above.

One patient with a benign sacral cyst was examined by both NMR and CT (Fig. 19). Only axial CT images were available on this patient, and only sagittal NMR images were obtained, but comparison of the two modalities was informative. The spatial resolution of CT was significantly better than that of NMR; however, the direct sagittal images provided by NMR were helpful in defining anatomic relationships. Additionally, NMR allowed differentiation between the two principal diagnostic possibilities, a fluid-filled cyst (dark appearance) and a lipoma (bright appearance).

One patient with thalassemia was examined. Of relevance to the musculoskeletal system, the vertebral marrow in this patient was of very low intensity (i.e., it produced virtually no NMR signal) on all spin echo images (Fig. 20). This is in marked contrast

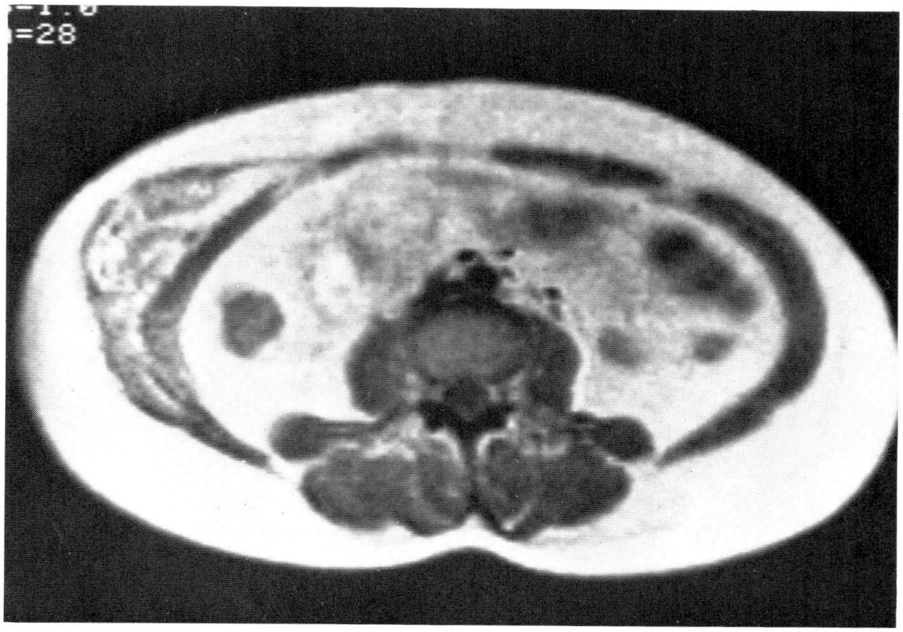

A.

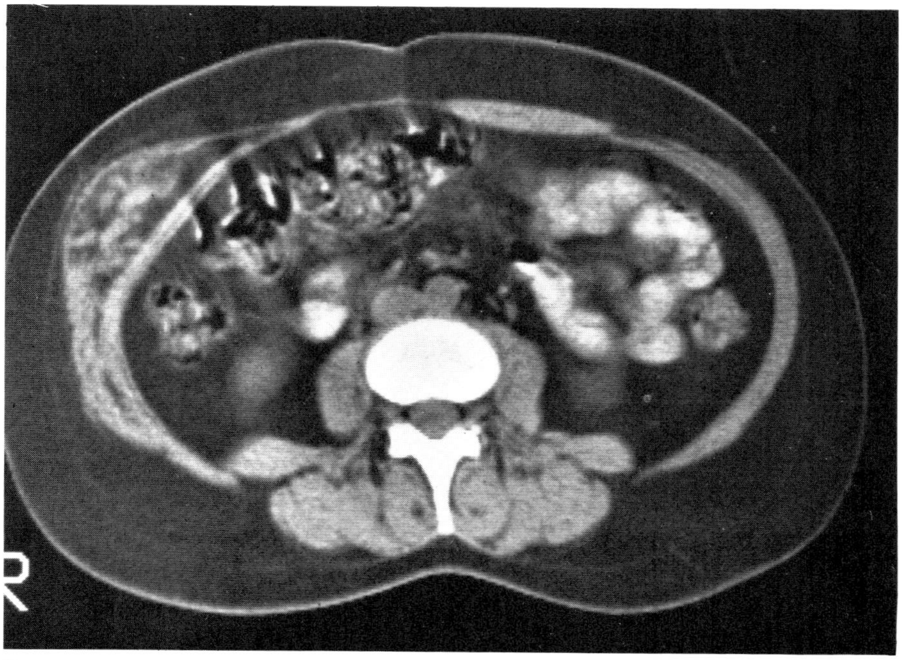

B.

Figure 17. Cavernous hemangioma of the abdominal wall.

A. Axial spin echo 1000/28 image through the hemangioma. The relatively high signal intensity within the malformed vessels (compare the aorta and mesenteric vessels) suggests that blood flow is minimal or absent. The blurred appearance of the intestine is typical of NMR images and is due to peristaltic motion occurring during the 8.5-min data acquisition period used for this image.

B. Axial CT image through the lesion.

(Reproduced, with permission, from Moon KL Jr, Genant HK, Helms CA, et al: Musculoskeletal applications of nuclear magnetic resonance. Radiology 147:161, 1983.)

A.

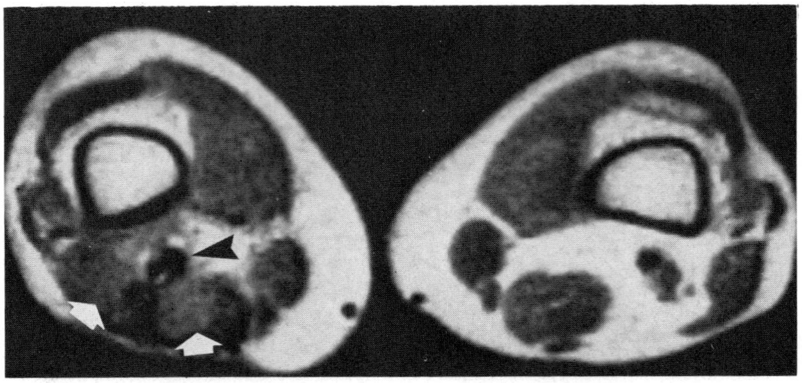

B.

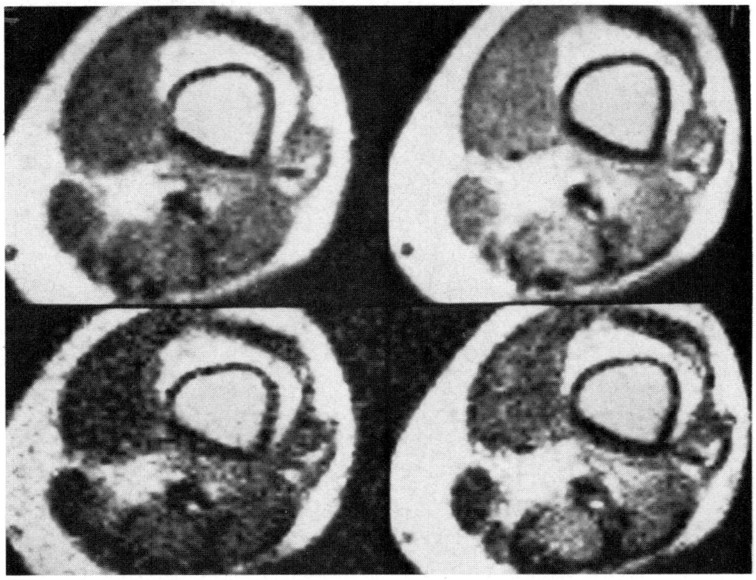

C.

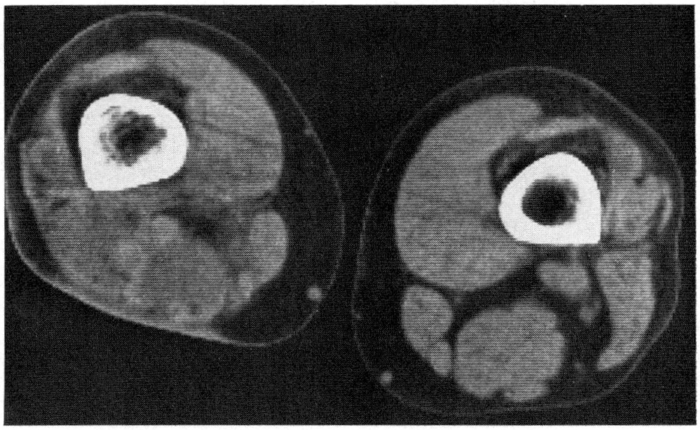

Figure 18. Recurrent liposarcoma of distal thigh.

A. Spin echo image through the tumor (arrows). Postoperative fibrosis is apparent as very dark tissue adjacent to the tumor. The popliteal vessels can be distinguished as dark round structures deep to the tumor (arrowhead).

B. Four spin echo intensity images of the tumor. From left to right on the top and on the bottom: 500/28, 1000/28, 500/56, 1000/56. Evaluation of all four images permits discrimination of tumor from fibrosis, blood vessels, fat, and adjacent muscle.

C. A CT image at a level similar to the image in A.

(Reproduced, with permission, from Moon KL Jr, Genant HK, Helms CA, et al: Musculoskeletal applications of nuclear magnetic resonance. Radiology 147:161, 1983.)

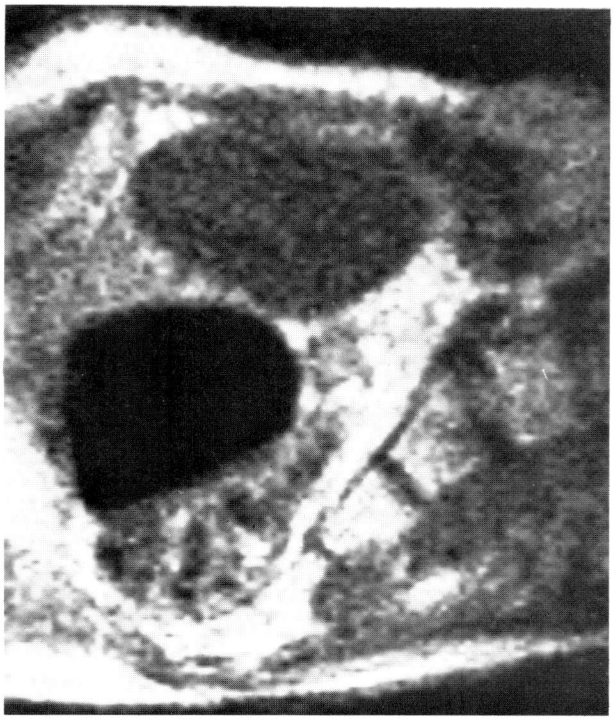

A.

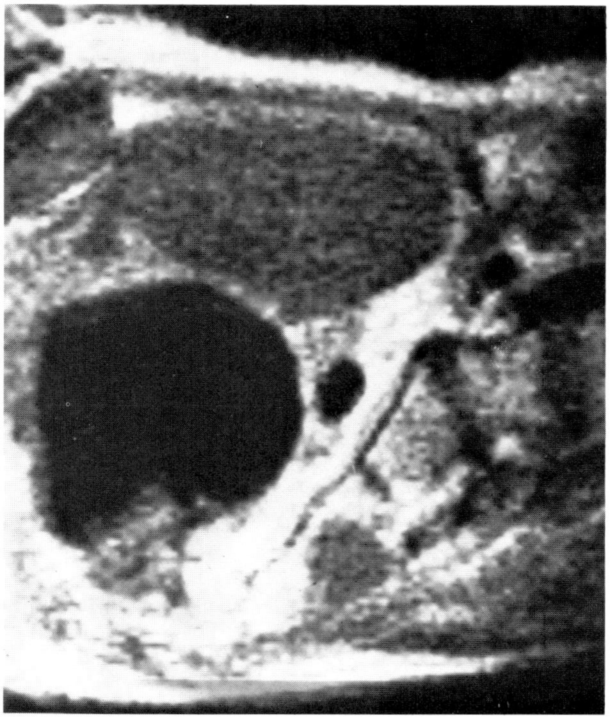

B.

Figure 19. Benign sacral cyst.
A and B. Spin echo 1500/28 sagittal images demonstrating low-intensity defect in sacrum.
C. Axial CT scan demonstrating cystic defect in sacrum adjacent to the spinal canal.
(Courtesy of Dr. David Spring. Reproduced, with permission, from Moon KL Jr, Genant HK, Helms CA, et al: Musculoskeletal applications of nuclear magnetic resonance. Radiology 147:161, 1983.)

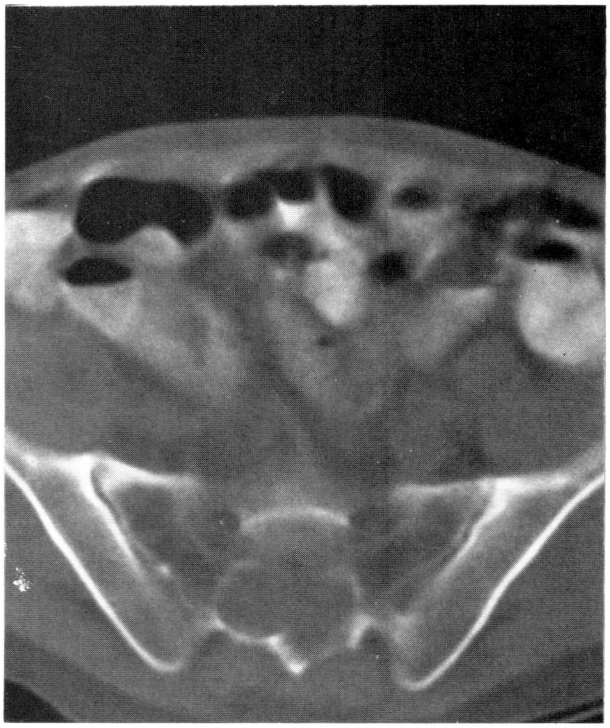

C.

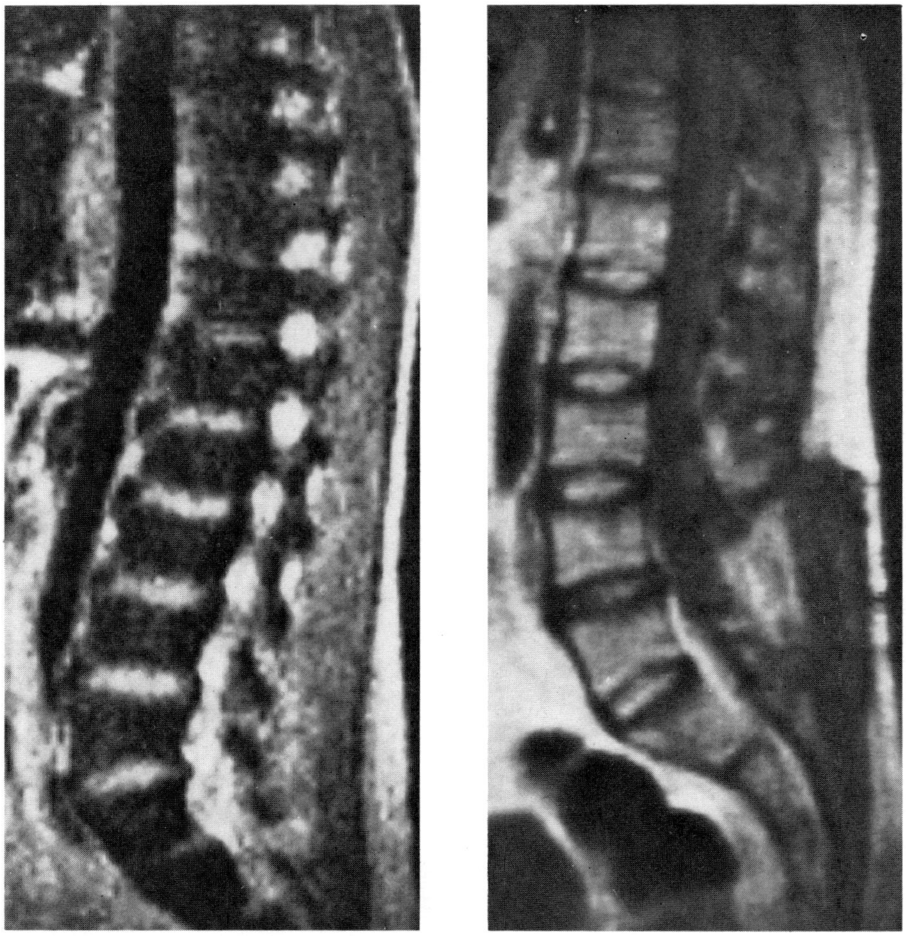

A.　　B.

Figure 20.
A and B. Beta-thalassemia with presumed excessive marrow from deposition. Sagittal spin echo 1000/28 image of thoracolumbar spine (A) showing dramatic loss of the normal high-intensity signal from the vertebral medullary cavities. Normal spine (B) for comparison.
(Courtesy of Drs. Robert Brasch and Charles Gooding.)

to the normal high-intensity signal seen in vertebral marrow in normal patients. Marrow biopsy demonstrated hemosiderosis, and the low signal intensity from marrow (and liver) was felt to reflect marked shortening of the T_1 and T_2 relaxation times of the tissue that resulted from the paramagnetic effects of the iron deposits.[11]

General Considerations

Our experience with NMR imaging in musculoskeletal disorders is still somewhat limited,[9,10] but it is sufficient to allow us to draw preliminary conclusions about the strengths and weaknesses of the technique in the several musculoskeletal applications that we have studied.

A general feature of NMR imaging that makes it well suited for a number of musculoskeletal applications is its superior soft-tissue contrast as compared with that of CT. Fat, muscle, fibrous structures (such as ligaments and tendons), nerves, and blood vessels have different NMR imaging characteristics (see Table III), whereas those same tissues (with the exception of fat) all have very similar CT attenuation values and are principally differentiated from each other on the basis of their anatomic appearances.

Nuclear magnetic resonance has demonstrated its ability to provide accurate definition of normal and pathologic anatomy. While NMR imaging devices are now competitive with CT scanners in spatial resolution (0.8-mm resolution for the University of California, San Francisco NMR imager in high-resolution mode, versus 0.5-mm resolution for current-generation CT scanners), imaging time is quite long (greater than 30 min) when the device is operated in this mode, and this option is usually selected only in special circumstances. When the standard imaging mode (1.7–2.5 mm) is used, imaging time is competitive with CT, but spatial resolution is inferior. Although standard-mode NMR imaging has a lower spatial resolution than that of CT, this difference becomes significant only in those applications requiring very high spatial resolution (such as in examination of the lumbosacral spine). In all other musculoskeletal applications studied to date, the difference does not appear to be sufficient to *impair* diagnostic utility.

Early experience in NMR examination of the lumbosacral spine has been promising. Despite its somewhat lower spatial resolution, NMR still demonstrates the important anatomic features in the region. Operated in the standard mode, it appears less sensitive than CT in demonstrating subtle abnormalities. On the other hand, its ability to create high-resolution images in the sagittal plane appears to be a significant advantage over CT. The ability of NMR to visualize the spinal cord without the need for intrathecal contrast material represents an exciting and important advance.

Intervertebral disc disease seems to be evaluated in a new fashion by NMR. Nuclear magnetic resonance usually differentiates nucleus pulposus from annulus fibrosus, while CT cannot make this distinction. In most patients with disc disease, the involved discs have demonstrated a loss of the normal high-intensity signal from the nucleus pulposus. It seems likely that NMR might be able to demonstrate abnormalities in disc material even in the absence of anatomic derangement (e.g., herniation). Differentiation between nuclear herniation and diffuse anular bulging also seems a reasonable possibility.

Since NMR differentiates dense fibrous tissue (such as scar tissue, ligament, and tendon) from other tissues, it may prove useful in the evaluation of the patient with previous back surgery. Computed tomography is frequently not capable of differentiating postoperative fibrosis from recurrent disc herniation because of the similar x-ray attenuation values of these tissues. While we have no direct experience with this problem to date, it is possible that NMR may be able to differentiate these tissues. We look forward to studying this problem in the near future.

Our preliminary experience with NMR in the evaluation of avascular necrosis of the femoral head suggests that NMR may be a useful means of making an early diagnosis in this condition. In two of the patients whom we have studied, rather striking abnormalities were present on the NMR images when conventional radiographs displayed minimal abnormalities. While early diagnosis of this condition can be made in some cases by using radionuclide bone scanning, NMR may be able to provide the same information with more precise anatomic localization and without exposing the patient to ionizing radiation. Further experience is required to determine the sensitivity and specificity of the technique in this application.

In the knee, the clear definition of the cruciate ligaments on axial and sagittal NMR images indicates that NMR may be useful as a noninvasive means of evaluating patients with suspected cruciate ligament disruption. Meniscal cartilage is less likely to be imaged successfully because of the difficulty in differentiating meniscal cartilage from articular cartilage and subchondral cortical bone. Nonetheless, optimal NMR examination techniques for visualization of cartilage have not been fully studied, and a better assessment of the ability of NMR to identify meniscal abnormalities must await further experience.

Our experience with NMR imaging of bone tumors is limited, but one characteristic of the technique seems likely to be of great utility. The superior soft-tissue contrast of NMR suggests that it will be a very sensitive technique for defining the extent of tumor in the medullary cavity. A recent report documents the ability of the technique to define intramedullary extension of giant cell tumors of bone.[7] Our experience with several patients with thalassemia also demonstrates the sensitivity of the NMR imaging technique to changes in the composition of the bone marrow. It should also prove sensitive in demonstrating encroachment of tumor upon important adjacent anatomic structures such as blood vessels. It is not known at present whether the NMR characteristics of different tumors are sufficiently distinct to allow specific tumor characterization.

Soft-tissue structures in the extremities are demonstrated in a superior fashion on NMR. Low-intensity muscle bundles are outlined by high-intensity fat. Blood vessels are clearly visible as black structures (the low signal intensity being due to the flow of blood within the vessels).[12] Slow or absent flow of blood is apparent as an increased signal from the involved vessels. This feature of NMR makes it a very promising modality for assessment of a variety of arterial and venous disorders, both peripherally and centrally. Soft-tissue tumors are shown to excellent advantage.

REFERENCES

1. Crooks L, Arakawa M, Hoenninger J, et al: Nuclear magnetic resonance whole-body imager operating at 3.5 kGauss. Radiology 143:169, 1982.
2. Alfidi RF, Haaga JR, El Yousef SF, et al: Preliminary experimental results in humans and animals with a superconducting, whole-body, nuclear magnetic resonance scanner. Radiology 143:175, 1982.
3. Young IR, Bailes DR, Burl M, et al: Initial clinical evaluation of a whole-body nuclear magnetic resonance (NMR) tomograph. J Comput Assist Tomogr 6:1, 1982.
4. Crooks LE, Mills CM, Davis PL, et al: Visualization of cerebral and vascular abnormalities by NMR imaging: The effects of imaging parameters on contrast. Radiology 144:843, 1982.
5. Kaufman L, Crooks L, Margulis AR (Eds): *Nuclear Magnetic Resonance Imaging in Medicine*. New York: Igaku-Shoin, 1981.
6. Smith FW: Clinical application of NMR imaging. In *NMR Imaging: Proceedings of an International Symposium on Nuclear Magnetic Resonance Imaging*, Witcofski RL, Karstaedt N, Partain CL (Eds). Winston-Salem: Bowman Gray School of Medicine, 1982.
7. Brady TJ, Gebhart MC, Pykett IL, et al: NMR imaging of forearms in healthy volunteers and patients with giant-cell tumor of bone. Radiology 144:549, 1982.
8. Newman RJ, Bore PJ, Chan L, et al: Nuclear magnetic resonance studies of forearm muscle in Duchenne dystrophy. Br Med J 284:1072, 1982.
9. Moon KL, Genant HK, Helms CA, et al: Musculoskeletal applications of nuclear magnetic resonance imaging. Radiology 147:161, 1983.
10. Moon KL, Genant HK, Davis PL, et al: Invited review: Nuclear magnetic resonance imaging in orthopaedics. Principles and applications. Orthop Res (in press).
11. Brasch RC, Gooding CA, Wesbey GE, et al: NMR imaging of iron deposition in children with thalassemia. Presented at the 68th Scientific Assembly of the Radiological Society of North America, Chicago, Nov. 28–Dec. 3, 1982.
12. Mills CM, Brant-Zawadzki M, Brooks LE, et al: Nuclear magnetic resonance imaging: Effect of blood flow. AJNR (in press).

17

Pediatric Applications

ROBERT C. BRASCH, M.D.
CHARLES A. GOODING, M.D.

Recent reports describing NMR efficacy for the diagnosis of human diseases are strongly favorable; however, these early reports on clinical NMR imaging have emphasized adult patients and adult diseases.[1–5] Yet this new imaging technique has particular appeal for the diagnostic evaluation of children, the lack of ionizing radiation being a most desirable feature. Further, NMR may have unmatched value for defining normal and abnormal tissue development in the early years of life. For example, NMR has the potential to demonstrate stages of myelinization within the brain of developing infants.[6]

This chapter describes our initial experience with NMR imaging in children. This experience is, as yet, too small to make conclusions about the ultimate clinical utility of pediatric NMR imaging. Rather, our aim is to indicate certain potential applications in children for which NMR may have unique diagnostic abilities. In virtually all cases, x-ray CT studies were available for correlation to the NMR imaging data, thereby permitting certain advantages and disadvantages of NMR imaging as compared with an advanced, computed-assisted radiologic technique to be assessed.

Techniques for Pediatric Magnetic Resonance Imaging

Our patients ranged in age from 4 days to 18 years. For all pediatric subjects, signed informed consent was obtained from parents and, when appropriate, from the patients themselves. The methods of immobilization and sedation were similar to those employed for pediatric CT examinations. Specifically, infants were restrained by means of blankets and Ace bandages upon a wooden board; no sedation was used. One small infant was not restrained and moved sufficiently to produce blurred, nondiagnostic images. Children 1 to 6 years old generally required sedation, which was given as intramuscular pentobarbital 5 mg/kg 30 min before the NMR examination. An alternative sedation regimen is intravenous diazepam 0.2–0.4 mg/kg, given slowly 10 min before imaging. Children over 6 years old required no sedation; cooperation was elicited by words of explanation and assurance.

Patients with internal metallic devices, including recently placed surgical clips, intravenous catheters, and pacemaker devices were excluded. Patients were dressed in special gowns without metallic snaps; all jewelry was removed prior to NMR examination.

After seeing the NMR imaging unit, no child refused the examination, and generally there was little or no apparent anxiety. Some patients expressed annoyance with the continuous noise that occurs during image acquisition.

The NMR imager is an 85-cm bore, 3.5-kGauss (0.35-T) superconducting magnet that operates at 15-MHz hydrogen resonance frequency.[7] A special 25-cm diameter radio-frequency coil was used for infant imaging. A 50-cm coil was employed for larger patients.

For all patients, at least four separate spin echo intensity images were obtained by varying the pulse interval (TR) between 0.5 and 2.0 sec and the echo delay after the 90° pulse (TE) between 28 and 56 msec. Images were variably obtained in axial, sagittal, and coronal planes. In most instances, 15 to 45 sections, each 7 mm thick, were obtained for each patient. Total time in the scanner ranged from 45 to 60 min; each 15-section data acquisition period was approximately 17 min long.

No adverse reaction to the entire NMR procedure was observed in these children.

Magnetic Resonance Imaging of Pediatric Neurologic Diseases

Nuclear magnetic resonance was able to demonstrate certain abnormalities within the brain that were not revealed by CT scanning. These included vascular malformations, subarachnoid collections of blood, and intracerebral petechial hemorrhages.

A 13-year-old girl was evaluated for a mild left-sided neurosensory hearing loss of 1 year's duration. This patient had multiple hemangioendotheliomas in early childhood involving the skin, mucous membranes, liver, and bone; numerous tissue specimens confirmed this diagnosis. In later childhood, during a 7-year course of steroid therapy, many of the hemangioendotheliomas disappeared, yet a central nervous system hemangioma was suspected as the cause of her recent hearing deficit. Computed tomography of the brain failed to show any lesions near the VIIIth nerve. Relatively mild streak artifacts in the posterior fossa marred the CT images. Small calcified foci in the supratentorial region of the brain identified on CT images were presumed to be old, healed vascular lesions.

In the same patient NMR spin echo images through the posterior fossa (Fig. 1) defined a lesion with approximately a 0.5-cm diameter, located in the region of the lateral cerebellar hemisphere near the internal auditory meatus. This lesion on first-echo images (TE of 28 msec) demonstrated a low intensity, but the same lesion on second-echo images (TE of 56 msec) had high intensity. Such a shift in NMR appearance between the first and second echo images is characteristic of slowly moving blood. On spin echo images obtained with a long echo delay time (TE of 56 msec), the signal produced by blood flowing at low velocities (1 to 5 cm/sec) may be even greater than that produced by stationary blood, and then decreases linearly with increasing velocity, reaching zero signal level at approximately 15 cm/sec (see Chaps. 7, 12). In this case, NMR imaging data demonstrated not only the lesion, but also revealed its vascular nature, strongly supporting the suspected diagnosis of hemangioendothelioma.

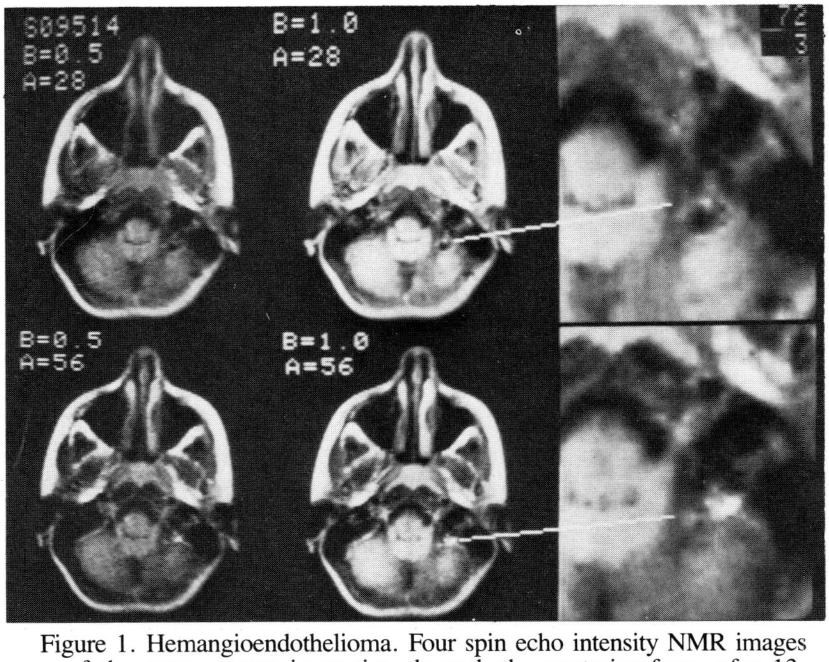

Figure 1. Hemangioendothelioma. Four spin echo intensity NMR images of the same anatomic section through the posterior fossa of a 13-year-old girl with cutaneous hemangioendotheliomas demonstrate an abnormal focus in the left cerebellar hemisphere near the internal auditory meatus. Detail of the abnormal region is better appreciated on photographic enlargements on the right side of the figure. Rapidly flowing blood yields no signal, as evidenced by the internal carotid artery. Slowly flowing blood yields no signal on the first-echo images (TE of 28 msec), but yields a strong signal on second-echo images (TE of 56 msec). Thus the abnormal focus contains slowly moving blood, compatible with cerebral hemangioendothelioma. This lesion, identified by NMR without contrast medium, was not evident on CT images.

This case demonstrated the potential of NMR imaging to define small intracerebral lesions and to characterize blood flow without contrast medium enhancement. The vascular lesion was not defined on the CT scan, probably because of streak artifacts from the nearby bone.[8] Such bone-related streak artifacts are not encountered on NMR images. Further, the formation of the NMR image uses a two-dimensional Fourier transform rather than the back-projection algorithms employed with CT that are prone to produce beam-hardening and streak artifacts at high-contrast interfaces.

Although angiography has been considered the "gold standard" for imaging vascular lesions, angiography may on occasion not demonstrate an arteriovenous malformation (AVM). Such a lesion was well demonstrated by NMR. A 6-year-old boy presented after a single focal seizure in the left arm. At age 1 year the patient had a small AVM surgically resected from the right temporal cortex and had remained asymptomatic for 5 years prior to the focal seizure. Computed tomographic images obtained before and after contrast medium administration demonstrated enhancing lesions in the right parietal and left temporal lobes without mass effect or calcification. Bilateral internal carotid arteriograms failed to reveal any vascular abnormality, a behavior of certain AVMs previously termed "cryptic."

Nuclear magnetic resonance images obtained in transverse and sagittal planes (Fig. 2) demonstrated the same two abnormal foci in the brain as shown by CT. Additionally, NMR images demonstrated both low-intensity and high-intensity areas within each lesion. The lesions were diagnosed as AVMs, with the high-intensity portions thought to represent clotted blood, and the low-intensity portions thought to be rapidly flowing blood. These features on NMR images were confirmed surgically. This case shows the potential of NMR imaging for differentiating static blood elements within a clot from flowing blood.

Of course, the search for tumors in children and their characterization are important potential applications of NMR imaging. Our small experience indicates that inherent contrast is high between pediatric brain tumors and surrounding normal cerebral tissue. A 2-year-old boy presented with signs of cerebellar dysfunction, including ataxia and nystagmus. A CT examination performed without and with contrast enhancement showed an isodense midline posterior fossa mass that differed in intensity from surrounding normal tissues only after contrast administration. Also noted was marked dilatation of the ventricles.

The NMR images demonstrated the same midline cerebellar tumor, a proven medulloblastoma, on multiple spin echo sequence images. No extrinsic contrast medium enhancement was necessary (Fig. 3). The NMR images also showed ventricular dilatation and a zone of decreased intensity in the periventricular white matter. This case illustrated the ability of NMR to clearly depict lesions that may be isodense on CT and that require radiographic contrast enhancement for their CT definition. Again, no posterior fossa streak artifacts were evident on the NMR images.

For delineation of the spinal cord, subarachnoid injection of contrast medium may be avoided in certain patients by using NMR. Sagittal images of the spine have demonstrated the spinal cord since it has a higher NMR intensity than the normally low intensity of cerebrospinal fluid. A 7-year-old girl with a 20-month history of dysarthria, ataxia, and other signs of pancerebellar dysfunction was treated empirically with prednisone for a clinical suspicion of multiple sclerosis. Computed tomographic images had suggested atrophy of the cerebellum with an exaggerated sulcal pattern. An NMR examination was performed because of its reported sensitivity for the plaques of multiple sclerosis.[9] The NMR images revealed no periventricular plaques, but did confirm marked cerebellar atrophy. The sagittal NMR images (Fig. 4) also demonstrated narrowing of the cervical spinal cord, findings indicative of spinocerebellar atrophy. This case illustrates the value of direct sagittal NMR imaging and the potential of NMR to eliminate contrast myelography in the evaluation of the spinal cord.

A disadvantage of NMR for the delineation of various diseases is the lack of sensitivity for calcium. A 17-year-old girl presented with tremor, loss of intellectual function, and dysarthria. Several family members had previously been institutionalized for "cerebral-muscular" failure. Physical examination and chemical examination of the patient's serum revealed classic signs of Wilson's disease, including Kayser-Fleischer rings and elevated serum copper. Computed tomographic images of the brain showed decreased density in the basal ganglia, a common finding in Wilson's disease, and punctate calcifications in the thalamus, an unusual finding. Predictably, these calcifications were not evident on NMR images. This case demonstrates the lack of NMR sensitivity to high-atomic-number, nonparamagnetic substances such as calcium and barium. Yet NMR may still be sensitive for Wilson's disease by demonstrating abnormally low spin echo intensity

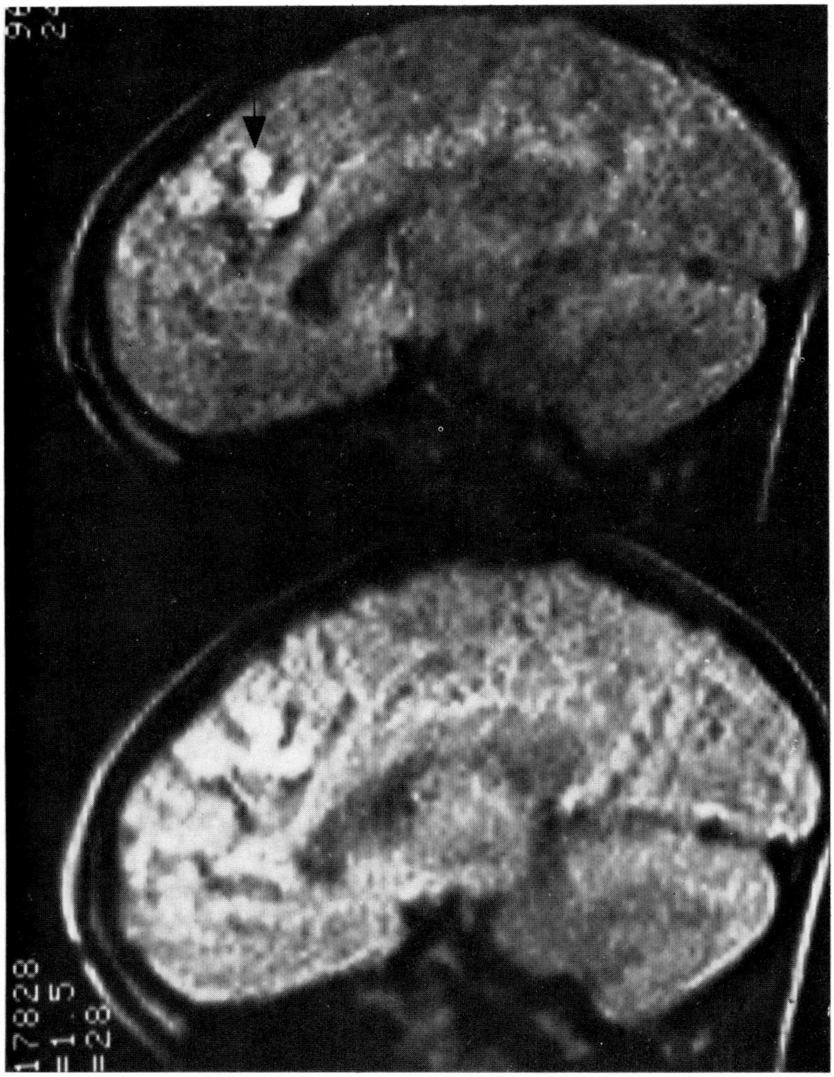

Figure 2. Partially clotted arteriovenous malformation (AVM). Two NMR spin echo intensity images (TE of 28 msec, TR of 1.5 sec) in a 6-year-old boy with seizures demonstrate an abnormal focus in the frontoparietal region. The core of the abnormality (black arrow) has a high intensity, suggesting static, clotted blood. The periphery of the lesion has low intensity, like the cerebral arteries, suggesting rapidly flowing blood. This diagnostic impression from the NMR study was confirmed surgically. A second AVM was identified in the left cerebrum on other NMR sections.

in the basal ganglia; this region shows abnormally high signal intensity when an alternative NMR technique such as inversion recovery is used.[10] The chemical changes responsible for these alterations in the magnetic relaxation factors of this tissue are unknown.

For the evaluation of the developing nervous system in the infant, NMR may be uniquely valuable. The NMR images of the brain (Fig. 5) of a 4-day-old girl with a known lipomeningocele of the sacrum demonstrated a relative lack of myelin on spin echo (TR of 1.0 sec) images. As opposed to the normal

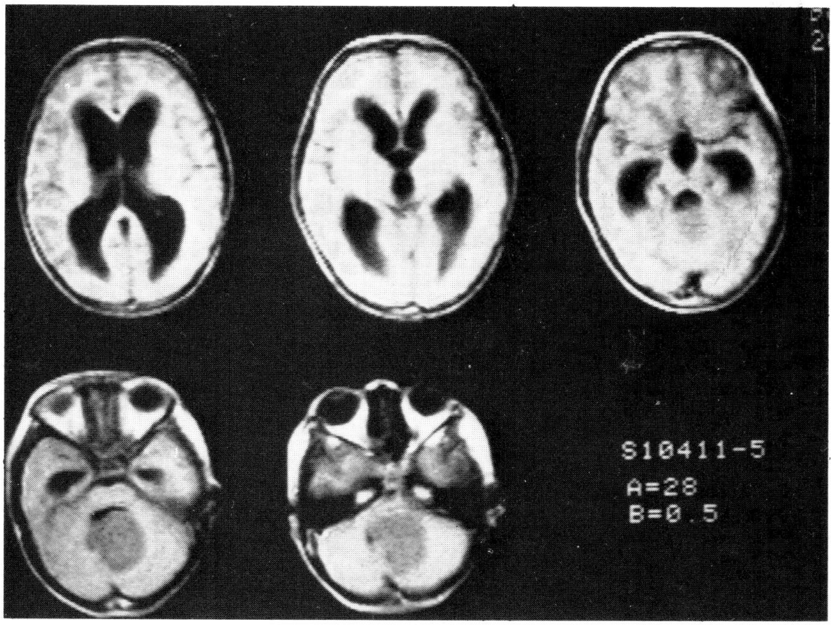

Figure 3. Medulloblastoma and obstructive hydrocephalus. Five contiguous spin echo NMR images of a 2-year-old boy demonstrate a large low-intensity mass in the midline of the posterior fossa. This mass was isodense with the cerebral hemispheres on nonenhanced CT images. A zone of low intensity surrounding the dilated ventricular system suggests cerebral edema and is consistent with obstruction as a cause of hydrocephalus.
(Reproduced, with permission, from Brant-Zawadzki M, Davis PL, Crooks LE, et al: NMR demonstration of cerebral abnormalities. AJR 140:847, 1983.)

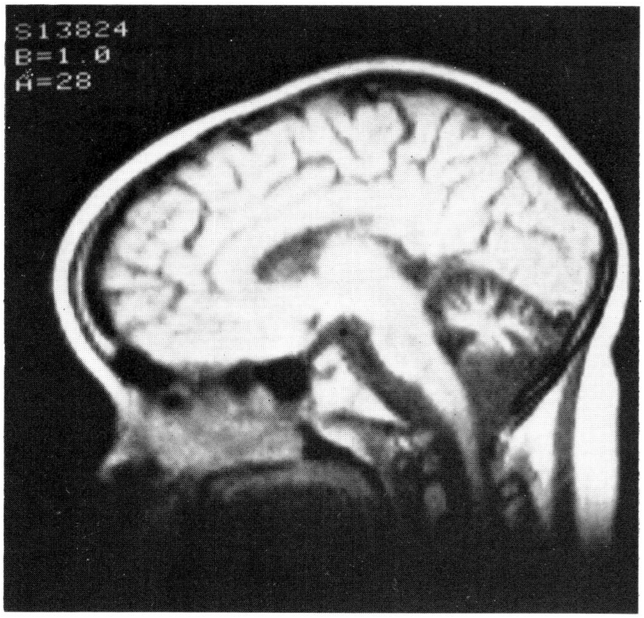

Figure 4. Spinocerebellar atrophy. Sagittal NMR image of a 7-year-old girl with signs of pancerebellar dysfunction demonstrates an atrophied cerebellum and a diminutive cervical spinal cord. Notice the sharp contrast between cerebellar white and gray matter and the ability to delineate the cord without contrast medium.

distribution of myelin throughout the white matter of the fully developed brain, this apparently normal brain of an infant had high NMR intensity only in the region of the basal ganglia. The ability to monitor the deposition of myelin with NMR may be a sensitive measure of neurologic development in the neonatal period and thus may contribute to the early diagnosis of abnormalities of the evolving brain.[6] This case further demonstrates that NMR can be used to image even very young children; this infant was restrained but not sedated for the NMR procedure.

Nuclear magnetic resonance imaging appears to be particularly sensitive for detecting tissue edema. The increased water content within the brains of several patients has been easily detected as areas of increased intensity on spin echo images. This observation can be largely attributed to lengthening of T_2 relaxation times. A 13-year-old girl was suspected of having a brain abscess following surgery and pulmonary infection. The brain CT examination clearly depicted a low-density lesion with marked rim enhancement after contrast administration. A larger region of low density surrounding the abscess cavity

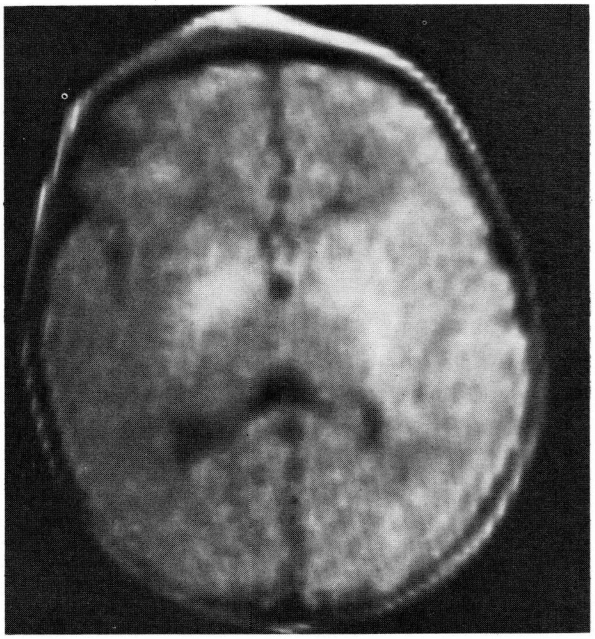

Figure 5. An NMR spin echo image of a normal 4-day-old brain demonstrates a paucity of myelinization. Only the regions of the basal ganglia show the characteristically high intensity of myelin. The potential to quantitatively evaluate brain development in the neonatal period may have useful clinical application. The slight asymmetry of the NMR image is attributable to asymmetric positioning of the infant.

suggested localized edema. The corresponding NMR sections vividly portrayed the edema as a region of normal intensity on first-echo images (TE of 28 msec) and very high intensity on second-echo images (TE of 56 msec) (Fig. 6). Such tissue magnetic characteristics are compatible with water deposition because water has a longer T_2 than myelin. The relative intensities of tissues with longer T_2 values are accentuated on images with TE of 56 msec.

The sensitivity of NMR imaging for small cerebral foci of hemorrhage was demonstrated in a 15-year-old girl with idiopathic bone marrow hypoplasia and thrombocytopenia. She acutely developed seizures and physical findings suggesting subarachnoid hemorrhage. Two separate CT examinations of the head failed to reveal either subarachnoid blood or a focus of hemorrhage. Nuclear magnetic resonance spin echo images (Fig. 7A) revealed three small foci of high intensity in the left occipital region. Intensity was higher than that of surrounding white matter on both first and second echoes, suggesting static blood rather than edema. Blood could also be identified in the subarachnoid space (Fig. 7B). The intracerebral blood was more apparent on short TR (0.5-sec) than on long TR (1.5-sec) images.

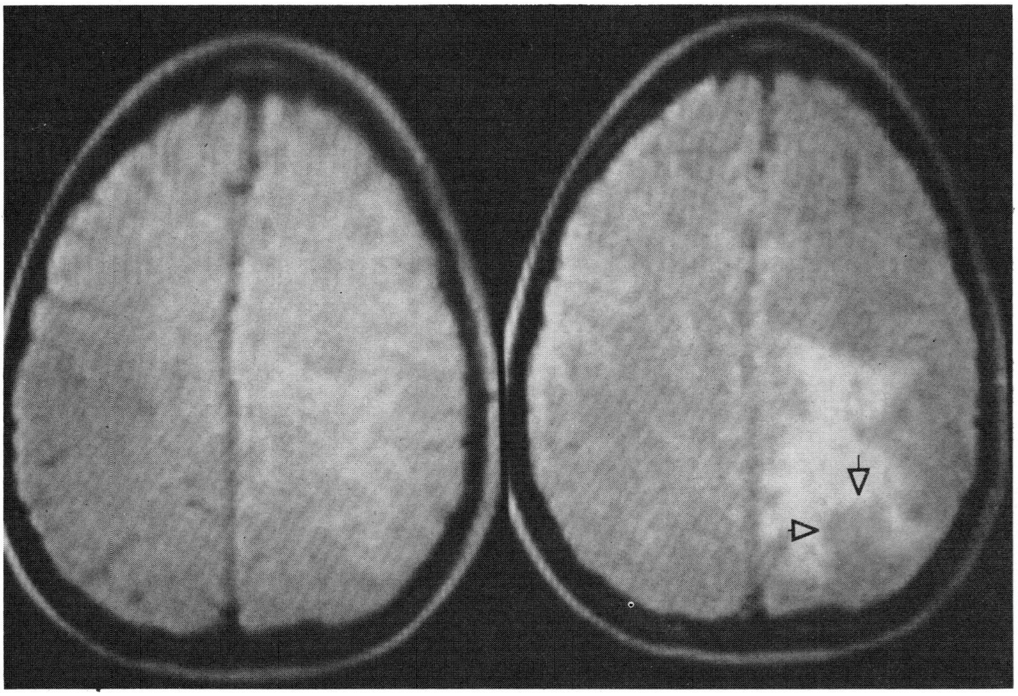

Figure 6. Brain abscéss. Spin echo NMR sections show an abscess as having relatively low NMR intensity (arrows), but the surrounding brain edmea is revealed as having higher intensity on the second echo (TE of 56 msec) (right) than on the first echo (TE of 28 msec) image (left).

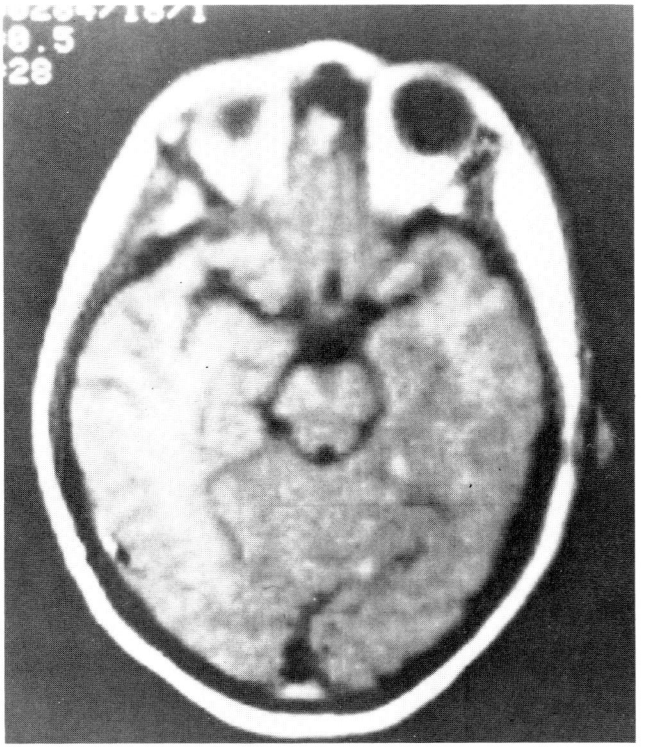

A.

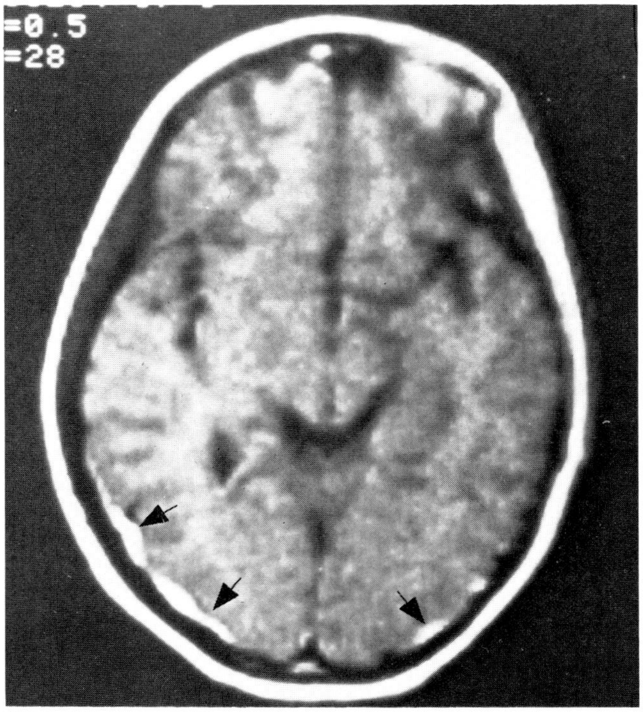

B.

Figure 7. Subarachnoid hemorrhage. Nuclear magnetic resonance image (A) reveals three small areas of petechial hemorrhage in a thrombocytopenic 15-year-old girl after CT examination showed no abnormality. Subarachnoid blood was seen best on more cephalic NMR section (B) as a high-intensity zone (arrows) between the calvarium and the cerebral cortex.

Magnetic Resonance Imaging of the Pediatric Chest and Abdomen

For three children with hemosiderosis, the NMR images demonstrated striking abnormalities of intensity for the liver, bone marrow, and kidneys (Fig. 8). For all three children the liver and bone marrow had extremely low intensities, appearing black on spin echo images, while the kidneys had relatively higher intensity, appearing white. These NMR intensity changes may be useful to diagnose and roughly quantitate iron deposition within various organs. These cases particularly well illustrate the dependence of NMR on the microchemical environment of hydrogen within tissues. All three children with thalassemia required frequent blood transfusions over many years to maintain adequate levels of hemoglobin. Because the body is poorly equipped to excrete excess iron from hemolysed blood, iron accumulates in the form of hemosiderin within various organs, particularly the liver, spleen, and bone marrow. Hemosiderin is a poorly understood, degraded form of the protein transferrin; the iron salts complexed within this protein molecule are primarily in the oxidized ferric (Fe^{+3}) state.

Ferric ions contain five unpaired electrons, making them strongly paramagnetic. Paramagnetic substances can be defined as those having a magnetic moment that aligns with an external magnetic field but that orients randomly when the external magnetic field is removed (refer to Chap. 18). Paramagnetic substances cause a proton relaxation effect on neighboring hydrogen nuclei; T_1 and T_2 relaxation times decrease. From in vitro studies we have learned that with high concentrations of paramagnetic ferric salts, both T_1 and T_2 are markedly shortened. The recorded NMR will be very weak because of the very short T_2. Because of the very high concentrations of ferric salts within the bone marrow and liver of our patients (greater than 100 mM), T_1 and T_2 times observed in these organs were extremely short, and the NMR intensity signal was low, as predicted; in vitro experiments had shown that ferric concentrations in excess of 20 mM produce a low NMR intensity.

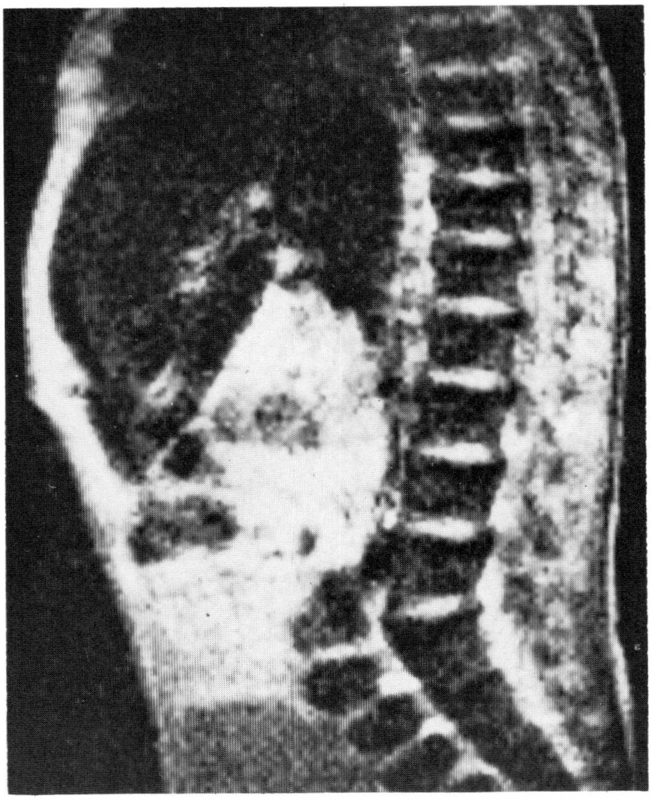

A.

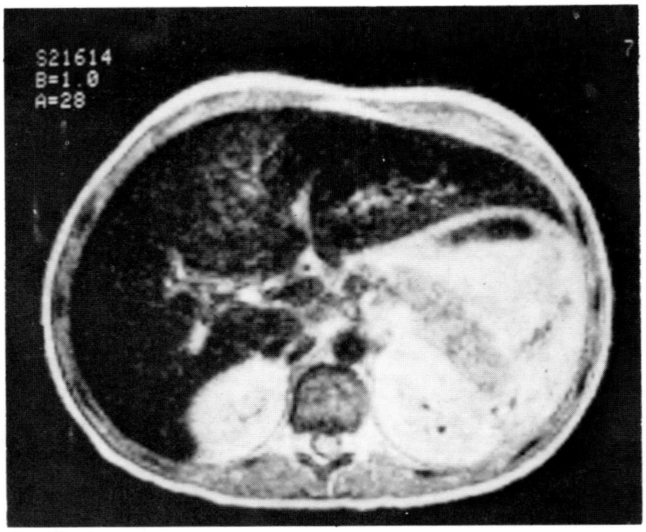

B.

Figure 8.

A. Hemosiderosis. Because of the deposition of large amounts of ferric salts contained within hemosiderin in the bone marrow and liver, these organs yield an unusually low signal on this sagittal spin echo image. This 14-year-old boy with thalassemia had been treated for many years with biweekly blood transfusions. The accumulated iron salts, which are strongly paramagnetic, dramatically shortened T_1 and T_2 relaxation times of nearby protons and thus reduced NMR intensity.

B. A transverse spin echo NMR image through the kidneys of the same patient. This appearance can also be attributed to deposition of paramagnetic iron salts. The high intensity observed for the kidneys suggests lesser concentrations of iron than within the liver, which has a very low-intensity signal.

With concentrations of ferric salts less than 20 mM, the proton relaxation effect in vitro from iron results in a relatively greater shortening of T_1 than T_2. Thus for relatively low concentrations of iron, T_1 is considerably shortened and T_2 values are relatively unchanged. This combination will increase, rather than decrease, the observed intensity signal. Thus the high-intensity NMR appearance of the kidneys in hemosiderosis was suggestive of a concentration of iron greater than normal but less than 20 mM.

It thus appears that NMR imaging may be useful to roughly estimate iron deposition in various organs. The same principles may also be applied to deposition of other paramagnetic substances such as manganese and copper.

The observed NMR signal abnormalities observed in these patients with hemosiderosis exemplify the sensitivity of NMR for the detection of tissue chemical imbalances. Further, the imaging assessment of bone marrow demonstrated in these children is another useful advantage of the NMR technique. It may be possible with NMR to detect a variety of bone marrow disorders, including hypoplasia, hyperplasia, and metastatic marrow replacement.

Nuclear magnetic resonance body imaging in children has been capable of demonstrating small anatomic abnormalities. The NMR images of a child with adrenal hyperplasia demonstrated bilateral glandular enlargement (Fig. 9). Similar changes were observed on CT images of this patient. Adrenal glands on NMR spin echo images characteristically appear

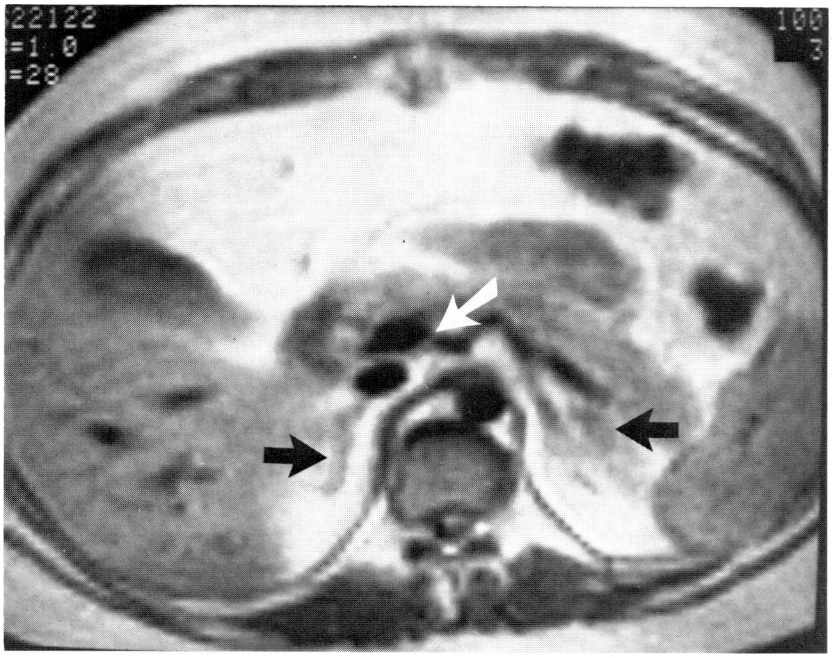

Figure 9. Adrenal hyperplasia. Bilaterally enlarged adrenal glands (black arrows) were identified in spin echo intensity NMR images as low-intensity structures surrounded by retroperitoneal fat. Fat characteristically yields a high-intensity signal. Incidentally noted was the junction of the splenic vein and the portal vein (white arrow).
(Reproduced, with permission, from Moon KL Jr, Hricak H, Crooks L, et al: Nuclear magnetic resonance imaging of the adrenal gland: A preliminary report. Radiology 147:155, 1983.)

as low-intensity structures surrounded by higher-intensity retroperitoneal fat. Although this case reveals no advantage of NMR over CT, it does indicate the ability of NMR to detect relatively subtle anatomic abnormalities within the abdomen.

Within the chest of children, NMR was found to be sensitive for the detection of pleural fluid, pulmonary nodules, bronchial abnormalities, and masses, particularly if the soft-tissue abnormality closely approximated a large vascular structure. An 11-year-old boy receiving chemotherapy for an osteogenic sarcoma arising in the lower extremity developed three pulmonary nodules. The same nodules were identified clearly on both CT and NMR spin echo images (Fig. 10). The NMR contrast difference was large between the air-filled lung that was low in hydrogen content and the metastatic nodules that were composed primarily of water. Even nodules located centrally in the pulmonary hilum should be readily detectable by NMR because of the low intensity of rapidly moving blood in the pulmonary arteries and veins, producing sharp contrast with the high intensity of a hilar mass lesion.

The experience with NMR for mediastinal masses suggests that clear definition of tumors may be possible without contrast enhancement. A 15-year-old girl with known Hodgkin's disease had a residual mediastinal mass despite radiation therapy and two courses of different chemotherapy combinations. A CT scan of the mediastinum demonstrated a large mediastinal mass with considerable calcification (Fig. 11A). The calcifications were not identified on NMR images; NMR defined a generally high-intensity mediastinal mass (Fig. 11B). However, NMR more sharply delineated the soft-tissue mediastinal mass from nearby vascular structures than was evident on the contrast-enhanced CT scan. Thus NMR may be particularly useful in defining small mediastinal and hilar masses because of the inherent sharp contrast differences between soft tissues and vessels.

The potential of NMR to add to the evaluation of children with abdominal tumors is illustrated by an 11-year-old girl who had a large pelvic rhabdomyosarcoma. This tumor, demonstrated by CT and at surgery, was being treated with a combination of antitumor drugs and radiotherapy. Computed tomography depicted the extent of a large mass extending cephalad from the vagina. The tumor had a uniform CT density before and after iothalamate administration. By comparison, the NMR images of

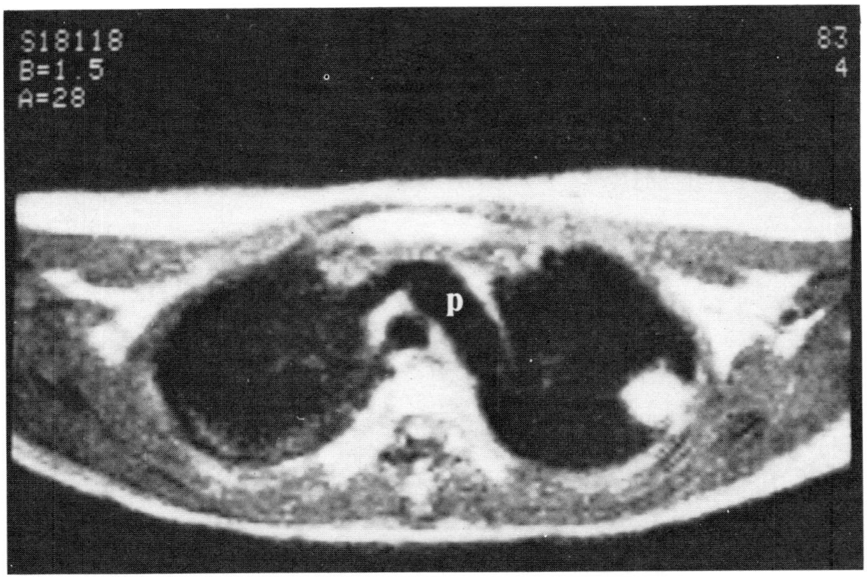

Figure 10. Pulmonary metastases. Nuclear magnetic resonance images of the lung of an 11-year-old boy with osteogenic sarcoma and known pulmonary metastases demonstrated a sharp contrast difference between the nodules and the surrounding normal lung parenchyma and normal hilar vessels. (p, pulmonary artery)

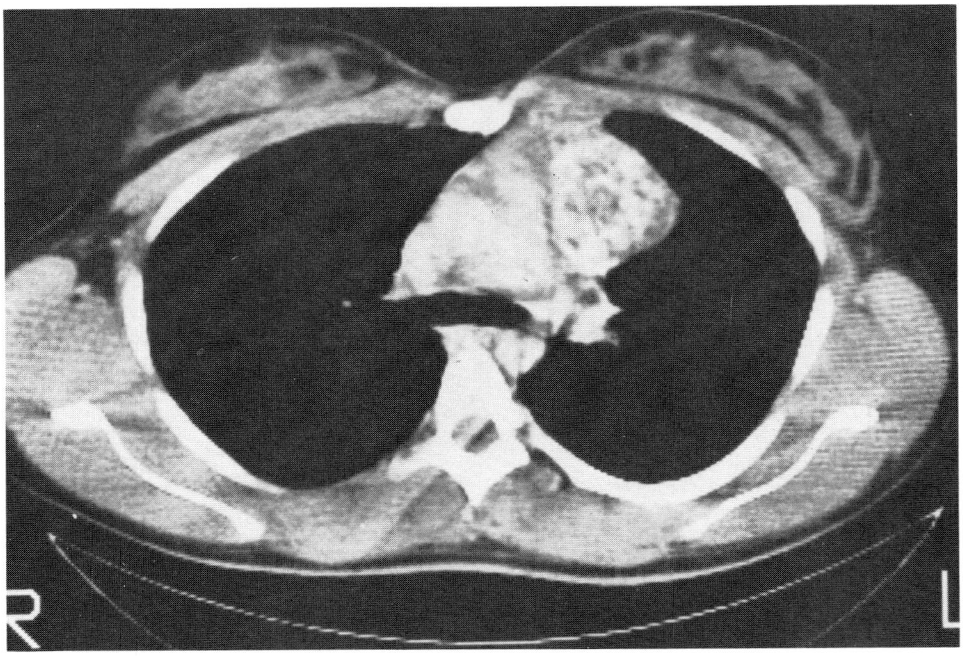

A. Figure 11. Hodgkin's disease. A 15-year-old girl with a residual mediastinal mass after therapy for Hodgkin's disease was noted on CT (A) to have a densely calcified left paramediastinal mass. The calcifications were not evident on the corresponding NMR section (B), but the differentiation of mass from mediastinal vessel was more easily appreciated on NMR than on CT.

(continued on next page)

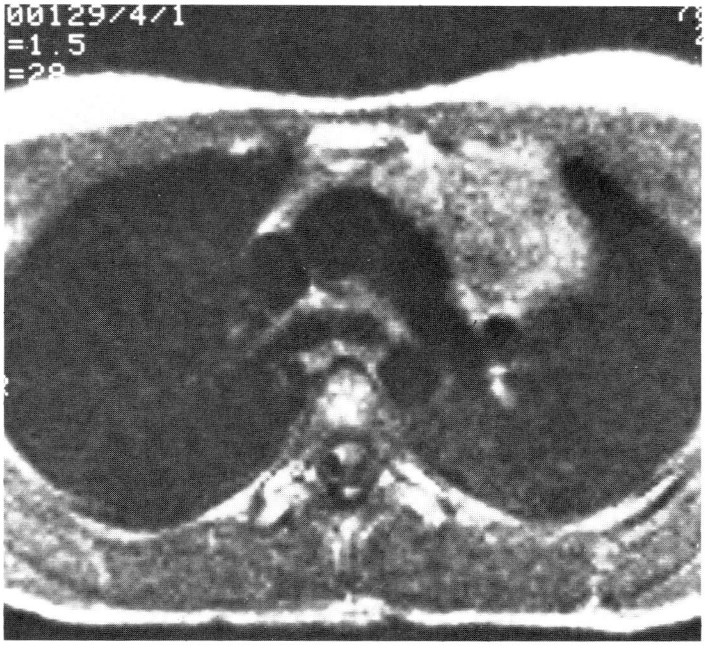

B.

this tumor revealed a central high-intensity core, with the remainder of the tumor having an intermediate intensity (Fig. 12). This inhomogeneity in the NMR appearance remains unexplained; perhaps the high-intensity focus represented a central hemorrhage in the tumor, a nidus of tissue not responding to therapy, or a difference in perfusion. One can speculate that different NMR appearances will be observed in treated tumors depending on their responsiveness. Extensive experience will be required before this supposition can be confirmed.

General Considerations

Our relatively small initial series in pediatric patients indicates certain advantages of NMR over CT.[11-13]

Besides the lack of ionizing radiation, excellent contrast was observed in NMR images without contrast medium administration. Bone artifacts were not observed, making NMR images of the posterior fossa uniquely diagnostic. Semiquantitative data reflecting movement of blood and deposition of paramagnetic elements could be extracted from the NMR intensity studies. Brain edema was readily apparent. Inhomogeneity of a treated tumor could be detected on NMR but not on CT. Nuclear magnetic resonance permits a direct imaging assessment of the bone marrow.

These apparent advantages should be weighed against the disadvantages of NMR that were revealed in our experience. As with CT, vigorous motion

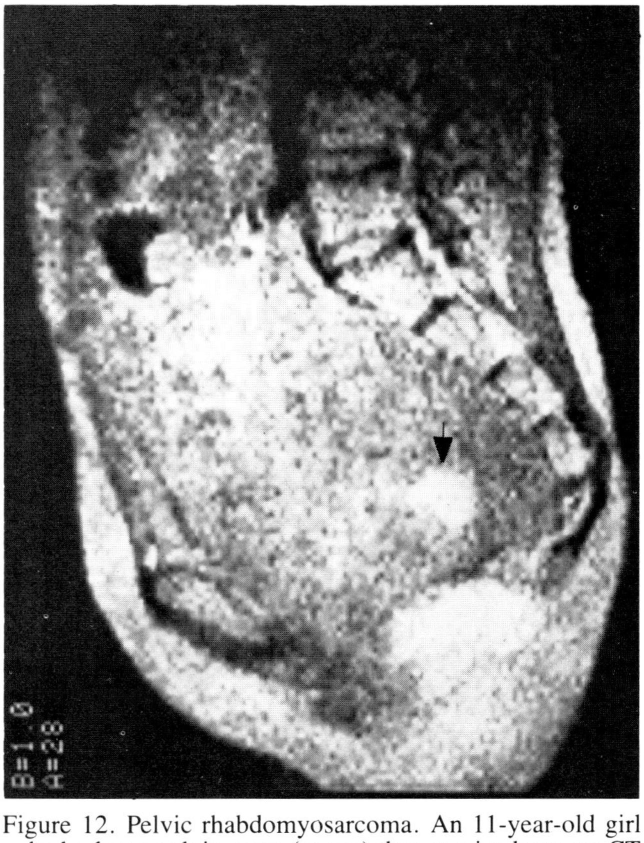

Figure 12. Pelvic rhabdomyosarcoma. An 11-year-old girl had a large pelvic mass (arrow) that was isodense on CT exam both before and after contrast administration. Nuclear magnetic resonance demonstrated inhomogeneity of the large pelvic mass with a high-intensity central core, a finding that could represent a difference in perfusion, a difference in tumor responsiveness to therapy, or hemorrhage into the center of the tumor.

during the acquisiton of data will significantly degrade NMR image quality. The time necessary to obtain a series of 15 sections was relatively lengthy, 13 to 17 min. Patients needed to remain relatively motionless. Patients with pacemakers and certain types of vascular clips may have to be excluded from the powerful magnetic fields of NMR.[14] As yet, nonmagnetic ventilating equipment that can be employed for patients within the scanner is not available. Further, NMR is insensitive to calcification.

The merits of NMR imaging in children must be viewed in the light of the very recent advent of clinical imaging devices. It is our hope that many of the present problems can be overcome by future innovation, making NMR imaging even more appealing for the diagnosis of pediatric disease. New applications and clinical benefits are likely to yet emerge.

REFERENCES

1. Holland GN, Moore WS, Hawkes RC: Nuclear magnetic resonance tomography of the brain. J Comput Assist Tomogr 4:1, 1980.
2. Bydder GM, Steiner RE, Young IR, et al: Clinical NMR imaging of the brain: 140 cases. AJR 139:215, 1982.
3. Doyle FH, Pennock JM, Banks LM, et al: Nuclear magnetic resonance imaging of the liver: Initial experience. AJR 138:193, 1982.
4. Hricak H, Crooks L, Sheldon P, et al: NMR imaging of the kidney: I. Radiology 146:425, 1983.
5. Young IR, Bailes DR, Burl M, et al: Initial clinical evaluation of a whole body nuclear magnetic-resonance (NMR) tomography. J Comput Assist Tomogr 6:1, 1982.
6. Levine MI, Whitelaw A, Dubowitz LMS, et al: Nuclear magnetic resonance imaging of IVH and related disorders. In *Proceedings of the Ross Laboratory Conference on Perinatal Intracranial Hemorrhage* 1:595, 1982.
7. Crooks LE, Arakawa M, Hoenninger J, et al: Nuclear magnetic resonance whole-body imager operating at 3.5 kGauss. Radiology 143:169, 1982.
8. Young IR, Burl M, Clarke GJ, et al: Magnetic resonance properties of hydrogen: Imaging the posterior fossa. AJR 137:895, 1981.
9. Young IR, Hall AS, Pallis CA, et al: Nuclear magnetic resonance imaging of the brain in multiple sclerosis. Lancet 2:1063, 1981.
10. Brant-Zawadzki M, Davis PL, Crooks LE, et al: NMR demonstration of cerebral abnormalities: Comparison with CT. AJR 140:847, 1983.
11. Gooding CA, Brasch RC, Brant-Zawadzki MN, et al: NMR imaging of the brain in children. (Submitted for publication.)
12. Moon KL Jr, Hricak H, Crooks L, et al: Nuclear magnetic resonance imaging of the adrenal gland: A preliminary report. Radiology 147:155, 1983.
13. Moon KL Jr, Genant HK, Helms CA, et al: Musculoskeletal applications of nuclear magnetic resonance. Radiology 147:161, 1983.
14. Budinger TF: Nuclear magnetic resonance (NMR) in vivo studies: Known thresholds for health effects. J Comput Assist Tomogr 5:800, 1981.

GENERAL AND ANCILLARY CONSIDERATIONS

18

Contrast Media

ROBERT C. BRASCH, M.D.

General Considerations

The inherently superior soft-tissue contrast afforded by proton NMR imaging, facilitated by the proper selection of imaging sequences, may render contrast agents unnecessary. However, as with any diagnostic study, instances of low specificity or sensitivity will arise in NMR imaging. Contrast agents can provide additional diagnostic information by the dynamics of their organ biodistribution, and thus can potentially affect the efficacy and general acceptance of NMR into contemporary diagnostic imaging.

Inherent differences in the observed nuclear resonance signals from various tissues produce differences in intensity and, hence, contrast on NMR images. The inherent contrast between tissues on NMR, which generally is quite high,[1] depends both on physical characteristics, such as viscosity and temperature, and chemical characteristics, such as hydrogen density and local magnetic environment. Inherent contrast can also be observed on images based upon computer-assisted mathematical analyses of the NMR signals, namely, the T_1 and T_2 images. (For a review of basic NMR concepts, including T_1 and T_2, the reader is referred to one of several excellent reports [2–4] and Chaps. 2, 5 and 6 of this book.) However, the intensity images for a particular body section may differ greatly from the T_1 or the T_2 image since there are at least four variables for each tissue; hydrogen density, flow, T_1, and T_2. Contrast between tissues may be greatest along any of these parameters (Fig. 1).

The use of pharmaceuticals as "extrinsic" NMR contrast agents adds a fifth variable. Contrast agents, by enhancing the signal differences observed between magnetically similar tissues, may produce NMR images with the best contrast and the most diagnostic information.

In several respects, NMR contrast agents are analogous to iodinated contrast media for diagnostic radiology. For example, NMR contrast agents excreted in the urine may yield useful information about renal function and anatomy. An intravascular NMR contrast agent may aid in the evaluation of tissue perfusion[5] and vascular anatomy. As with radiographic contrast agents, NMR contrast agents may be administered

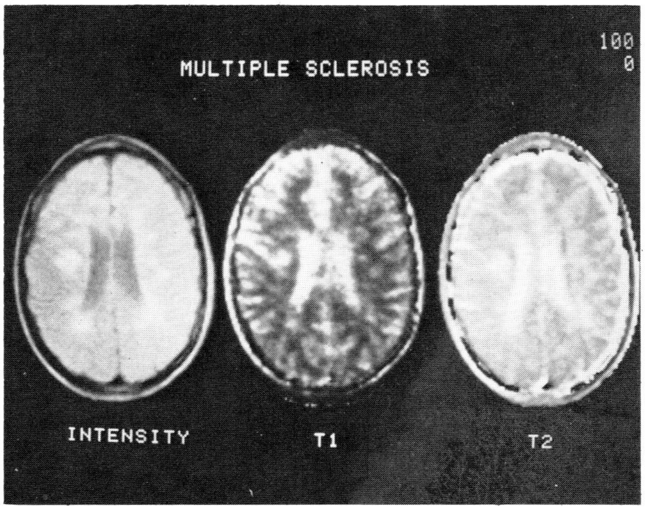

Figure 1. Nuclear magnetic resonance intensity, T_1, and T_2 images of the same brain section from a patient with multiple sclerosis readily demonstrate periventricular plaques of demyelination. The plaques are best seen on the intensity image as light areas as compared with the relatively gray background of the brain substance. Gray-white matter differentiation is best appreciated on the T_1 image. Contrast between different types of tissue may be best appreciated on any of the different types of NMR images.

by intravascular injection, ventilation, oral ingestion, or subarachnoid instillation.

Yet there are important differences between NMR contrast agents and iodinated radiographic media. The principles of contrast enhancement differ. Radiographic contrast agents, having high electron density, absorb x-rays to produce their contrast effect, whereas several types of NMR contrast agents, namely, paramagnetic agents, function by altering local magnetic environments.[6] Thus, NMR contrast agents may not be observed directly on the images; rather, the magnetic effects of the material on neighboring hydrogen nuclei may be the means of contrast enhancement. Alternatively, NMR contrast enhancement may be achieved by changes in hydrogen concentration, temperature, viscosity, or motion of a given tissue. Detailed explanations and considerations of feasibility for these different means of contrast enhancement will be the focus of this discussion.

Principles of Magnetic Resonance Contrast Enhancement

Paramagnetism and Proton Relaxation

All substances and all forms of matter possess magnetic properties to some degree, and three major types of behavior have been recognized: paramagnetism, diamagnetism, and ferromagnetism. A paramagnetic substance may be defined as one that is attracted toward the stronger portion of a magnetic field and aligns with it, whereas a diamagnetic substance is typically repelled by the stronger portion of a magnetic field and aligns against it. A paramagnetic substance has its own magnetic moment, which aligns in an external magnetic field, but it loses this alignment and orients randomly when the field is removed. Ferromagnetic substances also align with an external magnetic field but tend to retain that alignment after the external field is removed.

Paramagnetic substances, by possessing their own local magnetic fields, will act as "relaxation centers" for other nuclei in their microenvironment. The local magnetic field of a paramagnetic molecule will serve to increase the magnetic relaxation rates of surrounding hydrogen nuclei, a phenomenon called *proton relaxation enhancement* (PRE).[7] Proton relaxation enhancement shortens T_1 and T_2 values for neighboring hydrogen nuclei. This relaxation effect is dependent on many variables, two of which are the concentration of the paramagnetic agent and the distance from the paramagnetic center to the hydrogen nucleus. Proton relaxation enhancement can cause a higher NMR signal intensity (Fig. 2).

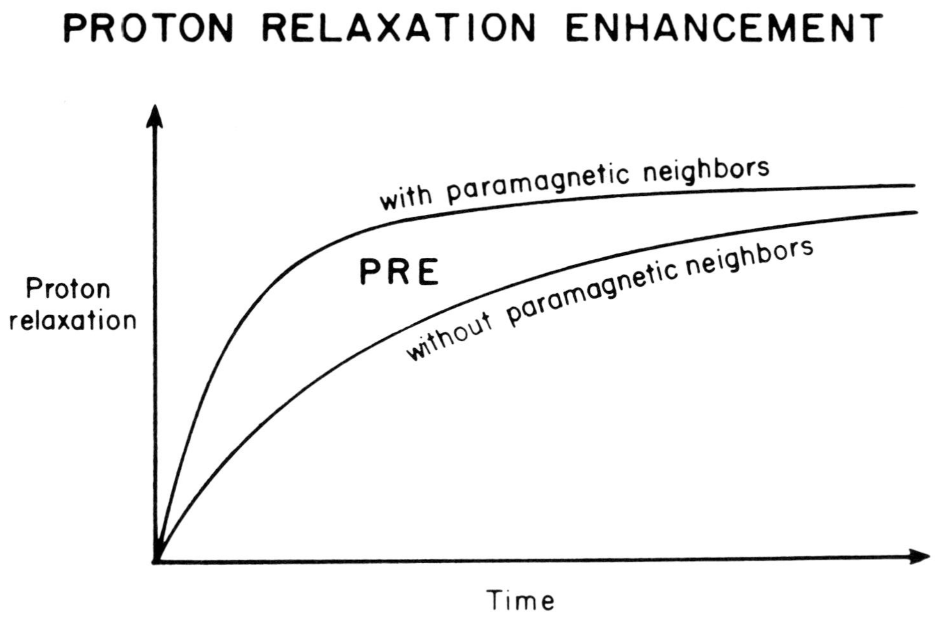

Figure 2. This graph indicates that after a radio-frequency pulse the subsequent relaxation of protons will be more complete in the presence of a neighboring paramagnetic molecule than in the absence of a paramagnetic molecule. This effect is termed proton relaxation enhancement (PRE).

(Reproduced, with permission, from Brasch RC: Work in progress: Methods of contrast enhancement for NMR imaging and potential applications. A subject review. Radiology 147:781, 1983.)

A review of the mathematical approximation between NMR spin echo intensity (I), T_1, and T_2 may be helpful to understand how changing T_1 and T_2 can influence intensity:[1]

$$I = N(H)f(v)\exp(-TE/T_2)[1-\exp(-TR/T_1)] \qquad \text{Eq. 1}$$

where *N(H)* is the local spin density; *TE* is the elasped time interval (in milliseconds) between the application of the 90° radio-frequency pulse and the reception of a spin echo signal; *TR* is the radio-frequency pulse repetition rate (in seconds); and f(*v*) is a function of both the speed with which the hydrogen nuclei move through the region being imaged and of the fraction of the total number of nuclei that are moving.

Intensity is measured directly by the NMR imaging unit; T_1 and T_2 parameters can be calculated from the intensity data. Note from Equation 1 that decreasing T_1 increases image intensity, whereas decreasing T_2 reduces image intensity.

Substances are paramagnetic by virtue of possessing one or more fundamental particles (proton, neutron, or electron) with a spin that is not canceled by a like particle with opposite spin. There are numerous paramagnetic substances, but only a few of these have promise as clinical NMR contrast media. The degree or strength of paramagnetism can differ markedly between substances. In Table I certain paramagnetic substances are grouped into nuclear and electron species. The magnetic moments of unpaired

TABLE I

Types of Paramagnetic Substances

Unbalanced Nuclear Spins

Element	No. of Protons	No. of Neutrons	Spin Quantum Number
H-1	1	0	½
C-13	6	7	½
N-14	7	7	1
0-17	8	9	5/2
F-19	9	10	½
P-31	15	16	½

Unbalanced Electron Spins

Unpaired Electrons
- Nitric oxide (NO)
- Nitrogen dioxide (NO_2)

Paired Electrons with Parallel Spins
- Molecular oxygen (O_2)

Ions Containing Unpaired Electrons
- A. Transition metal series
  - Mn^{+2}, Mn^{+3}
  - Fe^{+2}, Fe^{+3}
  - Ni^{+2}
  - Cr^{+3}
  - Cu^{+2}
- B. Lanthanide series
  - Gd^{+3}
  - Eu^{+2}
- C. Actinide series
  - Pa^{+4}

Stable Free Radicals
- A. Nitroxides
- B. Triphenylmethyl

electrons are about 700 times greater than those of proton or neutron magnetic moments; local magnetic fields generated by unpaired electrons are proportionately much stronger. Therefore, the PRE of the nuclear paramagnetic substances is generally too weak to substantially influence contrast on NMR images. It is the stronger paramagnetic compounds containing unpaired electrons that deserve our attention as NMR contrast enhancers.

Manipulation of Physical Characteristics of Tissue

Tissue accumulation of paramagnetic agents is just one of several potential methods of manipulating the NMR signal to enhance contrast. Manipulation of physical properties including temperature and viscosity will also change the NMR signal intensity. For example, a warm tissue generally has a longer T_1 than the same tissue at a lower temperature. The effects of heating and freezing water on the NMR signal intensity are depicted in Figure 3. The various solutions in test tubes were cross-sectionally imaged. Heating water from 25° to 40°C produced no significant change in spin echo intensity. However, freezing water to form ice markedly lengthened T_1 and shortened T_2, reducing intensity. However, because such dramatic changes in temperature (15° to 25°C) or viscosity (liquid to solid) cannot be feasibly employed for biologic systems, manipulation of these physical characteristics by temperature changes seems an unlikely choice for clinical NMR contrast enhancement. A more clinically feasible approach to PRE by altering proton viscosity may be through the use of ethanol or its derivatives.

Manipulation of Chemical Characteristics of Tissue

Alteration in the chemical characteristics, particularly hydrogen concentration, of a tissue will also change the NMR signal and may be a more practical means to enhance contrast. From Equation 1 we can recognize that changing hydrogen concentration (spin density) within a tissue will directly influence the NMR intensity. Because water is rich in hydrogen, the relative hydration of a tissue should influence NMR image appearance. However, increasing the hydration within a tissue not only will influence the hydrogen concentration N(H) but will tend to increase both T_1 and T_2 for that tissue. The resultant change in NMR signal intensity will depend on the imaging algorithm employed. T_1-weighted pulse sequences (saturation recovery, inversion recovery) will demonstrate a decrease in signal intensity (accentuating the longer T_1). Spin echo sequences, accentuating T_2 changes, will show an increase in signal intensity (longer T_2), which is most noticeable at longer echo delay times.

Changes in tissue hydration can also be achieved using parenteral hypertonic agents. Radiographic contrast media, approximately 1,500 milliosmolar, are concentrated in the kidney after intravenous injection and cause an osmotic diuresis. The resulting water shift into the renal tubules might be predicted to increase NMR intensity in a fashion parallel to overhydration. This NMR contrast-enhancing potential of radiographic contrast media is not attributable to iodine content but rather to hypertonicity.[8] Any other solute, such as urea, when administered in hypertonic concentrations and excreted in the kidneys, should have similar effects on NMR intensity. To date, studies of tissue hydration and hypertonic pharmaceuticals used as NMR contrast-enhancing agents are only preliminary,[8] and the mechanisms and potential diagnostic utility have not been defined.

The proportion of fat within a tissue is also an important determinant of NMR characterization. The hydrogen nuclei of fatty substances have much shorter T_1 than those of water.[9] Accordingly, the intensity signal from fatty tissues is higher than that of water. A 1.0 molar solution of sodium dodecylsulfate (a 12-carbon lipid) imaged in Figure 3 shows a moderate increase in intensity when compared with water. This contrast-enhancement potential using lipids may have clinical utility. In their animal studies Newhouse et al. were able to opacify bowel loops on NMR images after the oral administration of mineral oil.[10]

Another possible fatty contrast agent is the liposome. Liposomes are rounded, fatty vesicles composed of one or many layers of lipid membrane. These fatty vesicles may be generated in varying sizes and thereby may be targeted to specific organs such as the reticuloendothelial cells of the liver, spleen, and bone marrow. Because liposomes are composed almost entirely of lipid, they may be used as NMR contrast enhancers. In comparison with the NMR contrast-enhancing potential of paramagnetic agents (Fig. 3), however, the effects of fat appear to be weaker. Liposomes may yet be useful NMR contrast agents if their hollow cores can be filled with paramagnetic species.

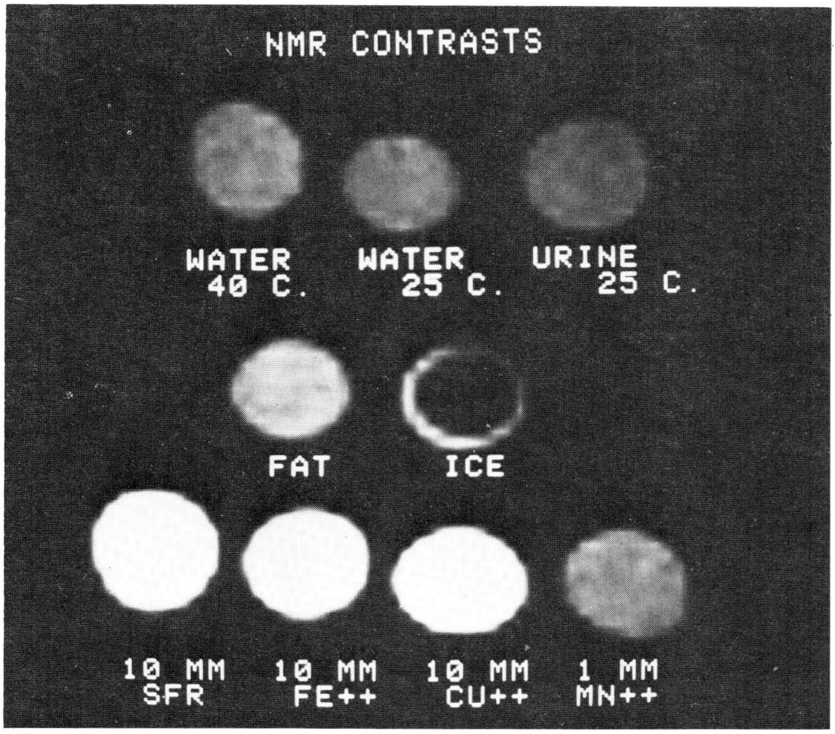

Figure 3. Nuclear magnetic resonance contrasts. An NMR intensity image of a transverse section through multiple test tubes containing different solutions demonstrates several potential means of NMR contrast enhancement. Distilled water at 40°C and 25°C has an NMR intensity signal that is virtually the same as that of urine. Freezing water to produce ice dramatically reduces the NMR intensity signal, an effect attributable primarily to T_2 shortening. A 1.0 M solution of sodium dodecysulfate (fat) produces an obvious increase in NMR intensity. From animal imaging studies it is known that fat has a short T_1. The solutions of stable free radical (SFR), ferrous ion (Fe^{+2}), and cupric ion (Cu^{+2}), all in 10 mM concentrations, produce a dramatic increase in NMR intensity compared with water or urine. These potential NMR contrast enhancers, as well as manganous ion (Mn^{+2}), are paramagnetic substances which alter the NMR intensity by PRE. The paramagnetic substances are generally more powerful contrast enhancers than the other demonstrated manipulations.

(Reproduced, with permission, from Brasch RC: Work in progress; Methods of contrast enhancement for NMR imaging and potential applications. A subject review. Radiology 147:781, 1983.)

Potential Clinical Applications and the "Ideal" Magnetic Resonance Contrast Enhancer

Before defining theoretical "ideal" characteristics for an NMR contrast agent, we must first define the goals for NMR contrast manipulation.

Identification of an isointense abnormal tissue focus surrounded by normal tissues of similar NMR characterization is one potential application for an NMR contrast enhancer. A well-known parallel in radiography is the use of contrast agents for differential enhancement of tumors which cannot be identified without contrast enhancement on CT scans. As with x-ray absorption, certain tissues have inherently similar NMR characteristics.[11] For example, the T_1 and T_2 values of normal muscle overlap with the values of liver,[11] and normal brain may have similar NMR characteristics to those of intracerebral hematoma, as Young et al. demonstrated.[13]

Another potential application of NMR was revealed by London and coworkers.[13] Three hours after the investigators experimentally produced renal ischemia by ligating the left renal artery of a rat, NMR images failed to demonstrate any intensity difference between the normal right kidney and the ischemic left kidney

(Fig. 4). For this and other instances of magnetically similar but histologically dissimilar tissues, contrast enhancement may prove valuable in increasing the intensity and relaxation differences between normal and pathologic tissues on NMR images.

Another application for an NMR contrast agent is the direct assessment of tissue function. The functional integrity of a tissue such as the brain or kidney may not be directly reflected on the nonenhanced NMR image. For example, renal incorporation of a contrast agent dependent on glomerular filtration would permit the identification and quantification of renal functional integrity. An NMR contrast agent incorporated by organs actively engaged in protein synthesis, such as the pancreas, might permit the identification of nonfunctioning foci, including tumors and abscesses.

Diseases of the central nervous system that disrupt the integrity of the blood-brain barrier may not be identifiable on nonenhanced NMR images. The regions of functional abnormality may become identifiable if one administers an NMR contrast agent that passes into the brain parenchyma only in areas of blood-brain barrier breakdown.

Tissue-specific targeting of NMR contrast agents could have numerous diagnostic applications. For example, NMR-active fibrinogen derivatives or platelet labels might help to identify thrombi or emboli. Contrast-enhanced protein aggregates may be useful for evaluation of pulmonary perfusion in a manner analogous to the use of radio-labeled albumin aggregates in nuclear lung scintigraphy. Monoclonal antibodies tagged with paramagnetic contrast agents may be useful in tumor identification. Moreover, NMR-active derivatives of the body's metabolic substrates may permit the assessment of enzyme pathways and substrate distribution. For example, an NMR-active glucose derivative might help to define regional metabolic activity within the brain.

Intravascular NMR contrast agents may be useful clinically despite the inherently strong negative contrast between rapidly moving blood and vessel walls. The apparent absence of an NMR intensity signal from rapidly moving blood perpendicular to the imaging plane is due to the movement of protons through the plane of section, disrupting the nuclear excitation and signal acquisition sequence. Yet positive contrast enhancers for the vascular system may still have utility in the assessment of organ microvascular perfusion, blood flow, and the measurement of blood volume. The spin echo NMR signal intensity of flowing blood in multisection imaging is dependent on the T_1 of blood.

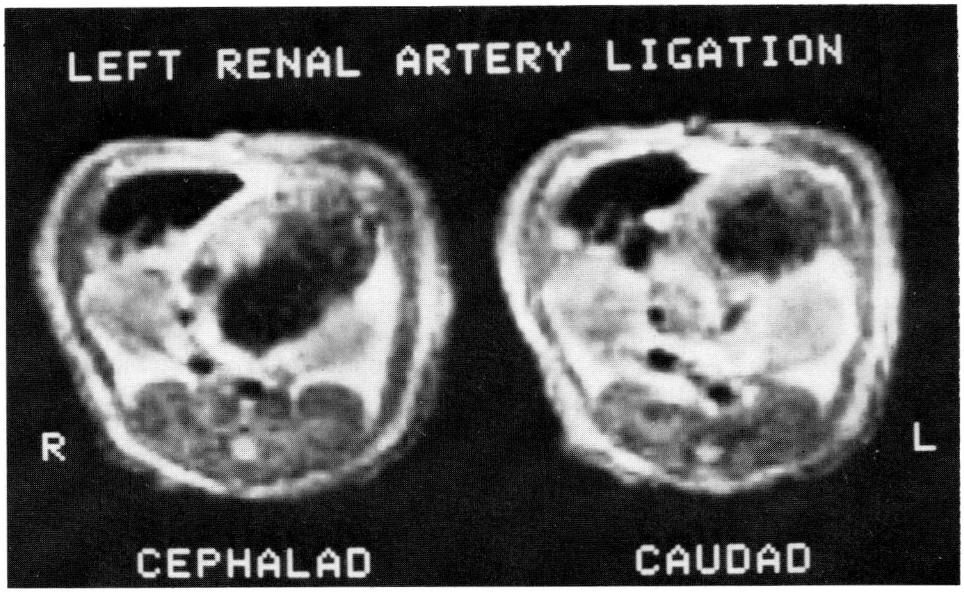

Figure 4. Two spin echo intensity images through the cephalic and caudal portions of rat kidneys were obtained 4 hr after surgical ligation of the left renal artery in a rat. No apparent differential in NMR intensity or renal size is noted between the normal right and the infarcted left kidney. This animal experiment suggests a need for an NMR contrast enhancement to identify disease and to quantitate renal functional integrity.

Having postulated several potential applications for NMR contrast enhancement, we can define criteria for the ideal agent. Ideal characteristics might include the following:

1. The effects of the contrast agent on the NMR signal should be reproducible from one examination to another.
2. The agent should be chemically versatile so that it can be bound to other compounds as a biologic probe and thereby permit selective tissue targeting.
3. The substance should have relatively strong NMR activity (paramagnetism) so as to increase local magnetic fields and thus reduce proton relaxation times.
4. The contrast agent should be chemically stable and easily stored without precautions, preferably in a form suitable for immediate administration.
5. The substance should be relatively easily manufactured from inexpensive starting materials.
6. The substance should be nonreactive in vivo and should be nontoxic in diagnostic doses.
7. The NMR agent should be quickly deactivated or excreted (within hours).

Using these criteria, we might examine in greater detail certain paramagnetic agents that have already been or might in the future be tested as NMR contrast enhancers.

A Closer Look at Paramagnetic Contrast Agents

Nitric Oxide and Nitrogen Dioxide

Nitric oxide (NO) and nitrogen dioxide (NO_2) are both paramagnetic molecules because they possess an odd number of electrons with uncanceled spins in valence molecular orbitals. These small molecules apparently have not been tested as NMR contrast agents. They both have certain disadvantages as pharmaceuticals, including limited solubility in biologic tissues. Another disadvantage is that they are gases at room temperature. Both agents have significant toxicity in relatively low doses; inhaled NO_2 causes an increase in airway resistance and a decrease in arterial oxygen. Nitric oxide produces methemoglobinemia. These molecules also lack the chemical versatility to permit conjugation to tissue-specific carrier molecules.

Molecular Oxygen

Oxygen (O_2) is also a paramagnetic molecule by virtue of containing two electrons of highest energy which occupy separate orbitals and have parallel spins that do not cancel. Inversion recovery NMR images of human subjects breathing 100% oxygen demonstrate an increase in the NMR signal from blood within the left ventricle (higher O_2 concentration) as compared with blood within the right ventricle (lower O_2 concentration).[12] This vascular contrast of O_2 can be attributed primarily to T_1 shortening. Yet a problem in evaluating the contrast effects of O_2 on blood in living subjects is that it is difficult to control for changes that might also result from changes in blood flow. Axel studied O_2 as an NMR contrast agent in vitro by bubbling 100% oxygen through static water, serum, and heparinized blood.[14] For water he found marked shortening of T_1 after oxygen saturation, but there was a relatively smaller effect of oxygen on serum, and even less on heparinized blood. Molecular oxygen has certain disadvantages as a contrast agent, including its general inability to be chemically complexed to carrier molecules without the loss of paramagnetism. Also, because oxygen is a gas at room temperature, it is difficult to direct within the body.

Paramagnetic Ions

Ions from the first transition series, which includes the elements of iron, manganese, cobalt, nickel, and copper, contain unpaired electrons within inner (3d) electron orbitals. There are five 3d orbitals capable of accommodating a total of 10 electrons. For the transition series the first five electrons fill the 3d orbitals singly, without pairing, and it is not until the sixth electron is reached—the ions of Fe^{+2} and Co^{+3}—that the electrons with opposite spins begin to pair. Maximum spin is achieved in this series with Fe^{+3} and Mn^{+2}, both of which have five unpaired 3d electrons and a spin quantum number of 5/2.

Manganous ions (Mn^{+2}) have been used extensively as paramagnetic probes for in vitro biochemical studies. In vivo use of manganous ions for NMR imaging has been more limited. Aqueous manganous chloride has been injected intravenously (0.05 mM/kg) in dogs having experimentally induced myocardial infarctions.[5] In vitro NMR images of the dog hearts demonstrated a lack of perfusion and reduced Mn^{+2} concentration in the infarcted tissue and an observable contrast differential from the normally perfused myocardium. No toxicity was reported from this in vivo use of manganous ions in low concentration.

Iron salts and, in particular, ferric salts can be used safely and effectively to produce positive contrast enhancement of the gastrointestinal tract (Fig. 5).[15] The same iron compounds would not be safe for

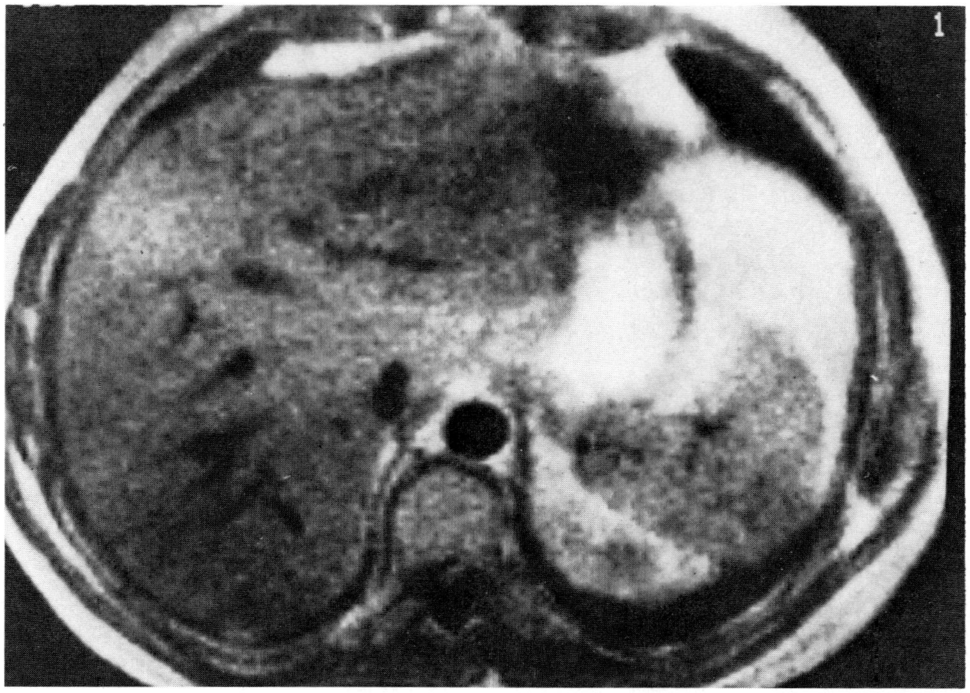

Figure 5. Positive contrast in the human gastrointestinal tract from oral administration of dilute ferric ammonium citrate.
(Reproduced, with permission, from Christiensen HE (Ed): *Registry of Toxic Effects of Chemical Substances*. Rockville, Md: U.S. Department of Health, Education, and Welfare, NIOSH, 1975.)

intravenous administration because of prolonged retention and storage of iron within the body.[16]

The lanthanide series also includes paramagnetic ions, those containing unpaired electrons in 4f orbitals. There are seven 4f orbitals; hence, maximum spin is achieved when ions in this group contain seven electrons with unpaired spin. This is the electron configuration of Gd^{+3} (gadolinium), the best known lanthanide paramagnetic ion, having a spin quantum number of 7/2. Gadolinium can be complexed in DPTA, a complexing agent commonly used for technetium in nuclear medicine, and injected intravenously. The resulting molecule is rapidly excreted in the urine and has strong paramagnetic properties in vivo. Gadolinium DPTA was shown by Fobben and Wolf to have contrast-enhancing properties for myocardial imaging in the rabbit;[17] in our laboratory the same complex, injected intravenously, has been shown in animals to enhance renal structures, subcutaneous inflammatory lesions, and brain abscesses.

The inorganic cations in the transition and lanthanide series are powerful paramagnetic compounds, which by virtue of their PRE are capable of enhancing contrast on NMR images. Figure 3 shows NMR intensity images of ferric (10 mM), cuprous (10 mM), and manganous (1 mM) salts and water for comparison. The contrast differences between water and these ionic solutions are striking. Potential disadvantages of these inorganic cations as intravenous NMR contrast media are their relative lack of chemical versatility, their delayed biologic excretion, and their toxicity.[16,18] On the basis of their toxicity, it seems questionable that these ions, if used alone, will achieve wide use in vivo as NMR contrast media. We may see the development of nontoxic complexes of these ions, such as gadolinium-DPTA, that are rapidly excreted and nontoxic but retain a strong PRE.

Stable Free Radicals

Free radicals are a group of paramagnetic molecules having one or more unpaired electrons in a molecular orbital. Unlike the paramagnetic inorganic ions, the unpaired electron in a stable free radical is contained in an outermost molecular orbital and is readily available for electron pairing with reducing agents (electron donors). The search for a stable free radical in which the electron is protected from pairing led in 1965 to the synthesis of a group of compounds that have come to be known as "stable free radicals" (SFRs) or "spin labels." The term spin label is now used to describe those SFRs that have been used as "probes" in the study of biochemical systems; unlike other free radical molecules, they can persist in biologic systems for minutes or longer without electron pairing. Many of these spin labels are chemically versatile and can be covalently attached to specific groups of biomolecules; moreover, the magnetic effects of these probes have been studied in vitro by electron spin resonance (ESR) spectroscopy. Further, the proton relaxation effect of spin labels on the relaxation values of neighboring hydrogen nuclei can be observed in vitro by NMR spectroscopy.

Stable free radicals deserve special consideration as NMR contrast agents because they demonstrate many of the characteristics that approximate the ideal characteristics of a pharmaceutical to be used for NMR contrast enhancement. At the University of California, San Francisco we have used SFRs in NMR imaging tests in animals, in particular, nitroxide stable free radicals (NSFRs). There are many paramagnetic NSFR compounds, but the major subgroups are the piperidine and pyrrolidine NSFR derivatives (Fig. 6).

Nitroxide stable free radical compounds are stable by virtue of (1) steric hindrance to electron transfer provided by four bulky methyl groups on the alpha carbons and (2) delocalization of the unpaired electron from the oxygen atom to the nitrogen atom. The stability of NSFRs makes them ideal probes, and by the suitable choice of a derivative it is possible to meet the specific requirements of many biologic problems. The nitroxide moiety is relatively nonreactive under most conditions; for example, it remains stable when heated up to 100°C and higher and over the pH range from 3 to 10. It can be stored at room temperature as a solid or in solution for extended periods without loss of activity.

Nitroxide stable free radicals have relatively strong PRE as a result of their electron paramagnetism. Like the transition series ions, the strong magnetic moment of an unpaired electron promotes both spin lattice and spin-spin relaxation of surrounding protons; T_1 and T_2 decrease. Because of these effects, relatively low concentrations of NSFR (1 to 10 mM) result in

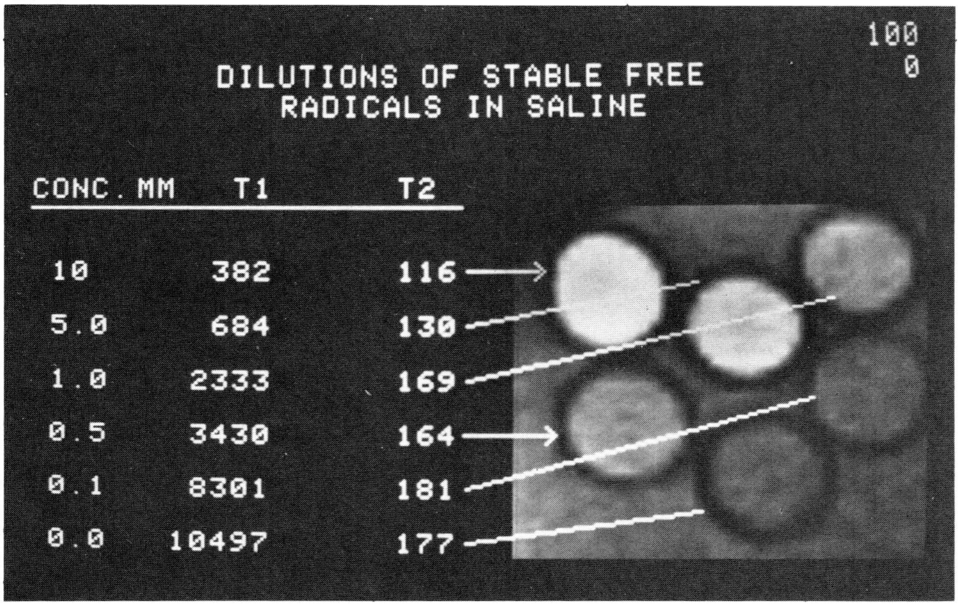

Figure 6. An NMR echo intensity image demonstrating transverse sections through six test tubes containing water and five dilutions of a piperidine NSFR derivative, ranging in concentration from 0.1 to 10.0 mM, demonstrates a dramatic increase in NMR intensity for the higher concentrations of NSFR as compared with water. Also tabulated are the calculated values of T_1 and T_2 given in milliseconds for each solution. Notice a general trend for T_1 and T_2 shortening with higher concentrations of NSFR. Because of an algorithmic problem in the calculation of T_1, values greater than 3,000 msec are relatively imprecise. By altering the "window width," which for this image is maximum, concentrations of NSFR as low as 0.5 mM (MM in figure) or less may be significant.

(Reproduced, with permission, from Brasch RC: Work in progress: Methods of contrast enhancement for NMR imaging and potential applications. A subject review. Radiology 147:781, 1983.)

an obvious and marked decrease in intensity within tissues on the NMR image. Figure 6 shows an NMR intensity image of NSFR solutions ranging from zero to 10 mM and computed numerical values of T_1 and T_2 relaxation values. The PRE of NSFRs is in the range of that of Cu^{+2} and Fe^{+2} in similar concentrations.

But unlike the paramagnetic ions, the potential for chemical conjunction of NSFRs makes them well suited for labeling of target-specific biomolecules. Nitroxide stable free radicals have been covalently attached to numerous drugs such as propranolol and steroids for the in vitro study of drug metabolism. By complexing NSFRs to substances of known specificity it seems possible to make them selective for different tissues and receptors. We have performed preliminary tests in animals with a piperidinyl NSFR (which we call "TES") used as a central nervous system and urographic contrast enhancer.[19,20] The normal excretion of TES is via the kidneys, probably by glomerular filtration; the half-life in the serum is approximately 40 min. Because TES is paramagnetic, it increases the observable NMR intensity image within the kidneys of living rats and dogs (Fig. 7). TES-enhanced spin echo renal images clearly demonstrated the presence of disease and functional abnormalities in diseased kidneys; these abnormalities were either not evident or only indirectly suggested by nonenhanced images (Fig. 8).[20]

Because TES is excluded from the brain by the normal blood-brain barrier, this NMR contrast agent has been useful in animals to detect focal blood-brain barrier defects. Dogs with experimentally induced unilateral bacterial cerebritis and radiation brain necrosis were imaged by NMR before and after intravenous administration of TES.[19] The normal brain showed no change in NMR appearance after contrast was given, but foci of disease demonstrated a dramatic increase in spin echo intensity and a decrease in T_1 relaxation times (Fig. 9). The areas of disease defined by TES enhancement were either not evident on the nonenhanced NMR images or were better defined after contrast administration. In-depth tests of toxicity, stability, and metabolism of TES and other NSFR derivatives are now in progress.

Because paramagnetic compounds show considerable potential to improve the diagnostic accuracy and utility of NMR imaging, our hope is that laboratory studies will soon lead to clinical trials and safe, diagnostically useful NMR pharmaceuticals.

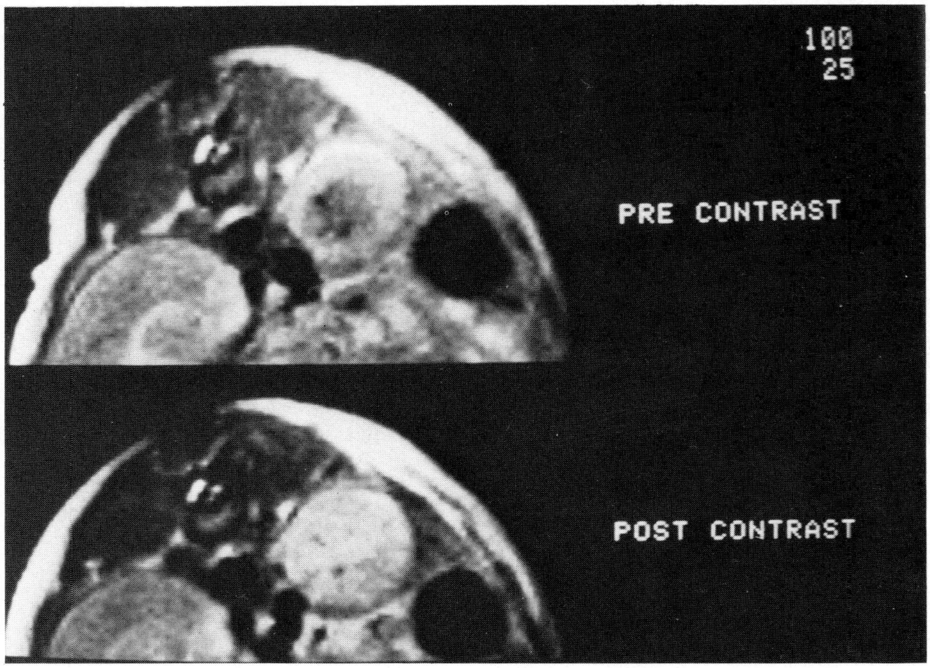

Figure 7. Nitroxide stable free radical contrast enhancement of the normal dog kidney.
(Reproduced, with permission, from Brasch RC, London DA, Wesbey GE, et al: Work in progress: Nuclear magnetic resonance study of a paramagnetic nitroxide contrast agent for enhancement of renal structures in experimental animals. Radiology 147:773, 1983.)

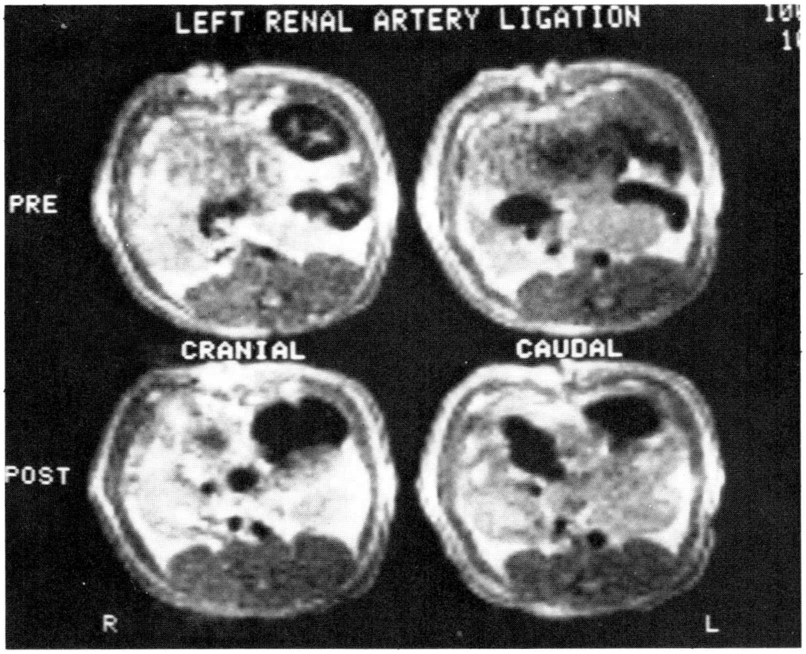

Figure 8. Nitroxide stable free radical contrast enhancement of the right kidney in rat with left renal artery ligation. Only the parenchyma and pelvocaliceal system of the right kidney are enhanced.
(Reproduced, with permission, from Brasch RC, London DA, Wesbey GE, et al: Work in progress: Nuclear magnetic resonance study of a paramagnetic nitroxide contrast agent for enhancement of renal structures in experimental animals. Radiology 147:773, 1983).

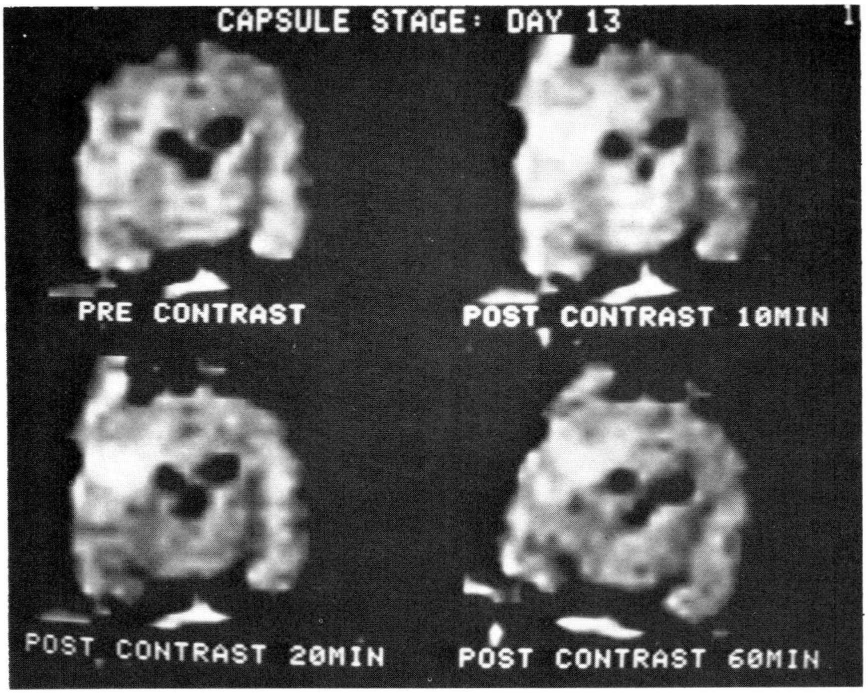

Figure 9. Nitroxide stable free radical contrast enhancement of a cerebral abscess in the right hemisphere of a dog.
(Reproduced, with permission, from Brasch RC, Nitecki DE, Brant-Zawadzki M, et al: Brain nuclear magnetic resonance imaging enhanced by a paramagnetic nitroxide contrast agent. Preliminary report. AJNR [in press]).

Acknowledgment

Portions of the text are reproduced, with permission, from Brasch RC: Work in progress: Methods of contrast enhancement for NMR imaging and potential applications. A subject review. Radiology 147:781, 1983, and from Brasch RC, London DA, Wesbey GE, et al: Work in progress: Nuclear magnetic resonance study of a paramagnetic nitroxide contrast agent for enhancement of renal structures in experimental animals. Radiology 147:773, 1983.

REFERENCES

1. Herfkins R, Davis P, Crooks L, et al: Nuclear magnetic resonance imaging of the abnormal live rat and correlations with tissue characteristics. Radiology 141:211, 1981.
2. Pykett IL: NMR imaging in medicine. Sci Am 246:78, 1982.
3. Kaufman L, Crooks L, Margulis AR (Eds): *Nuclear Magnetic Resonance Imaging in Medicine*. New York: Igaku-Shoin, 1981.
4. James AE Jr, Partain CL, Holland GN, et al: Nuclear magnetic resonance imaging: The current state. AJR 138:201, 1982.
5. Brady TJ, Goldman MR, Pykett IL, et al: Proton nuclear magnetic resonance imaging of regionally ischemic canine hearts: Effect of paramagnetic proton signal enhancement. Radiology 144:343, 1982.
6. Brasch RC: Work in progress: Methods of contrast enhancement for NMR imaging and potential applications. A subject review. Radiology 147: 781, 1983.
7. Slichter CP: *Principles of Magnetic Resonance,* pp 1–50, 137–183. New York: Harper and Row, 1963.
8. Wolf GL, Leigh J, Conrad B: The influence of solution of concentrated approved pharmaceuticals upon relaxation time in NMR. Presented at the Annual Meeting of the Association of University Radiologists, Baltimore, 1982.
9. Hansen G, Crooks LE, Davis P, et al: In vivo imaging of the rat anatomy with nuclear magnetic resonance. Radiology 136:695, 1980.
10. Newhouse JH, Brody TJ, Gebhardt M, et al: NMR imaging: Preliminary results in the upper extremities of man and the abdomen of small animals (abstract). Radiology 142:246, 1982.
11. Davis PL, Kaufman L, Crooks LE, et al: Detectability of hepatomas in rat livers by nuclear magnetic resonance imaging. Invest Radiol 16:354, 1981.
12. Young IR, Bailes DR, Burl M, et al: Initial clinical evaluation of whole-body nuclear magnetic resonance (NMR) tomography. J Comput Assist Tomogr 6:1, 1982.
13. London DA, Davis PL, Williams RC, et al: Nuclear magnetic resonance imaging of induced renal lesions. Radiology 148:167, 1983.
14. Axel L: Unpublished observation.
15. Wesbey GE, Brasch RC, Englestad BL, et al: Gastrointenstinal contrast enhancement for NMR imaging using nontoxic oral iron solutions. Radiology (in press).
16. Christiensen HE (Ed): *Registry of Toxic Effects of Chemical Substances*. Rockville, Md: U.S. Department of Health, Education, and Welfare, NIOSH, 1975.
17. Fobben E, Wolf GL: Gadolinium DTPA: A potential NMR contrast agent. Effects upon tissue proton relaxation and cardiovascular function in the rabbit. Presented at the Annual Meeting of the Association of University Radiologists, Mobile, Alabama, March 1983.
18. Browning E: *Toxicity of Industrial Metals*. New York: Appleton-Century-Crofts, 1969.
19. Brasch RC, Nitecki DE, Brant-Zawadzki M, et al: Brain nuclear magnetic resonance imaging enhanced by a paramagnetic nitroxide contrast agent. Preliminary report. AJNR (in press).
20. Brasch RC, London DA, Wesbey GE, et al: Work in progress: Nuclear magnetic resonance study of a paramagnetic nitroxide contrast agent for enhancement of renal structures in experimental animals. Radiology 147:773, 1983.

19

Biophysical Phenomena and Health Hazards of In Vivo Magnetic Resonance

THOMAS F. BUDINGER, M.D., PH.D.
CHRISTOPHER CULLANDER

The effects on the health of patients exposed to NMR in vivo measurement techniques are considered in six categories: (1) effects of static magnetic fields up to 2 T (20,000 Gauss); (2) effects of induced currents associated with rapidly varying magnetic fields used for spatial localization in imaging instruments; (3) radio-frequency (RF) heating effects from RF magnetic fields associated with the RF coil used in the NMR imaging or sampling systems; (4) effects of uniform and varying magnetic fields on cardiac pacemakers, surgical clips, and prostheses; (5) epidemiologic studies; and (6) prospective experimental directions.

Previous discussions of potential hazards of NMR imaging[1–13] have addressed only parts of these problems, and previously suggested guidelines[2] have been substantially modified in light of reinterpretation of experimental results on which those conservative guidelines were based.[12,13] There is a vast literature on biomagnetic effects, much of it contradictory and confusing. Usually the interpretation of experimental results is difficult because there are insufficient data to warrant a statistically valid conclusion. The circumstances of each experiment, such as temperature, lighting, and mechanical vibrations, are frequently neglected. For example, in a recent article that concluded that there were nonthermal RF effects on frog eggs,[14] the authors noted that small mechanical disturbances would produce the same result; however, there was no mention of efforts to isolate the experiment from the mechanical vibrations of the RF generating apparatus. Another example of an incorrect conclusion regarding the detrimental effects of static magnetic fields is the finding of glial hyperplasia and hypertrophy in rabbits and cats after exposure to 20–30 mT.[15] Because the same histopathology is commonly found in unexposed rabbits as a result of the endemic encephalitozoonosis,[16] the causal connection between central nervous system lesions and magnetic fields cannot be made. A final example of how environmental variables have led to erroneous results is the discovery by Tucker and Schmitt that audible cues were biasing experimental results on human detection of low-frequency oscillating magnetic fields.[17]

This chapter endeavors to set in some perspective present knowledge about physiologic effects caused by static magnetic and electromagnetic fields. At present there are no substantiated harmful effects from the NMR experiment, nor is there substantial evidence that one should expect effects when using currently available commercial and experimental instruments for NMR in vivo imaging and spectroscopy, with the exception of effects on prostheses, surgical clips, and pacemakers. As instruments go beyond 2 T (1 T equals 10,000 Gauss) for static fields, heating from the higher frequency and increased power of the RF to be used is an important question that requires continuing evaluation.

Static Magnetic Fields

With the exception of the well-established low-level magnetic field effects on the orientation of bacteria and birds[18–20] and the electrocardiographic (EKG) abnormalities associated with an induced electromotive force (EMF) on blood flow,[21] there is no substantiated scientific evidence for biologic effects of static magnetic fields. Conflicting reports, however, can be found in reviews by Barnothy,[22] Busby,[23] Kholodov,[24] Silver and Tobias,[25] Sheppard and Eisenbud,[26] Phillips et al.[27] and Tenforde.[28] From the vast literature on experiments with cell cultures, animals, and man, no experimental protocol has been found that when repeated by other investigators gives similar positive results. For example, a major experimental result reported in 1976 by Malinin and coworkers gave strong evidence for morphological transformation of mammalian cells (fibroblasts) in fields of 0.35 T.[29] It was this work as well as repeatable experiments of T-wave abnormalities in the EKG of animals exposed to fields greater than 0.3 T that led to the previous suggestion of a prudent threshold in the range of only 0.3 T for the static magnetic field in NMR imaging studies.[2] That suggestion was subsequently modified to 2 T because we learned that the EKG signals were not physiologic[3] (discussed below) and that the experiments reported by Malinin et al.[29] cannot be replicated under well-controlled conditions unless some slight modification in the method of treating control versus exposed groups is employed. This apparently was the case in the studies reported previously (Mahlum, personal communication, 1981). Extensive studies by Frazier[30] and other studies on tumor cells[31] show no effects of high fields if exposed and control biologic cells are handled in the same manner.

Experiments showing static field effects on the drug-induced behavior of *Spirostomum*[32] have not been repeated in another laboratory. Genetic effects have not been shown in chromosome studies by Wolff and coworkers at fields of 0.3 T,[4] by Baum and coworkers on the plant *Tradescantia* at fields up to 0.8 T,[33] on *Drosophila melanogaster* at 3.7 T,[34] and, recently, on cultured human blood cells at fields up to 1.0 T.[35]

Recently, some electrophysiologic effects on the pineal cells of the guinea pig were noted on exposure to fields comparable to earth's field (0.1 mT),[36] and Raybourn has reported a suppression of ionic current fluxes in the turtle retina,[37] but these results have not yet been evaluated in terms of a harmful health effect.

A possible magnetic phenomenon of great interest was the report by Baker of human homing ability,[38] but this result could not be repeated in an extensive study by other investigators.[39]

Four proposed mechanisms whereby magnetic fields might influence biologic processes or organismal behavior are: (1) macromolecule orientation changes, leading to changes in chemical kinetics and membrane permeability; (2) changes in enzyme kinetics by quenching superconductivity, which might be present in biologic molecular systems, or by conformational changes; (3) reduction of nerve conduction velocity; and (4) superposition of low potentials (induced EMF) on natural biopotentials. Other theoretical biomagnetic interactions have been suggested by Mansfield and Morris.[8]

Thresholds for the first three mechanisms are 1 to 20 T or above, as will be explained below. Measurable flow potentials of a few millivolts are induced by fields above 0.3 T. These potentials have not produced physiologic effects.

Magnetic Interaction Energy and Cell Orientation Effects

The magnetization energy or interaction energy of a field H with biologic molecules of diamagnetic susceptibility is given as:

$$W = -1/2 \int_V \vec{H}\cdot\overline{\overline{X}}\cdot\vec{H}\ dV \qquad \text{Eq. 1}$$

where $\overline{\overline{X}}$ is the magnetic susceptibility tensor and the integration is carried out over the volume of the biologic structure. This equation predicts that the magnetization energy will increase with the square of the field; thus, when going from fields of 1.0 T to 6.0 T, where human exposure experience will be gained in the near future, the magnetization energy will increase 36-fold. Therefore, biologic experiments at fields of 2 T and above are necessary. For fields up to 2 T, the magnetization energy for small biologic subcellular and cellular structures is on the order of 10^{-12} to 10^{-14} erg. This is comparable to the room temperature thermal energy of 4×10^{-14} erg.

Molecules and some organized cellular structures have magnetic susceptibilities that vary with direction (e.g., retinal rods DNA, liquid crystals, and sickled red cells).[40–42] If this anisotropy in susceptibility is sufficiently large, then a turning torque can be expected. The torque, or force, will be proportional to the difference of the susceptibilities, orientation of the molecule relative to the field, and the magnetic field squared, as can be seen from Equation 2.

$$W = -(V/2)(|\vec{X}_r| + |\vec{X}_l - \vec{X}_r|\cos^2 - \theta)|\vec{H}^2| \qquad \text{Eq. 2}$$

where $X_l >> X_r$ and θ is the angle between $\vec{X_l}$ and $\vec{H}$. Note that if n molecules or subcellular units are aligned, the torque will be proportional to the number of connected units in a cooperative fashion. With a sufficiently large anisotropy of each unit and a sufficient number of units, large turning torques can be expected at a field strength below 2 T. Such is the case in repeatable experiments on the orientation effects of retinal rods[41, 42] in 1-T fields and the alignment of sickle cells at 0.35 T.[40] Even the observations on orientation of bacteria and some animals[19] might be explained by the physical torque rather than the sensing of the turning torque by nervous tissues.[20] The susceptibility difference or magnetic polarization of the magnetite chains in the bacterium *Magnetotactum* appears to be ample for an orientation effect at earth field strengths. No mutation or lethal effects from static magnetic fields of 1 T were found with various bacterial strains.[43]

Effects on Enzyme Kinetics and Quenching of Superconductivity

Superconductivity is present in some organic molecules, and it has been postulated that superconductivity is involved in reaction kinetics. Magnetic fields can revert a superconductor to a normal state; thus, it has been proposed that enzyme reaction rates might be affected by this mechanism. The critical magnetic field required to effect this transition might be 20 T. This possibility and the reports of reduction in reaction kinetics at fields of 2 T[44–46] led to an investigation of the effects of fields from 5 T to 20 T on a number of different enzyme reactions.[47] They found no effects, contrary to previous reports, even though they carefully reproduced the circumstances of others' experiments. Also, as recently noted, the linear dependence of chemical shift dispersion on field strength implies that there is no magnetic field dependence on enzyme kinetics.[8] Other evidence for no enzyme effects from very high fields is from in vivo P-31 energy metabolism experiments[48–50] wherein no changes in brain and heart energy metabolism were noted during studies in animals, nor were late sequelae seen in exposed animals after the experiments.

Other recent experiments on the infectivity of DNA after exposures to 2 T show no effect (Tenforde, personal communication, 1981).

Effects on Nerve Conduction

The theoretical basis for expecting a change in conduction velocity of nerves is that magnetic fields act on current paths in a way similar to the effect of a magnetic field on a charge moving in that field:

$$\vec{F} = q(\vec{v} \times \vec{H}) \qquad \text{Eq. 3}$$

where q is the charge, $\vec{v}$ is velocity, and $\vec{H}$ is the magnetic field. This force causes a distortion that is perpendicular to both the magnetic field and the original direction of the ion movement. A magnetic field perpendicular to the current loop pairs associated with the propagation along the nerve will result in the maximum distortion of the current loops and a consequent reduction in current velocities (Fig. 1). Theoretical examinations of the field strengths necessary for a reduction in conduction velocities of action potentials argue that fields of 24 T are required to produce a 10% reduction in nerve impulse conduction velocity.[51,52]

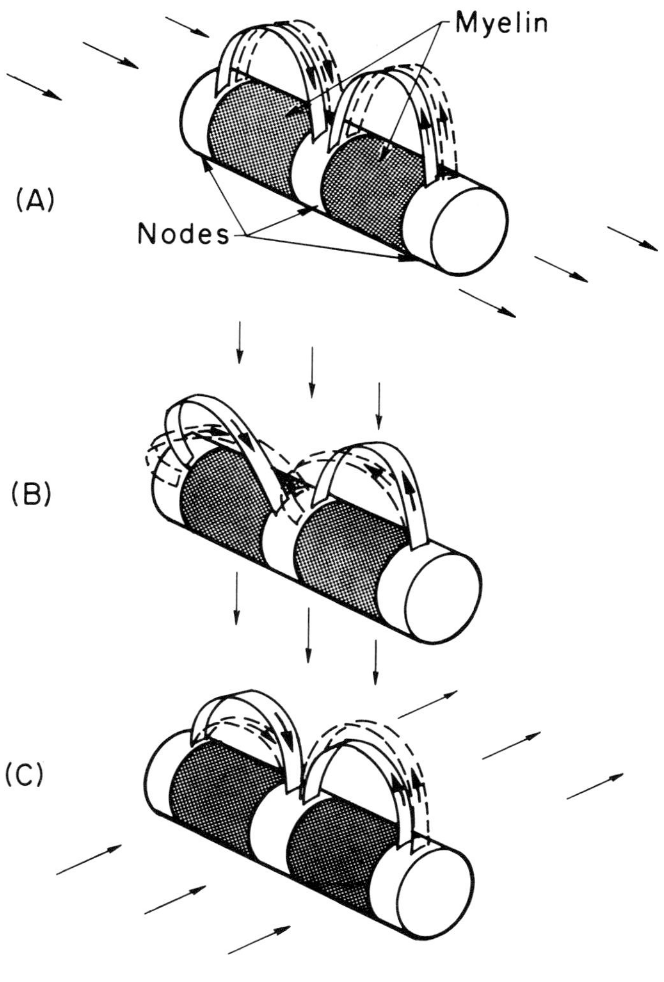

Figure 1.
A–C. Strong magnetic fields will affect current paths depending on the field strength and the direction of the current paths relative to the field. The normal current paths are shown by the open arrows, and the distorted paths are shown by the shaded arrows.

Observations that demonstrated large changes in conduction velocity in 1-T fields,[53] and other measurements in a constant inhomogeneous field[54] appear to be changes associated with experimental conditions other than magnetic field effects. These experiments could not be duplicated by Schwartz[55] with a lobster nerve preparation, nor more recently, in 1980, by Gaffey and Tenforde[56,57] with the frog sciatic nerve. In the latter studies, exposures as long as 4 hr at 2 T did not show a change in the action potential amplitude, the absolute refractory period, or in the conduction velocity. Small changes in submaximal action potentials were found to be associated with a very slight change in temperature (2%–4% change in submaximal action potential amplitude for a change of 0.1° C).

Induced Potentials and Currents Due to Flow

High static magnetic fields are associated with abnormalities in the electrocardiograms of the heart.[58,59] These changes in the T-wave at high magnetic fields are related to induced EMF potentials resulting from the flow of blood through the field. Recently, Gaffey and coworkers completed a comprehensive study of the T-wave changes in rats, dogs, and monkeys.[21,60,61] An alteration in the amplitude of the T-wave is noted with a threshold near 0.3 T (Fig. 2). The T-wave abnormality increases linearly with field strength to 2 T (highest level in the experiments). These observations are consistent with the hypothesis that the EKG changes are from the induced EMF associated with high-velocity blood flow perpendicular to the static magnetic field. The essential features of their observations that are relevant to NMR imaging are the following: No arrhythmias or changes in heart rate are induced; the normal EKG appears immediately on cessation of field exposure; and blood pressure and chemistry of the animals are not altered. What is being observed is an effect quite separate from any physiologic effect.

The physical basis of this phenomenon is the well-known voltage generation phenomenon associated with motion of a conductor in a magnetic field. The induced voltage is derived from Equation 3 and is

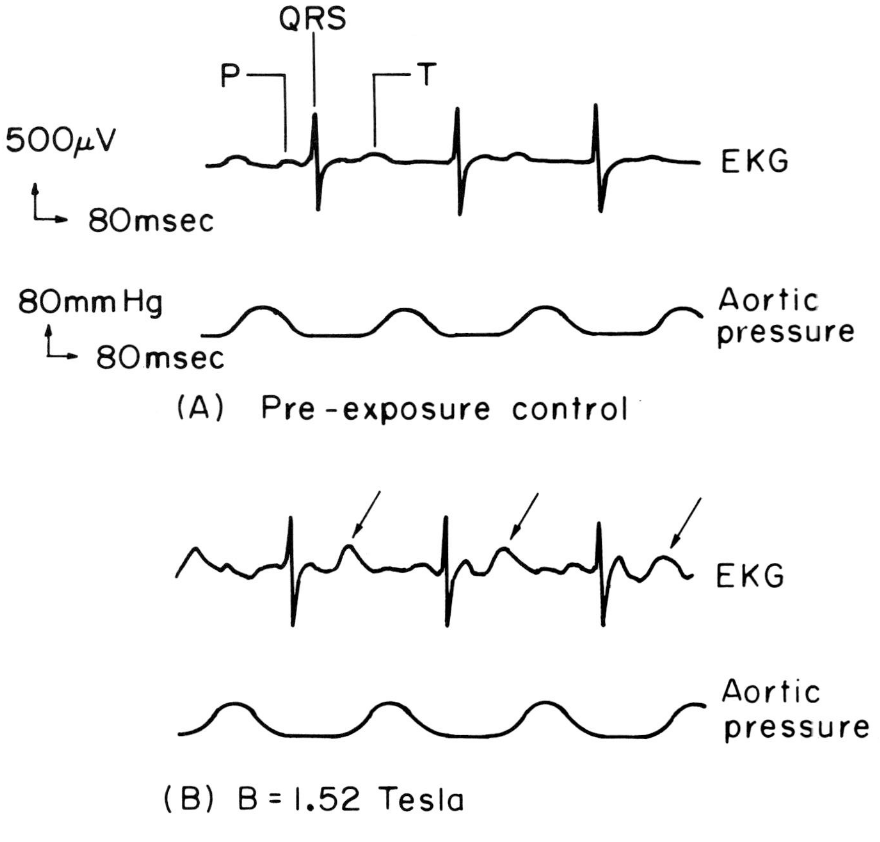

Figure 2. The normal EKG and the aortic blood pressure in the monkey (A) and the changes in the EKG (B) (arrows) at 1.52 T without any corresponding change in the aortic pressure, as would be expected from theory.

given by

$$\epsilon(\text{mvolts}) = \frac{\vec{H} \times \vec{v} d}{10} \qquad \text{Eq. 4}$$

where ϵ is in millivolts, $\vec{H}$ is in T (10^4 Gauss), and d is the diameter (cm) of a vessel through which there is a uniform velocity of $\vec{v}$ in cm/sec (Fig. 3). A simple conversion of Equation 4 to volumetric flow gives

$$\epsilon(\text{mvolts}) = \frac{\text{flow (cm}^3\text{/sec)}H}{2.5\pi d} \qquad \text{Eq. 5}$$

The magnitude of these potentials can be appreciated from the example below. The maximum flow in a dog aorta is about 240 cm^3/sec (180 beats/min; stroke volume 80 cm^3), and the aorta diameter is approximately 1.8 cm. At a field of 0.3 T (3,000 Gauss) perpendicular to flow, the induced EMF is 5 mV, which, neglecting the effects of volume conduction, will give a signal larger than the T-wave. The temporal pattern of the induced voltage depends on the position of the heart chambers and great vessels relative to the static magnetic field. Because of volume conduction effects, the potential measured on the surface of a body will be 10 or more times less than

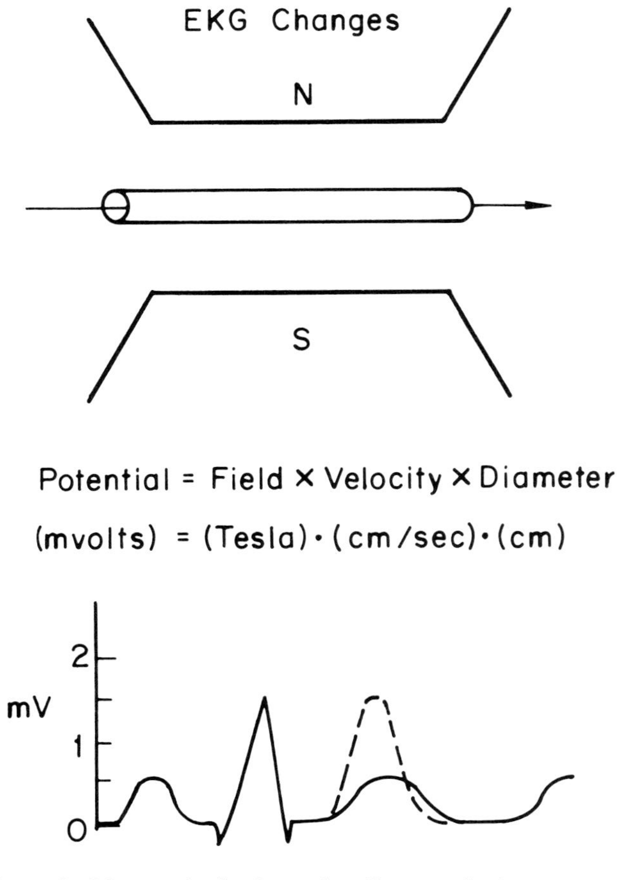

Figure 3. Magnetohydrodynamic effect results in a superposition of a potential over the normal physiologic EKG signals.

the induced potential change. The induced voltage recorded from monkeys is about 3 mV at 1.8 T. The presence of an induced EMF from normal flow through the human leg immersed in fields approaching 1 T has also been observed.[62]

A second effect that can contribute an EKG artifact in high fields is Faraday induction:

$$\epsilon(\text{volts}) = -10^{-4}|\vec{H}|(da/dt) \qquad \text{Eq. 6}$$

where a is the area perpendicular to the field. The chest diameter changes slightly as the heart forces blood through the thoracic vasculature. Suppose a 20-cm diameter chest changes by 2 mm during a 100-msec interval because of the pulse of blood through the thoracic vasculature during a heart contraction. This can cause a change of 10 cm^2/sec in the area perpendicular to the field. In a field of 0.3 T, a voltage of 0.3 mV will be induced. At 2 T, the induced voltage in the surface electrode loop will be nearly 2 mV. The area change can be expected to oscillate, and a ripple of time varying induced potentials is expected. This effect is sensitive to electrode orientation, but not to the flow direction relative to the magnetic field, as is the case for the magnetohydrodynamic effect phenomenon of Equation 5.

A third phenomenon of pressure and flow change due to the effect of a static field perpendicular to the direction of flow of a conductor has been predicted.[63–65] This effect is a decrease in internal pressure in the major arteries as the result of an increase in viscous force and a slight decrease in the timing of the pressure pulses in the artery. A recent analysis has shown that this change in effective viscosity would result in a pressure increase in large-diameter vessels of only 0.002 mm Hg for a 1-T field.[66] At 6 T the predicted change would be only 0.2 mm Hg; thus cardiovascular risk from pressure changes in the vascular system is not significant.

Induced Currents from Rapid Magnetic Field Changes

Rapidly changing magnetic fields are superimposed on the main static field to obtain spatial information in NMR imaging.[67] These time-varying magnetic fields generate internal electric currents by the well-known Faraday induction phenomenon. Rapid changes in magnetic fields of 3 T/sec can induce currents of 3 $\mu A/cm^2$, and if this current density is maintained for a sufficient time duration in one direction, nonthermal biologic effects (not necessarily harmful) can be expected. Consider the arrangement shown in Figure 4, wherein the subject is exposed to a changing field dB/dt. The induced EMF around each loop J is

$$\epsilon(\text{volts}) = -10^{-8}(dB/dt)\pi r^2 \qquad \text{Eq. 7}$$

where r is the radius of the loop in which the EMF is induced. The E field is given by

$$|\vec{E}| = \frac{\epsilon}{loop} = \frac{\epsilon}{2\pi r} \qquad \text{Eq. 8}$$

and the current density is

$$\vec{J} = \vec{E}\sigma \qquad \text{Eq. 9}$$

where σ is the tissue conductivity. Thus, the current density is

$$\vec{J} = -10^{-4}(dB/dt)(r\sigma/2) \qquad \text{Eq. 10}$$

where J is in A/m^2, dB/dt is in T/sec, r is in meters, and σ is in S/m.

The absorbed power per unit of mass of tissue is derived from observing that the energy per time imparted to charged particles by an E field is $\vec{E} \cdot \vec{J}$. Thus, the average absorbed power per unit mass or the specific absorbed power[67] for the induced average electric field $|\vec{E}|/\sqrt{2}$ is

$$SAR = (|E|/\sqrt{2})^2(\sigma/\rho) = (\sigma/2\rho)|\vec{E}|^2 \qquad \text{Eq. 11}$$

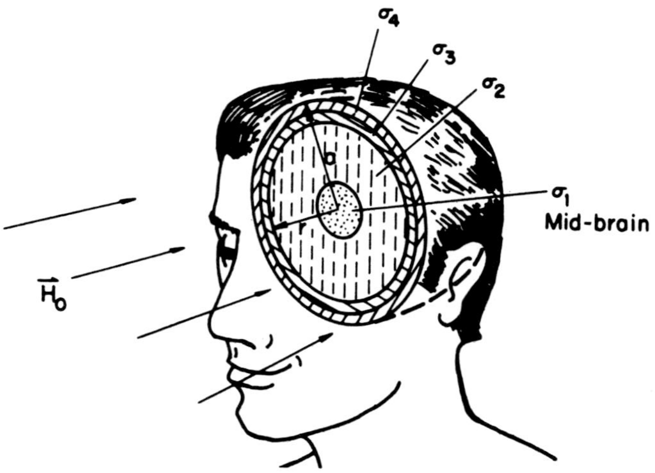

Figure 4. The currents induced in different regions of the body from a rapidly changing field depend on the conductivity of that portion of the body and the radius of the effective loop. The current density in region σ_4 will be greater than that for other regions of similar conductivity.

where ρ is the tissue density. The energy absorbed is, of course,

$$W = SAR \cdot \text{time} \qquad \text{Eq. 12}$$

Using Equation 12, we can calculate the energy available to induce biologic effects through electrical

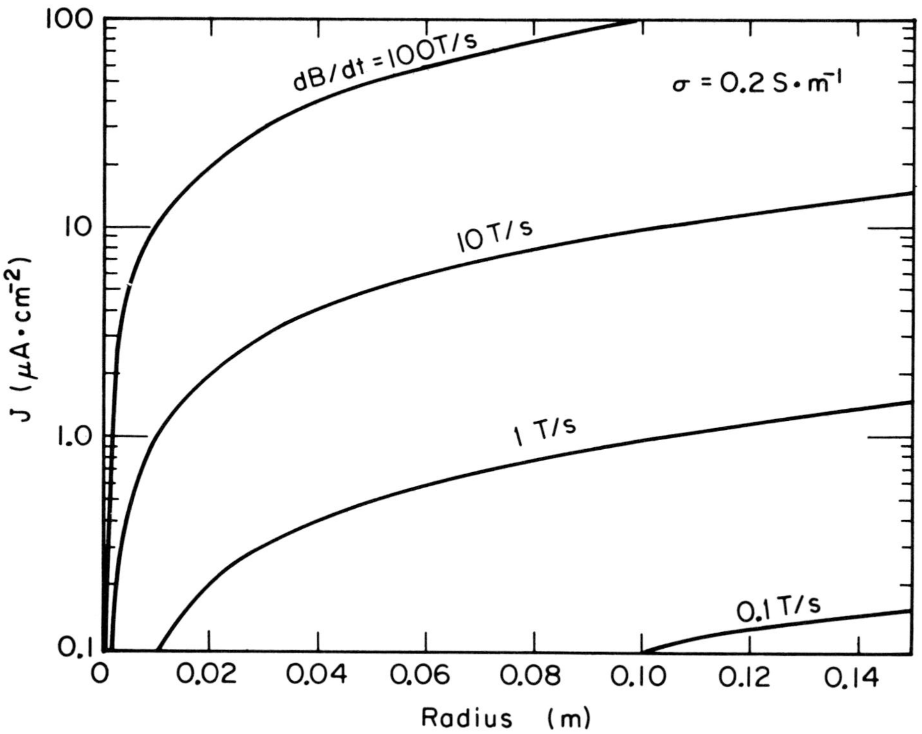

Figure 5. The current density in μ/A/cm^2 is shown as a function of the radius of the loop diameter for various rates of field change. Conductivity of 0.2 S/m is assumed. For a different conductivity the ordinate scale can be modified in proportion to the conductivity change.

phenomena or through heating. Heat production is the main mechanism for biologic effects of the sinusoidal magnetic field changes discussed in a later section.

The threshold current densities required for known biologic effects were evaluated previously[2] using the assumption of a 20-cm diameter loop and the above equations with a tissue conductivity of 0.2 S/m. A working relation of 1 $\mu A/cm^2$ for a dB/dT of 1 T/sec was used.[2] This working relationship between induced current density and rate of change of magnetic field has two important defects: First, the effective loop size changes by large factors from one experiment to another; thus, the current density will also vary (Fig. 5). Second, the biologic effect is believed to be strongly dependent on the duration of the EMF, that is, some minimum energy is necessary for an effect. Nevertheless, the conversion, 1 $\mu A/cm^2$ for 1 T/sec, gives an order of magnitude for the levels of current densities that will be experienced in the changing fields used in NMR imaging strategies.

The current densities involved in induction of visual light flashes by the application of electrodes across the human head are only 17 $\mu A/cm^2$ as inferred from experiments published in 1947 by Barlow and coworkers[69] and corroborated by induction of light flash sensations in partially blind children at frequencies of 10–30 Hz.[70] Perhaps the best documented biologic effect of magnetic fields is the induction of visual light flashes by changing magnetic fields. These visual sensations known as magnetic phosphenes were first described by d'Arsonval in 1896.[71] Other studies of note on this phenomenon were by Barlow and coworkers in 1947[72] and extensive studies by Swedish investigators published in 1980.[73] From their data, our experiments, and studies at the Francis Bitter National Magnet Laboratory, the threshold field change for phosphene induction is 2–5 T/sec. However, it is important to note that a minimum magnetic field amplitude of approximately 10 mT (100 Gauss) is necessary for this effect at a frequency of 20–30 Hz.[72] In other words, at a dB/dt of 5 T sec, the duration of current density in a particular direction must be 2 msec (the durations for NMR instruments is 0.2 m/sec or less). Thus, it is not surprising that in the experiments of Tucker and Schmitt,[17] magnetophosphenes were not induced using 60-Hz fields with a maximum amplitude of 5 mT. Experiments conducted by the author at 210 T/sec with pulses of ~3 μsec duration did not induce phosphenes. Figure 6 shows a composite of experiments by Barlow and coworkers,[72] Lovsund and coworkers,[73] Tucker and Schmitt,[17] and experiments by the author. Though the dependence of the induction of magnetophosphenes on both the rate of change of the field and the time duration has been clearly stated in previous discussions,[2,3] some misunderstanding has persisted with regard to suggested standards of 3T/sec without an accompanying qualification concerning the time duration.[12] A qualification was given by the British

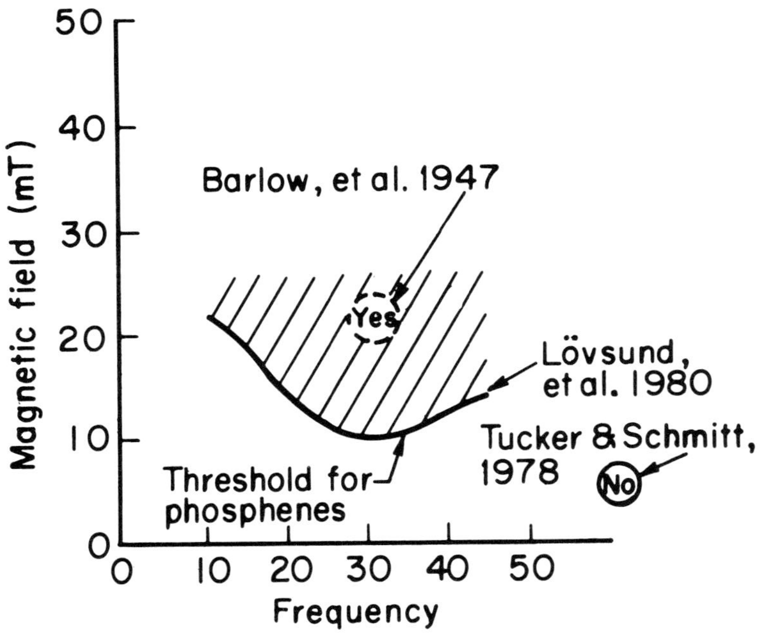

Figure 6. Observed relation between the maximum amplitude and frequency for the induction of visual light sensation (phosphenes).

Bureau for Radiation Protection by the statement, "The maximum rate of change (at any part of the body exposed to the field) should not exceed 20 tesla per second for pulses of 10 msec or longer."[13] This level is in the range where experiments are necessary, particularly for long-duration exposures of a developing fetus and with a prosthesis with a large area.

A second area of known biologic effects at low-current densities is stimulation of bone healing by induced electric fields using a coil near the fracture site. This technique has recently become an accepted method of therapy.[74–76] A current density of 10 $\mu A/cm^2$ for each pulse with a pulse repetition rate of 30–60/sec can be inferred from studies performed by Bassett and coworkers.[75] Pulses are asymmetric, with a rise time of about 1 μsec. Positive bone healing is reported when these pulses are applied repeatedly over days or weeks. Of relevance to the NMR situation is the fact that sinusoidal waveforms have little or no effect in the range of 30–65 Hz (Bassett, personal communication, 1980), whereas asymmetric waveforms give remarkable changes both in in vitro experiments and in patients. The biophysical phenomenon of electrophoresis is the most likely underlying mechanism for those effects, and experiments are presently under way to measure cell and molecular behavior under induced current conditions.

Three other areas of biologic and medical effects of induced currents are electroshock therapy, induction of heart fibrillation using capacitive coupling, and the stimulation of nerve action potential. Our study of the electrode configurations and currents applied to the head in the work of Fink[77] shows that current densities of 3 mA/cm^2 are applied for 300 msec in order to cause seizures. Animal experiments have shown that current densities of 1 A/cm^2 cause reversible nerve damage.

Magnetic stimulation of muscles by means of time-varying magnetic fields has been reported by several investigators.[78–82] Nerve and muscle stimulation have been induced by varying magnetic fields in two configurations.[81] Of interest in studies by Ueno and coworkers is that the strength-duration (dB/dt versus time) curve shows a chronaxy of 30 μsec for single-burst experiments. This means that below a stimulation amplitude of some threshold value (perhaps in the range of 3 T/sec), the nerve-muscle preparation will not be excited no matter what the pulse duration might be. In another series of experiments Ueno and coworkers found that a pulsed magnetic field of 6×10^5 T/sec for 20 μsec or of 900 T/sec for 1.5 msec did not alter the action potential of the giant squid axon, nor did it produce an action potential itself.[82] Because the induced EMF and thus the current density is proportional to the effective loop size (area) in accord with Equations 8–10, and the equivalent loop diameter was only 1 cm in their experiments, the current density was less than 1 mA/cm^2. Current densities of 1 mA/cm^2 or greater are needed to induce an action potential in the squid axon.

The current densities involved in mammalian nerve action potential are about 3.0 mA/cm^2, and those for induction of heart fibrillation are 0.1 to 1.0 mA/cm^2. Current densities involved in nerve and muscle stimulation are nearly 10^6 times greater than those being used in NMR imaging strategies. Pulsed fields on the order of 10^4 T/sec have been used to stimulate peripheral nerve tracts.[83] The field strength used was 2.2 T, and the risetime was 180 μsec. However, the current density at the periphery of a 10-cm diameter stainless steel object ($\sigma = 2 \times 10^3$ S/m) will be 50 mA/cm^2 for a field change of 10 T/sec.

Perhaps the most important area of experimental results with regard to the potential hazards of induced currents is that of heart fibrillation. Extrapolation of experimental data obtained by Watson et al.[84] from human hearts using 1.8-cm^2 electrodes gives a lower threshold for fibrillating the human heart of 300 $\mu A/cm^2$. The minimum fibrillation current density for the dog heart with an electrode area of 0.9 cm^2 is 550 $\mu A/cm^2$. Data obtained from a review by Roy[85] and the above observations are summarized in Figure 7. Unfortunately, the time during which voltage was applied across the heart was not varied in these experiments. The method of inducing fibrillation involved application of sinusoidal voltages through electrodes of various sizes. The state of fibrillation was reached by slowly increasing the voltage. It can be concluded from previous work that the duration of application of a sinusoidal field is 3 sec or less. There was no evidence of damage at these current densities.

At Donner Laboratory, we explored the EKG behavior of the heart and the blood-brain barrier integrity in anesthetized rats exposed for varying periods up to 10 min to field changes of 24 T/sec at repetition rates of 30/sec. The rise time was about 1 μsec. No abnormalities were observed in the EKG pattern or in the animal behavior after anesthesia.[86] Because the current density is very dependent on the loop diameter as discussed above, a dB/dt of 24 T/sec might induce a current of only 5 $\mu A/cm^2$, 100 times less than that required for fibrillation.

A plot of hypothetical relations between the induced

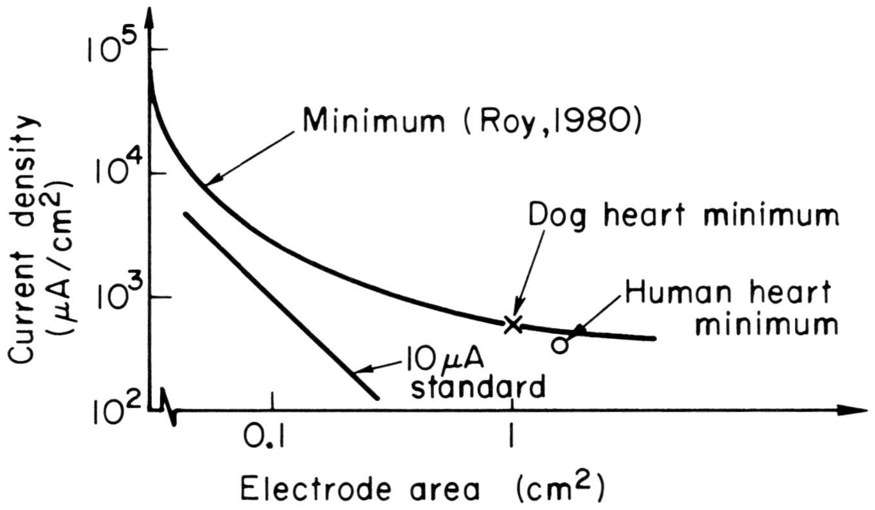

Figure 7. Relationship between the current density and electrode area for induction fibrillation in dog and human hearts based on experimental data. The change in field with time necessary for the induction of fibrillation is probably over 500 T/sec assuming 1 $\mu A/cm^2$ for 1 T/sec.

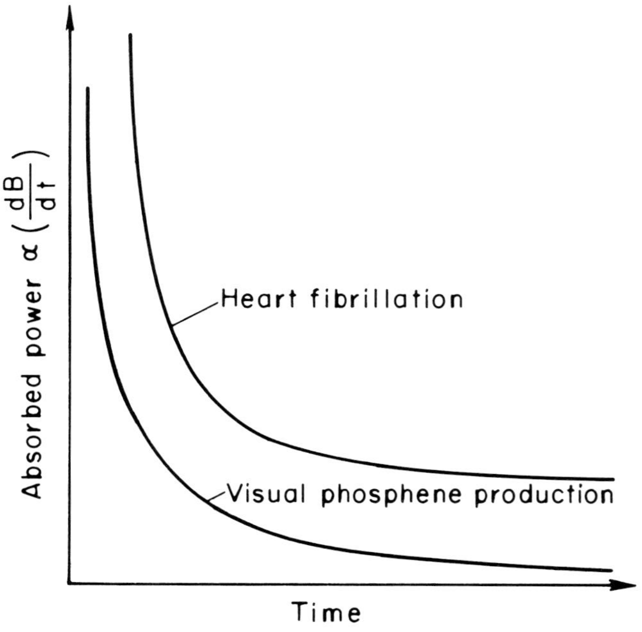

Figure 8. Proposed relationship between the absorbed power or induced currents and the biologic changes of the heart fibrillation and visual phosphene production.

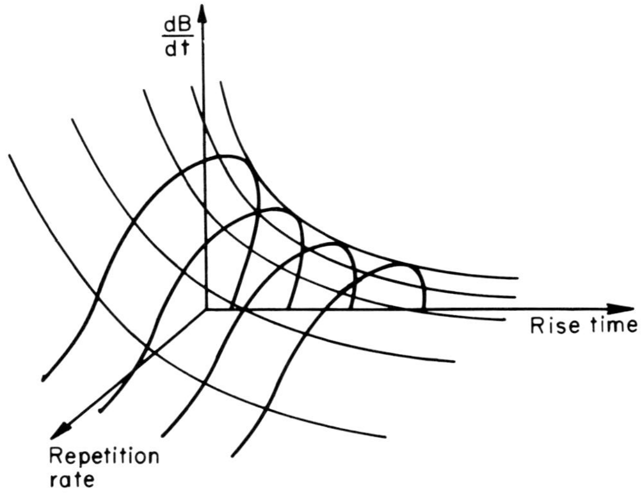

Figure 9. Biologic effects from induction of currents is dependent on the risetime, the repetition rate of each wave form, the duration of exposure, and the rate of change of field.

current density and time is given in Figure 8. The safe threshold, as has been emphasized previously,[2] is a complex function of waveform, amplitude, time duration, and repetition rate (Fig. 9). Experimental observations are not extensive enough to warrant a comprehensive guideline at this time. An interim guideline for design of systems with sinusoidally switching fields at frequencies of <1–200 Hz is to keep the amplitude of the oscillating magnetic field below 5 mT.

Radio-Frequency Magnetic Radiation Effects

Heating Effects

The RF heating problem from propagating electromagnetic radiation is conventionally approached by calculation of the absorbed energy resulting from field exposures to incident electromagnetic power densities.[68, 87, 88] In the past, tissue heating was evaluated based on measurements of the incident power density, frequency of the radiation, and the size of the object. This led to an appreciation of the fact that there is as much as a 100-fold difference between absorbed power in a rat and that in adult human subjects for the same frequency and incident power density because of the size of the object relative to the wavelength of the RF radiation. Theoretical estimates have appeared recently that show a relationship between power deposition in the body exposed to various magnetic RF frequencies and pulse sequences.[6] Estimations using the incident power density as ascertained from the exposed surface of the object and the average power of the RF system are not appropriate for the NMR imaging situation, since the RF radiation is magnetic and the object is in the near field.

Perhaps the best method for estimating the potential heating hazards in NMR studies is to measure the absorbed power using the difference of the Qs of the RF coil for the loaded and the unloaded condition. That is,

$$SAR = [\alpha\omega E/Q_L][1-(Q_L/Q_u)][T/\tau][1/V_p]\text{w/kg} \qquad \text{Eq. 13}$$

where α is the fraction of absorbed power not radiated or reflected out of the system, ω is the frequency, E is the energy, Q_L is the measured Q with the subject in the coil, Q_u is that of the unloaded coil, T is the time duration of each pulse, $1/\tau$ is the number of pulses per unit time, and V_p is the mass in kilograms of the object within the effective limit of the coil. The equation with $\alpha = 1$ should give a reasonable measure of the total absorbed power. The reflectivity at the air/tissue interface might be as high as 60% at the NMR frequencies used. This will of course decrease the amount of absorbed power and, therefore, heating. Most of the heating will occur at the body's surface, and because of the geometry and dielectric constant variations at tissue interfaces, hot spots[89,90] can be expected to occur with high frequencies used in P-31 NMR. The skin depth δ, the depth at which the power is reduced by $(1/e)^2$, is shown as a function of frequency for biologic tissues in Table I.[88]

TABLE I

Frequency (MHz)	δ *(cm)*
1	91
10	22
41	11
100	7

Whereas the majority of proton NMR is being done at frequencies less than 15 MHz, commercial instruments that operate at 1.5 T and 1.9 T will use 70-MHz and 80-MHz frequencies if proton imaging is pursued or if proton decoupling is employed in C-13 studies. At the planned RF power and duty cycles, some heating is anticipated.

The prudent threshold value for NMR studies is 1.5 w/kg for long-duration studies and 4 w/kg for studies of 10 min or less. The basis for these suggested absorbed power guidelines is the fact that the basal metabolic rate of humans is 1.5 w/kg during sleep and 15 w/kg for very heavy exercise. The body has the ability to lose heat by radiation and evaporative (Joule-Thompson) mechanisms. The latter mechanism is most effective in low-humidity environments where the body can lose 500 kcal/hr. It is instructive to calculate the temperature rise using a soft-tissue specific heat of 0.83 kcal/kg/°C (3.5 kjoules/kg/°C), assuming the absorbed energy is not transported to other parts of the body. The expected temperature rise is

$$\Delta T(^\circ\text{C}) = (SAR\cdot t/4186\cdot h) \qquad \text{Eq. 14}$$

where SAR is in units of w/kg, t is the time in seconds, and h is the specific heat in kcal/kg/°C. For a specific absorbed power of 4 w/kg for 10 min, the expected temperature rise is

$$\Delta T(^\circ\text{C}) = \frac{4\ \text{w/kg}\cdot 600\ \text{sec}}{4186\ \text{joules/kcal}\cdot 0.83\ \text{kcal/kg/}^\circ\text{C}} \cong 0.7^\circ\text{C} \qquad \text{Eq. 15}$$

Conduction of heat away from the tissue volume will decrease this heat burden. The skin will receive the maximum power, and the maximum temperature elevation is expected in the subcutaneous tissues for proton NMR imaging.[90] Thus, a conservative guideline value for long-term exposures (e.g., 20 min) is the average metabolic waste energy of 1.5 w/kg. For exposures less than 10 min, 4 w/kg average absorbed power to the portion of the body in the RF coils should be tolerated at frequencies less than 10 MHz. The effects of the fields employed in NMR imaging on pacemakers and prostheses is discussed in more detail below, but it should be noted here that absorption of NMR RF emissions by large metallic implants may result in significant local heating if the prosthesis is large and blood circulation limited.[11] The housings of newer pacemakers are typically of stainless steel to shield against electromagnetic interference from direct RF exposure.[10] Implants containing a significant amount of ferromagnetic material may also change the RF homogeneity sufficiently to create imaging artifacts; however, distortion of the local magnetic field is probably a stronger source of artifact.[9]

Nonthermal Effects

The literature on nonthermal effects is speculative and controversial at present. Some effects are repeatable, but the mechanism for the effects is still unknown. For example, some investigations of the effects of electromagnetic fields on calcium transport show a 15%–20% change in the calcium bound to brain cells following a 20-min exposure to 1–75 Hz.[91–93] It has been proposed that the mechanism for these observations was a field-induced alteration in the binding of calcium ions, which would lead to conformational changes in nervous tissue membranes and a consequent shift in the threshold for excitation. A recent study by Tenforde[94] has demonstrated that the observed alteration in calcium ion binding does not occur through a thermal energy transfer mechanism. The field interaction may involve a resonant or cooperative energy transfer mechanism.[95]

At high field intensities and frequencies in the range of 1–100 mHz, a "nonthermal" effect is characterized by alignment of particles of charcoal, starch, milk, or blood cells to form chains parallel to the electric lines of force. This chain formation, known as the "bead chain effect," has been quantified, and the proposed mechanism is an attraction between particles in which dipole charges are induced by the RF fields. It appears, however, that this effect will *not* occur at levels below that at which thermal effects occur.[96]

Changes in blood-brain barrier permeability are a possible nonthermal effect noted at power densities less than 10 mw/cm^2 in small animals using GHz frequencies. It is noteworthy that the specific absorbed power at this frequency is about 5 w/kg and 100 times greater than SAR for a human under the same exposure condition of power density and wavelength in the far field.[2,68] This is an experimental area of great controversy. Experiments of Oscar and Hawkins[97] and Albert[98] showed blood-brain barrier permeability changes associated with GHz exposures in the rat using fluorescein, mannitol, and horseradish peroxidase in microscopic and tracer analytic studies These observations could not be repeated by others.[99–101]

The most recent study and review by Oscar and coworkers[102] suggests that previously reported small increases in blood-brain barrier permeability may be a secondary effect caused by microwave alteration of blood flow, blood pressure, or blood vessel area. In these latter experiments as well as in other experiments with GHz frequencies on small animals, and with incident power densities of 10 to 15 mw/cm^2, high local power absorption and possible "hot spots" are likely to occur, and we cannot discount local heating as the biophysical mechanism for observed effects.

Magnetic Effects on Pacemakers, Prostheses, and Potential Projectiles

The experiments of Davis and coworkers[11] indicate that there will not be significant heating of implanted surgical clips during an NMR experiment; however, the more important effect for both small and large implants is probably the forces and torques exerted on them by the fields. In a recent study, over two-thirds of surgical vascular clips tested had sufficient ferromagnetism to be moved by the fields used, and the effects on five types of aneurysm clips were sufficiently large to produce a risk of hemorrhage or injury by the dislocation of the clip.[9] Ferromagnetic implants also produced imaging artifacts in proportion to their size, while steel alloys with little ferromagnetic character distorted the imaging field to a lesser extent.[9]

Static and time-varying magnetic fields are potentially hazardous to the patient with an implanted pacemaker.[10] Field strengths greater than 10 Gauss may be sufficient to close the magnetic reed switch on a demand pacemaker, converting it to asynchronous mode. Current machines are capable of putting out sufficient electromagnetic interference to confuse the sensing circuitry of some demand pacemakers and may result in the inhibition of the device. This will continue to be a problem, particularly as magnets become larger. The time-varying magnetic fields are capable of generating voltage pulses at frequencies that mimic cardiac activity, and most pacemakers are sufficiently ferromagnetic to be moved within the chest wall by the changing fields. Thus, for the present, patients with pacemakers should be excluded from the NMR procedure and environment.

Finally, it is important not to underestimate the fact that ferromagnetic objects outside the patient will also be set in motion by a large field, and will tend to be attracted towards the center of the magnet. A hypodermic needle or a pen top which becomes a projectile is capable of causing significant injury to both patient and expensive equipment.

Epidemiologic Perspectives

Studies of the effects of NMR imaging on human health can be carried out by prospective epidemiologic studies or retrospective surveys. Well-controlled studies require years of effort on the part of dedicated and experienced researchers. Some small population studies have been accomplished in the past, but the observations have been for the most part anecdotal.

The first analysis of the experiences of physicists and technicians working in fields up to 2 T in nuclear physics laboratories was a survey by Beischer,[103] who concluded that no deleterious effects were experienced by humans even after cumulative exposures to fields of 0.5 T for 3 days per year per person. The sensations of tooth pain and taste reported in Beischer's study were due to movement of the head through a field gradient that produced an induced electric current effect as discussed above (see Eq. 5). The most alarming report suggesting an effect of static fields appeared initially 19 years ago from Russian scientists who presented data from 1,600 workers whose hands were exposed to fields of 35 to 350 mT.[104] In addition to general symptoms of headache and fatigue, some workers had hand swelling and skin desquamation. There were no adequate controls for the chemicals used, and no corroborating reports exist, although hand exposures at similar or higher levels have been experienced by NMR researchers, bubble chamber technicians, and technical support staff at magnet facilities.[105]

Wertheimer and Leeper[106] associated an increase in the incidence of leukemia in children with the geographic location of their dwellings relative to high-current electrical power delivery lines, and suggested that the leukemia might result from the extremely low-level magnetic fields. According to the authors, these results did not seem to be an artifact of neighborhood, street congestion, social class, or family structure. However, there has not been an independent investigation of the phenomenon, and other possible causes (e.g. transformer oil, industrial pollution) have not been ruled out.

In a cross-sectional study of workers exposed to magnetic fields of up to 20 mT from large electrolytic cells, Marsh et al.[107] reported only small effects on blood pressure and a slightly lowered white cell count. The statistical significance and interpretation of these data have not been carefully reevaluated. A recent report by Milham[108] suggests a causal connection between leukemia and industrial exposure to electric and magnetic fields. This correlation was supported by Wright et al.,[109] but rejected by Liburdy,[110] and a more extensive evaluation by Rockette and Arena[111] does not find the same result in aluminum plant workers whose exposure can be in the range of 80 mT for 8 hr per day.

The major problem in studies aimed at showing the presence or absence of effects by evaluating the change in incidence of some phenomenon or disease is the need to study a sufficiently large sample with adequately matched controls. For example, the prevalence of cancer is about 2.5/1,000, and for cardiovascular disease is 5/1,000. With an exposed population size of 1,000, a threefold increase in the prevalence of cardiovascular disease would have to have occurred before a firm conclusion (90% confidence) could be made that there is a health effect from magnetic field exposures.[112]

New Experimental Work on Health Effects

At Donner Laboratory there are three programs of investigation for the health effects from in vivo NMR. These are: (1) the effects of rapidly changing fields (high dB/dt), including the influence of induced currents from NMR procedures on embryogenesis and on the electrical behavior of the heart; (2) RF heating effects studies by phantoms; and (3) an ongoing epidemiologic study. At the University of Pennsylvania, Wolf is conducting studies of the effects of rapidly changing fields on the cardiovascular system.[113]

The potential changes and accompanying currents that occur in the growth and development of an organism[114] may be affected by the currents induced by the rapidly changing field of an NMR procedure. We have constructed a water-cooled coil that will accommodate a standard mouse cage (or a large beagle) and can provide pulses of up to 7 T/sec with a risetime of 5 msec (Fig. 10). Mice exposed continuously from conception to birth to this flux change are compared with animals kept under the same conditions in a duplicate (unpowered) coil. Similarly, while the dB/dt required to induce fibrillation is believed to be on the order of 500 T/sec, less intense switched fields may alter the fibrillation threshold, so that patients already at risk may be compromised by the fields in an NMR procedure. Dog studies are under way with the coil (Fig. 10), using a ventricular threshold determination technique.[115]

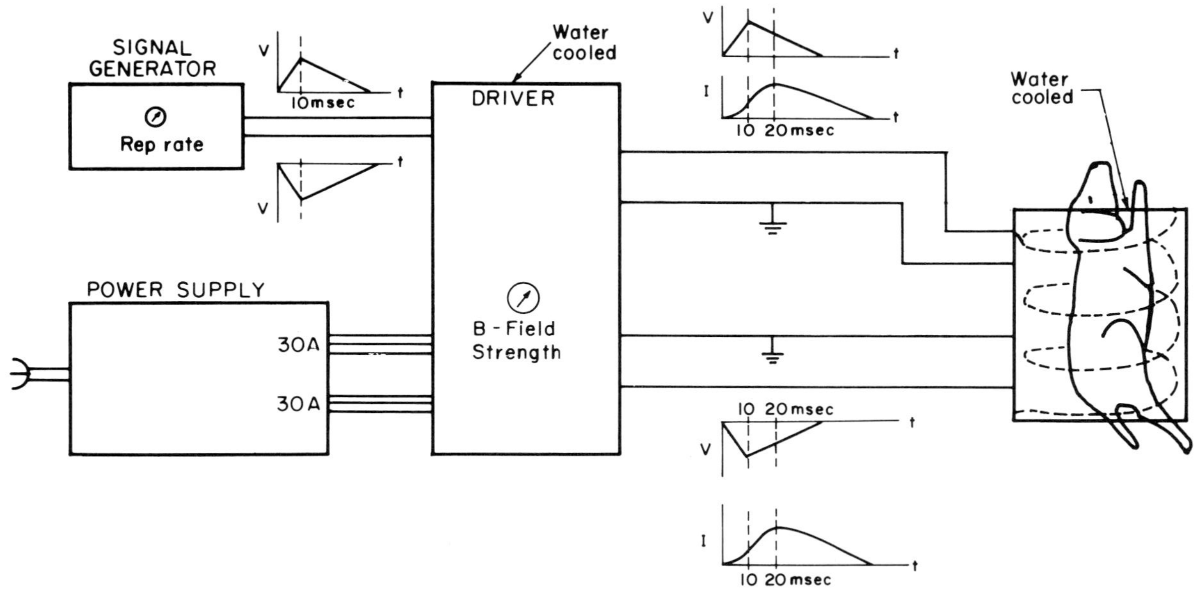

Figure 10. Experimental configuration for measurements of the effects of induced currents in biologic systems, including small mouse colonies and medium-sized dogs.

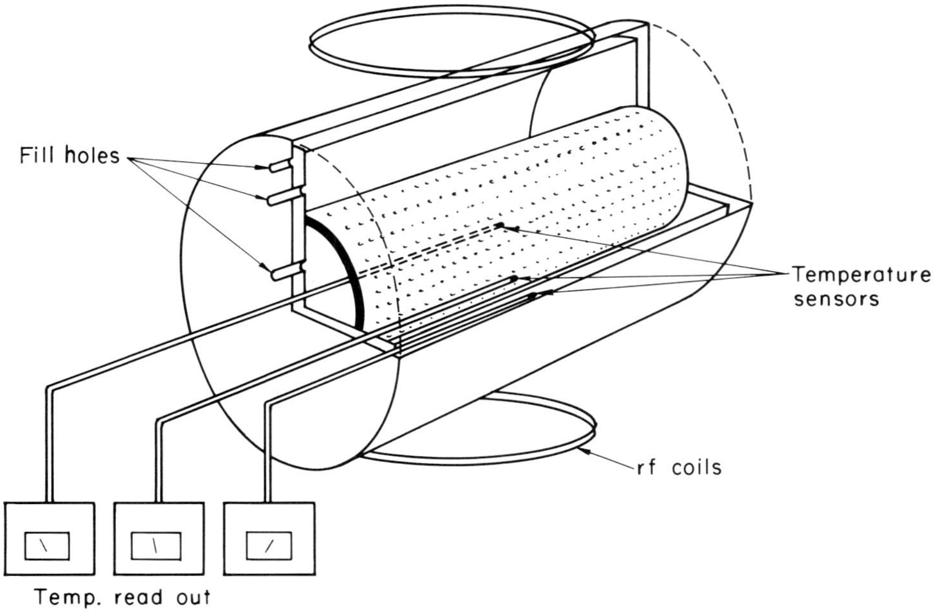

Figure 11. Overall configuration for measuring RF heating. Heat transfer conditions are controlled by changing the permeability of one compartment relative to another. Temperature sensors can be light pipes from liquid crystal sensors or very small thermistors, which will cause little perturbation in the temperature field for the RF frequencies under study.

Experimental studies of RF heating are planned using saline-filled arm and head phantoms (Fig. 11). The telescoping inner cylinders can be rotated to control the amount of convective flow and diffusion between the inner and outer volumes, and the miniature thermistors are easily slid out of the field. The device will be used in installations with field strengths from 0.15–6 T and with several RF frequencies and pulse sequences.

We are completing epidemiologic studies on workers at high-energy physics laboratories who have been exposed to fields as high as 1 T frequently during their working careers.[112] These studies include 800 exposed and 800 matched controls and will be complete in late 1983.

General Considerations

It is not clear at this time whether a prospective study of patients exposed to NMR imaging procedures is warranted because of the difficulty in separating RF effects from magnetic field effects (if any) and the fact that we have no hypothesis as to what effects one might expect. Two protocols that can easily be pursued involve a long-term follow-up study of 300 or more patients who have undergone head-imaging studies with magnetic fields less than 0.3 T but relatively high RF power exposures, and a study of at least 300 patients who have undergone high-field (> 1.5 T) in vivo P-31 spectroscopy measurements. Since these series will involve patients, disease- and age-matched controls are required.

Two biophysical effects of static fields above 2 T (20,000 Gauss) that have some potential for hazardous health effects are the magnetohydrodynamic induction of electric potentials and the turning torque on organized structures that have magnetic susceptibility anisotropy. Induced EMF (magnetohydrodynamic effects) from flow accounts well for perturbation in EKG signals, and this phenomenon is not considered to be of physiologic importance because the induced voltage is insufficient to perturb normal tissue excitation. The magnitude of the induced EMF is dependent on volume flow, and membrane depolarization from these electric potentials will not occur at 2 T or even 6 T for the normally encountered blood flow magnitude in humans.

Orientation distortion of molecular ensembles or organized subcellular structures as a result of prolonged exposure (hours) of field greater than 2 T requires further investigation.

The effects of static fields on surgical clips and metallic objects such as pacemakers can lead to hazards.

Electric fields are induced in the body by the rapid changes of magnetic field gradients used in some spatial imaging strategies. The repetition rate, waveform, and current density are important parameters that control the average energy absorbed by the exposed tissue. Under some conditions, visual phosphenes can be produced, and it is possible to cause heart fibrillation if field excursions are extremely high. Present experimental conditions give induced currents far below this threshold as discussed in this chapter. However, further investigations are required for effects on the cardiovascular system, memory, and embryogenesis.

Heating from RF-induced electric fields is likely to be observed in some imaging instruments and strategies. Since the average metabolic waste heat of humans is 1.5 w/kg, it is prudent not to exceed this level of absorbed power for studies of long duration. Short periods (< 10 min) of exposure wherein 4 w/kg is absorbed in the tissue within the RF coil can be easily tolerated in a low-humidity and cool environment. Metal prostheses will absorb power in proportion to the conductivity of the device; however, because of the capacity of surrounding tissues to absorb heat, problems are not expected unless very large devices are in the field. As the RF field is increased to correspond to an increase in magnetic field in order to achieve a better signal-to-noise ratio, we expect "hot-spot" heating to occur. This area is now being actively investigated.

Guidelines for NMR in vivo medical exposures have been proposed by the National Radiological Protection Board of England and the Bureau of Radiological Health of the U.S.F.D.A. Except for the lack of a qualification of the value for the duration of the switched fields used in imaging, both of these guidelines do not restrict currently available instrumentation. Guidelines for magnetic field exposure in industrial activities are necessarily based on different objectives and requirements.[116]

Acknowledgment

This paper benefited from the suggestions and criticisms of the following scientists: Drs. T. W. Athey, P. Czerski, H. Ho, and Mr. R. Schneider, Bureau of Radiological Health; Dr. P. Lauterbur, State University of New York at Stony Brook; Drs. P. Bottomley and W. Adelstein of General Electric, Inc.; Drs. A. Portis, T. Tenforde, and J. Singer of the University of California, Berkeley; Dr. D. Hoult, National Institutes of Health; Dr. Richard Saunders, British National Radiological Protection Board; and Dr. Peter Mansfield, University of Nottingham.

This work was supported by the Biology and Medicine Division of the U.S. Department of Energy under Contract No. DE-AC03-76SF00098, and by the National Heart, Lung, and Blood Institute Grant No. P01 HL25840-01.

REFERENCES

1. Bottomley PA, Andrew ER: RF magnetic field penetration, phase shift and power dissipation in biological tissue: Implications for NMR imaging. Phys Med Biol 23:630, 1978.
2. Budinger TF: Thresholds for physiological effects due to RF and magnetic fields used in NMR imaging. IEEE Trans Nucl Sci NS-26:2821, 1979.

3. Budinger TF: Nuclear magnetic resonance (NMR) in vivo studies: Known thresholds for health effects. J Comput Assist Tomogr 5:800, 1981.
4. Hoult DI, Lauterbur PC: The sensitivity of the zeugmatographic experiment involving human samples. J Magnet Reson 34:425, 1979.
5. Wolff S, Crooks LE, Brown P, et al: Tests for DNA and chromosomal damage induced by nuclear magnetic resonance imaging. Radiology 136:707, 1980.
6. Bottomley PA, Edelstein WA: Power deposition in whole-body NMR imaging. Med Phys 8:510, 1981.
7. Saunders RD, Orr JS: Biologic effect of NMR. In *Nuclear Magnetic Resonance (NMR) Imaging,* Partain CL et al (Eds), Chap 28. Philadelphia: W.B. Saunders, 1983.
8. Mansfield P, Morris PG: NMR imaging in biomedicine. In *Advances in Magnetic Resonance,* Waugh JS (Ed), Suppl. 2. New York: Academic, 1982.
9. New PF, Rosen BR, Brady TJ, et al: Potential hazards and artifacts of ferromagnetic and nonferromagnetic surgical and dental materials and devices in nuclear magnetic resonance imaging. Radiology 147:139, 1983.
10. Pavlicek W, Geisinger M, Castle L, et al: The effects of nuclear magnetic resonance on patients with cardiac pacemakers. Radiology 147:149, 1983.
11. Davis PL, Crooks L, Arakawa M, et al: Potential hazards in NMR imaging: Heating effects of changing magnetic fields and RF fields on small metallic implants. AJR 137:857, 1981.
12. Bureau of Radiological Health: *Guidelines for Evaluating Electromagnetic Exposure Risks for Trials of Clinical NMR Systems.* Washington, D.C.: Department of Health and Human Services, U.S. Public Health Service, Food and Drug Administration, 1982.
13. National Radiological Protection Board: Exposure to nuclear magnetic resonance clinical imaging. Radiography 47:250, 1981.
14. Mild KH, Sandstrom M, Lovtrup S: Development of Xenopus laevis embryos in a static magnetic field. Bio-electromagnetics 2:199, 1981.
15. Aleksandrivskaya M, Kholodov Y: Vozmoznaya role neyroglyyi v voznyknovenyi bioelectricheskoy reactsiyi golovnogo mozga na pastoyannoyo magnitoye polie. Rep Acad Sci USSR 170:482, 1966.
16. Friedman H, Carey RF: The effects of magnetic fields upon rabbit brains. Physiol Rev 4:539, 1969.
17. Tucker RD, Schmitt OH: Tests for human perception of 60 Hz moderate strength low frequency magnetic fields. IEEE Trans Biomed Bioeng BME-25:509, 1978.
18. Moore BR: Is the homing pigeon's map geomagnetic? Nature 285:69, 1980.
19. Blakemore R: Magnetotactic bacteria. Science 190:377, 1975.
20. Kalmijn AJ: Biophysics of geomagnetic field detection. IEEE Trans Magn 17:1113, 1981.
21. Tenforde TS, Gaffey CT, Moyer BR, et al: Cardiovascular alterations in Macaca monkeys exposed to stationary magnetic fields: Experimental observations and theoretical analysis. Bioelectromagnetics 4:1, 1983.
22. Barnothy MF (Ed): *Biological Effects of Magnetic Fields,* Vols 1 and 2. New York: Plenum, 1964.
23. Busby ED: Biomagnetics. NASA CR-889, 1967.
24. Kholodov YA (Ed): Influence of Magnetic Fields on Biological Objects. Joint Publ Res Service Rep 63038. Springfield, Virginia, National Technical Information Service, 1974.
25. Silver IL, Tobias CA: Magnetic fields and their biological effects. In *Space Radiation Biology and Related Topics,* Tobias CA, Todd PW (Eds), pp 258–292. New York: Academic, 1974.
26. Sheppard AR, Eisenbud M: *Biological Effects of Electric and Magnetic Fields of Extremely Low Frequency.* New York: New York University Press, 1977.
27. Phillips RD, Gillis MF, Kaune WT, et al (Eds): *Biological Effects of Extremely Low Frequency Electromagnetic Fields.* Proceedings of the 18th Annual Hanford Life Sciences Symposium at Richland, Washington, October 16–18, 1978, NTIS, DOE Symposium Series 50, CONF-781016, 1979.
28. Tenforde TS (Ed): *Magnetic Fields Effects on Biological Systems.* New York: Plenum, 1979.
29. Malinin GI, Gregory WD, Morelli, L, et al: Evidence of morphological and physiological transformation of mammalian cells by strong magnetic fields. Science 184:844, 1976.
30. Frazier ME: Biological effects of magnetic fields: A progress report. In *Health Implications of New Energy Technologies,* Rem WN, Archer VE (Eds), pp 679–685. Ann Arbor: Ann Arbor Scientific Publications, 1980.
31. Chandra S, Stefani S: Effect of constant and alternating magnetic fields on tumor cells *in vitro* and *in vivo.* In *Biological Effects of Extremely Low Frequency Electromagnetic Fields,* Phillips RD, Gillis MF, Kaune WT, et al (Eds), pp 436–446. NTIS, DOE Symposium Series 50, CONF-781016, 1979.
32. Ettienne E, Ripamonti A, Frankel RB: Effect of magnetic fields on the drug-induced contractility of the ciliate Spirostomum. In *Magnetic Field Effects on Biological Systems,* Tenforde TS (Ed), pp 54–56. New York: Plenum, 1979.
33. Baum JW, Schairer LA, Lindahl KL: Tests in the plant Tradescantia for mutagenic effects of strong magnetic fields. In *Magnetic Field Effects on Biological Systems,* Tenforde TS (Ed), pp 22–24. New York: Plenum, 1979.
34. Kale PG, Baum JW: Genetic effects of strong magnetic fields in Drosophila melanogaster: I. Homogeneous fields ranging from 13,000 to 37,000 Gauss. Environ Mutagen 1:371, 1979.
35. Cooke P, Morris PG: The effects of NMR exposure on living organisms: II. A genetic study of human lymphocytes. Br J Radiol 54:622, 1981.

36. Semm P, Schneider T, Vollrath L: Effects of an earth-strength magnetic field on electrical activity of pineal cells. Nature 288:607, 1980.

37. Raybourn M: The effects of direct-current magnetic fields on turtle retinas in vitro. Science 220:715, 1983.

38. Baker RR: Goal orientation by blindfolded humans after long-distance displacement: Possible involvement of a magnetic sense. Science 210:555, 1980.

39. Gould JI, Able KP: Human homing: An elusive phenomenon. Science 212:1061, 1981.

40. Murayama M: Orientation of sickled erythrocytes in a magnetic field. Nature 206:420, 1965.

41. Hong FT, Mauzerall D, Mauro A: Magnetic anisotropy and the orientation of retinal rods in a homogenous magnetic field. Proc Natl Acad Sci USA 68:1283, 1971.

42. Geacintov NE, Van Nostrand F, Becker JF, et al: Magnetic field induced orientation of photosynthetic systems. Biochim Biophys Acta 267:65, 1972.

43. Thomas A, Morris PG: The effects of NMR exposure on living organisms: I. A microbial assay. Br J Radiol 54:615, 1981.

44. Akoyunoglou G: Effect of a magnetic field on carboxydismutase. Nature 202:452, 1964.

45. Fardon JC, Poydock ME Sr, Basulto G: Effect of magnetic fields on the respiration of malignant, embryonic and adult tissue. Nature 211:433, 1966.

46. Levengood WC: Cytogenetic variations induced with a magnetic probe. Nature 209:1009, 1966.

47. Rabinovitch B, Maling JE, Weissbluth M: Enzyme-substrate reactions in very high magnetic fields. Biophys J 7:187, 1967.

48. Chance B, Nakase Y, Bond M, et al: Detection of ^{31}P nuclear magnetic resonance signals in brain by *in vivo* and freeze-trapped assays. Proc Natl Acad Sci USA 75:4925, 1978.

49. Radda GK, Seeley PJ: Recent studies on cellular metabolism by nuclear magnetic resonance. Ann Rev Physiol 41:749, 1979.

50. Nunnally RL, Bottomley PA: Assessment of pharmacological treatment of myocardial infarction by phosphorus-31 NMR with surface coils. Science 211:177, 1980.

51. Liboff RL: A biomagnetic hypothesis. Biophys J 5:845, 1965.

52. Wikwso JP Jr, Barach JP: An estimate of the steady magnetic field strength required to influence nerve conduction. IEEE Trans Biomed Eng BME-27:722, 1980.

53. Reno VR: Conduction velocity in nerve exposed to a high magnetic field. NASA Report No. NAMI-1089, 1969.

54. Kolta P: Strong and permanent interaction between per-ipheral nerve and a constant inhomogeneous magnetic field. Acta Physiol Acad Sci Hung 43:89, 1973.

55. Schwartz JL: Influence of a constant magnetic field on nervous tissue: I. Nerve conduction velocity studies. IEEE Trans Biomed Eng BME-25:467, 1978.

56. Gaffey CT, Tenforde TS: Electrical properties of conducting frog sciatic nerves exposed to high DC magnetic fields. Bioelectromagnetics 1:208, 1980.

57. Gaffey CT, Tenforde TS: Bioelectric properties of frog sciatic nerves during exposure to stationary magnetic fields. Radiat Environ Biophys 22:61, 1983.

58. Togawa T, Okai O, Oshima M: Observation of blood flow EMF in externally applied strong magnetic field by surface electrodes. Med Biol Eng 5:169, 1967.

59. Beischer DE, Knepton JC Jr:Influence of strong magnetic fields on the electrocardiogram of squirrel monkeys (Saimiri sciureus). Aerospace Med 35:939, 1964.

60. Gaffey CT, Tenforde TS, Dean EE: Alterations in the electrocardiograms of baboons exposed to DC magnetic fields. Bioelectromagnetics 1:209, 1980.

61. Gaffey CT, Tenforde TS: Alterations in the rat electrocardiogram induced by stationary magnetic fields. Berkeley: University of California, Lawrence Berkeley Laboratory Report LBL-12099, 1981.

62. Salles-Cinha SX, Battocletti JH, Sances A Jr: Steady magnetic fields in noninvasive electromagnetic flowmetry. IEEE Proc 68:149, 1980.

63. Belousouva L Ye: Possible braking or arrest of blood by a magnetic field. Biofizika 10:365, 1965.

64. Sud VK, Suri PK, Mishra RK: Effect of magnetic field on oscillating blood flow in arteries. Studia Biophys 46:163, 1974.

65. Kumar V: Exact analysis for calculating hemodynamic parameters in unsteady blood flow in arteries in the presence of static magnetic fields. Studia Biophys 72:43, 1978.

66. Easterly, CB: Cardiovascular risk from exposure to static magnetic fields. Am Ind Hyg Assoc J 43:533, 1982.

67. Hinshaw WS: Image formation by nuclear magnetic resonance: The sensitive point method. J Appl Physiol 47:3709, 1976.

68. Durney CH, Johnson CC, Barber PW, et al: Radiofrequency Radiation Dosimetry Handbook (ed 2), (Report SAM-TR-78-22). Salt Lake City: Departments of Electrical Engineering and Bioengineering, University of Utah, 1978.

69. Barlow HB, Kohn HI, Walsh EG: The effects of dark adaptation and of light upon the electric threshold of the human eye. J Physiol (London) 148:376, 1947.

70. Budinger TF: Indirect electrical stimulation of the visual apparatus. Berkeley: University of California, Lawrence Radiation Laboratory Report UCRL-18347, 1968.

71. d'Arsonval MA: Dispositifs pour la mesure des courants alternatifs à toutes fréquences. CR Soc Biol (Paris) 3:451, 1896.

72. Barlow HB, Kohn HI, Walsh EG: Visual sensations aroused by magnetic fields. Am J Physiol 148:372, 1947.

73. Lovsund P, Nilsson SEG, Reuter T, et al: Magnetophosphenes: A quantitative analysis of thresholds. Med Biol Eng Comput 18:326, 1980.

74. Brighton CT (Ed): *Electric and Magnetic Control of Musculo-Skeletal Growth and Repair.* New York: Grune and Stratton, 1979.

75. Bassett CAL, Pilla AA, Pawluk RJ: A non-operative salvage of surgically-resistant pseudarthroses and non-unions by pulsing electromagnetic fields. Clin Orthop 124:128, 1977.

76. Bassett CAL: Personal communication, 1983.

77. Fink M: *Convulsive Therapy: Theory and Practice.* New York: Raven, 1979.

78. Maas JA, Asa MM: Contactless nerve stimulation and signal detection by inductive transducer. IEEE Trans Mag 6:322, 1970.

79. Irwin DD, Rush S, Evering R, et al: Stimulation of cardiac muscle by time-varying magnetic field. IEEE Trans Mag 6:321, 1970.

80. Oberg PA: Magnetic stimulation of nerve tissue. Med Biol Eng 11:55, 1973.

81. Ueno S, Matsumoto S, Harada K, et al: Capacitive stimulatory effect in magnetic stimulation of nerve tissue. IEEE Trans Mag 14:5, 1978.

82. Ueno S, Lovsun P, Oberg A: On the effect of alternating magnetic fields on action potential in lobster giant axon. In *Proceedings of the 5th Nordic Meeting on Medical and Biological Engineering,* Linkoping, and the *24th Anniv. of the Swedish Society for Medical Physics and Medical Engineering,* Umea, Sweden, ISBN 91, 1981.

83. Poulson MJR, Barker AT, Freeston II: Stimulation of nerve trunks with time-varying magnetic fields. Med Biol Eng Comput 20:243, 1982.

84. Watson AB, Wright JS, Loughman J; Electrical thresholds for ventricular fibrillation in man. Med J Aust 1:1179, 1973.

85. Roy OZ: Technical note: Summary of cardiac fibrillation thresholds for 60 Hz currents and voltages applied directly to the heart. Med Biol Eng Comput 18:657, 1980.

86. Barghe-Sharghi V: "Effects of time-varying electromagnetic waves on blood platelets" (Master's thesis). University of California, Berkeley, 1981.

87. Schwan HP, Piersol GM: The absorption of electromagnetic energy in body tissues: Part I. Am J Phys Med 33:371, 1954.

88. Johnson CC, Guy AW: Nonionizing electromagnetic wave effects in biological materials and systems. Proc IEEE 60:692, 1972.

89. Guy AW: Analyses of electromagnetic fields induced in biological tissues by thermographic studies on equivalent phantom models. IEEE Trans Microwave Theory Tech (Special Issue of Biological Effects of Microwaves) MTT-19:205, 1971.

90. Kritikos HN, Schwan HP: Hot spots generated in conducting spheres by electromagnetic waves and biological implications. IEEE Trans Biomed Eng BME-19:53, 1972.

91. Bawin SM, Gavalas-Medici RJ, Adey WR: Effects of modulated very high frequency fields on specific brain rhythms in cats. Brain Res 58:365, 1973.

92. Hamer J: Effects of low level, low frequency electric fields on human reaction time. Commun Behav Biol A 2:217, 1968.

93. Adey WR: Tissue interactions with nonionizing electromagnetic fields. Physiol Rev 61:435, 1981.

94. Tenforde TS: Thermal aspects of electromagnetic field interactions with bound calcium ions at the nerve cell surface. J Theor Biol 83:517, 1980.

95. Schwan HP, Foster KR: RF-field interactions with biological systems: Electrical properties and biophysical mechanisms. IEEE Proc 68:104, 1980.

96. Schwan HP: Biophysics of diathermy. In *Therapeutic Heat and Cold* (ed 2), SH Licht (Ed), pp 63–125. New Haven: Elizabeth Licht, 1965.

97. Oscar KJ, Hawkins TD: Microwave alteration of the blood-brain barrier system of rats. Brain Res 126:281, 1977.

98. Albert EN: Light and electron microscopic observations on the blood-brain barrier after microwave radiation. In *Proceedings of the Symposium on Biological Effects and Measurements of Radiofrequency/Microwaves,* Hazard DG (Ed), pp 294–304. Rockville: DHEW, 1977.

99. Merritt JH, Chamness AF, Allen SJ: Studies on blood-brain barrier permeability after microwave radiation. Radiat Environ Biophys 15:367, 1978.

100. Spackman DH, Riley V, Guy AW, et al: Is the blood-brain barrier altered by RF irradiation (abstract)? Int Union Radiol Sci BEMS: 362, 1979.

101. Preston E, Vavasour EJ, Assenheim HM: Permeability of the blood-brain barrier to mannitol in the rat following 2450 MHz microwave irradaition. Brain Res 174:109, 1979.

102. Oscar KJ, Gruenau SP, Folker MT, et al: Local cerebral blood flow after microwave exposure. Brain Res 204:220, 1981.

103. Beischer DE: Human tolerance to magnetic fields. Astronautics 7:24, 1962.

104. Vyalov AM: Clinico-hygienic and experimental data on the effects of magnetic fields under industrial conditions. In *Influence of Magnetic Fields on Biological Objects,* Kholodov YA (Ed), pp 163–174. Joint Publ Res Service Report 63038. Springfield, Virginia: National Technical Information Service.

105. Rubin L: Personal communication, National Magnetic Laboratory, Massachusetts Institute of Technology, Boston, Massachusetts, 1977.

106. Wertheimer N, Leeper E: Electrical wiring configurations and childhood cancer. Am J Epidemiol 109:273, 1979.

107. Marsh JI, Armstrong TJ, Jacobson AP, et al: Health effects of occupational exposure to steady magnetic fields. Am Ind Hyg Assoc J 43:387, 1982.

108. Milham S: Mortality from leukemia in workers exposed to electrical and magnetic fields. N Engl J Med 307:249, 1982.

109. Wright WE, Peters JM, Mack TM: Leukemia in workers exposed to electrical and magnetic fields. Lancet 1:1160, 1982.

110. Liburdy RP: Carcinogenesis and exposure to electrical and magnetic fields. N Engl J Med 307:1402, 1982.

111. Rockette HE, Arena VC: Mortality studies of aluminum reduction plant workers: Potroom and carbon department. J Occup Med (in press).

112. Budinger TF, Wong PDC, Yen CK: Magnetic field effects on humans: Epidemiological study design. In *Biological Effects of Extremely Low Frequency Electromagnetic Fields,* Phillips RD, Gillis MF, Kaune WT, et al (Eds), pp 379–399. Technical Information Center, U.S. Dept. of Energy, DOE Symposium Series 50, CONF-7891016.

113. Wolf H: Personal communication, 1983.

114. Jaffe LF, Nuccitelli R: Electrical controls of development. Ann Rev Biophys Bioeng 6:445, 1977.

115. Moore EN, Spear JF: Ventricular fibrillation thresholds, its physiological and pharmacological importance. Arch Intern Med 135:446, 1975.

116. Alpen EL: Magnetic field exposure guidelines. In *Magnetic Field Effects on Biological Systems,* Tenforde TS (Ed), pp 25–32. New York: Plenum, 1979.

20

Siting

LEON KAUFMAN, Ph.D.
LAWRENCE E. CROOKS, Ph.D.
ALEXANDER R. MARGULIS, M.D.
JAMES O. PROSEUS

The siting of an NMR imager is a trivial task if the given area is high enough, large enough, and devoid of large iron structures. The challenge is to bring NMR into the typical hospital or private clinic. There are two different aspects of the siting problem that need to be considered: The magnet affects its immediate environment, and, in turn, the environment affects the magnet and imaging process. All other aspects of NMR imaging mirror the situation that exists for CT: the need for adequate temperature environments for computers, control rooms, space for viewing areas, reading rooms, patient holding, patient preparation, dressing rooms, supplies, and so on. In the following discussion we will concentrate on the siting of superconducting magnets. Resistive magnets are unlikely to be capable of performance that would make them a serious component of diagnostic work in the developed countries. Permanent magnets are either of low field or very large size and have yet to demonstrate characteristics suitable for routine clinical work. Conversely, superconducting magnets are based on well-understood technology, do not strain the state of the art, and are undergoing engineering changes that will further enhance their attractiveness.

Impact of the Magnet on Its Environment

Superconducting magnets arrive assembled, so that 8.5-ft high and 7-ft wide openings are needed to allow them to reach their destination within the building. The weight of these units, together with gradient coils and bed, is in the range of 10,000–14,000 lb, but future units will be lighter. Currently, magnet services (cryogen refill probe, exhaust, and current charging leads) require a minimum distance of 9 ft (and preferably distance of 10.5 ft) between the base of the magnet and the ceiling.

The magnet, once in operation, is surrounded by an approximately elliptical magnetic field that is directly proportional to its internal magnetic field strength. The strength of this field drops in proportion to the cube of the distance to the center of the magnet (Fig. 1). Ideally, the regions of stronger fringe field should be dedicated to storage, offices, and other activities not affected by this field. Only weaker fringe field areas can be allocated for cathode ray tubes, photomultipliers, and image intensifiers (in increasing order of sensitivity). If this condition cannot be met, the fringe fields can be drastically reduced by the use of a half-inch of iron forming a wall or floor at 2.5–3 m from the magnet. Any remaining effects can be reduced by more iron or by shielding of particular pieces of hardware. With adequate shielding, a 3.5-kGauss magnet can be used in the floor above or below an operating room, as demanding an environmental constraint as is likely to be found in a hospital (Fig. 2).

Magnets with stronger fields, like the 1.5-T imagers currently being tested by several companies, require larger volumes since the fringe fields are proportionately stronger (Fig. 1). Such units present severe siting problems in medical centers and usually must be introduced in new buildings. Shielding must be such that the field outside the NMR-controlled area will be at 5 Gauss or below as ordered by the FDA. This requirement protects people with pacemakers or metallic prosthetic heart valves. Warning signs at the border of the controlled area should inform personnel, patients, and visitors of the possible dangers of entering a magnetic field.

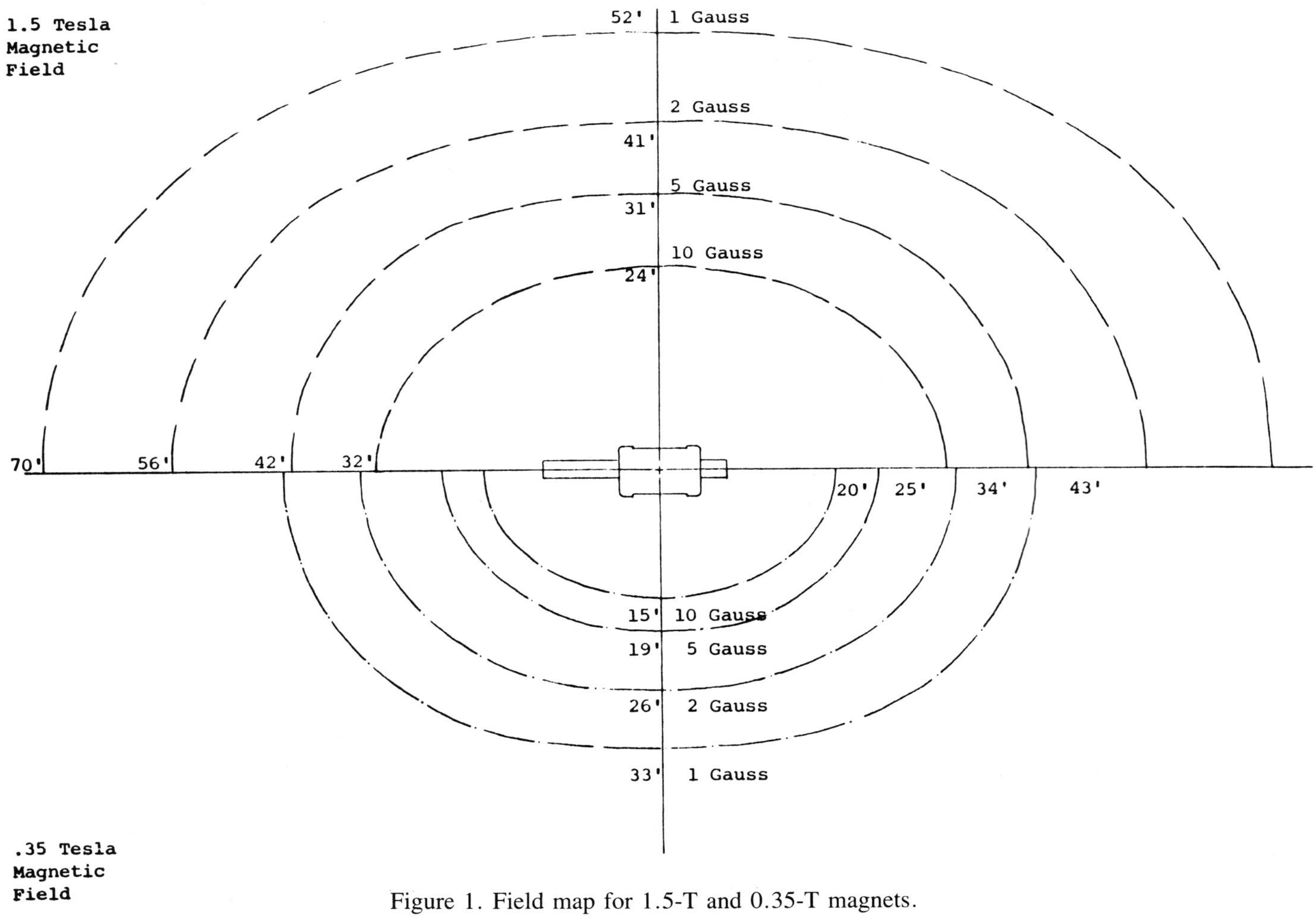

Figure 1. Field map for 1.5-T and 0.35-T magnets.

Impact of the Environment on the Imager

Large iron structures affect the uniformity of the magnetic field. Moving iron (elevators, cars) and large transformers may produce time-dependent variations, so that it is desirable to place distince (20–30 ft), iron shielding (which can be used to reduce this distance), or both between the magnet and the moving iron. The stationary iron used for shielding or in the structure of the building distorts the shape of the internal field of the magnet. These distortions can introduce linear or higher-order variations of the field along any one axis, in addition to complex cross-term variations. The degree of nonuniformity that can be afforded by the imaging technique in use varies among different manufacturers. In our case (Diasonics NMR), the specification calls for a uniformity of 100 ppm or better within a 30-cm-diameter sphere. A drastic environment may result in a uniformity of 2,000 ppm or worse, well below specification. Correction for this effect can be accomplished by using a set of shim coils mounted on a tube that fits within the inner bore of the magnet. Usually a set of 13 coils is provided, each independently driven by a direct current. Iron pieces can also be used together with these coils, or, instead of them, to distort the field in such a way that the distortion compensates for the structural iron effects. Our experience demonstrates that 3.5-kGauss magnets will be able to operate in most hospital environments when adequate measures are taken to compensate for site inadequacies.

Another environmental cause of interference is external radio-frequency (RF), produced by computers, welding units, radio stations, or other sources. The most widely used form of interference elimination consists of construction of an RF-shielded room around the magnet and patient bed. The size of this room is typically on the order of 25–30 ft long and 15–20 ft wide. The height has to meet the 9-ft minimum ceiling requirement over the center of the magnet. Materials used for nonmagnetic RF shielding are aluminum plate, copper sheet, or copper mesh,

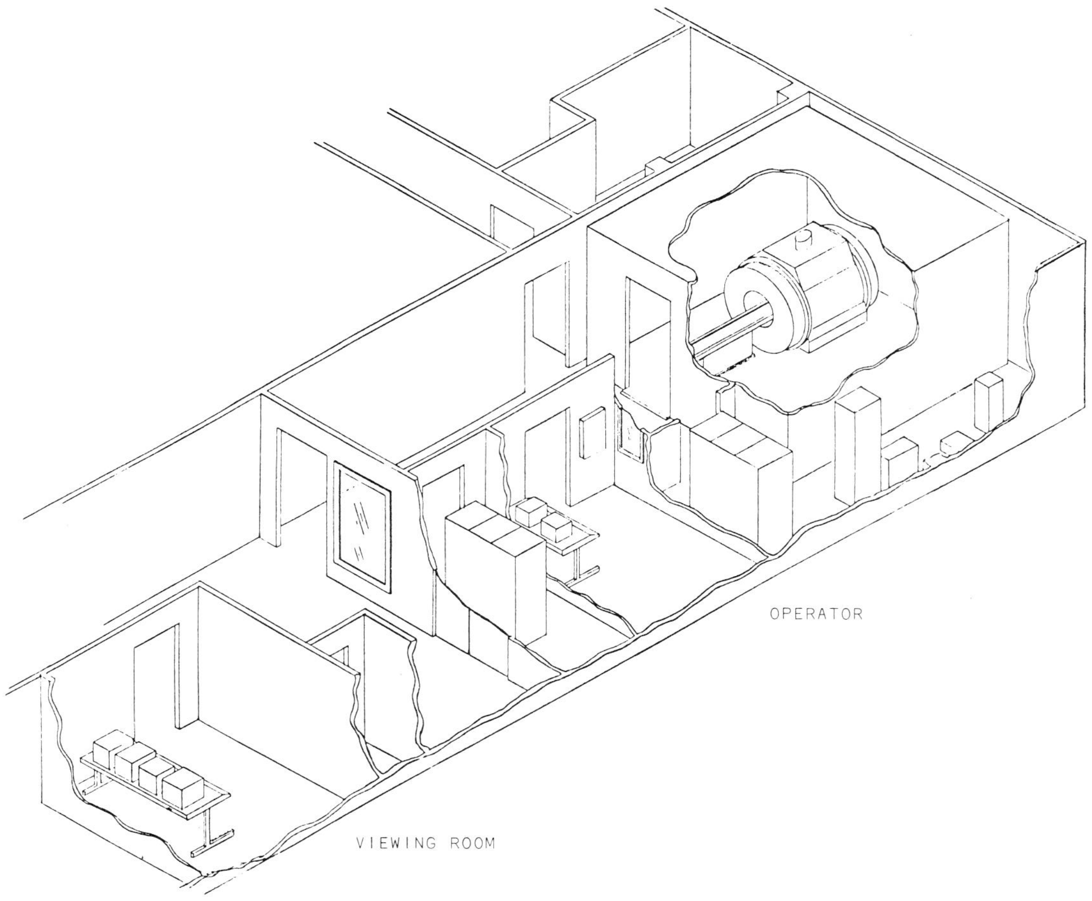

A.

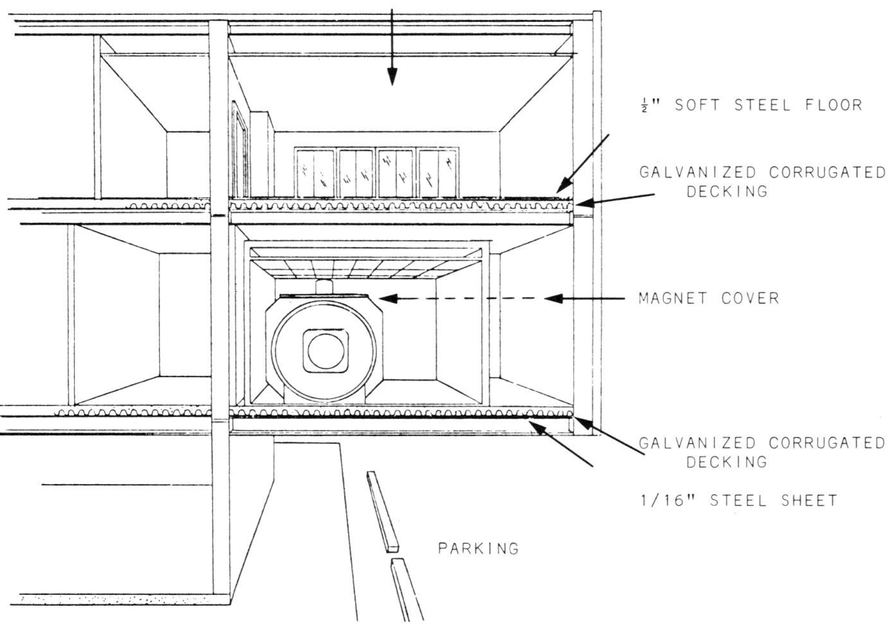

B.

Figure 2. Schematic drawing of the University of California, San Francisco clinical 0.35-T cyrogenic magnetic resonance imager.
A. Side view of facility.
B. Vertical cross-section showing open heart surgery suite above and loading dock below the magnet.

depending on manufacturing techniques. As opposed to x-ray shielding, where overlap of lead is sufficient, the metal in an RF room has to be in continuous *electrical* contact, which makes assembly of the latter a more exacting job. Doors also need to be part of the electrically continuous surfaces, thus requiring special construction of the door frame.

All entry points into the RF room are shielded or filtered to avoid introduction of RF interference. Radio-frequency wave guides and RF power filters accomplish this protection.

Continuing Sources of Interference

Completion of the installation of an NMR imager does not end the need to monitor sources of interference. The introduction of ferromagnetic objects in the RF room poses operational problems and hazardous conditions. Paper clips, hair pins, or small screws that enter the magnet can produce artifacts in the images that are interpreted as slight increases in noise levels. As described in Chapter 19, large objects can become projectiles that could injure patients, operating personnel, and the hardware. Although the danger area is within 2–4 ft from the magnet, it is prudent to consider the whole RF room as the area in need of absolute access control.

There exist two basic philosophies for site control. The first relies on automatic devices (metal detectors) for sceeening. Common experience at airports indicates the weakness of such an approach, which yields high false-positive and false-negative rates. The second philosophy for site control involves limiting access to the RF room to authorized personnel, and strict escorting procedures for visitors who have an absolute need to enter the room. Patients are dressed in hospital garb only, and gurneys, wheelchairs, and any additional patient support hardware must be nonmagnetic. After working hours, access to the machine can be controlled by combination locks, the code being known only to a limited number of staff. This control method depends on the quality of the personnel given this task and on the willingness of senior staff to support subordinates' decisions that may offend other hospital personnel or outside visitors.

Ancillary equipment that is necessary for patient care must be nonmagnetic. This is essential because a 2-lb metallic wrench is attracted into the magnet with a force of 18 lb when brought close to its opening. This means that a small oxygen tank when brought inadvertently into the room close to the magnet would fly into the machine with such a force that two weightlifters could not hold it back. The patient who becomes a medical emergency in the cryogenic magnet must be pulled out and taken care of or resuscitated in the room adjacent to the imager. A crash cart and suction devices should be available there; for safety reasons and to avoid predictable errors, these should be nonmagnetic. Because the market for such nonmagnetic appliances and equipment expands as magnetic resonance imaging becomes a clinical presence, the appliance industry will undoubtedly make available a whole line of products.

21

Economic Considerations

BARBARA J. McNEIL, M.D., Ph.D.
HERBERT L. ABRAMS, M.D.

The increase in the cost of medical care over the past few decades has clearly gone beyond the limits of inflation. Much of the increase has resulted from the cost of hospital care. (Of course, the quality of medical care—as evidenced by decreased morbidity and mortality—has increased, so that costs are not compared for equivalent services.) Several years ago the Department of Health, Education, and Welfare summarized rates of change for charges of various components of hospital services. These rates were variable, ranging from a low of 8% for oxygen to a high of 20.2% for operating room services.[1] Although the rate of change in charges for radiologic services was less than those for clinical laboratory services, radiologic procedures have nonetheless been subjected to intense scrutiny over the past decade, undoubtedly in part because of their high unit cost.

The most recent imaging method to attract the attention of the regulatory agencies was x-ray CT. The interest of review bodies was heightened by the belief that diffusion of CT was unaccompanied by a systematic and scientific evaluation of its efficacy. In contrast to therapeutic technologies such as gastric freezing, myocardial revascularization by internal mammary artery implantation, and hyperbaric oxygen chambers, which also diffused rapidly but ultimately proved of limited usefulness, CT constituted a revolutionary new addition to the armamentarium of health care. Impressive sets of data over the past 8 years have indicated that CT has played a significant role in improving the practice of many areas of medicine and surgery. Its costs have been substantial, however, and the tenuous balance between increased costs and increased benefit has kept CT in the limelight of discussion among health care planners and regulators.

It should come as no surprise, therefore, that NMR, a promising new diagnostic technology with applications not fully defined, will surely be subject to a strong demand for systematic exploration. National and regional committees are already being formed to deal with issues over its potential benefits and costs. Potential benefits from NMR include its ability to derive additional diagnostic information beyond that available from existing technologies, to provide data that will improve therapeutic planning, and—as a consequence—to effect changes in either morbidity or mortality. Its costs have several components and are the subject of this chapter. These cost figures must ultimately be compared with data on its medical impact so that appropriate cost-effectiveness ratios can be calculated.

In considering the cost of NMR, several factors must be analyzed: capital costs, maintenance costs, site preparation, and operating costs, including personnel and supplies. Average and total costs are both relevant and important.

Component Costs

Capital Costs

In terms of equipment, prices range from $1–$2.2 million per unit, with most vendors quoting a price of $1.5–$2.2 million. Low field strength resistive units tend to be at the lower end of this range, while higher field strength superconductive instruments are more costly. For example, Toshiba, Phillips, and Picker quote prices of $1–$1.5 million for 1.2-kGauss to 1.5-kGauss (0.12-T to 0.15-T) resistive whole body units. Siemens, Phillips, Diasonics, and Picker quote prices of $1.4–$1.7 million for 5-kGauss (0.5-T) superconductive units, and $2.2 million for 15-kGauss (1.5-T) units. Even if *permanently* magnetized ceramic blocks for head and body prove practical and become available in the future, both the capital and the maintenance costs of NMR may not be reduced in a significant manner.

There has been some conjecture that low field strength resistive systems may be economically produced and furnish the "work horse" units of community hospitals and private offices in the future. Reliable sources have suggested that such units might be produced for less than $500,000. Nevertheless, the patterns of production and pricing and the recovery of development costs, as reflected in the diffusion of CT, do not support the likelihood of such a development, nor does the quality of images suggest such a development.

Maintenance Costs

Service costs for resistive systems are estimated to be $80,000 a year, and for superconductive systems, $105,000 a year. Many individuals also add an additional security cost related to the unique effects of the fringe magnetic field.

Site Preparation

The presence of the magnetic field induces renovating costs greater than those for other technologies. Minimum space requirements for a 5-kGauss superconductive system, for example, are said to be 1,300 square feet and must allow for protection both of the magnet and of nearby individuals and pieces of equipment. Areas above and below must also be considered. Renovating existing structures costs at least $100–$300 per square foot, and most individuals estimate renovation costs from $300,000 to $500,000. New building would cost more, probably over $1 million.

The shift from a 1.5-, 3-, or 5-kGauss unit to a significantly higher field intensity—such as 15–18 kGauss—is accompanied by an increase in complexity of many orders of magnitude. It is not the square footage but the cubic footage that becomes more critical if the unit is sited as part of an existing structure with activities adjacent and above it, or if a new facility is constructed to house it. Although shielding is potentially feasible, it is not only costly, but may also seriously interfere with the homogeneity of the field and, therefore, with the proper functioning of the unit. For units of 1.5 T or above, site preparation costs of well over $1 million must be considered. For most consumers (hospitals), this will not be a serious consideration because commercially available units are currently designed to operate at 5-kGauss or less.

Operating Costs

Personnel requirements are in some sense similar to those required for CT. For example, for a 12-hr day it is estimated that one physician, three technologists, one and a half nurses, and one and a half receptionist-typists are necessary. During start-up activities, a full-time physicist will be essential in institutions interested in research. As additional experience is gained, about one-quarter of a physicist-engineer's time will still be necessary. These personnel costs would account for nearly $250,000 of the operating costs (Table I).

Other operating costs include electricity and water for resistive magnets: Phillips estimates a $20,000 a year cost for electricity and a $20,000 water cost. (Some cities do not allow dumping of water because of water shortages. Air conditioning systems for cooling are frequently used, and their cost probably approximates that of the electricity needed for heating the magnet.) For superconductive magnets the cost of electricity is significantly less than for resistive units. Superconductive instruments need cryogens, usually helium and nitrogen, and these could add another $35,000 or more per year. It is possible that newer devices that allow for refrigeration of nitrogen and helium will result in diminished cryogen costs. In the case of a catastrophic failure of the magnet, the gases of the superconductive system would need to be recharged, and this could amount to about $12,000 per year. Normal quenching is a far more benign process.

The cost of supplies (film, clerical supplies, perhaps contrast agents in the future, etc.) probably is close to that of CT. For an average machine operating 12 hr a day, an additional $65,000 (or $17 per patient) a year is reasonable. (This figure is a bit less than the $20 per patient estimate of the American Hospital Association.[2])

Indirect Costs

All of the foregoing costs are direct costs, and these must be modified on an individual basis by indirect costs of particular institutions. These costs will obviously include space, light, power, telephone, housekeeping, institutional record-keeping and billing functions, and so on.

Interpretation Fees

In addition to the technical components for NMR and the associated indirect costs, professional fees must be included. These costs are likely to be similar to those of CT. Evens's and Jost's recent study on CT suggests that in 1982, typical professional charges for CT of the head and body were $86 and $104, respectively.[3] Professional charges for NMR should be somewhat higher because of the increased complexity of interpretation. By including a full-time physician in the personnel budget (Table I), we have deliberately avoided the issue of charges.

Total Costs of One Machine

Table II summarizes the above data for the current year and the next 4 years. In constructing this table we have, by necessity, made approximations since we have not separated out costs of resistive *versus* superconductive systems. However, our numbers provide an estimate, probably an overestimate, for resistive systems and an underestimate for superconductive systems. In the case of both types of systems we have assumed outright purchase in 1983; leasing arrangements would increase direct scanner costs. Thus, for 1983, per machine direct costs run about $750,000; indirect costs, $525,000; and total costs, $1,275,000. In a 12-hr work day and a throughput of 15 patients per day (about 3,750 annually), the average cost per examination is $340 in 1983.

TABLE I

Estimated Personnel Costs (Including Fringe) for Nuclear Magnetic Resonance Machine

1	Physician	$100,000
3	Technologists	80,000
1.5	Nurses	27,000
0.25	Physicist-engineer	12,500
1.5	Receptionist-typists	21,000
		$240,500

Total Costs

Total costs of NMR to society will vary with two factors: its rate of diffusion and the extent to which prices decrease from either competition in the marketplace or because of major changes in construction (for example, recirculation of coolants, permanently charged magnets, etc.). The latter are impossible to estimate at this point, except to say that on the basis

TABLE II

Average Annual Costs for a Nuclear Magnetic Resonance Machine

	Year 1	*Year 2*	*Year 3*	*Year 4*	*Year 5*
Direct Costs⁺					
Scanner Costs*	$320,000	$320,000	$320,000	$320,000	$320,000
Site Costs*	80,000	80,000	80,000	80,000	80,000
Maintenance Contract**	0	80,000	88,000	95,000	103,000
Operating Costs**					
Personnel	240,000	260,000	280,000	302,000	326,000
Electricity and Coolants	45,000	49,000	53,000	57,000	62,000
Supplies	65,000	70,000	76,000	82,000	88,000
	$750,000	$859,000	$897,000	$936,000	$979,000
Indirect Costs (70%)⁺	$525,000	$601,000	$628,000	$655,000	$685,000
Total Costs	$1,275,000	$1,460,000	$1,525,000	$1,591,000	$1,664,000

⁺ The figures on power and supplies are derived from suggested costs from a group of manufacturers. The figures on personnel were arrived at using CT as a partial model, and incorporating data for existing NMR installations. Scanner cost and site preparation figures represent an average of estimates from both manufacturers and physicists. Indirect costs are computed from the experience of a group of hospitals.

*Depreciated over 5 years.

**Inflated at approximately 8% per year.

of the CT experience, costs of some units will increase, while others—of lesser quality—may decline. Although the diffusion rate is somewhat difficult to estimate, analogy with CT is instructive. Table III shows the rate of installation of new CT scanners in the United States by year: in 1973 there were six CT scanners; in 1976, 475; in 1977, 955; and in 1980, 1,733.[4] There are now over 2,000 scanners in the United States, and over 4,000 worldwide. (On a per million population basis in 1980, these numbers range approximately from less than one scanner per million population in France to nearly 12 per million in Japan. The United States had between six and seven per million.)

At this time, we believe that the CT diffusion curve is a reasonable starting point for estimating NMR costs. Others could view this assumption as either conservative or liberal. It might be considered conservative if the promise of NMR is so great and barriers to diffusion of new diagnostic technologies diminished enough that we would expect the diffusion of NMR to be more rapid than that of CT. On the other hand, it could be considered a liberal estimate if we assume that despite NMR's actual benefits, its *excess* price for capital over that for CT will lead to a slower rate of diffusion.

Before estimating total societal costs, one adjustment in Table III is necessary: It currently appears that there will be 15 NMR scanners in place by the end of 1983, leading us to estimate cumulative totals for 1984 to 1987 of 54, 205, 484, and 964, respectively, or nearly 1,000 by 1987. These numbers and the costs in Table III lead to yearly societal and cumulative costs shown in Table IV. Or, broadly speaking, if there are 1,000 scanners by 1987, total costs for NMR in that year will be about $1.66 billion.

Whether these yearly costs (and, by extension, cumulative costs) for NMR are large or small is not clear. Careful prospective studies linking diagnostic effectiveness with cost will be needed before such a judgment can be made. Our hope is that with the lessons learned from assessments of CT and other new technologies over the past decade, we should be able to provide timely and thoughtful answers to questions regarding the costs, risks, and benefits of NMR put to us by public policy makers and regulators, referring physicians, patients, manufacturers, and ourselves.

TABLE III

Installation of New Computed Tomography Scanners by Year

Year	*Number*	*Cumulative Total*
1973	6	6
1974	39	45
1975	151	196
1976	279	475
1977	480	955
1978	244	1,199
1979	204	1,403
1980	330	1,733
1981	435	2,168

TABLE IV

Yearly Societal Costs for Nuclear Magnetic Resonance

	Estimated No. of Scanners in Operation	*Total Costs That Year (millions)*
1983	15	$19.1
1984	54	78.8
1985	205	312.6
1986	484	770.0
1987	964	1,604.0

REFERENCES

1. Department of Health, Education and Welfare: *Medical Care Expenditures, Prices and Costs: Background Book*, p 23. No(SSA) 75-11909.
2. American Hospital Association: Financial considerations. In "NMR (Nuclear Magnetic Resonance): Guideline Report." AHA Hospital Technology Series 2:57, 1983.
3. Evens RG, Jost RG: Computed tomography utilization and charges in 1981. Radiology 145:427, 1982.
4. Market Scan. CT scanning redux. Diagn Imaging 51:15, 1982.

GLOSSARY of Nuclear Magnetic Resonance (NMR) Terms*

Prepared under the auspices of American College of Radiology:

Subcommittee on NMR Nomenclature and Phantom Development
Leon Axel, Ph.D., M.D., *Chairman*

Committee on NMR Imaging Technology and Equipment
Alexander R. Margulis, M.D., *Chairman*

Commission on Nuclear Magnetic Resonance
Thomas F. Meaney, M.D., *Chairman*

Acquisition time—see Image acquisition time.

ADC—see Analog to digital converter.

Adiabatic fast passage (AFP)—technique of producing rotation of the *macroscopic magnetization vector* by sweeping the *frequency* of an irradiating *RF* wave (or the strength of the *magnetic field*) through *resonance* (the *Larmor frequency*) in a time short compared to the *relaxation times*. Particularly used for *inversion* of the *spins*. A *continuous wave NMR* technique.

Adiabatic rapid passage—see Adiabatic fast passage.

AFP—see Adiabatic fast passage.

Analog to digital converter (ADC)—part of the *interface* that converts ordinary (analog) voltages, such as the detected *NMR signal*, into digital number form, that can be read by the *computer*.

Angular frequency (ω)—*frequency* of oscillation or rotation (measured, e.g., in radians/second) commonly designated by the Greek letter ω: $\omega = 2\pi f$, where f is frequency (e.g., in *hertz* (Hz)).

Angular momentum—a *vector* quantity given by the vector product of the momentum of a particle and its position vector. In the absence of external forces, the angular momentum remains constant, with the result that any rotating body tends to maintain the same axis of rotation. When a *torque* is applied to a rotating body, the resulting change in angular momentum results in *precession*. Atomic nuclei possess an intrinsic angular momentum referred to as *spin*, measured in multiples of Planck's constant.

Antenna—device to send or receive electromagnetic radiation. Electromagnetic radiation per se is not relevant to NMR, as it is the magnetic vector alone that couples the *spins* and the *coils*, and the term coil should be used instead.

Array processor—optional component of *computer* system specially designed to speed up numerical calculations like those needed in *NMR imaging*.

Artifacts—false features in the image produced by the imaging process.

B_0—a conventional symbol for the constant *magnetic (induction) field* in an *NMR* system. (Although historically used, H_0 (units of *magnetic field* strength, ampere/meter) should be distinguished from the more appropriate B_0 (units of magnetic induction, tesla).)

B_1—a conventional symbol for the *radiofrequency magnetic induction field* used in an *NMR* system (another symbol historically used is H_1). It is useful to consider it as composed of two oppositely rotating *vectors*, usually in a plane transverse to B_0. At the *Larmor frequency*, the vector rotating in the same direction as the *precessing* spins will interact strongly with the *spins*.

Bloch equations—phenomenological "classical" equations of motion for the *macroscopic magnetization vector*. They include the effects of *precession* about the *magnetic field* (static and *RF*) and the *T1* and *T2 relaxation times*.

Boltzmann distribution—if a system of particles which are able to exchange energy in collisions is in thermal equilibrium, then the relative number of particles, N_1 and N_2, in two particular energy states with corresponding energies, E_1 and E_2, is given by

$$\frac{N_1}{N_2} = \exp\left[-(E_1 - E_2)/kT\right]$$

where k is Boltzmann's constant and T is absolute temperature. For example, in *NMR* of protons at room temperature in a *magnetic field* of 0.25 *tesla*, the difference in numbers of *spins* aligned with the magnetic field and against the field is about one part in a million; the small excess of nuclei in the lower energy state is the basis of the net *magnetization* and the *resonance* phenomenon.

Carr-Purcell (CP) sequence—sequence of a *90° RF pulse* followed by repeated *180° RF pulses* to produce a train of *spin echoes*; useful for measuring *T2*.

Carr-Purcell-Meiboom-Gill (CPMG) sequence—modification of *Carr-Purcell RF pulse* sequence with 90° *phase* shift in the *rotating frame of reference* between the *90° pulse* and the subsequent *180° pulses* to reduce accumulating effects of imperfections in the 180° pulses. Suppression of effects of pulse error accumulation can alternatively be achieved by alternating phases of the 180° pulses by 180°.

Chemical shift (δ)—the change in the *Larmor frequency* of a given nucleus when bound in different sites in a molecule, due to the magnetic shielding effects of the electron orbitals. Chemical shifts make possible the differentiation of different molecular compounds and different sites within the molecules in high-resolution *NMR spectra*. The amount of the shift is proportional to *magnetic field* strength and is usually specified in parts per million (ppm) of the *resonance frequency* relative to a standard.

Coherence—maintenance of a constant *phase* relationship between rotating or oscillating waves or objects. Loss of phase coherence of the *spins* results in a decrease in the *transverse magnetization* and hence a decrease in the *NMR signal*.

Coil—single or multiple loops of wire (or other electrical conductor, such as tubing, etc.) designed either to produce a *magnetic field* from current flowing through the wire, or to detect a changing magnetic field by voltage induced in the wire.

*Reproduced, with permission, from the American College of Radiology.

Computer—as used for *NMR,* can be divided into central processing unit (CPU), consisting of instruction interpretation and arithmetic unit plus fast access memory, and peripheral devices such as bulk data storage and input and output devices (including, via the *interface,* the *spectrometer*). Under *software* control, the computer controls the *RF pulses* and *gradients* necessary to acquire data, and processes the data to produce *spectra* or images. (Note that devices such as the spectrometer may themselves incorporate small computers.)

Continuous wave NMR (CW)—a technique for studying *NMR* by continuously applying *RF* radiation to the sample and slowly sweeping either the RF *frequency* or the *magnetic field* through the *resonance* values; now largely superceded by *pulse NMR* techniques.

Contrast—contrast can be defined as the relative difference of the signal intensities in two adjacent regions. In a general sense, we can consider image contrast, where the strength of the image intensity in adjacent regions of the image is compared, or object contrast, where the relative values of a parameter affecting the image (such as *spin density* or *relaxation time*) in corresponding adjacent regions of the object are compared. If the two intensities are J_1 and J_2, a useful quantitative definition of contrast is $(J_1 - J_2)/(J_1 + J_2)$. Relating image contrast to object contrast is more difficult in *NMR imaging* than in conventional radiography, as there are more object parameters affecting the image and their relative contributions are very dependent on the particular imaging technique used. As in other kinds of imaging, image contrast in NMR will also depend on region size, as reflected through the modulation transfer function (MTF) characteristics.

CP—see Carr-Purcell.

CPMG—see Carr-Purcell-Meiboom-Gill.

CPU—see Computer.

Crossed-coil—*coil* pair arranged with their *magnetic fields* at right angles to each other in such a way as to minimize their mutual electromagnetic interaction.

Cryomagnet—See Superconducting magnet.

Cryostat—an apparatus for maintaining a constant low temperature (as by means of liquid helium). Requires vacuum chambers to help with thermal isolation.

CW—see Continuous wave.

DAC—see Digital to analog converter.

Data system—see Computer.

dB/dt—the rate of change of the *magnetic field* (induction) with time. Because changing magnetic fields can induce electrical fields, this is one area of potential concern for safety limits.

Demodulator—another term for *detector,* by analogy to broadcast radio receivers.

Detector—portion of the *receiver* that demodulates the *RF NMR signal* and converts it to a lower *frequency* signal. Most detectors now used are *phase* sensitive (e.g. *quadrature demodulator/detector*), and will also give phase information about the RF signal.

Diamagnetic—a substance that will slightly decrease a *magnetic field* when placed within it (its magnetization is oppositely directed to the magnetic field, i.e., with a small negative *magnetic susceptibility*).

Diffusion—the process by which molecules or other particles intermingle and migrate due to their random thermal motion. *NMR* provides a sensitive technique for measuring diffusion of some substances.

Digital to analog converter (DAC)—part of the *interface* that converts digital numbers from the *computer* into analog (ordinary) voltages or currents.

Echo—see Spin echo.

Echo planar imaging—a technique of *planar imaging* in which a complete planar image is obtained from one *selective excitation pulse.* The *FID* is observed while periodically switching the *y-gradient field* in the presence of a static *x-gradient field.* The *Fourier transform* of the resulting *spin echo* train can be used to produce an image of the excited plane.

Eddy currents—electric currents induced in a conductor by a changing *magnetic field* or by motion of the conductor through a magnetic field. One of the sources of concern about potential hazard to subjects in very high *magnetic fields* or rapidly varying *gradient* or main magnetic fields. Can be a practical problem in the *cryostat* of *superconducting magnets.*

Excitation—putting energy into the *spin* system; if a net *transverse magnetization* is produced, an *NMR signal* can be observed.

f—see Frequency.

Faraday shield—electrical conductor interposed between *transmitter and/or receiver coil* and patient to block out electric fields.

Fast Fourier transform (FFT)—an efficient computational method of performing a *Fourier transform.*

Ferromagnetic—a substance, such as iron, that has a large positive *magnetic susceptibility.*

FFT—see Fast Fourier transform.

FID—see Free induction decay.

Field gradient—see Gradient magnetic field.

Field lock—a feedback control used to maintain the static *magnetic field* at a constant strength, usually by monitoring the *resonance frequency* of a reference sample or line in the *spectrum.*

Filling factor— a measure of the geometrical relationship of the *RF coil* and the body. It affects the efficiency of irradiating the body and detecting *NMR signals,* thereby affecting the *signal-to-noise ratio* and, ultimately, image quality. Achieving a high filling factor requires fitting the coil closely to the body, thus potentially decreasing patient comfort.

Filtered back projection—mathematical technique used in *reconstruction from projections* to create images from a set of multiple *projection profiles.* It essentially involves "correcting" the projection profiles by convolving them with a suitable mathematical filter and then back projecting the filtered projections into image space. Widely used in conventional computed tomography (CT).

Flip angle—amount of rotation of the *macroscopic magnetization vector* produced by an *RF pulse,* with respect to the direction of the static *magnetic field.*

Fourier transform (FT)—a mathematical procedure to separate out the *frequency* components of a signal from its amplitudes as a function of time, or vice versa. The Fourier transform is used to generate the *spectrum* from the *FID* in *pulse NMR* techniques and is essential to most imaging techniques.

Fourier transform imaging—*NMR imaging* techniques in which at least one dimension is phase encoded by applying variable *gradient pulses* along that dimension before "reading out" the *NMR signal* with a *gradient magnetic field* perpendicular to the variable gradient. The *Fourier transform* is then used to reconstruct an image from the set of encoded NMR signals. An imaging technique of this type is *spin warp imaging.*

Free induction decay (FID)—if *transverse magnetization* of the *spins* is produced, e.g., by a *90° pulse,* a transient *NMR signal* will result that will decay toward zero with a characteristic time constant *T2* (or *T2**); this decaying signal is the FID. In practice, the first part of the FID is not observable due to residual effects of the powerful exciting *RF pulse* on the electronics of the *receiver.*

Free induction signal (FIS)—see Free induction decay.

Frequency (f)—the number of repetitions of a periodic process per unit time. For electromagnetic radiation, such as radio waves, the old unit, cycles per second (cps), has been replaced by the *SI* unit, *hertz*, abbreviated *Hz*. It is related to *angular frequency*, ω, by $f = \omega/2\pi$.

FT—see Fourier transform.

G—see Gauss.

G_x, G_y, G_z—conventional symbols for *gradient magnetic field*. Used with subscripts to denote *spatial direction* component of gradient, i.e., direction along which the field changes.

Gauss (G)—a unit of magnetic flux density in the older (CGS) system. The Earth's magnetic field is approximately one half gauss to one gauss, depending on location. The currently preferred (*SI*) unit is the *tesla* (T) (1 T = 10,000 G).

Golay coil—term commonly used for a particular kind of *gradient coil*, commonly used to create *gradient magnetic fields*, perpendicular to the main *magnetic field*.

Gradient—the amount and direction of the rate of change in space of some quantity, such as *magnetic field* strength.

Gradient coils—current carrying *coils* designed to produce a desired *gradient magnetic field* (so that the *magnetic field* will be stronger in some locations than others). Proper design of the size and configuration of the coils is necessary to produce a controlled and uniform *gradient*.

Gradient magnetic field—a *magnetic field* which changes in strength in a certain given direction. Such fields are used in *NMR imaging* with *selective excitation* to select a region for imaging and also to encode the location of *NMR signals* received from the object being imaged. Measured (e.g.) in *teslas* per meter.

Gradient pulse—briefly applied *gradient magnetic field*.

Gyromagnetic ratio (γ)—the ratio of the *magnetic moment* to the *angular momentum* of a particle. This is a constant for a given nucleus.

H_0—conventional symbol historically used for the constant *magnetic field* in an *NMR* system; it is physically more correct to use B_0. A magnet provides a field strength, H; however, at a point in an object, the *spins* experience the *magnetic induction*, B.

H_1—conventional symbol historically used for the *radiofrequency magnetic field* in an *NMR* system; it is physically more correct to use B_1. It is useful to consider it as composed of two oppositely rotating *vectors*. At the *Larmor frequency*, the vector rotating in the same direction as the *precessing spins* will interact strongly with the *spins*.

Hardware—electrical and mechanical components of *computer*.

Helmholtz coil—pair of current carrying *coils* used to create uniform *magnetic field* in the space between them.

Hertz (Hz)—the standard (*SI*) unit of *frequency*; equal to the old unit cycles per second.

Homogeneity—uniformity. In *NMR*, the homogeneity of the static *magnetic field* is an important criterion of the quality of the magnet. Homogeneity requirements for *NMR imaging* are generally lower than the homogeneity requirements for NMR spectroscopy, but for most imaging techniques must be maintained over a larger region.

Hz—see Hertz.

I—see Nuclear spin number.

Image acquisition time—time required to carry out an *NMR imaging* procedure comprising only the data acquisition time. The additional image reconstruction time will also be important to determine how quickly the image can be viewed. In comparing *sequential plane imaging* and *volume imaging* techniques, the equivalent image acquisition time per slice must be considered, as well as the actual image aquisition time.

Inductance—measure of the magnetic coupling between two current carrying loops (mutual) (reflecting their spatial relationship) or of a loop (such as a *coil*) with itself (self). One of the principal determinants of the *resonance frequency* of an *RF* circuit.

Inhomogeneity—degree of lack of *homogeneity*, for example the fractional deviation of the local *magnetic field* from the average value of the field.

Interface—set of devices that enables the interaction of the *computer* and the *spectrometer*. Particularly, this includes an *analog to digital converter* (ADC), which turns the analog voltages, such as the output of the RF *receiver*, into numbers that can be read by the *computer*. It also includes a *digital to analog converter* (DAC), which does the reverse, enabling the computer to produce control voltages.

Interpulse time—times between successive RF *pulses* used in *pulse sequences*. Particularly important are the *inversion time* (*TI*) in *inversion recovery*, and the time between a *90° pulse* and the subsequent *180° pulse* to produce a *spin echo*, which will be approximately one half the *spin echo time* (*TE*). The time between repetitions of pulse sequences is the *repetition time* (*TR*).

Inversion—a nonequilibrium state in which the *macroscopic magnetization vector* is oriented opposite to the *magnetic field*; usually produced by *adiabatic fast passage* or *180° RF pulses*.

Inversion-recovery (IR)—*pulse NMR* technique which can be incorporated into *NMR imaging*, wherein the nuclear magnetization is inverted at a time on the order of *T1* before the regular imaging pulse-gradient sequences. The resulting partial *relaxation* of the spins in the different structures being imaged can be used to produce an image that depends strongly on T1. This may bring out differences in the appearance of structures with different T1 relaxation times. Note that this does not directly produce an image of T1. T1 in a given region can be calculated from the change in the *NMR signal* from the region due to the inversion pulse compared to the signal with no inversion pulse or an inversion pulse with a different inversion time (TI).

Inversion time (TI)— time between *inversion* and subsequent *90° pulse* to elicit *NMR signal* in *inversion-recovery*.

Inversion transfer—see Saturation transfer.

IR—see Inversion recovery.

k—Boltzmann's constant: appears in *Boltzmann distribution*.

kHz—see Kilohertz.

Kilohertz (kHz)—unit of *frequency*; equal to one thousand *hertz*.

Larmor equation—states that the *frequency* of precession of the nuclear *magnetic moment* is proportional to the *magnetic field*.

$$\omega_0 = -\gamma B_0 \text{ (radians per second)}$$
$$\text{or } f_0 = -\gamma B_0/2\pi \text{ (hertz)}$$

where ω_0 or f_0 is the *frequency*, γ is the *gyromagnetic ratio*, and B_o is the *magnetic induction field*. The negative sign indicates the direction of the rotation.

Larmor frequency (ω_0 or f_0)—the frequency at which magnetic resonance can be excited; given by the *Larmor equation*. By varying the *magnetic field* across the body with a *gradient magnetic field*, the corresponding variation of the Larmor frequency can be used to encode position. For protons (hydrogen nuclei), the Larmor frequency is 42.58 MH_z/tesla.

Lattice—by analogy to *NMR* in solids, the magnetic and thermal environment with which nuclei exchange energy in *longitudinal relaxation*.

Line imaging—see Sequential line imaging.

Line scanning—see Sequential line imaging.

Line width—width of line in *spectrum*; related to the reciprocal of the transverse *relaxation time* (*T2** in practical systems). Measured in units of *frequency*, generally at the half-maximum points.

LMR—see Localized magnetic resonance.

Localized magnetic resonance (LMR)—a particular technique for obtaining *NMR spectra,* for example, of phosphorus, from a limited region by creating a *sensitive volume* with *inhomogeneous* applied *gradient magnetic fields,* which may be enhanced with the use of *surface coils.*

Longitudinal magnetization (M_z)—component of the *macroscopic magnetization vector* along the static *magnetic field.* Following excitation by *RF pulse,* M_z will approach its equilibrium value M_o, with a characteristic time constant *T1.*

Longitudinal relaxation—return of *longitudinal magnetization* to its equilibrium value after *excitation;* requires exchange of energy between the *nuclear spins* and the *lattice.*

Longitudinal relaxation time—see T1.

Lorentzian line—usual shape of the lines in an *NMR spectrum,* characterized by a central peak with long tails; proportional to $1/[(1/T2)^2 + (f - f_O)^2]$, where *f* is *frequency* and f_O is the frequency of the peak (i.e., central resonance frequency).

M—conventional symbol for *macroscopic magnetization vector.*

M_{xy}—see Transverse magnetization.

M_z—see Longitudinal magnetization.

M_o—equilibrium value of the *magnetization;* directed along the direction of the static *magnetic field.* Proportional to *spin density, N.*

Macroscopic magnetic moment—see Macroscopic magnetization vector.

Macroscopic magnetization vector—net *magnetic moment* per unit volume (a *vector* quantity) of a sample in a given region, considered as the integrated effect of all the individual microscopic nuclear magnetic moments. Most *NMR* experiments actually deal with this.

Magnetic dipole—north and south magnetic poles separated by a finite distance. An electric current loop, including the effective current of a spinning nucleon or nucleus, can create an equivalent magnetic dipole.

Magnetic field (H)—the region surrounding a magnet (or current carrying conductor) is endowed with certain properties. One is that a small magnet in such a region experiences a *torque* that tends to align it in a given direction. Magnetic field is a *vector* quantity; the direction of the field is defined as the direction that the north pole of the small magnet points when in equilibrium. A magnetic field produces a magnetizing force on a body within it. Although the dangers of large magnetic fields are largely hypothetical, this is an area of potential concern for safety limits.

Formally, the forces experienced by moving charged particles, current carrying wires, and small magnets in the vicinity of a magnet are due to *magnetic induction* (B), which includes the effect of *magnetization,* while the magnetic field (H) is defined so as not to include magnetization. However, both B and H are often loosely used to denote magnetic fields.

Magnetic field gradient—see Gradient magnetic field.

Magnetic induction (B)—also called magnetic flux density. The net magnetic effect from an externally applied *magnetic field* and the resulting *magnetization.* B is proportional to H ($B = \mu H$), with the *SI* unit being the *tesla.*

Magnetic moment—a measure of the net magnetic properties of an object or particle. A nucleus with an intrinsic *spin* will have an associated *magnetic dipole* moment, so that it will interact with a *magnetic field* (as if it were a tiny bar magnet).

Magnetic resonance—see Nuclear magnetic resonance (NMR). Another magnetic resonance phenomenon is electron spin resonance (ESR).

Magnetic susceptibility (χ)—measure of the ability of a substance to become magnetized.

Magnetization (see also Macroscopic magnetization vector)—the magnetic polarization of a material produced by a *magnetic field* (magnetic moment per unit volume).

Magnetogyric ratio—see Gyromagnetic ratio.

Maxwell coil—a particular kind of *gradient coil,* commonly used to create *gradient magnetic fields* along the direction of the main *magnetic field.*

Megahertz (MHz)—unit of *frequency,* equal to one million *hertz.*

Meiboom-Gill sequence—see Carr-Purcell-Meiboom-Gill sequence.

MHz—see Megahertz.

Multiple line-scan imaging (MLSI)—variation of *sequential line imaging* techniques that can be used if *selective excitation* methods that do not affect adjacent lines are employed. Adjacent lines are imaged while waiting for *relaxation* of the first line toward equilibrium, which may result in decreased *imaging time.* A different type of MLSI uses simultaneous excitation of two or more lines with different phase encoding followed by suitable decoding.

Multiple plane imaging—variation of *sequential plane imaging* techniques that can be used with *selective excitation* techniques that do not affect adjacent planes. Adjacent planes are imaged while waiting for relaxation of the first plane toward equilibrium, resulting in decreased *imaging time.*

Multiple sensitive point—*sequential line imaging* technique utilizing two orthogonal oscillating *magnetic field gradients,* an *SFP* pulse sequence, and signal averaging to isolate the *NMR spectrometer* sensitivity to a desired line in the body.

N (H)—see Spin density.

NMR—see Nuclear magnetic resonance.

NMR imaging (see also Zeugmatography)—creation of images of objects such as the body by use of the *nuclear magnetic resonance* phenomenon. The immediate practical application involves imaging the distribution of hydrogen nuclei (protons) in the body. The image brightness in a given region is usually dependent jointly on the *spin density* and the *relaxation times,* with their relative importance determined by the particular imaging technique employed. Image brightness is also affected by motion such as blood flow.

NMR signal—electromagnetic signal in the *radiofrequency* range produced by the *precession* of the *transverse magnetization* of the *spins.* The rotation of the transverse magnetization induces a voltage in a *coil,* which is amplified and demodulated by the *receiver;* the signal may refer only to this induced voltage.

Nuclear magnetic resonance (NMR)—the absorption or emission of electromagnetic energy by nuclei in a static *magnetic field,* after *excitation* by a suitable RF magnetic field. The peak *resonance frequency* is proportional to the magnetic field, and is given by the *Larmor equation.* Only nuclei with a non-zero *spin* exhibit NMR.

Nuclear signal—see NMR signal.

Nuclear spin (see also Spin)—an intrinsic property of certain nuclei that gives them an associated characteristic *angular momentum* and *magnetic moment.*

Nuclear spin quantum number (I)—property of all nuclei related to the largest measurable component of the nuclear *angular momentum.* Non zero values of nuclear angular momentum are quantized (fixed) as integral or half-integral multiples of $(h/2\pi)$, where h is Planck's constant. The number of possible energy states for a given nucleus in a fixed *magnetic field* is equal to $2I + 1$.

Nucleon—generic term for a neutron or proton.

Nutation—a displacement of the axis of a spinning body away from the simple coneshaped figure which would be traced by the axis during *precession.* In the *rotating frame of reference,* the nutation caused by an *RF pulse* appears as a simple precession, although the motion is more complex in the stationary frame of reference.

Orientation—a suggested standard orientation for the presentation of *NMR* images is: 1) transverse: patient's right on the left side of the image, anterior or ventral on top, 2) coronal: patient's right to left side of image, superior or head to the top, 3) sagittal: patient's head to the top, anterior to the left side of image. R, L, S and A should be shown on the screen, as appropriate. In displaying sagittal images, it is helpful to indicate whether a slice is to the left or right of the midline.

Paramagnetic—a substance with a small but positive *magnetic susceptibity* (magnetizability). The addition of a small amount of paramagnetic substance may greatly reduce the *relaxation times* of water. Typical paramagnetic substances usually possess an unpaired electron and include atoms or ions of transition elements, rare earth elements, some metals, and some molecules including molecular oxygen and free radicals. Paramagnetic substances are considered promising for use as contrast agents in *NMR imaging*.

Partial saturation (PS)—*excitation technique* applying repeated *RF pulses* in times on the order of or shorter than *T1*. In *NMR imaging* systems, although it results in decreased signal amplitude, there is the possibility of generating images with increased *contrast* between regions with different *relaxation times*. It does not directly produce images of T1. The change in NMR signal from a region resulting from a change in the *interpulse time, TR,* can be used to calculate T1 for the region. Although partial saturation is also commonly referred to as *saturation recovery,* that term should properly be reserved for the particular case of partial saturation in which recovery after each *excitation* effectively takes place from true *saturation*.

Permanent magnet—a magnet whose *magnetic field* originates from permanently magnetized material.

Permeability (μ)—tendency of a substance to concentrate *magnetic field*, $\mu = B/H$.

Phantom—an artificial object of known dimensions and properties used to test aspects of an imaging machine.

Phase—in a periodic function (such as rotational or sinusoidal motion), the position relative to a particular part of the cycle.

Phase sensitive detector—see Demodulator.

Pixel—acronym for a picture element; the smallest discrete part of a digital image display.

Planar spin imaging—one particular technique of *planar imaging* that creates an *NMR image* of a plane from one excitation sequence by selectively exciting a grid of points within the plane and then applying a *gradient magnetic field* so that each point has a different *Larmor frequency. Fourier transformation* of the *FID* can then be used to separate the signals from each selected point and create the image.

Planar imaging—imaging technqiue in which image of a plane is built up from signals received from the whole plane. See also Sequential plane imaging.

Point imaging—see Sequential point imaging.

Point scanning—see Sequential point imaging.

Precession—comparatively slow gyration of the axis of a spinning body so as to trace out a cone; caused by the application of a *torque* tending to change the direction of the rotation axis, and continuously directed at right angles to the plane of the torque. The *magnetic moment* of a nucleus with *spin* will experience such a torque when inclined at an angle to the *magnetic field,* resulting in precession at the *Larmor frequency.* A familiar example is the effect of gravity on the motion of a spinning top or gyroscope.

Precessional frequency—see Larmor frequency.

Probe—the portion of an *NMR spectrometer* comprising the sample container and the *RF* coils, with some associated electronics. The RF coils may consist of separate *receiver* and *transmitter* coils in a *crossed-coil* configuration, or, alternatively, a single coil to perform both functions.

Program—see Software.

Progressive saturation—see Saturation recovery.

Projection profile—*spectrum* of *NMR signal* whose *frequency* components are broadened by a *gradient magnetic field.* In the simplest case (negligible *line width,* no *relaxation* effects, and no effects of prior gradients), it corresponds to a one-dimensional projection of the *spin density* along the direction of the gradient; in this form it is used in *reconstruction from projections imaging*.

PS—see Partial saturation.

Pulse, 90° ($\pi/2$ pulse)—*RF pulse* designed to rotate the *macroscopic magnetization vector* 90° in space as referred to the *rotating frame of reference,* usually about an axis at right angles to the main *magnetic field.* If the *spins* are initially aligned with the magnetic field, this pulse will produce *transverse magnetization* and an *FID*.

Pulse, 180° (π pulse)—*RF pulse* designed to rotate the *macroscopic magnetization vector* 180° in space as referred to the *rotating frame of reference,* usually about an axis at right angles to the main *magnetic field.* If the *spins* are initially aligned with the magnetic field, this pulse will produce *inversion*.

Pulse, RF—see RF pulse.

Pulse length (width)—time duration of a *pulse*. For an *RF pulse* near the *Larmor frequency,* the longer the pulse length, the greater the angle of rotation of the *macroscopic magnetization vector* will be (greater than 180° can bring it back toward its original orientation).

Pulse NMR—*NMR* techniques that use *RF pulses* and *Fourier transformation* of the *NMR signal*; have largely replaced the older *continuous wave* techniques.

Pulse programmer—part of the *spectrometer* or *interface* that controls the timing, duration, and amplitude of the *pulses* (*RF* or *gradient*).

Pulse sequences—set of *RF* (and/or *gradient*) *magnetic field pulses* and time spacings between these pulses; used in conjunction with gradient magnetic fields and *NMR signal* reception to produce *NMR* images. See also Interpulse times.

Pulsed gradients—see Gradient pulse.

Q—see Quality factor.

Quadrature detector—a *phase* sensitive *detector* or *demodulator* that detects the components of the signal in phase with a reference oscillator and 90° out of phase with the reference oscillator.

Quality factor (Q)—applies to any electrical circuit component; most often the *coil* Q is limiting. Inversely related to the fraction of the energy in an oscillating system lost in one oscillation cycle. Q is inversely related to the range of *frequency* over which the system will exhibit *resonance.* It affects the *signal-to-noise ratio,* because the detected signal increases proportionally to Q while the noise is proportional to the square root of Q. The Q of a coil will depend on whether it is unloaded (no patient) or loaded (patient).

Quenching—loss of superconductivity of the current carrying coil that may occur unexpectedly in a *superconducting magnet.* As the magnet becomes resistive, heat will be released that can result in rapid evaporation of liquid helium in the *cryostat.* This may present a hazard if not properly planned for.

Radian—dimensionless unit of angular measure; $360° = 2\pi$ radians.

Radiofrequency (RF)—wave *frequency* intermediate between auditory and infra red. The RF used in *NMR* studies is commonly in the *megahertz* (MHz) range. The principal effect of RF *magnetic fields* on the body is power deposition in the form of heating, mainly at the surface; this is a principal area of concern for safety limits.

Readout delay—see TE.

Receiver—portion of the *NMR* apparatus that detects and amplifies *RF* signals picked up by the receiving *coil.* Includes a preamplifier, amplifier, and *demodulator.*

Receiver coil—*coil* of the *RF receiver;* "picks up" the *NMR signal.*

Reconstruction from projections imaging—*NMR imaging* technique in which a set of *projection profiles* of the body is obtained by observing *NMR signals* in the presence of a suitable corresponding set of *gradient magnetic fields.* Images can then be reconstructed using techniques analogous to those used in conventional computed tomography (CT), such as *filtered back projection.* It can be used for *volume imaging* or, with plane selection techniques, for *sequential plane imaging.*

Refocusing—see Spin echo.

Relaxation rates—reciprocals of the *relaxation times.*

Relaxation times—after *excitation,* the *spins* will tend to return to their equilibrium distribution, in which there is no *transverse magnetization* and the *longitudinal magnetization* is at its maximum value and oriented in the direction of the static *magnetic field.* It is observed that in the absence of applied RF, the transverse magnetization decays toward zero with a characteristic time constant *T2,* and the longitudinal magnetization returns toward the equilibrium value M_o with a characteristic time constant *T1.*

Repeated FID—a form of *NMR imaging* in which repeated *90° pulses* are applied. Results in *partial saturation* if interpulse times are of the order of or less than *T1.* Strictlv speaking, applies only if *NMR* signal is detected as an *FID.*

Repetition time—see TR.

Rephasing gradient—*gradient magnetic field* applied for a brief period after a *selective excitation* pulse, in the opposite direction to the *gradient* used for the selective excitation. The result of the gradient reversal is a rephasing of the *spins* (which will have gotten out of phase with each other along the direction of the selection gradient), forming an *echo* by *"time reversal,"* and improving the sensitivity of imaging after the selective excitation process.

Resistive magnet—a magnet whose *magnetic field* originates from current flowing through an ordinary (nonsuperconducting) conductor.

Resolution, spatial—although generally referring to the ability of the imaging process to distinguish adjacent structures in the object (an important measure of image quality), the specific criterion of resolution to be used depends on the type of test used (e.g. bar pattern or contrast-detail *phantom*). As the ability to separate or detect objects depends on their *contrast,* and the different *NMR* parameters of objects will affect image contrast differently for different imaging techniques, care must be taken in comparing the results of resolution *phantom* tests of different machines and no single simple measure of resolution can be specified.

Resonance—a large amplitude vibration in a mechanical or electrical system caused by a relatively small periodic stimulus with a *frequency* at or close to a natural frequency of the system; in *NMR* apparatus, resonance can refer to the NMR itself or to the tuning of the *RF* circuitry.

Resonant frequency—frequency at which *resonance* phenomenon occurs; given by the *Larmor equation* for *NMR*; determined by *inductance* and capacitance for *RF* circuits.

RF—see Radiofrequency.

RF coil—used for transmitting *RF pulses* and/or receiving *NMR signals.* Most commonly used in *saddle coil* or *solenoid* configurations for *NMR imaging.*

RF pulse—brief burst of *RF magnetic field* delivered to object by *RF transmitter.* For *RF frequency* near the *Larmor frequency,* it will result in rotation of the *macroscopic magnetization vector* in the *rotating frame of reference* (or a more complicated *nutational* motion in the stationary frame of reference). The amount of rotation will depend on the strength and duration of the RF pulse; commonly used examples are *90° (π/2)* and *180° (π) pulses.*

Rotating frame of reference—a frame of reference (with corresponding coordinate systems) that is rotating about the axis of the static *magnetic field* B_o (with respect to a stationary ("laboratory") frame of reference) at a *frequency* equal to that of the applied *RF* magnetic field, B_1. Although B_1 is a rotating *vector,* it appears stationary in the rotating frame, leading to simpler mathematical formulations.

Rotating frame zeugmatography—technique of *NMR imaging* that uses a *gradient* of the *RF excitation* field (to give a corresponding variation of the *flip angle* along the gradient as a means of encoding the spatial location of *spins* in the direction of the RF field gradient) in conjunction with a static *gradient magnetic field* (to give spatial encoding in an orthogonal direction). It can be considered to be a form of *Fourier transform imaging.*

Saddle coil—*RF coil* configuration design commonly used when the static *magnetic field* is coaxial with the axis of the coil along the long axis of the body (e.g. *superconducting magnets* and most *resistive magnets*) as opposed to solenoid or surface coil.

Saturation—a nonequilibrium state in *NMR,* in which equal numbers of *spins* are aligned against and with the *magnetic field,* so that there is no net *magnetization.* Can be produced by repeatedly supplying *RF* pulses at the *Larmor frequency* with interpulse times short compared to *T1.*

Saturation recovery (SR)—particular type of *partial saturation pulse sequence* in which the preceding pulses leave the *spins* in a state of *saturation,* so that recovery at the time of the next pulse has taken place from an initial condition of no magnetization.

Saturation transfer (or Inversion transfer)—nuclei can retain their magnetic orientation through a chemical reaction. Thus, if *RF* radiation is supplied to the *spins* at a *frequency* corresponding to the *chemical shift* of the nuclei in one chemical state so as to produce *saturation* or *inversion,* and chemical reactions transform the nuclei into another chemical state with a different chemical shift in a time short compared to the *relaxation time,* the *NMR spectrum* may show the effects of the saturation or inversion on the corresponding, unirradiated, line in the spectrum. This technique can be used to study reaction kinetics of suitable molecules.

SE—see Spin echo.

Selective excitation—controlling the *frequency spectrum* of an irradiating *RF pulse* (via *tailoring*) while imposing a *gradient magnetic field* on *spins,* such that only a desired region will have a suitable *resonant frequency* to be excited. Originally used to excite all but a desired region; now more commonly used to select only a desired region, such as a plane, for excitation.

Selective irradiation—see Selective excitation.

Sensitive plane—technique of selecting a plane for *sequential plane imaging* by using an oscillating *gradient magnetic field* and filtering out the corresponding time-dependent part of the *NMR signal.* The *gradient* used is at right angles to the desired plane and the magnitude of the oscillating gradient magnetic field is equal to zero only in the desired plane.

Sensitive point—technique of selecting out a point for *sequential point imaging* by applying three orthogonal oscillating *gradient magnetic fields* such that the local *magnetic field* is time dependent everywhere except at the desired point, and then filtering out the corresponding time dependent portion of the *NMR signal.*

Sensitive volume—region of the object from which *NMR signal* will preferentially be acquired because of strong *magnetic field inhomogeneity* elsewhere. Effect can be enhanced by use of a shaped *RF* field that is strongest in the sensitive region.

Sequence time—see TR.

Sequential line imaging (Line scanning, Line imaging)—*NMR imaging* techniques in which the image is built up from successive lines through the object. In various schemes, the lines are isolated by oscillating *gradient magnetic fields* or *selective excitation,* and then the *NMR signals* from the selected line are encoded for position by detecting the *FID* or *spin echo* in the presence of a gradient magnetic field along the line; the *Fourier transform* of the detected signal then yields the distribution of emitted NMR signal along the line.

Sequential plane imaging (Plane imaging)—*NMR imaging* technique in which the image of an object is built up from successive planes in the object. In various schemes, the planes are selected by oscillating *gradient magnetic fields* or *selective excitation.*

Sequential point imaging (Point scanning)—*NMR imaging* techniques in which the image is built from successive point positions in the object. In various schemes, the points are isolated by oscillating *gradient magnetic fields (sensitive point)* or shaped *magnetic fields.*

SFP—see Steady state free precession.

Shim coils—*coils* carrying a relatively small current that are used to provide auxiliary *magnetic fields* in order to compensate for *inhomogeneities* in the main magnetic field of an *NMR* system.

Shimming—correction of *inhomogeneity* of the *magnetic field* produced by the main magnet of an *NMR* system due to imperfections in the magnet or to the presence of external *ferromagnetic* objects. May involve changing the configuration of the magnet or the addition of *shim coils* or small pieces of steel.

SI (International System of Units)—the preferred international standard system of physical units and measures.

Signal-to-noise ratio (SNR or S/N)—used to describe the relative contributions to a detected signal of the true signal and random superimposed signals ("noise"). One common method to improve (increase) the SNR is to average several measurements of the signal on the expectation that random contributions will tend to cancel out. The SNR can also be improved by sampling larger volumes (with a corresponding loss of spatial *resolution*) or, within limits, by increasing the strength of the *magnetic field* used. S/N will depend on the electrical properties of the sample or patient being studied.

Simultaneous volume imaging—see Volume imaging.

Skin depth—time dependent electromagnetic fields are significantly attenuated by conducting media (including the human body); the skin depth gives a measure of the average depth of penetration of the *RF* field. It may be a limiting factor in *NMR imaging* at very high frequencies (high *magnetic fields*). The skin depth also affects the *Q* of the *coils.*

S/N—see Signal-to-noise ratio.

SNR—see Signal-to-noise ratio.

Software—the set of instructions, or programs, that controls the activities of the *computer.* Programs may be written in machine language (sequences of numbers directly interpretable by the computer), assembly language, or higher level languages such as Fortran or Basic. The software includes overall supervising "executive" programs, data acquisition programs, data processing programs (including image reconstruction), and display programs.

Solenoid coil—a coil of wire wound in the form of a long cylinder. When a current is passed through the coil, the *magnetic field* within the coil is relatively uniform. Solenoid *RF coils* are commonly used when the static magnetic field is perpendicular to the long axis of the body.

Spectrometer—the portions of the *NMR* apparatus that actually produce the NMR phenomenon and acquire the signals, including the magnet, the *probe,* the *RF* circuitry, etc. The spectrometer is controlled by the *computer* via the *interface* under the direction of the *software.*

Spectrum—an array of the frequency components of the *NMR signal* according to *frequency.* Nuclei with different *resonant frequencies* will show up as peaks at different corresponding frequencies in the spectrum, or "lines."

Spin—the intrinsic *angular momentum* of an elementary particle, or system of particles such as a nucleus, that is also responsible for the *magnetic moment;* or, a particle or nucleus possessing such a spin. The spins of nuclei have characteristic fixed values. Pairs of neutrons and protons align to cancel out their spins, so that nuclei with an odd number of neutrons and/or protons will have a net nonzero rotational component characterized by an integer or half integer quantum *"nuclear spin number"* (I).

Spin density (N)—the density of resonating *spins* in a given region; one of the principal determinants of the strength of the *NMR signal* from the region. The SI units would be moles/m^3. For water, there are about 1.1×10^5 moles of hydrogen per m^3, or .11 moles of hydrogen/cm^3. True spin density is not imaged directly, but must be calculated from signals received with different *interpulse times.*

Spin echo—reappearance of an *NMR signal* after the *FID* has died away, as a result of the effective reversal of the dephasing of the *spins* ("refocusing") by such techniques as reversal of a *gradient magnetic field* (often referred to as a form of *"time reversal"*), or by specific *RF pulse* sequences such as the *Carr-Purcell sequence* (applied in a time shorter than or on the order of *T2*). Multiple spin echoes or a series of spin echoes at different times can be used to determine T2 without contamination by effects of the *inhomogeneity* of the *magnetic field.*

Spin-echo imaging—any of many *NMR imaging* techniques in which the *spin echo NMR signal* rather than the *FID* is used. Can be used to create images that depend strongly on *T2.* Note that spin echoes do not directly produce an image of T2.

Spin-lattice relaxation time—see T1.

Spin number, nuclear—see Nuclear spin number.

Spin—Spin relaxation time—see T2.

Spin tagging—nuclei will retain their magnetic orientation for a time on the order of *T1* even in the presence of motion. Thus, if the nuclei in a given region have their spin orientation changed, the altered spins will serve as a "tag" to trace the motion of any fluid that may have been in the tagged region for a time on the order of T1.

Spin warp imaging—a form of *Fourier transform imaging* in which *phase* encoding *gradient pulses* are applied for a constant duration but with varying amplitude. This is distinct from the original *FT* imaging methods in which phase encoding is performed by applying gradient pulses of constant amplitude but varying duration. The spin warp method, as other Fourier imaging techniques, is relatively tolerant of nonuniformities (*inhomogeneities*) in the static or *gradient magnetic fields.*

SR—see Saturation recovery.

SSFP—see Steady state free precession.

Steady state free precession (SFP or SSFP)—method of *NMR excitation* in which strings of *RF pulses* are applied rapidly and repeatedly with interpulse intervals short compared to both *T1* and *T2.* Alternating the phases of the RF pulses by 180° can be useful in obtaining maximal signal strength.

Superconducting magnet—a magnet whose *magnetic field* originates from current flowing through a *superconductor.* Such a magnet must be enclosed in a *cryostat.*

Superconductor—a substance whose electrical resistance essentially disappears at temperatures near absolute zero. A commonly used superconductor in *NMR imaging* system magnets is niobium-titanium, embedded in a copper matrix to help protect the superconductor from *quenching.*

Surface coil NMR—a simple flat *RF receiver coil* placed over a region of interest will have an effective selectivity for a volume approximately subtended by the coil circumference and one radius deep from the coil center. Such a coil can be used for simple localization of sites for measurement of *chemical shift spectra,* especially of phosphorus, and blood flow studies. Some additional spatial selectivity can be achieved with *gradient magnetic fields.*

Susceptibility—see Magnetic susceptibility.

T—see Tesla.

T1 ("T-one")—spin-lattice or longitudinal *relaxation time;* the characteristic time constant for *spins* to tend to align themselves with the external *magnetic field.* Starting from zero *magnetization* in the z direction, the z magnetization will grow to 63% of its final maximum value in a time T1.

T2 ("T-two")—spin-spin or transverse *relaxation time;* the characteristic time constant for loss of *phase coherence* among *spins* oriented at an angle to the static *magnetic field,* due to interactions between the spins, with resulting loss of *transverse magnetization* and *NMR signal.* Starting from a non zero value of the *magnetization* in the xy plane, the xy magnetization will decay so that it loses 63% of its initial value in a time T2.

T2* ("T-two-star")—the characteristic time constant for loss of *phase coherence* among *spins* oriented at an angle to the static *magnetic field* due to a combination of magnetic field *inhomogeneities,* ΔB, and spin-spin transverse relaxation with resultant more rapid loss in *transverse magnetization* and *NMR signal.* NMR signal can still be recovered as a *spin echo* in times less than or on the order of *T2.* $1/T2^* = 1/T2 + \Delta\omega/2$; $\Delta\omega = \gamma\Delta B$.

Tailored excitation—see Selective excitation.

Tailored pulse—shaped pulse whose magnitude is varied with time in a predetermined manner. Affects the *frequency* components of an *RF pulse* in a manner determined by the *Fourier transform* of the pulse.

TE—*echo time.* Time between middle of *90° pulse* and middle of *spin echo* production. For multiple echoes, use TE1, TE2. . . .

Tesla (T)—the preferred (*SI*) unit of magnetic flux density. One tesla is equal to 10,000 *gauss,* the older (CGS) unit.

Thermal equilibrium—a state in which all parts of a system are at the same effective temperature, in particular where the relative alignment of the *spins* with the *magnetic field* is determined solely by the thermal energy of the system (in which case the relative numbers of spins with different alignments will be given by the *Boltzmann distribution).*

TI—*inversion* time. Time after middle of inverting *RF pulse* to middle of *90° pulse* to detect amount of *longitudinal magnetization.*

Time reversal—technique of producing a *spin echo* by subjecting *excited* spins to a *gradient magnetic field,* and then reversing the direction of the gradient field. All methods of spin echo production can be viewed as effective time reversal.

Torque—the effectiveness of a force in setting a body into rotation. It is a *vector* quantity given by the vector product of the force and the position vector where the force is applied; for a rotating body, the torque is the product of the moment of inertia and the resulting angular acceleration.

TR—repetition time. The period of time between the beginning of a *pulse sequence* and the beginning of the succeeding (essentially identical) pulse sequence.

Transmitter—portion of the *NMR* apparatus that produces *RF* current and delivers it to the transmitting *coil.*

Transmitter coil—*coil* of the *RF transmitter.*

Transverse magnetization (M_{xy})—component of the *macroscopic magnetization vector* at right angles to the static *magnetic field* (B_o). *Precession* of the transverse magnetization at the *Larmor frequency* is responsible for the detectable *NMR signal.* In the absence of externally applied *RF* energy, the transverse magnetization will decay to zero with a characteristic time constant of *T2* or *T2*.*

Transverse relaxation time—see T2.

Two-dimensional Fourier transform imaging (2DFT)—a form of *sequential plane imaging* using *Fourier transform imaging.*

Tuning—process of adjusting the *resonant frequency,* e.g., of the *RF* circuit, to a desired value, e.g., the *Larmor frequency.* More generally, the process of adjusting the components of the *spectrometer* for optimal *NMR signal* strength.

Tunnel—opening into *NMR imaging* machine to place patient into imaging region.

Vector—a quantity having both magnitude and direction, frequently represented by an arrow whose length is proportional to the magnitude and with an arrowhead at one end to indicate the direction.

Volume imaging—imaging techniques in which *NMR signals* are gathered from the whole object volume to be imaged at once, with appropriate encoding *pulse/gradient* sequences to encode positions of the *spins.* Many *sequential plane imaging* techniques can be generalized to volume imaging, at least in principle. Advantages include potential improvement in *signal-to-noise ratio* by including signal from the whole volume at once; disadvantages include a bigger computational task for image reconstruction and longer *image acquisition times* (although the entire volume can be imaged from the one set of data). Also called simultaneous volume imaging.

Voxel—volume element; the element of 3-D space corresponding to a *pixel,* for a given slice thickness.

x—dimension in the stationary (laboratory) frame of reference in the plane orthogonal (at right angles) to the direction of the static *magnetic field* (B_o), *z,* and orthogonal to *y,* the other dimension in this plane.

x^1—dimension in the *rotating frame of reference* in the plane at right angles to the direction of the static *magnetic field* (B_o), *z*; defined to be in the direction of the magnetic *vector* of the *RF* field (B_1).

y—dimension in the stationary (laboratory) frame of reference in the plane orthogonal to the direction of the static *magnetic field* (B_o and H_o), *z,* and orthogonal to *x,* the other dimension in this plane.

y^1—dimension in the *rotating frame of reference* in the plane orthogonal (at right angles) to the direction of the static *magnetic field* (B_o and H_o), *z,* and orthogonal to the other dimension in this plane, x^1.

z—dimension in the direction of the static *magnetic field* (B_o and H_o), in both the stationary and *rotating frames of reference.*

Zeugmatography—term for *NMR imaging* coined from Greek roots suggesting the role of the *gradient magnetic field* in joining the *RF* magnetic field to a desired local spatial region through *nuclear magnetic resonance.*

2DFT—see Two-dimensional Fourier transform.

γ—see Gyromagnetic ratio.

δ—see Chemical shift.

μ—see Permeabilty.

ζ—often used to denote different time delays between *RF pulses.* See Interpulse times.

χ—see Magnetic susceptibility.

ω—see Angular frequency.

ω_o—see Larmor frequency.

INDEX